BURNHAM'S
CELESTIAL HANDBOOK
An Observer's Guide to the Universe
Beyond the Solar System

O Night, sweet though sombre space of time . . .
  All things find rest upon their journey's end——
Our cares thou canst to quietude sublime;
  For dews and darkness are of peace the friend.

MICHELANGELO

BURNHAM'S
CELESTIAL HANDBOOK

An Observer's Guide to the Universe
Beyond the Solar System

ROBERT BURNHAM, JR.

Staff Member, Lowell Observatory, 1958–1979

IN THREE VOLUMES
Volume Three, Pavo–Vulpecula

DOVER PUBLICATIONS, INC.
NEW YORK

Frontispiece: Venus

For information about our audio products, write us at:
Newbridge Book Clubs, 3000 Cindel Drive, Delran, NJ   08370

Published in Canada by General Publishing
Company, Ltd., 30 Lesmill Road, Don Mills,
Toronto, Ontario.
Published in the United Kingdom by Con-
stable and Company, Ltd.

*Burnham's Celestial Handbook: Volume Three,
Pavo–Vulpecula* is a new work, first published by
Dover Publications, Inc., in 1978.

INTERNATIONAL STANDARD BOOK NUMBERS:
paperbound edition: 0-486-23673-0
clothbound edition: 0-486-24065-7
Library of Congress Catalog Card Number:
77-082888

Manufactured in the United States of America
Dover Publications, Inc.
31 East 2nd Street
Mineola, N.Y. 11501

PREFACE

This is the third and final volume of the Celestial Handbook, covering the constellations Pavo through Vulpecula.

The format and arrangement of material remains the same as in the two preceding volumes. A list of symbols and abbreviations used appears in the first volume, beginning on page 98.

Throughout the Handbook, all positions are given for the Epoch 1950.0 to permit the direct location and plotting of objects on *Norton's Star Atlas* and the Skalnate Pleso *Atlas of the Heavens*. A Constellation Index and Atlas Reference Table appears in the back of each of the three volumes.

In offering the complete Handbook to the world, the author wishes, once again, to thank all those who have offered assistance and encouragement, and have helped to make this book a reality.

Clear skies to you all !

Robert Burnham, Jr.

Flagstaff, Arizona
November, 1977

NOTE

Indexes covering all three volumes of the Celestial Handbook are to be found at the end of this volume:

ACKNOWLEDGMENTS

The author wishes to express his deep appreciation to Evered Kreimer, Alan McClure, Kent De Groff, and David Healy, for the professional-quality astronomical photographs which these observers have generously contributed to the *Celestial Handbook*.

Grateful acknowledgment is rendered to the American Association of Variable Star Observers for their kind permission to use AAVSO charts and comparison magnitudes in designing the variable star charts in this book.

The poem by Tu Fu of the T'ang Dynasty, *"Night Thoughts While Travelling"*, on page 1627, is a translation by Kenneth Rexroth, from *One Hundred Poems From the Chinese*. All rights reserved. Copyright © 1971 by Kenneth Rexroth. Reprinted by permission of New Directions Publishing Corporation, N.Y.

The quotations from *The Dream-Quest of Unknown Kadath* by Howard Phillips Lovecraft, on pages 1141 and 1787, are reprinted from *At the Mountains of Madness and Other Novels*, by H.P.Lovecraft, published by Arkham House Publishers; Copyright 1939, 1943, 1964. Reprinted by kind permission of Arkham House Publishers, Inc., Sauk City, Wisconsin.

Grateful acknowledgment is also extended to the Directors of the Lick Observatory of the University of California, the Yerkes Observatory of the University of Chicago, the Lowell Observatory of Flagstaff, Arizona, and the Mount Wilson and Palomar Observatories of the California Institute of Technology, for their kind permission to reproduce the many fine astronomical photographs which appear in this book.

The author takes great pleasure in offering his special thanks and appreciation to Herbert A. Luft, whose unflagging interest and support has helped immeasurably to make the *Celestial Handbook* a reality.

PAVO

LIST OF DOUBLE AND MULTIPLE STARS

NAME	DIST	PA	YR	MAGS	NOTES	RA & DEC
h5029	2.0	96	24	8 - 8	PA slow dec, spect G5	18108s5752
Hd 289	2.5	298	01	6½- 10	(L7574) spect A0	18127s6815
I 249	7.1	1	28	6 - 10½	Slight PA dec, spect G0	18148s6354
Cp 79	6.0	55	28	8½- 8½	spect F5	18183s6237
ξ	3.5	154	55	4½- 8	Slight PA inc, color contrast, spect K2	18186s6131
Hd 290	1.8	346	29	6- 12½	(Rst 987) gG8	18256s5734
	33.9	120	13	10½		
I 633	0.6	322	43	8½- 8½	PA inc, spect F5	18274s5937
Mlb 5	4.6	294	42	7½- 9½	relfix, spect G5	18292s6619
R 314	2.0	269	47	6½- 9	slow PA inc, cpm; spect A0	18436s7303
h5065	22.5	21	13	7½- 10	spect A0	18478s5800
h5069	0.8	82	42	8 - 8	spect F2	18518s6154
	15.2	92	35	-12		
Gale 3	0.3	259	59	6 - 7	PA inc, cpm; spect A2	19122s6645
I 114	0.5	287	30	7½-8½	spect G5	19152s6259
h5109	24.7	144	40	7- 9	relfix, spect K0	19248s6725
	36.3	13	40	-10		
I 117	1.0	184	47	7½- 8	spect A2	19277s6023
h5132	21.6	309	16	7½- 10	relfix, spect G0	19392s6625
I 120	0.4	103	26	8 - 8½	binary, 55 yrs; PA dec, spect G0, all cpm; (h5141)	19448s6157
	13.9	343	33	-10		
h5140	1.5	80	52	8 - 8	PA dec, spect G0	19452s6502
h5137	29.6	201	17	7- 11½	spect F8	19455s7256
	42.8	313	17	- 11		
I 121	0.4	133	59	5½- 7½	PA inc, spect A2	19466s5920
h5163	1.7	249	45	8 - 8½	relfix, spect A2	20007s6313
h5162	6.7	292	17	7½- 10	relfix, spect K0	20027s7058
I 1042	0.5	83	28	8 - 8½	PA dec	20032s7142
L8337	0.6	256	52	7 - 7½	(Hd 295) PA inc, spect A0	20071s5740
h5167	0.2	338	42	8 - 8	(Rst 5153) spect A0	20072s6345
	7.2	34	40	-8½		

LIST OF DOUBLE AND MULTIPLE STARS (Cont'd)

NAME	DIST	PA	YR	MAGS	NOTES	RA & DEC
h5171	17.3	305	27	7 - 10	spect A2	20100s6435
	30.1	336	16	- 10		
Rmk 25	7.2	28	40	8 - 8	(h5177) spect F8	20109s5707
h5185	18.6	61	16	7½- 11	spect A0	20168s5853
h5194	4.3	256	17	6½- 12	spect A2	20254s6914
h5200	12.2	137	17	7½-10½	spect A0	20271s6832
Hu1615	0.4	22	59	8 - 8½	PA inc, spect A0	20305s6329
△231	57.2	288	17	7 - 9	spect A0	20315s7114
L8550	2.5	87	59	6½- 6½	(Rmk 26) PA dec,	20475s6237
					spect A2; cpm	
h5231	7.1	116	17	7½- 8	relfix, spect F2	20539s7037
h5231b	1.0	325	27	8½- 9	(I668)	
L8625	0.9	350	00	6½- 6½	AB uncertain, not	21040s7322
	8.0	131	31	- 14	seen after 1900;	
					AC PA dec, spect G3	
					(I379)	

LIST OF VARIABLE STARS

NAME	MagVar	PER	NOTES	RA & DEC
κ	3.9--4.8	9.065	Cepheid, W Virginis type;	18518s6718
			Spect F5--G5	
λ	3.4--4.3	Irr	Spect B2	18476s6215
R	7.3--13.8	230	LPV. Spect M4e	18081s6338
S	6.6--10.4	386	Semi-reg; spect M7e	19510s5920
T	7.2--14.0	244	LPV. Spect M4e	19451s7154
U	8.6--12..	290	LPV. Spect M4e	20514s6254
V	8.0--10..	225	Semi-reg; Spect N	17390s5742
W	8.1--14.6	283	LPV. Spect M4e--M7e	17458s6224
X	8.0--10..	397	Semi-reg; spect M	20076s6005
Y	7.8--9.0	233	Semi-reg; spect N	21198s6957
Z	8.0--9..	136	Semi-reg; spect M	19309s6252
RR	9.8--13..	240	LPV. Spect M4e	20022s6334
RT	8.5--10..	85	Semi-reg; spect M	18310s6956

LIST OF VARIABLE STARS (Cont'd)

NAME	MagVar	PER	NOTES	RA & DEC
RZ	9.8--14..	288	LPV.	17445s5844
SU	8.7--13..	245	LPV. Spect M4e--M6e	20135s6013
XZ	8.4--15..	332	LPV.	17541s5911
BL	9.5--11..	135	Semi-reg	19150s6713
BM	9.0--10..	62	Semi-reg	19237s6256
BO	9.3--10.1	19.23	Ecl.bin.	19455s6555
BQ	9.8--12..	112	LPV.	19551s7001
BR	8.7--15..	251	LPV.	20053s5729
BU	9.3--10.5	71	Semi-reg	21181s6333
KZ	7.4--8.0	1.900	Ecl.bin; spect F2	20537s7037

LIST OF STAR CLUSTERS, NEBULAE, AND GALAXIES

NGC	OTH	TYPE	SUMMARY DESCRIPTION	RA & DEC
----	I.4662	⊘	I or SBp; 11.8; 2.0' x 1.2' F,pS,1E	17421s6439
----	I.4710	⊘	SBc; 12.8; 4.0' x 2.5' vF,vS,1E,bM	18235s6701
6630		◎	Mag 15; diam 19" x 15" F,S,R	18277s6319
----	I.4721	⊘	SBc; 12.9; 3.5' x 1.3' F,cL,E	18301s5832
----	I.4723	◎	Mag 15; diam 19" vF,vS,R	18311s6326
6684		⊘	SB; 11.7; 2.0' x 1.5' vB,pL,R,psvmbM	18441s6514
6699		⊘	Sb; 12.4; 1.2' x 1.2' pF,pS,1E,1bM	18478s5723
6721		⊘	E0; 13.1; 1.2' x 1.2' pF,cS,R,mbM	18565s5751
6744	△262	⊘	SBc; 10.6; 9.0' x 9.0' cB,cL,R,vmbM (*)	19050s6356

LIST OF STAR CLUSTERS, NEBULAE, AND GALAXIES (Cont'd)

NGC	OTH	TYPE	SUMMARY DESCRIPTION	RA & DEC
6752	△295	⊕	!! Mag 7, diam 15', class VI B,vL,vmC, stars mags 11... Fine cluster (*)	19064s6004
6753			SB; 11.7; 2.4' x 2.0' pB,pL,R	19072s5708
6769			SBb; 12.7; 2.0' x 1.2' vF,S,1E,1bM; group with 2 F spirals 6770 & 6771	19139s6035
6776			E2; 12.8; 0.7' x 0.6' pB,S,R,pmbM	19207s6359
6782			S0; 12.8; 1.5' x 0.8' cF,cS,R,1bM	19195s6002
6808			Sa; 13.0; 1.1' x 0.4' pB,E,bN	19385s7046
6810			Sa; 12.4; 2.5' x 0.8' pS,R,vgbM	19394s5847
6876			E3; 12.7; 1.5' x 0.8' pB,S,1E,BN	20131s7101
6943			SB; 12.5; 3.5' x 1.8' pF,L,mE	20398s6855
----	I.5052		Sd; 12.3; 4.8' x 0.8' F,L,mE	20475s6925
7020			S ; 13.1; 2.0' x 0.9' pB,S,1E,gbM	21073s6415

DESCRIPTIVE NOTES

ALPHA   Name- The "Peacock Star", honoring the bird
sacred to Juno; the constellation name is of
no great antiquity, however, as it was introduced by Bayer
in the early 17th century. Magnitude 1.93; spectrum B3 IV;
position 20217s5654. The computed distance is about 310
light years, and the actual luminosity about 1200 times the
Sun. Alpha Pavonis shows an annual proper motion of 0.09";
the radial velocity is 1.2 mile per second in recession.
Spectroscopic studies show that the star is a close binary
with a period of 11.753 days.

BETA   Magnitude 3.42; spectrum A5 IV; position
20405s6623. The distance is about 160 light
years and the actual luminosity about 90 times the Sun.
The annual proper motion is 0.05"; the radial velocity is
6 miles per second in recession.

NGC 6752   Globular star cluster. Position 19064s6004,
about 10° WSW from Alpha Pavonis. One of the
finest of the globular clusters, though almost unknown to
observers in North America or Europe owing to its far
southern location. It was probably first observed by J.
Dunlop in 1828, and with a total integrated magnitude of
7.2 ranks as the 7th brightest globular in the sky. In
apparent size it possibly holds third place, only Omega
Centauri and 47 Tucanae appear to exceed it. The total
diameter is about 42' on the best photographs, though the
visual size is about 15' or 16' in most telescopes. "One
of the gems of the sky" states E.J.Hartung. "On a clear
dark night this is a most lovely object......a moderately
condensed type of globular cluster, the central region
about 3' wide and the unusually bright outliers extending
over 15', involving an elegant pair (7.7+9.3, 3".0, 238°).
Many of the brighter stars of the cluster are in curved and
looped arms, and look distinctly reddish.." The cluster is
among the nearer globulars with a computed distance of
about 20,000 light years and a true luminosity of close to
100,000 suns. NGC 6752 contains only two known variable
stars, one of which has been known since 1897. The total
integrated spectral type is about F6; the radial velocity
is a very moderate 23 miles per second in approach.

GLOBULAR STAR CLUSTER NGC 6752 in PAVO. This object ranks among the dozen finest globular clusters in the sky.
                                        Radcliffe Observatory

SPIRAL GALAXY NGC 6744 in PAVO. This is one of the largest of the barred spirals. The photograph was made with the 74-inch reflector at Mt.Stromlo Observatory in Australia.

PEGASUS

LIST OF DOUBLE AND MULTIPLE STARS

NAME	DIST	PA	YR	MAGS	NOTES	RA & DEC
Σ2767	2.1	30	62	8 - 8½	relfix, spect F5	21082n1945
β681	2.7	239	45	7 - 11	relfix, spect K2	21110n1643
1	36.3	312	54	4 - 8	AB cpm, relfix,	21198n1935
	75.0	20	21	- 12	spect K1, K0	
Σ2797	3.5	216	54	6½- 8	relfix, spect A2	21243n1328
Σ2799	1.8	276	66	7 - 7	PA dec, spect F2;	21264n1052
	136	336	12	-9½	Globular cluster	
					M15 is 1° to NNE	
h3032	17.3	93	29	8 - 12	relfix, spect F0	21300n0439
Σ2804	3.0	347	59	7½- 8	Dist dec, spect F5,	21306n2029
	97.1	106	24	-11½	F8; PA inc.	
Σ3112	7.3	239	52	7½- 9½	relfix, spect G0	21320n0917
β374	1.3	329	56	7½- 9½	PA inc, spect F8	21329n2110
Hu 371	0.2	286	68	6½- 7	Binary, about 175	21332n2414
					yrs; PA inc, spect	
					A3	
OΣ443	8.3	349	37	8 - 8½	cpm pair, spect K2	21351n0629
3	39.2	349	34	6½- 8½	(ΣI56) cpm, spect	21352n0624
	88.3	119	59	-13½	A0; OΣ443 in field	
OΣ444	8.1	277	16	7½-10½	relfix, spect F5	21367n2023
OΣ445	0.8	113	59	8 - 8½	relfix	21370n2030
Ho 165	0.5	64	60	8 - 8	relfix, spect A2	21397n1846
Hu 280	0.2	286	66	8 -8½	Binary, about 115	21398n0541
					yrs; PA inc, spect	
					A5	
ε	81.8	325	10	2½-11½	(S798) optical,	21417n0939
	143	321	14	- 8½	spect K2 (*)	
Ho 166	0.3	118	69	8 - 8	Binary, about 80	21417n2737
					yrs; PA dec, spect	
					dF5	
κ	0.2	273	69	4½- 5	(Σ2824) (β989)	21424n2525
	13.8	292	58	- 10	AB binary, 11½ yr;	
					PA dec, spect F5;	
					AC dist inc. (*)	
Ho 465	42.6	246	23	7 - 9	spect A2,	21441n2157
Ho 465b	4.7	79	23	9 - 11	BC dist inc.	
Σ2828	30.8	141	63	8 - 9	Dist inc, PA dec,	21470n0310
Σ2828b	3.7	39	63	9 -10½	spect G5	
Σ2829	17.4	16	24	8 - 9	relfix, spect A0,A2	21472n3031
13	0.4	232	69	5½- 7	Binary, 31 yrs, PA	21478n1703
					inc, spect F2	

LIST OF DOUBLE AND MULTIPLE STARS (Cont'd)

NAME	DIST	PA	YR	MAGS	NOTES	RA & DEC
β692	2.6	9	34	7½- 11	spect K0	21479n3137
	37.6	297	12	- 11		
h1697	9.1	258	28	7½- 11	relfix, spect K5	21480n3435
Ho 467	2.2	214	69	8 - 10	Dist & PA inc,	21483n2201
	40.1	338	05	- 12	spect K2	
h947	19.4	95	24	6½- 10	optical, spect B6	21492n1935
	23.8	320	24	- 12		
Σ2834	4.3	295	54	7½-10½	slow PA inc, spect F2	21493n1904
Σ2841	22.7	110	58	6½- 8	cpm pair, relfix, spect K0	21520n1929
β75	0.3	236	70	8 - 8½	Binary, about 300 yrs; PA inc, spect G5	21531n1039
0Σ452	1.0	180	56	7½- 8½	relfix, spect F8	21532n0701
Σ2848	10.8	56	63	7 - 7½	relfix, spect A2	21555n0542
Σ2850	3.0	260	43	7 - 11	relfix, spect M3	21575n2342
A306	1.2	306	29	7½- 13	spect B9	21589n2635
0Σ228	22.4	160	25	7½- 12	Dist inc, spect G5	21594n0432
	78.3	26	27	- 9		
Σ2854	2.2	82	63	7½- 8	Dist dec, spect F5	22020n1324
h953	21.4	106	34	6½-11½	Spect G5	22024n3242
Σ2856	1.3	199	67	8 - 8½	relfix, spect G5	22030n0437
h1721	10.6	270	55	7½- 8½	Dist inc, PA slow dec, spect M	22035n2939
Σ2861	7.1	221	66	7½- 8	relfix, spect A3	22036n2033
Σ2857	20.4	113	25	7- 8½	cpm pair, relfix, spect A2	22037n0951
Es 386	6.2	35	25	8½-12½		22046n3421
27	27.5	323	34	5½- 12	(π¹) PA inc,	22070n3256
	70.2	263	23	- 10	spect gG8	
Σ2868	1.2	353	67	8½- 9	PA slow inc, 'spect F5	22070n2218
Σ2867	10.5	209	51	8 - 9	relfix, spect G0	22076n0742
Σ2867b	8.9	253	04	9- 13½		
Σ2869	21.5	254	34	6 -11½	spect K0	22079n1423
0Σ463	4.0	358	48	7½- 11	PA inc, spect G0	22079n1330
β698	10.5	338	26	7 - 12	spect F0	22094n0639
	108	289	13	- 11		
Ho 179	0.9	276	67	8 - 9	PA inc, spect F5	22104n2958

LIST OF DOUBLE AND MULTIPLE STARS (Cont'd)

NAME	DIST	PA	YR	MAGS	NOTES	RA & DEC
Σ2877	16.0	13	40	6½- 9½	PA & dist inc,	22119n1657
	86.9	45	33	-10½	spect K2, optical	
Σ2879	1.4	122	58	6½- 8	PA dec, spect A0	22120n0744
	66.0	122	37	- 9½		
	124	274	37	- 11		
Σ2881	1.2	84	67	7½- 8	cpm, PA dec,	22123n2920
					spect F5	
OΣ467	23.0	274	14	6½-10½	spect K0	22124n2217
OΣ468	12.9	162	07	7 - 11	cpm pair, relfix,	22140n3329
					spect F8	
30	6.1	19	58	5½- 11	(h962) relfix, AC	22180n0532
	14.8	225	58	- 12	dist inc, PA slight	
					inc, spect B5	
OΣ469	28.7	287	45	7 - 8½	Optical, dist dec,	22183n3452
	53.1	52	08	-12½	PA inc, spect F0	
β1217	0.4	226	69	7½-10½	relfix, spect K0	22187n3102
32	72.5	127	24	5 - 9½	(Ho 615) spect B8	22190n2805
	42.3	307	24	- 12		
	60.3	116	24	- 12		
32b	2.4	18	24	9½-10½		
Ho 474	44.5	112	06	7 - 11	Spect G0	22202n3006
Ho 474b	4.2	36	15	11½-11½		
Ho 292	4.1	64	35	8 - 11	Perhaps slight PA	22207n0524
					inc, spect A2	
33	0.3	173	58	6 - 9	(Σ2900) PA & dist	22213n2036
	78.4	312	56	- 8	dec, spect dF4,A4;	
					AC PA dec, dist inc	
Σ2901	3.3	148	47	8½- 9	relfix, cpm, spect	22219n0334
					F0	
34	3.5	224	37	6- 11½	(β290) AB cpm, PA	22240n0408
	103	272	24	- 13	& dist slow inc,	
					spect dF5, K4	
Σ2905	3.5	281	44	8½- 8½	cpm pair, relfix,	22248n1454
					spect G5	
β701	0.9	217	69	7 - 10	PA dec, spect K0	22256n1200
	119	129	24	- 11		
Σ2908	9.0	116	37	7 - 8½	relfix, spect K0	22258n1700
Σ2910	5.5	336	66	8½- 9	Slow PA dec, spect	22258n2316
					K0	

LIST OF DOUBLE AND MULTIPLE STARS   (Cont'd)

NAME	DIST	PA	YR	MAGS	NOTES	RA & DEC
37	1.0	118	69	6 - 7	(Σ2912) Binary, edge-on orbit, dist inc; about 150 yrs; spect dF5 (*)	22274n0411
Hu 388	0.4	223	69	8 - 8½	PA inc, spect F0	22278n2213
Σ2915	13.7	134	66	8½- 8½	PA dec, spect F0	22301n0709
h1779	22.5	216	26	8 - 9	Spect A5	22303n3358
Ho 293	1.4	324	69	8 - 12	PA dec, spect G5	22310n3341
Σ2920	13.7	144	36	7½- 8½	relfix, spect A0	22320n0358
	21.3	63	07	- 14		
h966	13.3	268	26	7 - 11	relfix, spect K2	22327n3032
	36.8	276	06	-12½		
Hu 982	0.7	223	65	7- 10½	Slight PA inc; spect K2	22328n1421
Ho 618	6.4	236	21	7½-12½	PA inc, spect G0	22348n2629
40	1.8	236	48	6- 11½	(Kui 113) cpm, spect gG7	22365n1916
Ho 296	0.4	66	69	6 - 6½	Binary, 21 yrs; PA dec, spect dG3	22384n1417
	72.2	235	24	-11½		
ζ	62.0	140	25	3½- 11	(42 Peg) Optical, spect B8 (*)	22390n1034
Σ2934	0.9	85	68	8 - 9	Binary, about 500 yrs; PA dec, spect G0; A= 0".1 pair, PA inc, about 80 yrs.	22394n2110
η	91.0	339	25	3 - 9	Spect G8 (*)	22406n2958
η b	0.2	83	49	10- 10	(β1144) no certain change	
ξ	11.8	101	54	4½- 12	(46 Peg) (h301) AB cpm, PA dec, spect F7, dM1	22442n1155
	145	15	24	- 11		
β1146	0.2	137	61	7½- 8½	Binary, about 145 yrs; PA dec, spect B9	22461n3050
Σ2945	4.2	296	53	8½- 8½	relfix, spect F0	22473n3103
Ho 482	0.3	54	69	7 - 7	Binary, about 125 yrs; PA dec, spect A3	22490n2608
	51.1	198	38	- 9		
Ho 191	3.2	90	07	7 - 13	spect A?	22512n3029
	24.3	280	23	- 10½		

LIST OF DOUBLE AND MULTIPLE STARS   (Cont'd)

NAME	DIST	PA	YR	MAGS	NOTES	RA & DEC
Σ2952	17.5	136	25	7½- 10	Spect G0	22518n2745
Σ2958	3.8	13	65	6½- 9	cpm, PA inc, spect A3	22543n1135
0Σ536	0.3	153	59	7 - 7½	Binary, 27 yrs; PA dec, nearly edge-on orbit, spect dG1	22560n0906
	233	84	24	-9½		
52	0.7	285	69	6 - 7½	(0Σ483) Binary, about 290 yrs; PA inc, spect F0	22567n1128
Σ2968	3.4	91	41	7 - 9½	relfix, spect A0	22583n3048
Σ2974	2.7	163	54	8 - 8	relfix, spect A0	23026n3306
0Σ488	14.4	337	25	7 - 11	spect K0	23050n2019
Σ2978	8.4	145	58	6 - 7½	cpm, relfix, spect A3	23051n3233
β378	18.8	55	15	7 - 11	spect A2	23055n3111
	48.8	62	15	- 11½		
A1238	0.2	301	65	7½- 8	AB binary, 72 yrs; PA dec, spect F5	23063n1041
	70.7	294	26	- 11		
A1238c	1.4	121	61	11½-11½		
57	32.9	198	23	6 - 10	(Σ2982) relfix, spect gM4	23070n0824
Σ2986	31.4	271	55	7½- 9½	cpm pair, relfix, spect G0	23075n1409
β385	0.4	101	59	7 - 8	(h5532) PA dec, spect B9	23079n3213
	57.9	77	51	- 9		
β852	58.3	283	13	7 - 10	Spect A2	23083n2614
β852b	1.2	347	54	10-11½	BC PA dec.	
Σ2990	2.2	58	65	8½- 8½	PA dec, spect F2	23108n2148
Σ2991	33.6	359	59	6 - 10	relfix, spect G5	23109n1048
Hu 400	0.4	157	68	7½- 8½	Binary, about 170 yrs; PA dec, spect A8	23151n1802
Σ3000	3.3	51	64	8½- 8½	relfix, spect F5	23163n2455
0Σ494	3.2	82	59	7½- 8	relfix, spect F0	23183n2140
64	0.5	133	58	5½- 9	(β718) PA inc, spect B3	23195n3132
	113	147	12	-13½		
Σ3007	5.9	91	66	6½- 9½	AB cpm, PA inc, AC dist inc, PA dec spect G0, K4	23203n2017
	88.3	311	56	- 8½		
66	0.1	56	54	5 - 5	(Ho 300) spect K3 PA inc.	23206n1202

LIST OF DOUBLE AND MULTIPLE STARS (Cont'd)

NAME	DIST	PA	YR	MAGS	NOTES	RA & DEC
β719	1.5	346	67	8 - 11	PA dec, spect F8	23219n1412
	108	106	16	- 10		
Σ3012	3.0	192	64	8½- 9	relfix, spect G0;	23251n1621
	52.0	66	58	- 8½	Star C= Σ3013	
Σ3013	3.2	275	53	8½- 9	relfix	23251n1621
β1266	0.1	38	62	8 - 8	Binary, 49 yrs;	23279n3034
	18.9	203	24	-9½	PA dec, ABC cpm, spect dF5 (Σ3018)	
0Σ497	1.3	214	65	8 - 8½	relfix, spect F8	23284n0912
Σ3020	2.1	104	55	7½- 9½	PA slow dec, spect A3	23286n1830
Σ3021	8.6	308	63	7½- 9	relfix, spect F8	23289n1557
	119	24	31	-10½		
Σ3023	1.9	282	55	7 - 9½	relfix, spect F2	23299n1708
72	0.5	252	69	5 - 5	(β720) Binary, PA inc, about 220 yrs; spect gK4	23315n3102
Ho 303	0.7	236	54	8 - 11	PA inc, spect K2	23383n2005
β858	0.8	236	65	7½- 9	PA dec, spect A0;	23388n3218
	23.3	52	09	-12½	AC= β389	
0Σ503	1.4	133	69	7 - 7½	relfix, spect F8	23395n2001
0Σ504	7.7	176	20	7 - 10	relfix, spect K0	23400n1823
β994	1.5	313	70	8 - 11	Slow PA inc, spect K0	23410n2449
78	0.9	245	70	5 - 8	(AGC 14) cpm pair; PA inc, spect K0	23415n2905
0Σ505	2.3	60	69	7 - 10	relfix, spect K0	23430n2008
Σ3041	61.0	352	36	7½- 8	Spect A0, G	23453n1647
Σ3041b	3.4	178	66	- 8		
A424	0.2	76	61	7½- 8	Binary, 85 yrs; PA inc, spect F0	23473n2724
Σ3044	17.8	282	51	7 - 7½	relfix, spect F0	23504n1139
h321	21.0	133	11	7 - 10½	Spect A0	23552n1112
Σ3048	8.8	313	64	7½- 8½	cpm pair, relfix;	23555n2404
	45.5	246	18	-10	AC PA inc, dist dec Spect G5	
Ho 208	1.1	203	70	8 - 10	PA dec, spect G0	23589n3028
85	0.8	160	70	6 - 11	AB binary, 26 yrs;	23595n2649
	75.5	330	32	-8½	PA inc, spect G3;	
	109	296	21	-13	C & D optical (*)	

LIST OF DOUBLE AND MULTIPLE STARS (Cont'd)

NAME	DIST	PA	YR	MAGS	NOTES	RA & DEC
Σ3055	5.7	358	43	7 - 11	Spect F0	00014n1152
A1250	0.8	58	70	8- 10½	PA dec, spect F5	00023n2949
ADS 42	23.1	122	07	6½-10½	spect K2	00024n2622
Σ3061	7.6	146	56	8½- 8½	relfix, cpm pair; spect F2	00032n1734
Σ3060	3.6	127	59	8½- 8½	cpm, PA inc, spect K0	00034n1748
	76.5	265	33	- 11		
β1027	1.6	190	69	7 - 10	spect K2	00124n2116

LIST OF VARIABLE STARS

NAME	MagVar	PER	NOTES	RA & DEC
β	2.1--3.0	Irr	Spect M2 (*)	23013n2749
γ	2.84 ±.03	.1517	β Canis Majoris type; Spect B2 (*)	00107n1454
R	7.2--13.6	378	LPV. Spect M6e--M9e	23041n1016
S	7.5--13..	319	LPV. Spect M5e--M8e	23180n0839
T	8.5--14..	374	LPV. Spect M6e--M7e	22065n1218
U	9.2--9.9	.3748	Ecl.bin; W Ursa Maj type; Spect F3+F3 (*)	23554n1540
V	8.0--15..	302	LPV. Spect M3e--M6e	21585n0553
W	7.6--13.4	345	LPV. Spect M6e--M8e	23174n2600
X	8.8--14..	201	LPV. Spect M2e--M5e	21186n1414
Y	9.6--15..	207	LPV. Spect M3e	22092n1407
Z	7.7--13.6	325	LPV. Spect M6e--M8e	23576n2537
RR	8.6--14..	264	LPV. Spect M4e--M6e	21422n2447
RS	8.4--14..	412	LPV. Spect M6e--M7e	22098n1418
RT	9.4--14..	215	LPV. Spect M3e--M6e	22020n3453
RU	9.7--13.0	Irr	SS Cygni type; spect sdBe +G8	22116n1227

PEGASUS

LIST OF VARIABLE STARS (Cont'd)

NAME	MagVar	PER	NOTES	RA & DEC
RV	9.0--14.5	389	LPV. Spect M6e	22233n3013
RW	8.8--14..	208	LPV. Spect M3e--M6e	23017n1502
RX	8.0--9.5	630	Semi-reg; spect N	21541n2237
RZ	7.7--13.5	439	LPV. Spect N	22037n3316
SS	8.0--13..	416	LPV. Spect M7e	22316n2418
ST	8.5--10..	136	Semi-reg; spect M6	22467n2706
SU	9.2--11.5	198	Semi-reg; spect M3e	23345n3225
SV	8.0--10..	145	Semi-reg; spect M7	22035n3506
SW	8.0--14..	396	LPV. Spect M4e	21202n2147
SX	9.0--14..	307	LPV. Spect S4e	22480n1738
SY	9.6--10..	Irr	Spect M0	22036n3440
TT	9.3--11.2	153	Semi-reg; spect M6e	00039n2649
TU	8.2--13.8	322	LPV. Spect M7e--M8e	21427n1228
TV	9.0--14..	247	LPV. Spect M0e	21244n1622
TW	7.0--9.2	956	Semi-reg; spect M6--M7	22017n2806
TX	8.6--10..	120	Semi-reg; spect M5e	22159n1321
TZ	9.0--13..	213	LPV. Spect M3e	21066n1550
UW	8.7--9.9	106	Semi-reg; spect M5	22156n0229
UY	9.5--11..	Irr	Spect M1	22424n3002
VX	9.0--10.5	935	Semi-reg; spect M7	21405n2215
AF	8.8--9.8	65:	Semi-reg; spect M5	22489n1751
AG	6.4--8.2	Irr	Erratic, Z Andromedae type; spect Be+M (*)	21486n1223
AK	7.8--10..	195	Semi-reg; spect M5e	23007n1106
AM	9.0--11.0	137	Semi-reg; spect M2e--M3e	21077n1215
AN	9.6--13..	275	LPV. Spect M5	21090n1310
AP	9.0--13..	300	LPV. Spect M6e	21268n1757
AS	9.0--14..	329	LPV. Spect M2e--M6	21107n1844
AT	9.5--10.3	1.1461	Ecl.bin; spect A0	22109n0811
AU	8.8--9.4	2.398	Cepheid; W Virg type; spect F8	21217n1804
AV	9.7--11.1	.3904	Cl.Var; Spect A7--F5	21498n2219
AW	7.2--8.0	10.62	Ecl.bin; spect A3+F5	21500n2347
AZ	9.0--10..	102:	Semi-reg; spect M5e	22043n2858
BC	9.3--10.3	125	Semi-reg; spect M6	22394n2054
BD	8.5--10..	78	Semi-reg; spect M6	22406n2754
BE	9.5--11..	74	Semi-reg; spect M5	22413n2255
BI	8.8--10..	500	Semi-reg; spect M6e	22554n1745
BK	9.5--10.0	2.745	Ecl.bin; spect F8	23446n2617

LIST OF VARIABLE STARS   (Cont'd)

NAME	MagVar	PER	NOTES	RA & DEC
DF	9.5--11.0	14.70	Ecl.bin; spect A2	21523n1419
DH	9.0--10.0	.2555	Cl.var.	22129n0634
DI	8.8--9.5	.7118	Ecl.bin; spect K0	23297n1442
DK	9.7--10.5	1.632	Ecl.bin; spect A5	23390n0956
DL	9.0--13..	181	LPV. Spect M0	23458n1523
DX	9.4--10.4	Irr	Spect M8	21449n2337
EE	7.0--7.5	2.627	Ecl.bin; spect A4	21376n0857
EI	9.0--10.5	61	RV Tauri type; spect M4	23193n1219
EM	9.0--10..	Irr	Spect M5	21367n0805
EO	8.5--9.5	Irr	Spect M7	23143n1020
EP	9.5--10.3	340	Semi-reg; spect M7	23576n1958
EQ	9.8--10.5	Irr	Red dwarf flare star; spect dM4e	23293n1940
EZ	9...10	Irr	Uncertain type; spect G5	23144n2527
FF	9.7--14..	252	LPV.	23336n0932
FZ	8.4--9..		Semi-reg; spect M5	21326n2803
GH	9.0--9.5	2.556	Ecl.bin; spect A3	21485n1501
GI	9.3--10..	Irr	Spect M6	22044n2444
GK	9.3--10..		Semi-reg; spect M6	22113n2511
GO	7.4--8..	Irr	Spect M4	22526n1917
GR	9.0--10..	Irr	Spect M8	23486n2654

LIST OF STAR CLUSTERS, NEBULAE, AND GALAXIES

NGC	OTH	TYPE	SUMMARY DESCRIPTION	RA & DEC
16	15[4]	⊘	E3; 13.1; 1.0' x 0.7' pB,pS,R	00065n2727
23	147[3]	⊘	SBa? 12.9; 1.0' x 0.6' pB,pS,1E	00073n2539
7078	M15	⊕	Mag 6.5; diam 10'; class IV ! vB,vL,vmbMN, stars mags 13.... fine object (*)	21276n1157
7137	261[2]	⊘	Sc; 13.1; 0.9' x 0.9' F,pS,R,vglbM	21459n2156

LIST OF STAR CLUSTERS, NEBULAE, AND GALAXIES   (Cont'd)

NGC	OTH	TYPE	SUMMARY DESCRIPTION	RA & DEC
7177	247[2]	⊖	Sa/Sb; 12.0; 2.3' x 1.3' pB,pS,E,bMN	21583n1729
7217	207[2]	⊖	Sb; 11.3; 2.7' x 2.4' B,pL,vlE,gbM (*)	22056n3107
7331	53[1]	⊖	Sb; 10.4; 10.0' x 2.4' B,pL,mE,sbM (*)	22348n3410
7332	233[2]	⊖	E7; 12.0; 2.1' x 0.4' cB,S,mE,smbMN; 7339 is 5' E = edge-on spiral or lenticular S0	22350n2332
7448	251[2]	⊖	Sc; 11.8; 2.0' x 0.9' pB,L,E,vgbM	22576n1543
7454	249[2]	⊖	E2; 13.1; 0.5' x 0.4' F,cS,1E,1bM	22586n1607
7457	212[2]	⊖	E5; 12.5; 3.0' x 1.3' cB,cL,1E,gmbM	22586n2953
7469	230[3]	⊖	S/pec; 12.8; 1.3' x 1.0' vF,vS,vsmbM, Faint outer ring	23007n0836
7479	55[1]	⊖	SBb; 11.8; 3.2' x 2.5' pB,cL,mE, S-shape spiral (*)	23024n1203
7619	439[2]	⊖	E1; 12.6; 0.8' x 0.6' cB,pS,R,psbM, On Peg-Pisces border; 7626 is 6.9' E	23178n0755
7625	250[2]	⊖	E/pec; 13.0; 0.7' x 0.6' pB,cS,1E,mbM	23180n1657
7626	440[2]	⊖	E2; 12.7; 0.9' x 0.7' cB,pS,R,psbM. Pair with 7619; brightest members of small galaxy group	23182n0756
7678	226[2]	⊖	Sc; 12.7; 1.5' x 1.1' vF,pL,R,1bM; compact spiral	23261n2209
7741	208[2]	⊖	SBc; 12.1; 3.1' x 2.5' cF,cL,iR; S-shape spiral with thick central bar (*)	23414n2548
7742	255[2]	⊖	E0/pec or S/pec; 12.4; 0.8' x 0.8'; cB,cS,mbM	23418n1029
7743	256[2]	⊖	Sa; 12.5; 1.7' x 1.4' pF,pS,R	23418n0939

NGC	OTH	TYPE	SUMMARY DESCRIPTION	RA & DEC
7769	230^2	⊖	Sb/Sc; 12.8; 1.0' x 0.8' pF,pS,R,mbM; 7771 is 5.3' ESE= F,E spiral	23485n1952
7772		⠿	Group of 7 small stars mags 11...14; diam 1.6' (*)	23492n1559
7814	240^2	⊖	Sa/Sb; 12.0; 1.0'x 5.0' cB,cL,mE,vgmbM; edge-on, equatorial dust lane (*)	00007n1551

DESCRIPTIVE NOTES

ALPHA   Name- MARKAB or MARCHAB, from the Arabian word for Saddle, though the term might also refer to a ship. Other Arabic names were *Matn al Faras*, the Horse's Shoulder, and *Yed Alpheras*, the Horse's Forearm or Hand. Magnitude 2.50; spectrum B9 or A0 III; position 23023n1456. The star marks the southwest corner of the "Great Square of Pegasus", the huge squarish figure about 18° X 14° that outlines the body of the Horse. Alpha Pegasi lies at a distance of about 110 light years, and has about 95 times the solar luminosity (absolute magnitude about -0.1). The annual proper motion is 0.07"; the radial velocity is 2.2 miles per second in approach, with slight but definite variations. The star lies in a rather blank part of the sky, lacking in faint stars; the interesting spiral galaxy NGC 7479, however, will be found about 2.9° almost directly south. (See photograph on page 1392)

Pegasus is, of course, the famed *Flying Horse* of Greek mythology, one of the most curious, but also one of the loveliest concepts created by the ancient myth-makers of the Greek world. In legend he was born from the blood of the Medusa, when that monster had been slain by Perseus, and his name, it is thought, comes from the Greek Πηγαι or

DESCRIPTIVE NOTES   (Cont'd)

"Pegae", the "Springs of the Ocean" at the place of his birth. The word πηγός or "strong" has also been suggested as a possible source of the name. After his creation, the Winged Horse made his first landing on the rocky heights above Corinth, where the blow of his hoof caused the famous spring of Peirene to gush forth; the spot was sacred to the Corinthians, and Pegasus was held in special reverence by the inhabitants of the city. A similar tradition credited Pegasus with having produced the Fount of Hippocrene on Mt.Helicon. Pegasus appears on coins of Corinth as early as 550 BC, where he is shown in a curiously archaic style (Fig.1); some of the later issues, in high classic style, are among the finest of coin creations of the ancient world and are eagerly sought by collectors:

1   2   3   4

The coin shown in Figure 2 was struck in the early 4th Century BC at Ambracia, while the specimen shown in Fig.4 dates to about 320 BC and was minted at Lokroi in Bruttium. The style was adopted by many cities of the Greek world; the obverse of virtually all of these coins shows a classic head of Athena, wearing the traditional Corinthian helmet. Pegasus also is featured on some of the large bronze coins of Carthage, probably minted in Sicily during the final years of the city's existence, before its total destruction by the Romans in 146 BC. (Fig.3)

Pegasus was tamed by Athena or Minerva according to Greek legend, and given to the Muses, in whose service he became the symbol of poetic inspiration; in another tradition he carried the thunder and lightning for Zeus. In another classic tale he became the steed of the Greek hero Bellerophon, Prince of Corinth, and slayer of the fearsome Chimaera, a most unlikely combination of lion, serpent, and goat. Bellerophon tamed the fabulous Flying Horse with the aid of Athena, after spending a night in prayer in her temple, and had many other fabulous adventures with the great horse. Eventually, however, Bellerophon became so bold as to attempt to fly to Olympus itself; the wiser Pegasus refused to attempt the flight and threw his rider to Earth. The tradition which connects Pegasus with the hero Perseus is of more modern origin, and is not supported by the ancient myths. The famous painting by Rubens, which depicts Pegasus present at the Rescue of Andromeda, is a part of this modern mythos; Shakespeare also refers in *Troilus and Cressida* to "Perseus' horse", evidently an allusion to Pegasus. In Greek writings Pegasus is often called simply "The Horse" or occasionally "The Divine Horse"; the Romans called it *Equus Gorgoneus* or *Equus Ales*, the "winged Horse"; another popular title was *Alatus* or "The Winged One" which appears in the *Alfonsine Tables*. In other Latin manuscripts it is called *Equus Medusaeus* which requires no translation. In the 1551 edition of Ptolemy's *Almagest* it is given as *Equus Pegasus*.

According to R.H.Allen, the constellation is identified as the *Horse of Nimrod* by ancient Jewish writers; the identification with the Archangel *Gabriel* is relatively modern, and has been attributed to Julius Schiller.

In the sky Pegasus appears turned over on his back with his body outlined by the Great Square (Alpha, Beta, Gamma, and the Alpha of Andromeda). His front legs are marked by Eta and Iota Pegasi, and his head by Epsilon; the great wings are not clearly indicated, but would lie more or less at the position of the "Circlet of Pisces", some 10° below the southern edge of the Great Square.

BETA Name- SCHEAT, from the Arabic *Al Sa'id* or *Sa'd*, the "Upper Part of the Arm" or possibly "The Foreleg". Riccioli has it labeled *Scheat Alpheraz* while Bayer has *Seat Alpheras*. Schickard's title appears to be very corrupted Arabic: *Saidol-Pharazi*. Beta Pegasi is magnitude 2.50 (variable); spectrum M2 II or III; the position is 23014n2749. The star marks the northwest corner of the Great Square of Pegasus.

Scheat is an irregular red variable star, similar in behavior to Betelgeuse, but much less extreme in size and luminosity. It varies from magnitude 2.1 to about 3.0 in an irregular period. The star was one of the first to be

measured with the beam interferometer on the 100-inch
reflector at Mt.Wilson; the apparent angular size was found
to be about 0.021". At the computed distance this corresponds to about 145 times the diameter of the Sun. Beta
Pegasi, like Betelgeuse, varies somewhat in size during
the course of the light cycle; the maximum diameter may be
about 160 times the diameter of our Sun. The true luminosity varies from about 240 suns up to about 500. If Beta
Pegasi should replace our sun, the star would not quite
fill the Earth's orbit. E.J.Hartung (1968) comments on the
fine appearance of the spectrum of this star, "with broad
dark bands in red and orange, and a series of narrower
bands in green, blue and violet."

The computed distance is about 210 light years, the
surface temperature about 3100°K, the mass about 5 solar
masses, and the average density about one millionth that of
the Sun. The star shows an annual proper motion of 0.23" in
PA 54°; the radial velocity is 5 miles per second in
recession.

Two faint field stars are listed in the ADS catalogue
as companions to Beta Pegasi, but these are optical atten-
dants only, and do not share the proper motion of the
bright star:

Mag 11   at   108.5" in PA 211°   (1924)
     9   253.1"   98°   "

The AB separation is slowly increasing from about 80" in
1828, while the AC distance is diminishing; both changes
are the result of the proper motion of the bright star.

GAMMA   Name- ALGENIB, probably from the Arabic *Al
            Janb*, "The Side", though some authorities
derive it from *Al Janah*, "The Wing". Magnitude 2.84,
spectrum B2 IV, position 00107n1454. The star marks the
southeast corner of the Great Square of Pegasus. Gamma Pegasi is a giant star, at a computed distance of about 570
light years, and with an actual luminosity of about 1900
times that of the Sun (absolute magnitude -3.4.) The annual
proper motion is about 0.01"; the radial velocity is 2.5
miles per second in recession.

One of the Beta Canis Majoris variables, the star has
an unusually short period of 0.1517495 day, or about 3 hours

and 38 minutes. As in all the stars of this type, the light variations are very slight, only a few hundredths of a magnitude.

The variable radial velocity of this star was first detected by K.Burns at the Lick Observatory in 1911. Owing possibly to the small amplitude of the variations, no further studies of the star were made until 1952 when D.H. McNamara and A.D.Williams at the University of California found the period to be remarkably short, only 3.63 hours. At the time this was the shortest period known for any Beta Canis star, but the star Theta Ophiuchi has since been found to have a period of 3h 22m. Light curves for Gamma Pegasi were measured in different colors by M.Jerzykiewicz in 1970, and it was found that the B magnitude shows slight but regular variations of less than 0.01 magnitude in a period of about 44 minutes, close to 1/5 the main period of the star. The V-B color index becomes smallest around the time of maximum light. Stars of this type are believed to be rather massive, young stars which are beginning to evolve away from the main sequence. (Refer also to Beta Canis Majoris, page 435)

The interesting eclipsing variable U Pegasi lies about 3.8° to the west and slightly north. (See page 1378)

EPSILON Name- ENIF, from *Al Anf*, "The Nose". Medieval Arabian charts sometimes label it *Fum al Faras* or "The Horse's Mouth". Magnitude 2.31, spectrum K2 Ib, position 21417n0939. The star is at a computed distance of about 780 light years, giving an actual luminosity of about 5800 times that of the Sun, and an absolute magnitude of -4.6. The annual proper motion is 0.025"; the radial velocity is slightly under 3 miles per second in recession. From the position on the H-R diagram, the estimated mass of the star is close to 10 solar masses.

Two faint stars in the field are not true physical companions to Epsilon; the further one, at 143", has been mentioned in many observing books owing to the statement by Herschel that the star exhibits a curious optical phenomenon: "the apparent pendulum-like oscillation of a small star in the same vertical as the large one, when the telescope is swung from side to side.." Herschel suggested

that the seemingly larger arc traversed by the small star
was due to the greater time required for its faint light
to affect the eye, so that "the reversal of motion is first
perceived in the larger object".

The fine globular star cluster M15 may be found in
binoculars by sweeping an area about 4° to the northwest.
(Refer to page 1383)

ZETA   Name- HOMAM, probably from the Arabic phrase
              Sa'd al Humam, the "Lucky Star of the Hero",
though Thomas Hyde derived it from *Al Hammam*, which seems
to mean "The Whispering One". According to R.H.Allen, the
names *Sa'd al Na'amah*, "The Lucky Star of the Ostriches"
and *Na'ir Sa'd al Bahaim*, "The Bright Fortunate One of the
Two Beasts" were also in use among the Arabs. The Chinese,
for some unknown reason, connected the star with thunder.
Zeta Pegasi is magnitude 3.46, spectrum B8 V, position
22390n1034. The computed distance is about 210 light
years, the actual luminosity about 145 times the sun, the
annual proper motion is 0.08", and the radial velocity
about 4 miles per second in recession.

The 11th magnitude companion at 62" was first noted
by S.W.Burnham in 1879, but appears to have no real connection with the primary. The distance between the two stars
is slowly decreasing from the proper motion of Zeta itself.

ETA   Name- MATAR, from the Arabic *Al Sa'd al Matar*,
              "The Fortunate Rain". Magnitude 2.96, spectrum
G8 II + F?, Position 22407n2958. Eta Pegasi is some 360
light years distant; the actual luminosity is about 630
times that of the Sun (absolute magnitude -2.2). The star
shows an annual proper motion of 0.03"; the radial velocity
is 2.5 miles per second in recession.

W.W.Campbell at Lick Observatory in 1898 found the
star to be a spectroscopic binary with a period of 818 days
and an eccentricity of about 0.155. The primary star is
close to 1 AU from the center of gravity of the system
and the spectroscopic companion appears to be an F-star of
uncertain class.

A visual companion at 91" is itself a very close
pair of about 0.2", but probably does not form a true

physical system with the bright star. The faint pair was
first resolved by S.W.Burnham with the 36-inch refractor
at Lick in 1889, and has shown no definite change in PA
or separation since that time.

KAPPA   Magnitude 4.27; Spectrum F5 IV, Position
             21424n2525. The parallactic distance is close
to 100 light years, giving a total luminosity of about 16
suns. The annual proper motion is 0.033"; the radial velocity is about 5 miles per second in approach.

    The 11th magnitude companion at about 14" has been
known since 1776 when it was recorded by William Herschel;
the PA has been decreasing slowly and the separation has
increased about 2" in the last century. The observed change
can be accounted for by the known proper motion of the
primary; thus it appears that the two stars form an optical
pair only, and are not physically associated.

    Kappa itself is a very close binary in rapid motion,
discovered by S.W.Burnham in August 1880. Burnham wrote in
1891: "The extreme difficulty of measuring so close a pair
seems to have deterred other observers, with a single

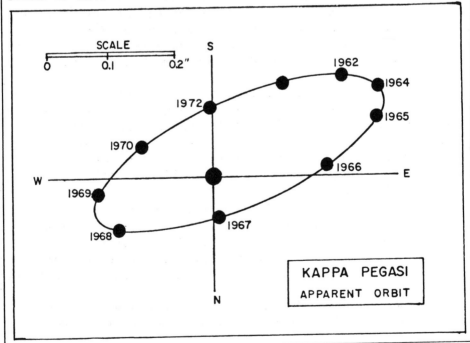

KAPPA PEGASI
APPARENT ORBIT

exception, from doing anything with it.... Since I have
been at Mt.Hamilton I have measured the close pair each
year with the 36-inch refractor. During the measures of the
past year it was extremely difficult, and was a severe test
of the power of the great telescope with the very best
atmospheric conditions."

T.Lewis, at the Royal Observatory at Greenwich, in a
paper written in November 1894, called attention to the
fact that the star was the most rapid binary then known,
and stated that Burnham's note "naturally induced an inspection with the 28-inch refractor of the Royal Observatory, Greenwich, which showed them distinctly separated
with a power of 1030, and our measures appear to confirm
the remarkably short period".

Modern measurements give the period as 11.53 years;
periastron was in late 1955, and the separation reaches
about 0.3" at maximum, as in 1964.   Individual magnitudes
are 4.8 and 5.2; spectral types are F5 (subgiant) and about
K0.   From the computed orbit the masses of the two stars
are 1.6 and 1.5; the absolute magnitudes are +2.3 and +2.7.
Orbital elements, according to W.J.Luyten, are: Semi-major
axis= 0.22" or about 7.5 AU; Eccentricity= 0.30; Inclina-
tion= 109°; motion retrograde with periastron= 1909.86. In
addition, the brighter star is a spectroscopic binary with
a period of 5.9715 days.

MU   Magnitude 3.50; Spectrum G8 III; Position
            22476n2420, about 4.5° SW from Beta Pegasi.
The parallactic distance is about 100 light years, which
gives the star an actual luminosity of about 30 suns. The
annual proper motion is 0.15"; the radial velocity is about
8.5 miles per second in recession.

37   Magnitude 5.47; Spectrum F5 IV; Position
            22274n0411, about 4.5° north of Zeta Aquarii
which is the central star of the "Water Jar" in Aquarius.
37 Pegasi is a rather close binary system with a period of
about 150 years, and a much elongated comet-like orbit
which is inclined at only about 2° from the edge-on position. The magnitudes of the two stars have been measured as
5.79 and 6.95, and the apparent separation varies from less
than 0.1" at closest approach (1917) to about 1.1" (1960).

DESCRIPTIVE NOTES (Cont'd)

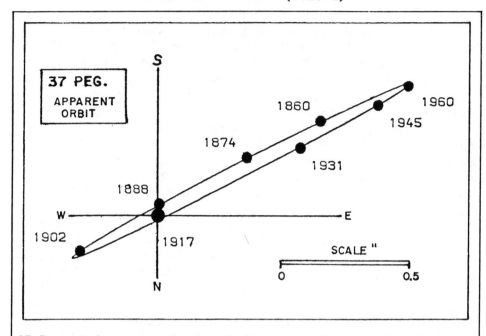

37 Pegasi lies at a computed distance of about 165 light years, which gives a total absolute magnitude of about +2 and individual luminosities of about 10 and 4 suns. The system shows an annual proper motion of 0.15"; the radial velocity is less than one mile per second in recession.

S.W.Burnham, in the BDS catalogue (1906) wrote that "this system has been difficult to measure with ordinary apertures, and many of the observations are discordant and uncertain". John Herschel, however, found it not difficult with 6-inch aperture and a power of 320, when near maximum separation. The orbital computations of three different authorities are compared below; the agreement is fairly good considering the difficulties of measurement.

	Period	Semi-M axis.	Eccen.	Incl.	Peri.
V.Biesbroeck	150 yr	0.81"	0.51	90	1908
G.F.Knipe	140	0.75	0.51	89	1908
T.Jastrzebski	143	0.75	0.56	88	1911

Incidentally, the true dimensions of the 37 Pegasi system are fairly comparable to the size of the Solar System; the computed mean separation is close to 40 AU.

85   Magnitude 5.75; Spectrum G3 V; position
23595n2649, about 1.8° SSW from Alpha Andromedae in the NE corner of the Great Square of Pegasus. This
is one of the best studied and most interesting of the
binary systems, but unfortunately an object suited for
observation only by large telescopes. It was discovered by
S.W.Burnham with the 18-inch refractor at Dearborn in 1878.
The period is 26.27 years with periastron occurring in 1962
and widest separation about 1970. The star is always a
frustratingly difficult object because of the closeness of
the pair and the brightness difference of at least three
magnitudes.
     Discrepancies in the reported magnitude of the faint
star are probably the result of these observational difficulties. The ADS Catalogue gives the small star a magnitude
of 11.0, Burnham has 12.5, and the Yale Catalogue of Bright
Stars lists a magnitude difference of 2.7 for the two stars
which would give the B component a value of 8.45. Orbital
elements according to R.G.Hall (1949) are as follows:
Period= 26.27 yrs; Semi-major axis= 0.83"; eccentricity=

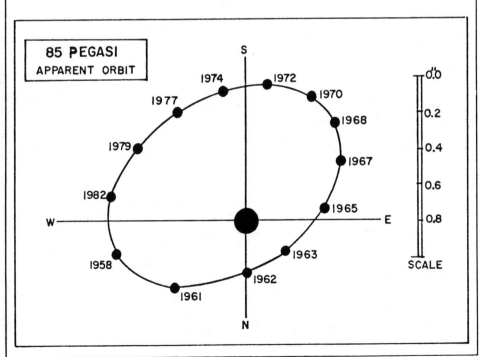

85 PEGASI
APPARENT ORBIT

DESCRIPTIVE NOTES   (Cont'd)

0.38; inclination= 50°; perihelion= 1910.11. At the known
distance of about 40 light years the true separation works
out to about 9.5 AU, fairly comparable to Saturn and the
Sun.   85 Pegasi is thus one of the fairly close binary
systems in space, and shows a large proper motion of 1.29"
annually in PA 140°; the radial velocity is 21.5 miles per
second in approach.   Very similar in type to our own Sun,
the primary star has about 70% the solar luminosity; the
computed absolute magnitude is +5.2.

The great peculiarity of the system is that the two
masses are nearly equal, as shown by the observed orbit,
and the fainter star thus violates the mass-luminosity
relation to an unusual degree. O.Struve and V.Zebergs in
1959 found masses of 0.82 and 0.80 suns for the two stars;
approximately the same masses were derived by A.A.Wyller
in 1956 in a study made with the 24-inch telescope at
Sproul. R.G.Hall in 1948, adopting a slightly different
parallax, found a smaller total mass of about 1.3 suns,
but again found the individual masses nearly equal. Hall
suggested that the anomalously large mass of the faint star
might be partly explained by the supposition that the star
is a close double; current studies suggest that the large
discrepancy cannot be completely explained in this way;
the faint star is underluminous by a factor of at least 15
and must be an extreme sub-dwarf. No spectrum of the faint
component has been obtained, so this is another of those
intriguing systems about which astronomers can say "more
work needs to be done".

Two faint stars in the field at 75.5" and 109" are
merely optical companions and do not share the large proper
motion of the bright system.

U   Position 23554n1540; Spectra dF3 + dF3. This
is one of the best known dwarf eclipsing
binary systems, closely resembling W Ursae Majoris, and
easily located about 3.8° west and slightly north from
Gamma Pegasi. It was discovered by S.C.Chandler at Harvard
in 1894, and the first accurate light curve was obtained
by E.C.Pickering as early as 1898; G.W.Myers at that time
pointed out that "the distance of centers does not materi-
ally differ from the sum of the radii of the components",
implying that the two stars were close to actual contact.

In 1915 the first orbital elements of the star were derived by H.Shapley, using a series of observations obtained by O.C.Wendell in 1909. The period was shown to be 0.3747819 day, or 8h 59m 41.1s, just a few seconds short of 9 hours. During each revolution there are two eclipses, the primary minimum being only 0.1 magnitude fainter than the secondary one. The visual range of the system is 9.2 to 9.9; when observed photographically it is 9.7 to 10.3.

The U Pegasi system consists of two dwarf stars, both of spectral type dF3, separated by about 1.2 million miles center to center. The components are of very nearly equal size and mass, each star being about 60% the solar diameter and close to the solar mass. O.Struve (1949) derived a total mass of 1.98, with individual masses of 1.10 and 0.88; his study of the star at McDonald Observatory showed that the more massive star is in front at time of primary eclipse. The orbital period of the system has been slowly decreasing since discovery, suggesting that the masses of the stars are gradually being altered through an exchange of material. Slight but definite changes in the period of this system have been detected by S.Gaposchkin (1932), by

R.LaFara (1951) and Z.Kopal (1956). In 1958 an extensive
series of photoelectric observations was made by L.Binnen-
dijk at the Flower and Cook Observatory of the University
of Pennsylvania. The light curve of the star was found to
repeat itself over a two-month period when measured both in
the yellow wavelengths and the blue, but showed definite
differences when compared with the light curve obtained in
1949 and 1950. A binary system of this type presents some
interesting problems; the two stellar surfaces are so nearly in contact that considerable exchange of material must
occur; not only does this slowly change the relative masses
of the two stars and alter the orbital elements, but it
undoubtedly affects the pattern of stellar evolution. In
1951, sudden flares of up to 0.3 magnitude were detected at
the Tokyo Observatory, implying that outbursts of some sort
are occurring on at least one of the components.

An object of the U Pegasi type is regarded by R.P.
Kraft (1961) as the probable ancestor of the erratic dwarf
novae of the SS Cygni class. From studies by Kraft and W.J.
Luyten it is known that the two classes of objects have a
very similar distribution in the Galaxy, and are comparable
in masses and periods. In the U Pegasi system it appears
that the primary star is being accelerated in its evolution
toward the hot subdwarf state, which in the future will
cause it to more and more resemble systems of the SS Cygni
type. The observed flares may indicate that this process
is now well underway.

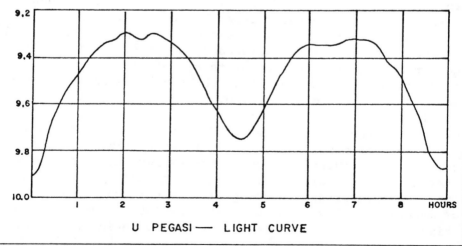

U PEGASI — LIGHT CURVE

The distance of the U Pegasi system is not accurately known as no measurable parallax has been detected. From the assumed total luminosity, the expected distance must be close to 360 light years. The annual proper motion is 0.09" and the radial velocity (strongly variable) implies that the orbital velocities of the two stars are about 100 and 123 miles per second. The whole system shows a recession velocity of some 15 miles per second. (Refer also to W Ursae Majoris, SS Cygni, and U Geminorum. Some speculations concerning the relationship of these stars to the classical novae will be found in the article on Nova Aquilae 1918)

The sparse grouping of stars called NGC 7772 lies about 1.4° to the west and slightly north. Although plotted on Norton's Atlas this little group does not appear to be a true cluster, and the members may not even form a physically attached group. The apparent size is about 1.6' and the individual magnitudes range from 11 to about 14. (Refer to photograph on page 1393)

AG   Position 21486n1223, about 3.3° NE from the bright star Epsilon Pegasi. AG Pegasi is a curious and erratic variable of uncertain class, probably to be regarded as a member of the "symbiotic" or "Z Andromedae" type. The star first came to the attention of the astronomical world in 1894 when bright hydrogen lines were discovered in its spectrum by W.Fleming at Harvard. Later studies showed spectral peculiarities resembling those of the "permanent nova" P Cygni. Light variations of the star were first noticed about 1920, and a check of photographic sky survey plates has made it possible to trace the light changes back to 1890, with a few early visual estimates available for the period 1821- 1890. The star seems to have been a 9th magnitude object until about 1850, brightening over the next 20 years to about 6th magnitude. From studies by A.J.Cannon (1912) it appears that no definite changes in the spectrum were detected from 1893 to 1912. But in 1920 there began a series of remarkable progressive changes, and within two years the spectrum had developed the peculiar composite appearance which is the identifying characteristic of the symbiotic stars. The original spectrum was of type Be, but since 1922 the additional spectrum of an M-star has gradually developed. Bands of titanium

DESCRIPTIVE NOTES (Cont'd)

COMPARISON MAGNITUDES: 20 Pegasi= 5.55; 17 Pegasi= 5.59
A= 6.08; B= 6.61; C= 6.73; D= 6.75; E= 7.59; F= 7.75;
G= 7.91; H= 8.05; J= 8.52.

oxide, first noticed in 1930, have increased in intensity
over the last 40 years, and have become one of the major
features of the star's spectrum. According to P.W.Merrill
in 1929, the bright lines of various elements show a cyclic
change in intensity and position in a period of about 800
days. This may represent the orbital period of the system,
though it is not known to which component the various spectral features should be attributed. The spectrum is further
complicated by gas streams around the star and the evident
presence of one of more gaseous shells expanding at different rates. Merrill in 1959 reported that at times 50 or
more of the bright lines have been accompanied by dark
absorption components on the side toward the violet; the
typical "P Cygni feature". "Change rather than stability
is the rule in this odd star" wrote Merrill in 1959. "The
velocity curve of a particular element may vary from time

to time. For example, the average velocity of the bright
hydrogen lines changed from a recession of 16 km/sec in
1915 to an approach of 27 km/sec in 1927. So far, however,
all velocity curves have maintained the 800-day period.
Since 1919 this has been about the only stable feature of
AG Pegasi."

In a study made in 1967, A.A.Boyarchuk of the Crimean
Astrophysical Observatory found that the spectrum of the
star showed typical Wolf-Rayet features combined with the
spectrum of an M-star. He derived spectral types of WN6 and
M3 III, and reported that the W-star resembles the nucleus
of a planetary nebula. The star evidently began to show
strong flare activity about 1850, and the resulting gaseous
shells are now beginning to resemble a true planetary. The
similar star FG Sagittae has shown the same type of activity during the last 80 years, and the surrounding nebula can
now be seen on photographs. R Aquarii is another object in
which the nebulosity is directly observable with modern
telescopes.

The light variations of AG Pegasi have not been large
in recent years. The general trend has been a slow fading
from a maximum of about 6th magnitude in 1870, and the
brightness of the star in 1961 was about 8.5. No sudden
outbursts of the Z Andromeda type have ever been observed.
The true distance and luminosity of the AG Pegasi system
is not well known, as the star is too remote to show either
a measurable parallax or proper motion. On the assumption
that the M-star emits about half the light of the system,
the total absolute magnitude may be in the range of -1.5
to -2; the distance must than be something like 3000 light
years. (Refer also to Z Andromedae, R Aquarii, BF Cygni,
and T Corona Borealis)

M15   (NGC 7078) Position 21276n1157. A beautiful
globular star cluster, discovered by Maraldi
in September 1746 during the search for the de Cheseaux
Comet of that year; the cluster was refound by Messier in
1764, and is generally considered to rank among the dozen
finest objects of its type in the northern sky. Easily
found about 4° NW of Epsilon Pegasi, the cluster has a
total magnitude of about 6½ and may be glimpsed as a fuzzy
star-like object in binoculars. Partial resolution is

GLOBULAR STAR CLUSTER M15 in PEGASUS. This is one of the richer and more compact globular clusters. U.S. Naval Observatory photograph made with the 61-inch reflector.

DESCRIPTIVE NOTES (Cont'd)

achieved in modest telescopes, though both Messier and Bode
described the cluster as a round "nebula" containing no
stars. Sir William Herschel, in 1783, was probably the
first to realize the true nature of M15.

Admiral Smyth found the cluster "not exactly round...
under the best circumstances it is seen with stragglers
branching from the central blaze..." "A very fine specimen
of a completely insulated cluster" says T.W.Webb, "bright
and resolvable, blazing in the center...a glorious object
in a 9½" mirror..." H.Shapley confirmed the slightly
elliptical shape of the cluster in about PA 35°; the oblate
form is most evident in the rich central mass, the distribution of the stars becoming very nearly spherical in the
outer portions. Long exposure photographs show several ill-
defined dark patches similar to those seen in the great M13
cluster in Hercules. According to Webb, one of these, near
the center, was originally discovered by Buffham with a 9-
inch mirror.

M15 is one of the richer and more compact globulars,
remarkable for the intense brilliance of its central core
where the countless stars seemingly crowd together into a
blazing nuclear mass; the central condensation is some 20"
in diameter. According to H.B.Sawyer (1947) M15 has a total
apparent width of about 12' and an integrated spectral type
of F3. Photographically, the total integrated magnitude has
been measured as 7.33, which puts it in 12th place in
brightness in the list of known globulars. The cluster has
an approach radial velocity of about 66 miles per second.

A large number of variable stars populate M15, 112
of these stars having been discovered up to 1973. The great
majority of these stars are short period variables of the
RR Lyrae class, all close to magnitude 15.9 (pg). About a
fourth of these stars have been known since the studies by
S.I.Bailey in 1897. One object, however, appears to be a
Cepheid of type II, with a period of 17.109 days. Another
unusual feature of M15 is the presence of a small planetary
nebula (K648) on the northeast side of the cluster. This
object, of photographic magnitude 13.8, and about 1" in
diameter, was found by F.G.Pease on plates made with the
100-inch telescope at Mt.Wilson in 1927, and in 1928 was
found to show a continuous spectrum of type O with a few
bright lines. From radial velocity measurements it appears

SPIRAL GALAXY NGC 7331 in PEGASUS. A system similar in type
and size to our own Milky Way Galaxy. Palomar Observatory
photograph made with the 200-inch reflector.

1386

DESCRIPTIVE NOTES (Cont'd)

certain that the nebula is a true member of the cluster.
M15 was identified in 1974 as a source of X-ray energy,
which suggests that the cluster may contain one or more
supernova remnants, or possibly one of those completely
collapsed stars or "black holes" so popular among writers
on speculative cosmology. As of 1975, the clusters NGC
1851 in Columba, NGC 6441 in Scorpius, and NGC 6624 in
Sagittarius are also known to be X-ray sources, so there
is growing evidence that a very massive collapsed body of
some sort may exist in the core of such clusters.

Shapley's original studies gave a distance of about
42,000 light years for M15, a value which has not been too
radically revised since the 1930's. H.C.Arp in 1965 found
the best modern value to be 10.47 kiloparsecs or about
34,000 light years. A study of the two color-magnitude
diagrams suggests that M15 is about 1.7 times more remote
than the Hercules cluster M13; this would give a distance
of about 39,000 light years. From these results the total
luminosity of the group is close to 200,000 times the Sun;
the cluster diameter is about 130 light years. The most
luminous members are red giants of absolute magnitude -2.
(Refer also to the Hercules star cluster M13, page 978)

NGC 7331   Position 22348n3410. Spiral galaxy, located
about 4.3° north and slightly west from Eta
Pegasi. NGC 7331 is a galaxy whose photograph is often used
in astronomy texts to illustrate the probable appearance
of our own Milky Way system if seen from intergalactic
space. It is a highly tilted spiral of type Sb, appearing
in moderate telescopes as an oval mass about 10' x 2½' and
with a total magnitude of about 10. The orientation is 15°
or 20° from the edge-on position, and the heavy dust lanes
bordering the west rim indicate that this is the closer
side to our own galaxy. Many bright nebulous regions mark
the outer arms, and A.Sandage at Palomar finds that the
spiral pattern can be traced to within 12" of the nucleus.
From photometric studies the total integrated magnitude
(pg) is 10.27; the integrated spectral type is G8. As seen
in the sky the tilt of the long axis is toward PA 163°.

A detailed study of NGC 7331 has been made by V.C.
Rubin, E.M. and G.R.Burbidge, D.J.Crampin, and K.H.Prender-
gast (1964). They describe the galaxy as similar in type

SUPERNOVA IN NGC 7331. This exploding star (arrow) appeared
in the galaxy in the summer of 1959.

Lick Observatory

and structure to the great Andromeda Galaxy M31, but about
23 times more distant. The apparent diameter is about 1/20
the Andromeda system, which implies a distance close to
50 million light years. Humason, Mayall and Sandage report
a corrected red shift of about 656 miles per second. From
a study of the radial velocity at various positions in the
system, the total mass appears to be about 80 billion solar
masses, out to a radius of 10 kiloparsecs (32,000 light
years) from the center. The faint outer portions probably
increase the total mass to about 140 billion suns. From the
same studies the total absolute magnitude (pg) is about
−21.2, and the total luminosity about 50 billion suns.

A supernova of apparent magnitude 12.7 (pg) appeared
in the galaxy in 1959; it was located in the most prominent
spiral arm, to the west of the nucleus. At the computed
distance the derived luminosity of this star was about −18.

Half a degree from NGC 7331, to the SSW, is the tight
little group of remote galaxies called "Stephan's Quintet",
consisting of NGC 7317, 7318A & B, 7319, and 7320.   This
is the best known case of an apparent cluster of galaxies
in which measurements show widely discordant red shifts.
NGC 7320 has a red shift of about 480 miles per second, but
the other four members show a mean value of about 3600. The
obvious explanation would be that 7320 is simply a foreground system, but photographs seem to show that it is
connected to the other members by faint tidal streamers.
In addition, the degree of photographic resolution seems
to imply that the other four members are not as remote as
their red shift values would indicate. One tentative theory
is that we may have here an expanding or exploding group;
an analysis by G.R. & E.M.Burbidge (1959) shows that the
Quintet must be disintegrating unless the individual masses
are in excess of one trillion suns. The scatter in the
individual velocity measurements seems to support the idea
that this group may be expanding or disintegrating:

NGC 7317	E4	Mag 15.3	+319 km/sec (with
7318A	E2	14.8	+221 respect to the
7318B	SBb	14.9	−779 mean)
7319	SBb	13.7	+240

STEPHAN'S QUINTET. This small group of galaxies lies some 30' south from NGC 7331. The largest member of the group is NGC 7320.   Lick Observatory

1390

GALAXY NGC 7814 in PEGASUS. An example of a precisely edge-on galaxy, showing a prominent equatorial dust-lane. Mt.Wilson Observatory photograph.

SPIRAL GALAXY NGC 7479 in PEGASUS. This S-shape barred
spiral lies about 3° south of Alpha Pegasi. U.S. Naval
Observatory photograph with the 61-inch reflector.

1392

DEEP SKY OBJECTS IN PEGASUS. Top: The small cluster NGC 7772 photographed at Lowell. Below: The many-armed spiral galaxy NGC 7217 as photographed at Palomar.

PERSEUS

LIST OF DOUBLE AND MULTIPLE STARS

NAME	DIST	PA	YR	MAGS	NOTES	RA & DEC
A1266	0.1	246	33	$7\frac{1}{2}$- $8\frac{1}{2}$	PA inc, spect A0	01361n5421
β1104	2.9	197	23	7- $11\frac{1}{2}$	Spect A3	01405n5238
Σ162	1.9	205	66	7 - $7\frac{1}{2}$	PA dec, spect A2,	01462n4739
	20.7	178	56	- 9	A2	
Σ213	1.9	322	36	$8\frac{1}{2}$- 9	Relfix, spect B9	02059n5051
	7.1	64	06	- 12		
5	5.6	270	35	$6\frac{1}{2}$- 12	(β874) spect B5	02080n5725
Σ230	24.1	258	25	8 - 9	relfix, spect B9	02114n5815
Σ235	1.9	46	46	$8\frac{1}{2}$- 9	Spect F8	02137n5541
7	69.9	354	11	6 - 11	(β1170) In field	02146n5717
7b	0.3	313	58	$11\frac{1}{2}$-$11\frac{1}{2}$	with Double Cluster	
					Prim.spect = gG6	
9	11.7	164	24	$5\frac{1}{2}$- 12	(β875) relfix;	02188n5537
					spect A2	
β---	8.3	122	06	$6\frac{1}{2}$- 14	spect G0	02203n5508
Hn 7	2.0	188	43	8 - 10	relfix, spect A2	02218n5759
Σ260	6.5	346	32	8 - $8\frac{1}{2}$	relfix, spect F5	02230n5405
Σ268	2.8	131	66	7 - 8	relfix, spect A2	02259n5519
Σ270	21.5	304	36	$7\frac{1}{2}$- $9\frac{1}{2}$	relfix, AB cpm;	02273n5520
	38.3	338	26	-$10\frac{1}{2}$	spect F5; AC dist	
	42.4	270	27	-$12\frac{1}{2}$	inc.	
β1314	3.6	120	04	7- 13	Field of cluster	02283n5729
	6.8	121	04	- 14	NGC 957	
	13.4	334	21	- $11\frac{1}{2}$		
	14.3	162	04	- 14		
	25.2	268	21	- $11\frac{1}{2}$		
Σ272	1.7	35	55	8 - 8	PA slow dec, spect	02295n5815
					A2	
OΣ42	0.2	278	67	7 - $7\frac{1}{2}$	Binary, PA inc,	02299n5205
					spect A2	
h2143	23.4	19	21	8 - $8\frac{1}{2}$	Field of cluster	02299n5719
					NGC 957	
A1279	2.0	307	59	9 - 9	relfix, spect A0	02368n5506
h1123	20.0	248	16	$8\frac{1}{2}$- $8\frac{1}{2}$	In cluster M34;	02387n4235
					spect both A0	
OΣ44	1.4	55	59	8 - $8\frac{1}{2}$	In cluster M34, PA	02390n4229
	86.3	290	08	-$9\frac{1}{2}$	slt.dec; spect B9	
Σ292	22.8	212	25	$7\frac{1}{2}$- 8	relfix, spect B9	02393n4003
β521	5.8	151	25	6 - 11	relfix, spect G5	02397n4803

LIST OF DOUBLE AND MULTIPLE STARS   (Cont'd)

NAME	DIST	PA	YR	MAGS	NOTES	RA & DEC
θ	18.3	303	53	4- 10	(13 Persei) (Σ296) cpm, binary, about 2700 yrs; spect F7, dM2; PA inc.	02407n4901
h654	32.2	45	24	7½- 10	spect B5	02411n3454
Σ297	15.7	278	26	8 - 8½	relfix, spect A0	02418n5621
	28.4	106	04	-10½		
Σ301	8.1	16	38	7½-8½	relfix, spect A0	02440n5344
β9	1.4	181	66	6½- 8½	PA inc, spect F2	02440n3521
	31.6	138	13	- 13		
Σ304	24.9	290	25	7½- 9	spect B9	02454n4859
η	28.4	301	32	4 -8½	(Σ307) AB cpm,	02470n5541
	66.3	269	25	-10	relfix; spect K3; yellow & blue pair	
η c	5.2	114	25	10-10½		
Σ316	14.3	135	57	8½- 8½	relfix, spect F8	02489n3705
Σ314	0.2	117	55	7 - 9	(A2906) spect B9;	02493n5248
	1.5	309	66	-7	AC PA inc.	
0Σ48	6.8	317	36	6½- 10	relfix, spect K0	02499n4822
20	0.2	239	62	5½- 6½	(Σ318) AB binary,	02506n3808
	14.0	237	29	- 10	31.6 yrs; PA dec, Spect F4	
β1293	2.0	351	70	7- 10½	Slight PA dec, spect B9	02506n4657
τ	51.7	106	23	5= 10½	(18 Persei) Spect G5, A5	02507n5234
τ b	3.5	87	25	10½-11	slow PA inc	
Σ325	15.0	160	57	8 - 9½	Optical, dist inc, PA dec, spect M0	02527n3417
Σ331	12.2	86	54	5½- 6½	relfix, cpm; spect B6 + B9	02572n5209
A1529	0.1	142	46	7½- 8½	PA dec, spect A2	02572n4741
Σ336	8.5	8	38	6½- 8	relfix, spect G5 color contrast	02584n3213
γ	0.4	49	54	4 - 4	(23 Persei)(h2170)	03012n5319
	56.9	325	39	- 11	binary, spect G8, A3 (*)	
Es 558	8.4	358	17	8- 9½	spect B9	03034n4534

LIST OF DOUBLE AND MULTIPLE STARS   (Cont'd)

NAME	DIST	PA	YR	MAGS	NOTES	RA & DEC
β	57.5	156	20	2- 12½	(β526) ALGOL, spect	03049n4046
	67.2	145	24	- 12½	B8; primary is ecl.	
	81.9	192	24	- 10½	var (*)	
Σ352	3.5	2	40	8 - 10	spect A0	03056n3516
Σ360	2.5	128	59	8 - 8	Dist inc, PA dec,	03090n3702
					spect G0	
OΣ51	0.9	324	66	8 - 8	PA inc, spect G0	03096n4406
Σ364	11.9	311	29	8½- 8½	relfix, cpm; spect	03104n3847
					F0	
Hu 544	1.2	99	54	6½- 9	spect A0	03122n5046
Σ369	3.4	30	58	6½- 8	relfix, spect A0	03139n4018
A1704	74.0	231	17	7½- 12	spect F0	03142n4230
A1704b	0.6	247	31	12-13		
OΣ53	0.8	271	69	7½- 8½	Binary, 118 yrs;	03145n3827
					PA dec, spect G0	
Ho 319	12.0	46	14	8,- 12	spect B8	03181n4512
	18.1	308	14	- 15		
Σ382	4.5	155	58	6 - 10	cpm, slight dist	03214n3322
	27.8	173	03	- 12½	inc, spect B9	
Ho 321	1.5	36	39	7½- 10	spect B8	03222n4520
Σ388	2.9	209	38	8 - 9	relfix, spect F0	03251n5016
AG67	23.6	360	25	7½- 10	spect G5	03256n4001
	52.8	354	15	- 11		
34	0.7	155	59	5- 10½	(β1179) PA dec,	03258n4920
					spect B5	
Σ391	3.7	95	49	7½- 8	relfix, spect B3	03258n4453
OΣ55	27.9	294	16	6- 10½	spect B5	03259n4646
OΣ55b	3.5	234	32	10½-13½		
β787	3.8	282	58	7½- 11	PA inc, dist inc,	03308n4827
	35.6	176	58	- 11	spect A0	
Σ413	2.5	129	51	8½- 8½	relfix, spect F0	03322n3331
β533	1.0	41	65	7 - 7	PA dec, dist inc,	03325n3131
					spect F0	
S430	41.1	95	25	8 - 8	spect A0	03348n4439
	53.3	360	04			
Σ425	2.0	79	56	7½- 7½	PA & dist slow dec,	03370n3357
					spect F5	
OΣ59	2.5	252	59	7½- 8	relfix, spect G5	03372n4552

LIST OF DOUBLE AND MULTIPLE STARS   (Cont'd)

NAME	DIST	PA	YR	MAGS	NOTES	RA & DEC
Σ426	19.8	342	30	7½- 9	relfix, spect A3	03375n3857
Σ426b	12.6	37	04	9- 13½		
40	20.0	238	25	5 - 9½	(Σ431) relfix, spect B1	03392n3348
β1182	18.6	244	10	6½- 13	relfix, spect K0	03405n4822
	4.4	260	34	- 14		
Σ434	31.0	85	25	7 - 8	Slight dist inc; spect K5	03407n3813
O	1.0	37	58	4 - 8½	(38 Persei) (β535) PA dec, spect B1; in neb IC 348. A= slight var.	03412n3208
Σ439	0.6	8	58	8½- 10	(β880) PA inc, spect B5, A0	03414n3200
	23.8	39	58	-9		
Ho 504	1.0	192	66	8 - 8	PA slight inc, spect A0	03414n3542
β1183	6.5	138	35	6 -14	spect B9	03425n4532
Σ443	8.0	50	54	8 - 8½	cpm, PA inc, spect G5; AC optical, AC dist dec, PA inc.	03435n4119
	15.3	178	15	- 11		
Σ443c	4.4	142	15	11-13½		
OΣ63	6.9	270	40	6 - 11	relfix, spect B8	03446n5035
Σ448	3.5	17	34	7- 9½	relfix, spect B3; Zeta Persei group	03447n3327
Σ446	8.5	255	34	7 - 9	relfix, spect B0	03457n5230
	11.9	39	10	- 12		
	66.5	336	09	- 10½		
Σ446d	2.6	232	09	10½-11		
OΣ516	2.3	48	21	7½- 9½	relfix, spect B9	03460n3207
Es 277	30.2	143	06	7 - 10	spect F0	03470n3440
Es 277b	7.4	290	06	10-14		
OΣ66	0.9	140	54	7½- 8	PA slow inc, spect A2	03487n4039
ζ	12.9	209	57	3- 9½	(Σ464) AB cpm, spect B1; AD dist inc, optical (*)	03510n3144
	33.0	287	23	- 11		
	94.2	195	57	-9½		
β263	0.8	96	70	8- 8½	PA inc, spect F8	03533n3302
ε	8.8	9	38	3 - 8	(Σ471) relfix cpm; spect B0,B8 (*)	03545n3952
	78.3	11	12	- 14		
OΣ69	1.5	326	50	6½- 9	PA dec, spect A0	03563n3841

LIST OF DOUBLE AND MULTIPLE STARS (Cont'd)

NAME	DIST	PA	YR	MAGS	NOTES	RA & DEC
Σ476	21.8	287	50	7½- 8½	Dist inc, spect K2	03582n3832
Σ477	3.1	214	27	8½- 9½	relfix, spect A2	03586n4143
Es 2085	3.9	266	51	7½- 10	spect A0	04005n3750
Σ483	0.8	99	67	8 - 9½	binary, about 400 yrs; PA dec, spect dG5	04007n3923
A1710	0.2	9	62	8 - 8	binary, about 80 yrs; PA dec, spect G5	04030n4317
0Σ71	0.8	206	57	7 - 9	(AG Persei) PA inc, spect B4. Primary is ecl.bin.	04038n3319
	36.0	118	48	-12½		
0Σ531	1.5	18	69	6½- 8	binary, about 700 yrs; PA & dist dec, spect dK2	04042n3757
Σ492	97.0	200	21	6½- 10	spect A2, B=5" pair	04049n4122
A998	0.3	300	55	8 - 8	PA dec, spect A0	04053n4606
Σ3114	2.8	161	48	7½- 10	PA dec, spect F8	04058n4002
Ho 327	15.9	310	06	6½- 12	both PA dec, spect F5	04064n3131
	14.8	160	27	- 14		
Σ500	4.2	80	36	8½- 9½	relfix, spect G0	04081n4008
μ	14.6	350	34	4½- 12	(51 Persei) (0Σ73)	04112n4817
	83.8	232	25	- 10	relfix, spect G0	
Σ512	5.2	223	50	8½- 8½	relfix, spect G5	04122n4517
0Σ77	0.8	269	70	8 - 8	binary, about 200 yrs, PA inc, spect dG0	04128n3134
	56.3	42	33	- 9		
0Σ76	3.9	210	03	7½- 12	relfix, spect A3	04129n3445
0ΣΣ44	58.4	321	24	6½- 7½	wide pair, spect A2, B	04137n4606
Σ519	18.2	347	26	7½- 9	spect K2	04171n5016
0ΣΣ47	72.6	327	19	7 - 8	dist dec, spect K2	04172n5008
b	22.8	258	13	8- 11		
Σ521	2.1	257	33	7 - 9	relfix, spect G0	04180n4955
β310	19.3	172	27	8 - 12	spect F8	04191n3949
0Σ80	0.5	166	58	6½- 7	cpm, PA dec, spect B9	04201n4219
Σ533	19.6	61	32	6- 7½	relfix, spect B9	04212n3412
56	4.3	30	47	6- 8½	(0Σ81) cpm, PA dec spect dF5	04214n3351

LIST OF DOUBLE AND MULTIPLE STARS   (cont'd)

NAME	DIST	PA	YR	MAGS	NOTES	RA & DEC
Hu 609	0.3	192	67	8½- 9	binary, about 95 yrs; PA dec, spect F5	04230n3436
Es 568	5.1	304	08	7½- 12	spect G5	04242n4305
Σ552	9.1	115	49	6½- 6½	neat pair, relfix, spect B8	04280n3954
β789	0.9	318	66	8 - 9	PA slight dec, spect F8	04282n3733
Σ551	13.7	126	15	8½- 9	relfix, spect A	04283n5206
57	115	199	13	5 - 6	(0ΣΣ50) both spect	04299n4258
	76.4	354	08	- 12	F0	
S451	58.4	199	24	7½- 7½	spect F5, F	04326n4715
Σ563	11.8	32	18	8 - 9½	(65H) relfix	04332n4059
Σ565	1.4	172	43	7½- 8½	PA dec, spect K0;	04346n4201
	29.5	109	14	-11½	AC dist dec.	
	71.6	52	06	-11		
Σ577	1.6	36	59	7½- 7½	binary, about 700 yrs; PA dec, spect F8	04388n3725
Σ582	5.7	23	34	7½- 10	relfix, spect B9	04405n4220
	97.2	140	04	-10½		
Σ582c	7.3	340	13	11-11	(Σ 581)	

LIST OF VARIABLE STARS

NAME	MagVar	PER	NOTES	RA & DEC
β	2.1--3.4	2.867	ALGOL. Spect B8; typical eclipsing binary (*)	03049n4046
ο	3.82 ±.02	4.419	(38 Persei) Ellipsoidal var. Spect B1. Also visual double star	03412n3208
ρ	3.3--4.0	40:	Semi-reg; spect M4	03020n3839
φ	4.3--4.4	Irr	Gamma Cass type? Spect B1e	01405n5026
48	4.0--4.1	55:	(MX Per) Semi-reg; B3p	04050n4735

PERSEUS

LIST OF VARIABLE STARS (Cont'd)

NAME	MagVar	PER	NOTES	RA & DEC
b[1]	4.6--4.7	1.527	Ellipsoidal binary, spect A2	04145n5010
R	8.1--14.6	210	LPV. Spect M2e--M5e	03269n3530
S	7.3--11..	810:	Semi-reg; spect M3e--M4e	02193n5822
T	8.2--9.1	326	Semi-reg; spect M2	02157n5844
U	7.7--12.0	321	LPV. Spect M6e--M7e	01562n5435
V	9....17	---	Nova 1887	01585n5630
W	8.5--11.8	470	Semi-reg; spect M5	02469n5647
X	6.0--6.6	Irr	RW Aurigae type or nebular var. Spect 0e	03523n3054
Y	8.1--10.9	252	LPV. Spect N	03243n4400
Z	9.4--12.0	3.056	Ecl.bin; spect A0	02368n4159
RR	8.1--15..	390	LPV. Spect M6e--M7e	02251n5103
RS	9.0--11..	152	Semi-reg; spect M4	02188n5653
RT	9.9--11.4	.8494	Ecl.bin; spect F2	03202n4624
RU	9.9--12.0	310:	Irr or semi-reg; spect M4e	03273n3929
RV	9.9--12.4	1.973	Ecl.bin; spect A2	04074n3408
RW	9.4--11.6	13.198	Ecl.bin; spect A5e+gG0	04168n4212
RX	9.0--13..	422	LPV. Spect M3e	03481n3253
RY	8.3--10.0	6.864	Ecl.bin; spect B8+F8	02423n4756
RZ	8.7--14..	354	LPV. Spect S4	01266n5036
ST	9.0--10.9	2.648	Ecl.bin; spect A3+K	02569n3858
SU	7.9--9.0	470	Semi-reg; spect M3	02186n5623
SV	8.6--9.6	11.129	Cepheid; spect F6--G1	04463n4212
SW	8.3--9.8	102	Semi-reg; spect M5	04074n4205
SY	9.5--12.5	476	Semi-reg; spect Ne	04128n5030
SZ	10....?	---	Nova? (*)	03439n3410
TT	8.2--9.6	82:	Semi-reg; spect M5	01472n5330
TV	9.9--13..	363:	Semi-reg; spect K0	02407n3602
TW	9.8--12..	335	LPV. Spect M2e	03168n3258
TX	9.8--12.5	78	Semi-reg; spect G5e--K2	02449n3645
UZ	8.0--10..	928	Semi-reg; spect M5	03170n3151
VX	9.0--9.8	10.89	Cepheid; spect F6--G1	02043n5812
XX	8.6--9.4	Irr	Spect M3	01598n5500
XY	9.2--10.6	Irr	RW Aurigae type? Spect B6 + A2	03463n3850
YY	9.0--10..	Irr	Spect M5	04329n4453
YZ	9.3--10.4	378	Semi-reg; Spect M1--M3	02348n5650

LIST OF VARIABLE STARS   (Cont'd)

NAME	MagVar	PER	NOTES	RA & DEC
AA	9.0--10.0	130	Semi-reg; spect M6	03118n4624
AD	9.0--10..	320	Semi-reg; spect M2--M3	02169n5646
AE	9.3--11.4	115	Semi-reg; spect M5	02568n4356
AF	9.5--10.7	89	Semi-reg; spect M6	03391n3621
AG	6.5--6.8	2.029	(0Σ71) Ecl.bin; spect B4 + B5. Also double star	04038n3319
AW	7.5--8.2	6.463	Cepheid; spect F6--G1	04444n3638
AX	9.4--13.5	Irr	Z Andromedae type, period averages ±680 days; spect gM3e	01331n5400
AY	9.7--10.6	11.777	Ecl.bin; spect A0	03068n5045
BU	9.3--11..	365	Semi-reg; spect M4	02153n5712
BW	9.0--10..	117	Semi-reg; spect M4	02438n4551
DM	7.7--8.5	2.728	Ecl.bin; spect B6	02224n5553
FZ	8.7--9.5	Irr	Spect M1	02174n5655
GK	0.2----14	---	NOVA PERSEI 1901 (*)	03278n4344
GQ	9.0--10..	Irr	Spect M6	04201n3606
IQ	7.5--8.0		Ecl.bin; spect B9	03561n4800
IR	9.2--11..	175	Semi-reg; spect M7	04166n4057
IW	5.8--5.9	.9172	Ellipsoidal binary, spect A3	03303n3944
IX	6.7 ±.01	1.326	Ellipsoidal var. Spect F2	03319n3152
IZ	7.8--9.0	3.688	Ecl.bin; spect B8	01289n5346
KK	6.6--7.6	Irr	Spect M1--M3	02068n5619
KP	6.3--6.4	.200	β Canis Major type; spect B2	03292n4441
KS	7.6--7.7	35:	Semi-reg; spect A5e	04453n4311
LT	5.08 ±.05	1.729	α Canum type, spect A0p	02543n3144
LX	7.6--8.6	8.038	Ecl.bin; spect G4	03098n4755
PP	9.2--10.3	Irr	Spect M0--M1	02135n5818
PR	8.9--9.6	Irr	Spect K5e--M2	02181n5738
V400	8.0--19..	---	Nova 1974	03042n4656

PERSEUS

LIST OF STAR CLUSTERS, NEBULAE, AND GALAXIES

NGC	OTH	TYPE	SUMMARY DESCRIPTION	RA & DEC
650	M76	◎	Mag 11; diam 140" x 70" with 16½^m central star; "Little Dumbbell" or "Cork Nebula". vB,E, irregular planetary (*)	01388n5119
869	33^6	⁙	!! vvL,vRi, diam 35'; stars mags 7.... class F. "Double Cluster" with NGC 884 (*)	02155n5655
884	34^6	⁙	!! vL,vRi; diam 35'; stars mags 7.... class E. "Double Cluster" with NGC 869 Beautiful field (*)	02189n5653
957		⁙	pL,pRi, diam 10'; about 40 stars mags 11...15; class E; incl. double stars β1314 and h2143	02289n5718
1023	156^1	⊖	E7/pec; 11.0; 4.5' x 1.3' vB,vL,vmE, vvmbM, lens-shape E-W, with faint tuft or satellite galaxy on E tip	02372n3852
1039	M34	⁙	B,vL,lC, Mag 6; diam 20'; 80 stars mags 8....class D. Incl double stars h1123 and 0 Σ44. Fine cluster (*)	02388n4234
1058	633^2	⊖	Sc; 12.5; 2.3' x 2.1' pF,cL,R,glbM. Compact spiral; supernovae in 1961 & 1969	02402n3708
1169	620^2	⊖	Sb; 13.0; 2.2' x 1.9' pF,pS, iR	03001n4612
1175	607^2	⊖	S ; 13.1; 1.1' x 0.2' F,cL,E	03013n4208
1220		⁙	S,F, Diam 3', mag 11; Irr; class E. About 25 F stars. 1° E from Gamma Persei	03078n5310
----	Kg 5	⁙	L,F, diam 7'; about 40 stars mags 13....	03110n5232
1245	25^6	⁙	pL,Ri,C,iR; diam 20'; mag 9; class E; about 100 stars mags 11..... (*)	03112n4703

LIST OF STAR CLUSTERS, NEBULAE, AND GALAXIES   (Cont'd)

NGC	OTH	TYPE	SUMMARY DESCRIPTION	RA & DEC
1270		⊖ (galaxy)	E2; 12.7; 0.6' x 0.5' vF,S,R, NGC 1275 nf 5'	03156n4118
1275		⊖ (galaxy)	S? 13.0; 0.7' x 0.6' vF,S, eruptive galaxy= radio source Perseus A (*)	03164n4120
1333		□ (nebula)	F,L, 5' x 9' with 11^m B9 star on N edge; in L dark neby.	03261n3112
1342	88[8]	⣉ (open cluster)	vL, Irr scattered group; diam 15', class C. about 50 stars mags 8.....	03284n3709
----	I.348	□ (nebula)	vvL neby 60' x 120' surrounds 4^m star Omicron Persei	03420n3200
----	I.351	◎ (planetary nebula)	Mag 11; diam 8" x 6" with 15^m central star	03443n3454
1465		⊖ (galaxy)	E7/S0; 14.5; 1.0' x 0.2' vS, edge-on spindle shape; 0.6° N from Zeta Persei	03505n3221
----	I.2003	◎ (planetary nebula)	pB,eS, Mag 12; diam 5" with 18^m central star	03532n3344
1491	258[1]	□ (nebula)	vB,S,Irr,bM; 3' x 3' with F outer wisps. Incl 11^m B0 star	03595n5110
1499		□ (nebula)	vvL,vvF,mE,Irr; 145' x 40'; "California Nebula" 0.6° N from Xi Persei, spect 07e (*)	04001n3617
1513	60[7]	⣉ (open cluster)	L,vRi,pC, diam 12'; mag 9; class D; about 40 stars mags 11.... 2° SSW from NGC 1528	04062n4923
1528	61[7]	⣉ (open cluster)	B,vRi,cC; Mag 6; diam 25'; class E; about 80 stars mags 8.... 1° NNW from b¹ Persei; NGC 1513 2° to SSW (*)	04114n5107
1579	217[1]	□ (nebula)	pB,pL,iR, diam 8' x 12' with dark lanes and 12^m star	04269n3510
1624	49[5]	□ (nebula)	F,cL,R, 3' x 3'; surrounds S group of 7 stars with 12^m Oe star	04365n5021

DESCRIPTIVE NOTES

ALPHA   Name- MIRFAK or MARFAK; the star also appears
           on various star-maps as ALCHEMB or ALGENIB,
evidently through some age-old confusion, as the latter
name is also given to both Gamma Persei and Gamma Pegasi.
Magnitude 1.79; spectrum F5 Ib; position 03207n4941. The
star lies in the center of the constellation, and dominates the curving row of stars extending from Eta Persei to
Delta and 48 Persei, a pattern often referred to as the
"Segment of Perseus". Alpha is the brightest star in the
constellation; the name signifies "The Elbow".

     Like Hercules and Orion, the star pattern of Perseus
was identified in many ancient cultures with prominent
national heroes or gods. According to Lalande, the Egypti-
ans regarded it as their god *Khem*, while to the Persians
it represented *Mithras*; Biblically minded star-watchers
saw it as *David* with the Head of Goliath, or as St.George
slaying the Dragon. The modern name was in use in the days
of the classical Greeks, who also referred to the constellation as "The Champion", "The Rescuer", or occasionally
'Ιπποτης, "The Horseman". The Moorish name *Almirazgual* is
from the Arabic *Hamil Ras al Ghul*, the "Bearer of the
Demon's Head". Another Arabic name, *Killab*, refers to the
hero's weapon, and is probably the source of the *Celeub* or
Chelub of the *Almagest* and Bayer's *Uranometria*.

     Perseus in mythology was one of the greatest of the
ancient Greek heroes, the great-grandfather of Hercules and
traditionally the ancestor of the Persians. Like many other
ancient heroes, Perseus was of partly divine parentage; his
mother Danae had been visited by Zeus in the form of a
shower of gold. Owing to a prophecy that the son of Danae
would slay his grandfather, King Acrisius of Argos, both
mother and son were imprisoned in a wooden chest and cast
into the sea. The chest, however, floated safely to the
island of Seriphus, where it was found by the fisherman
Dictys, father or brother of King Polydectes of the island.
Here Perseus grew to manhood, and to eventually fulfill his
destiny.

     It was at the request of Polydectes that Perseus
undertook his most famous exploit, the slaying of the fearsome Medusa, one of the three Gorgons whose glance turned
men to stone. With the aid of Athena and Hermes, who
supplied him with winged sandals, sword, and a helmet of

invisibility, Perseus found his way to the realm of the
Gorgons on the farthest shores of Oceanus, near the isles
of the Hesperides, and slew the monster while looking at
her reflection in his polished shield. Another tradition
places the lair of the Gorgons in a "sea-girt cave" near
Tartessus, the Biblical Tarshish on the southern coast of
Spain, west of Gibraltar near modern Cadiz. It was while
Perseus was returning from this adventure that he found
the princess Andromeda chained to a rock on the Ethiopian
coast as a sacrifice to the sea-monster Cetus, sent by the
sea god to punish the kingdom for the boastful vanity of
Queen Cassiopeia. Gods, even the best of them, seem to
lack the most elementary sense of justice; they dispense
punishments at random and appear to be quite satisfied as
long as someone suffers sufficiently in order to "atone"
for the original offense. Perseus rescued Andromeda,
destroyed the monster, and, it is to be hoped, warned the
vain Cassiopeia to be more careful in the future. Although
these events presumably took place on the Ethiopian coast
somewhere along the southern shore of the Red Sea, a great
rock formation on the Israeli coast near Jaffa (Joppa) is
also identified as the site of Andromeda's rescue. Returning to the court of Polydectes, the hero turned the schem-
ing king and all his noblemen to stone by showing them the
Gorgon's head. Dictys became king of the island, and the
head of Medusa was presented to Athena who set it in the
center of her shield.

        According to another tradition, Perseus released the
titan Atlas from his wearisome task of holding up the
Heavens; he showed the giant the Gorgon's head, and transformed him into a great rock (the Atlas Mountains) in
southwest Morocco. Later, while attending the funeral
games of the king of Larissa in Thessaly, Perseus fulfill-
ed the prophecy made at his birth when he unintentionally
struck Acrisius with a discus and killed him.

        Many of the characters of the Perseus legend appear
in the sky as constellations, including Perseus himself,
Andromeda, Cetus the Sea-monster, Queen Cassiopeia, and
King Cepheus. Perseus is one of the most extensive constellations, stretching from the borders of Cassiopeia to
almost the Pleiades in Taurus. He is depicted standing,
holding aloft the Medusa's head with one arm and grasping

THE ALPHA PERSEI GROUP. This scattered cluster forms a
splendid group for binoculars. Alpha is the bright star at
upper left. Lowell Observatory 5-inch camera photograph.

DESCRIPTIVE NOTES   (Cont'd)

his sword with the other. In art he usually appears in the
same pose, as in the great statue by Benvenuto Cellini,
completed about 1554 after nine years of labor, and now in
the Piazza della Signoria in Florence. A similar composition by Antonio Canova, created about 1806, is one of the
treasures of the Metropolitan Museum of Art in New York.
Ancient coin portraits of Perseus are much rarer than those
of Hercules; the most striking surviving specimens are the
coins of King Perseus of Macedonia, last king of the line
of Alexander, and who lost his throne to the Romans in 168
BC.   King Perseus evidently regarded himself as a descend-
ant or incarnation of the ancient hero, and had himself so
depicted on his coins:

Alpha Persei is a giant star, with an actual luminosity of over 4000 suns; the computed distance is about
570 light years and the absolute magnitude about -4.4. Both
Mary Proctor and R.H.Allen have called the color "lilac"
though to most modern observers it is merely white or even
slightly yellowish. The star shows an annual proper motion
of 0.04" and the radial velocity is about 1 mile per second
in approach.   Small periodic shifts in the spectral lines
have been detected in a cycle of about 4 days; it is not
certain whether this indicates that the star is a close
binary. In view of the very short period for so large a
star, it seems more likely that the cyclic shifts result
from pulsations in the atmosphere of a single star.
Lying directly in the Milky Way, Alpha Persei is the
center of a fine field of stars which form a rich and
brilliant group for small telescopes; the region appears
as a large and scattered cluster even without optical aid,
and is truly splendid in good binoculars. Modern studies

DESCRIPTIVE NOTES   (Cont'd)

confirm the reality of this cluster as a true moving group
in space. According to O.Heckmann and K.Lubeck (1958) 124
members are now recognized, though the membership of perhaps a dozen are uncertain. In a study made in 1971 by
W.W.Morgan, W.A.Hiltner, and R.F.Garrison, 106 stars were
accepted as members; the resulting color-magnitude diagram
(below) is plotted from their measurements made at the
McDonald Observatory. With the exception of Alpha itself,
the cluster stars appear to form a well defined main sequence in which the earliest type stars are spectral class
B3 with absolute magnitude -1.0, and the latest are about
G3 with absolute magnitude about +4.6. The entire group
lies about 175 parsecs from the Sun, and is moving about
10 miles per second in the general direction of Beta Tauri
or PA about 140°.

    After Alpha itself, the most prominent members are
Psi, 30, 34, 29, and 31 Persei; the bright stars Delta and
Epsilon show approximately the same motion. At the present
rate of motion, however, the group will require about
90,000 years to change its position by 1° in the sky.

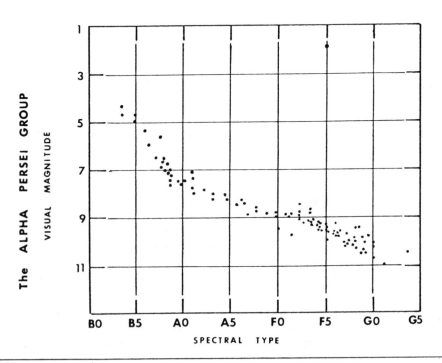

BETA   Name- ALGOL, the "Demon Star", the most famous
of the eclipsing variables. Magnitude 2.15
(variable); Spectrum B8 V; Position 03049n4046. The name is
from the Arabic *Al Ra's al Ghul*, "The Demon's Head". To
the writers of classical times the star represented the
head of Medusa held by Perseus in the mythological outline
of the constellation. This is the *Gorgoneum Caput* of
Vitrivius and the *Caput Gorgonis* of Hyginus; Manilius, in
the days of Augustus, called it *Gorgonis Ora*. The Hebrews,
according to R.H.Allen, knew the star as *Rosh ha Satan* or
"Satan's Head", but in some other traditions it is identified with the mysterious and sinister *Lilith*, the legendary
first wife of Adam. On 17th Century maps the star often
appears with the label *Caput Larvae*, "The Spectre's Head".
Ancient and medieval astrologers considered Algol the most
dangerous and unfortunate star in the heavens, which seems
to suggest that its strange variability might have been
noticed in antiquity; this reasonable conjecture, however,
remains unsupported by any other real evidence.

In Greek and Roman tradition, Medusa was the most
famous of the three *Gorgons*, the serpent-haired sisters
whose glance was literally petrifying, and who, as the
bright schoolboy wrote, "looked like women, only more
horrible". The gorgon legend is of very great antiquity, as
we find them mentioned in Book II of the *Odyssey*, which is
thought to date from the period of the 9th - 6th centuries
BC. The three sisters, it would seem, had been guilty of
the same sin as Cassiopeia, and had angered the gods by
their excessive vanity. For the next few centuries they
found themselves with a great deal of leisure time in which
to ponder the mysteries of divine retribution as they rest-
ed on the wet sea-rocks of the farthest shores of Oceanus,
thoughtfully combing their snakes. Medusa, according to a
curious feature of the legend, was the only one of the
three who could be killed; her two sisters were immortal.
She was eventually slain by the hero Perseus, who avoided
the monster's poisonous glance by looking at her reflection
in his polished shield; from the gorgon's blood sprang
forth the winged horse *Pegasus*, whose strange origin has
suggested a flock of allegorical interpretations. Medusa's
head was later presented to Athena, who had it set, as a

(1) Coins from Neapolis, c.450 BC. (2) Greek Sculpture from Syracuse. (3) Gorgon from Temple on Corfu, 6th Cent. BC. (4) Roman Bronze Pyxis Lid, 1st Cent. AD. (5) The Rondanini Medusa. (6) "Dying Medusa", National Museum, Rome.

rather impressive centerpiece, in her shield. The gorgon's head, about six times natural size, was one of the striking features of the great statue of Athena which once graced the Parthenon at Athens. Although this particular statue vanished centuries ago, the Corfu Museum still displays its own carving of the Medusa (Fig.3) from the west pediment of the Temple of Artemis on Corfu, dating from the early 6th Century BC. This relief resembles most of the early coin portraits of Medusa (Fig.1) which are done in a very stiff and archaic style and portray the snaky enchantress as a hideous monstrosity, complete with fangs and a long protruding tongue. These grotesque representations must be classed among the few real failures of Greek art; to the modern mind they seem about as frightening as a Halloween mask designed by a six-year-old child.

In time, however, the Greeks seem to have learned. By Hellenistic times the Medusa has become a figure of strange and sinister beauty, the prototype perhaps of the supernaturally lovely and infinitely terrifying "witch queen" or "white goddess" who haunts the pages of H.Rider Haggard, A.Merritt, and other writers in the classic "lost civilization" tradition. Edgar Allen Poe, in many of his somber tales, followed the ancient tradition of what might be called the "Medusa syndrome"; that curious but infinitely fascinating juxtaposition of beauty and horror, love and death, and the shadow of inexorable fate; we find it again in the eternal appeal of the story of Cleopatra, and in the *Tristan and Isolde* legend, the "most splendid love story of the western world". Shelley, after seeing the famous painting of the Medusa in the Uffizi Gallery, put it into words perhaps as effectively as any poet has ever done:

> *"Its horror and its beauty are divine.*
> *Upon its lips and eyelids seem to lie*
> *Loveliness like a shadow, from which shine,*
> *Fiery and lurid, struggling underneath,*
> *The agonies of anguish and of death...."*

The painting which inspired these lines was in Shelley's time attributed to Leonardo da Vinci, but is now thought to be the work of some unidentified member of the school of Caravaggio. Another widely known painting of the Medusa,

certainly one of the most repellent works of art in existence, was created by Peter Paul Rubens in the early 17th
Century and now graces - or disfigures - a wall of the
Kunsthistorisches Museum in Vienna. The ancient Greek vase
painters often showed their warriors carrying shields
adorned with the figure of the gorgon; Achilles, Patroclus,
and other heroes of the Trojan War are depicted in this
way.

The much-admired "Rondanini Medusa" (Fig.5) in the
Glyptothek in Munich, is one of the most effective "modern"
interpretations of the theme, though its actual age is not
definitely agreed upon; usually thought to be Hellenistic
in style, it could be a later Roman or Renaissance copy.
Equally mysterious is the "Dying Medusa" (Fig.6) in the
National Museum in Rome, possibly created, according to
some scholars, for the great Altar of Zeus at Pergamum in
the 3rd Century BC. Even the identification of this haunt-
ing portrait as Medusa has been questioned, since the
expected serpentine hair is only subtly suggested, if at
all. This, however, is equally true of the Rondanini portrait.

The gorgon's head was an immensely popular subject in
Greek and Roman art, and appears on ancient amulets and
talismans, wall reliefs, coins, household objects, and on
ornate sarcophagi well up into the Middle Ages. Joan Evans
of St.Hugh's College, Oxford, in the book *Magical Jewels of
the Middle Ages and the Renaissance* (1922) stated that "the
majority of extant Roman camei are carved with the prophy-
lactic head of Medusa", and that English manuscripts of the
High Gothic period refer to the use of the gorgon's head on
magical engraved gems. A 12th Century manuscript in the
collection of the Bodleian Library contains the following
intriguing passage:

*"Borallus vel Corallus vocatur lapis quidam qui valet
ad incantaciones et derisiones et iras si in eo sit nomen
hoc nocticula scultum et serpens scilicet gorgon et valet
scilicet ad omnes inimicos, et ad victoriam et plagam et
pavorem..."*

One who engraves the name and form of the gorgon on
a gem of coral, according to this text, will receive magical protection against injury and the attack of enemies.

Algol is one of the most famous variable stars in the sky, and its name seems to suggest that the light changes were known to the medieval Arabs. The first definite statement on the matter, however, was made by the Italian astronomer Geminiano Montanari of Bologna, about 1667. Although both Maraldi and Palitzsch confirmed the occasional fading of the star, the regularity of the period was first determined in 1782 by John Goodricke; he suggested that the periodic dimming might be attributed to the partial eclipse of the star by a dark companion revolving about it. Thus did astronomers become aware of the existence of "eclipsing

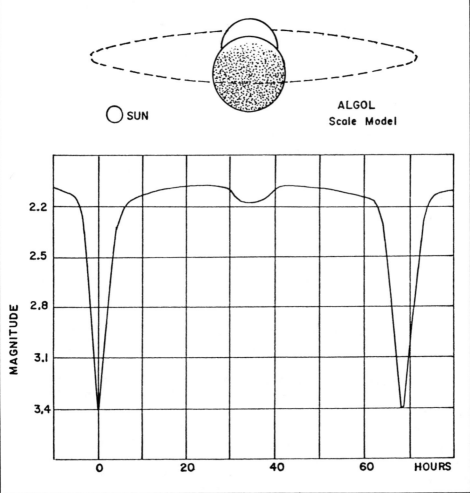

ALGOL
Scale Model

SUN

binaries". The theory was generally accepted, but remained a hypothesis until 1889 when H.C.Vogel at Potsdam proved it to be true by spectroscopic analysis. Algol is the most suitable object of its type for amateur observations, as the light changes may be readily followed with no optical aid whatever. The star is normally of magnitude 2.1 but at intervals of 2.86739 days it fades away to magnitude 3.4 and then slowly brightens again. The entire eclipse lasts some 10 hours; the exact period between minima is 2d 20h 48m 56s.

At a distance of about 100 light years, Algol is one of the nearest of all the eclipsing binaries, and has certainly been one of the most thoroughly studied. Yet the star still presents certain puzzles which await solution. One difficulty is the invisibility of the eclipsing component; completely lost in the glare of the bright primary, it has never been seen visually, and was detected spectroscopically for the first time as recently as 1978, at the McDonald Observatory. From the combined results of studies of the light curve, radial velocity measurements, and the combined spectra, our present picture of the system may be summarized as follows:

The primary star is a white B8 main sequence star about 100 times the solar luminosity, and close to 2.6 million miles in diameter. The mass of the star is not definitely determined, but is most probably between 3½ and 4 times the solar mass. The "dark" companion is known to be relatively dark only by comparison to the primary; actually it must be a half magnitude or so brighter than the Sun. The diameter is slightly over 3 million miles. From the computed size and luminosity the spectral type is most probably late G or early K. The mass is uncertain, but is not likely to be much greater than 1 solar mass. This fainter star seems to have the characteristics of a typical subgiant.

The orbit of the bright star is well determined from spectroscopic studies, and it is found that the bright star is just over a million miles from the center of gravity of the system. The faint star, being less massive, has a larger orbit around the gravitational center, and the true separation of the two stars is probably about 6½ million miles, center to center. From spectroscopic studies the

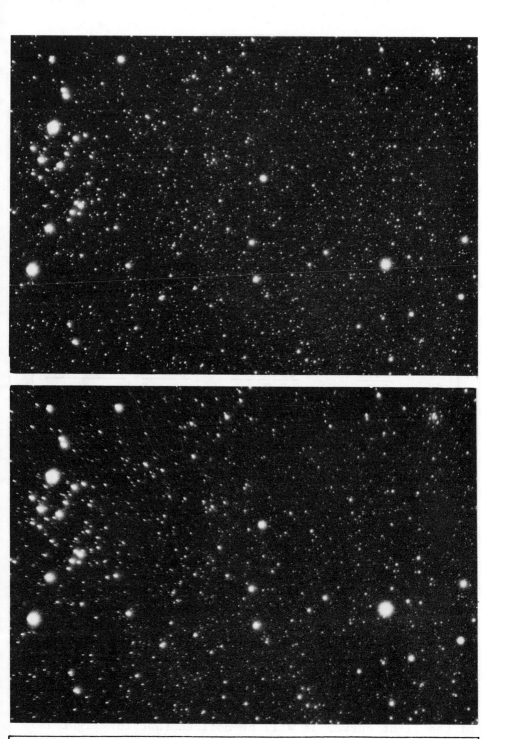

ECLIPSE OF ALGOL, March 15, 1977. The star appears near minimum (top) and at normal light (below). Photographed by R.Burnham, Jr., with a 1.7-inch Xenar lens.

eccentricity of the system is about 0.033, and the inclination about 82°. At primary eclipse, approximately 79% of
the bright star is hidden by the larger companion. A slight
secondary minimum midway between the eclipses, is noticeable on the light curve, and is produced when the companion
is partly eclipsed by the bright primary. Thus the relative
luminosities may be deduced from the derived orbit and the
relative depths of the two minima; this makes it evident
that the companion is not really a "dark" star at all.

  According to a study by G.Hill, J.V.Barnes, J.B.
Hutchings, and J.A.Pearce of the Dominion Astrophysical
Observatory (1971), the masses and luminosities for the
components of the Algol system are close to the values in
the following table:

	Mass	Diameter	Abs.Mag.Bolometric
A	3.7	3.0	-0.4
B	0.8	3.4	+3.1
C	1.7	1.5	+2.3

  In another study, made with the 100-inch reflector,
A.S.Meltzer of Princeton University derived the results
given below:

	Mass	Spectrum	Abs.Mag.	Lum.
A	5.0	B8 V	-0.08	90
B	1.0	K0 IV	+3.4	3½
C	1.3	F2 V	+3.2	4

The discrepancy in the derived masses is partly due to the
uncertainty in the exact distance; measured parallaxes
range from 0.03" to about 0.041", giving a range of from
101 light years down to about 79. Algol shows a very small
annual proper motion, less than 0.01"; the radial velocity
is about 2.5 miles per second in recession with large variations.

  Some additional facts of interest appear from a study
of the light curve. It will be noticed that the light is
not constant at maximum, but continues to rise slowly after
the end of primary eclipse, reaching a peak just before the
slight secondary eclipse. This is explained as a reflection
effect; the faint star reflects light from the primary and

ALGOL, photographed with the 13-inch camera at Lowell Observatory on January 27 and 28, 1965. The star appears at normal light (top) and at primary minimum (below).

shows moon-like phases as it revolves in its orbit. The "full moon" phase must obviously come when the fainter star is seen beyond the brighter one, so that the whole illuminated face is turned toward the earth. This moment comes just before and after secondary eclipse, so it is naturally at this time that the combined light of the system is at a maximum. A study of this effect shows that the faint star appears about 1.7 times brighter on the side turned toward the primary.

Another effect, called limb darkening, also slightly alters the form of the light curve. Limb-darkening is the result of the greater thickness of the star's atmosphere lying along the line of sight at the rim of the disc; this causes the disc of any star to appear brighter in the center than at the edge. Since the limb of a star is occulted first during a stellar eclipse, the loss of light at first appears gradual, but grows increasingly more rapid as the brighter central portion of the disc is hidden. The light curve of an eclipsing binary shows this effect.

Algol has been identified as a source of radio energy by C.M.Wade and R.M.Hjellming at the National Radio Astronomy Observatory in Green Bank, Virginia. They state that the radio emission is "best described as erratic, with occasional periods of strong flaring and long quiescent periods". The authors suggest that the radio flares are possibly the result of sudden changes in the stellar structure, best described as "starquakes". "The sudden energization of a large and very hot thermal plasma would be the basic source of the typical radio flares". Radio emission has also been detected from red giant stars, as Betelgeuse and Antares, as well as from red dwarf flare stars of the UV Ceti type.

DISTANT COMPONENTS OF THE ALGOL SYSTEM. The fact that the Demon Star is not a simple two-star system was recognized as early as 1869. For nearly a century it has been noticed that the period of the eclipsing pair is subject to small but definite changes. Modern studies show that the major axis is gradually changing its orientation in space in a period of about 32 years; an analysis of this effect proves the existence of a third star in the system, and the possibility of a fourth companion has been suspected.

The third star, called "Algol C", has now been detected spectroscopically, and has the characteristics of an F-type star of the main sequence. It appears to be at least a magnitude brighter than our Sun, and is definitely more luminous than the "dark" star of the eclipsing pair. It revolves about the AB pair in a period of 1.862 years, at a distance of about 50 million miles. P.van de Kamp finds that the orbit of the third star is not in the same plane as the eclipsing pair; a study in 1951 gave the inclination as about 63° with considerable uncertainty. The orbit of the A-C system is also more eccentric (about 0.21) than the eclipsing pair.

Of the suspected fourth star, Algol D, little is known, except that the period about the system has been calculated at about 188 years. Studies in 1971, however, cast some doubt on the reported perturbations from which this figure was derived, and there appears to be no definite evidence that the fourth body actually exists.

EVOLUTION OF THE ALGOL BINARIES. A peculiar feature of the Algol system, shared by other binaries of the type, is that the fainter and less massive component has evolved to the subgiant stage, while the primary remains a main sequence object. This is an evolutionary paradox, for if the stars are of the same age, it should be the brighter and more massive star which evolves more rapidly.

Fred Hoyle has suggested an interesting and plausible solution. He assumes that the fainter star was originally the more massive and luminous of the pair. As it began its evolutionary expansion, it lost great quantities of matter to the close companion. It thus grew fainter as it evolved to the subgiant stage. At the same time the companion grew more brilliant as the result of its increased mass. The small separation of an Algol-type binary makes this interpretation seem quite logical, and it is fairly evident that a similar sequence of events is occurring also on systems of the Beta Lyrae type. (Refer also to Beta Lyrae, U Cephei, and U Sagittae)

Observers of Algol will find tables giving predicted times of minima in the yearly *Handbook of the British Astronomical Association;* predictions for each month are also carried in *Sky and Telescope* magazine.

GAMMA   Magnitude 2.91; Spectrum G8 III and A3 V;
Position 03012n5319. The star is labeled
"Algenib" on some star charts, but this name is also given
to both Alpha Persei and Gamma Pegasi. The star is thought
to be about 150 light years distant, but there is some
uncertainty, as the distance derived from a direct parallax
is about double the value found from the luminosity criteria. The total absolute magnitude, from the spectral types,
should be about -0.4; the annual proper motion is less than
0.01"; the radial velocity is about 1 mile per second in
recession.
        Gamma Persei is a very close and difficult binary
star with an orbital period of 14.647 years; the orbit was
originally determined spectroscopically, but it may be seen
as a visual pair in large telescopes when at greatest
separation, about 0.4". The widest separation occurs in
1954, 1969, 1983, etc.; the apparent orbit is a nearly
edge-on ellipse, with the true separation varying from 3 AU
(1947) up to about 20 AU (1954). D.B.McLaughlin, in his
detailed study of the system, found the following values
for mass, luminosity, etc:

	Spectrum	Mass	Diameter	Luminosity
A	G8 III	4.7	12	100
B	A3 V ?	2.7	1.8	25

The primary of the system thus resembles the brightest
component of Capella, while the fainter star closely matches Sirius. A third star, at 57" and magnitude 11, is not
known to be a true physical companion to the system.
        About 4° north of Gamma Persei is the radiant point
of the very dependable Perseid Meteors, which reach their
maximum each year on August 10--12. The Perseids are celestial debris from the Swift-Tuttle Comet (1862 III), but
seem to be so evenly distributed around the orbit that the
number remains very nearly the same from year to year; at
maximum about 60 per hour may be expected. The comet itself
has a computed period of about 120 years, with a probable
uncertainty of several years either way. The Perseids are
sometimes called *The Tears of Saint Lawrence* in honor of
the early Christian martyr whose feast day is August 10.

DELTA   Magnitude 3.00; Spectrum B5 III; position
03394n4738, about 3½° SE of Alpha Persei. The
star is a bluish giant over 1700 times more luminous than
our Sun (absolute magnitude -3.3) and lies at a computed
distance of about 590 light years. Delta Persei seems to
show about the same space motion as the Alpha Persei group
and may be a member of that aggregation. The annual proper
motion is 0.05"; the radial velocity is 5½ miles per second
in approach.

Variations in the ultraviolet emission of this star
have been detected by the OAO-2 satellite in July 1971.
According to M.R.Molnar of the Laboratory for Atmospheric
and Space Physics, "although these observations are insuf-
ficient to determine the exact period and nature of the
photometric variations, the amplitude is on the order of
0.03 magnitude and the period may be around one day".

EPSILON   Magnitude 2.88; Spectrum B0 or B1 V; position
03545n3952, about 8° east of Algol. Epsilon
Persei is another B-type giant star, some 2500 times more
luminous than our Sun (absolute magnitude -3.7) and lying
at a computed distance of about 680 light years. The star
shows an annual proper motion of 0.04"; the radial velocity
is less than 1 mile per second in approach.

For the small telescope, the star is a visual double
with an 8th magnitude companion lying about 9" distant.
Webb calls the colors "greenish and bluish-white" while
E.J.Hartung finds them "pale yellow and slate-coloured";
the spectral type of the faint star is B8. This is a common
motion pair, but has shown no definite change in separation
or PA since the early measurements of F.G.W.Struve in 1832.
The projected separation is about 2000 AU. "Not an easy
double" says C.E.Barns. Although lying some 10° from the
Alpha Persei group, Epsilon shows about the same space
motion, and is probably an outlying member of that large
aggregation. The primary star is also a spectroscopic
binary of uncertain period.

ZETA   Magnitude 2.83; Spectrum B1 Ib; position
03510n3144, in the Foot of Perseus, about 8°
north of the Pleiades cluster in Taurus. This is one of the
most luminous stars in Perseus and lies at an estimated

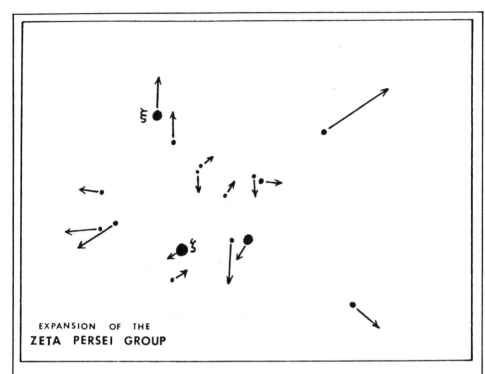

EXPANSION OF THE
ZETA PERSEI GROUP

distance of about 1000 light years; the actual luminosity
is about 6300 times that of the Sun, and the computed
absolute magnitude about -4.7. Zeta Persei shows an annual
proper motion of 0.02"; the radial velocity is 13 miles per
second in recession.

For the small telescope the star has several faint
companions; the 9th magnitude star at 12.9" seems to share
the proper motion of the primary, though the projected
separation is about 4000 AU. Webb calls the colors greenish
white and ashy. If the derived distance is correct, the
faint star has about 13 times the luminosity of the Sun; no
spectrum appears to have been obtained for the companion.

Zeta Persei is the brightest member of the stellar
association "II Persei", an expanding group of bright O and
B stars which appear to be moving outward from a common
center at a velocity of about 12 km/sec. It would seem that
this group must be exceedingly young on the astronomical
time-scale, since expansion to the present size would have
required scarcely more than a million years. The Zeta Persei group is thus of great interest to astronomers who are

ZETA PERSEI REGION. Zeta is the brightest star on the
print; Xi Persei and the nebula NGC 1499 are at the top.
Lowell Observatory 5-inch camera photograph.

DESCRIPTIVE NOTES   (Cont'd)

specializing in problems of stellar formation and evolution. Among the brighter members of the group are Omicron
Persei, Xi Persei, 40 and 42 Persei, the double star Σ448,
and about 15 other stars brighter than 6th magnitude. The
expansion of the group is illustrated by the diagram on
page 1422, where the length of the arrows represents the
motion for each star for the next half-million years; the
diagram is based on a study by A.Blaauw at Palomar. Known
members of the Zeta Persei group extend over an area about
100 light years in diameter.

Some very conspicuous nebulosity still exists in the
group, implying that star formation may not be completely
ended in this region of space. Omicron Persei is involved
in the faint cloud IC 438; the star itself is a B-type
ellipsoidal variable and also has a close visual companion.
Xi Persei, on the north edge of the association, is the
earliest type star in the group with a spectral class of
O7e; it also appears to be the illuminating star for the
huge "California Nebula" NGC 1499 which lies just to the
north and which evidently shares the outward expansion of
the group. A difficult object visually, the nebula extends
over a $2\frac{1}{2}°$ field, and shows much filamentary detail on red-
sensitive long exposure photographs.

The position of the mysterious disappearing star SZ
Persei is about 2.6° to the NNW of Zeta Persei. (See page
1426)

RHO   Magnitude 3.30; Spectrum M4 II or III;
position 03020n3839, about $2\frac{1}{4}°$ south of Algol
and slightly west. The star is a semi-regular variable
resembling Alpha Herculis; a primary cycle of about 33 days
is given in the Yale *Catalogue of Bright Stars* (1964) while
the Moscow Variable Star Catalogue reports a probable period of about 40 days. At least one much longer cycle of
about 1100 days seems to be involved. The variations, never
very sudden, have a total amplitude of about 0.7 magnitude
in the visual. From a direct parallax measurement, Rho Persei appears to be some 260- 300 light years distant, which
gives it a mean absolute magnitude of close to -1. The
annual proper motion is 0.17" in PA 129°; the radial velocity is about 17 miles per second in recession.

THE CALIFORNIA NEBULA. This is NGC 1499, which lies at the north edge of the Zeta Persei Association. Photographed in red light with the Lowell Observatory 13-inch camera.

Rho Persei shows very nearly the same space motion as the interesting binary Zeta Herculis. Dr.O.J.Eggen has made a search for other possible members of this group and and has identified about a dozen stars which appear to show the same motion, including Iota Reticuli, Beta Hydri, TV Piscium, 84 Virginis, Phi-1 Lupi, and Epsilon Octantis. As the presumed members are scattered well around the sky the reality of such a "group" remains questionable.

SZ   Suspected nova. Position 03439n3410, about 1° ENE from 40 Persei, and some 10° due north of the Pleiades in Taurus. This is the star BD+33°715, which appears on the BD charts (1862) as an object of magnitude 9.5. On a photograph made in October 1891 it could not be found, but was seen again at about 10th magnitude in 1894. In 1908 it had apparently vanished once more. Aside from these few observations, virtually nothing is known about this star, which may be a recurrent nova or an erratic type of variable. No star brighter than 14th magnitude exists at this position on plates made from 1959 to 1976.

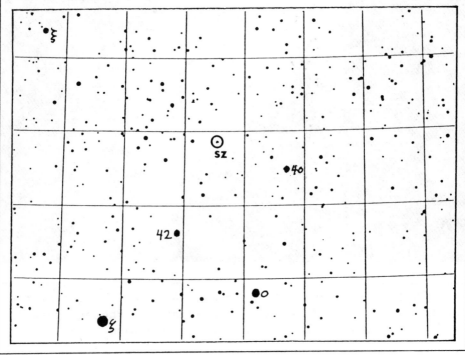

GK   Nova Persei 1901. Position 03278n4344, about
4½° NE of Algol. The first bright nova of the
20th Century, first noticed on February 21, 1901, by the
amateur astronomer T.D.Anderson of Edinburgh, Scotland. At
that time the star was of the 2nd magnitude, but within 2
days it had increased its light more than 6 times, and
attained its maximum brilliancy of magnitude 0.2 on February 23.

The new star began to fade almost immediately, and
had fallen to magnitude 2 by February 28. It dropped about
4 magnitudes in the first three weeks, then began a peculiar series of oscillations with a perid of about 4 days and
an amplitude of about 1.5 magnitudes. Sir Robert Ball of
the University Observatory at Cambridge gave an amusing
account of his experiences during the oscillation phase of
Nova Persei; one night he went out on his lawn to show some
visitors the new star but it had disappeared from view; two
nights later he invited other friends to "observe the star
which had disappeared" but when they looked, there it was
"as large as life". Two nights later other visitors were
invited to see the star that had apparently come to life
again so miraculously, but it had disappeared once more!

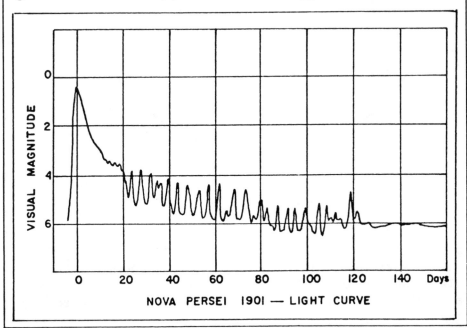

NOVA PERSEI 1901 — LIGHT CURVE

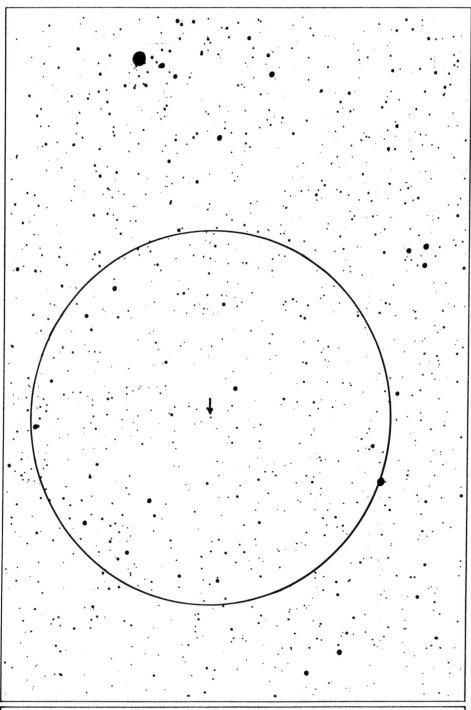

NOVA PERSEI Identification Chart, from a 13-inch telescope plate made at Lowell Observatory. The circle is 1° in diameter with north at the top; stars to about 14^m are shown.

DESCRIPTIVE NOTES (Cont'd)

About four months after maximum the oscillations ceased and the star, then at 6th magnitude, faded slowly from sight.

Previous photographs of the region of Nova Persei showed that the star had been a 13th magnitude object up to three days before the outburst. At the maximum it was some 200,000 times more luminous than our Sun, an increase of 13 magnitudes in only a few days. The computed distance of the star was about 1300 light years which implies an absolute magnitude of about −8.4 at the peak brightness.

A few months after the appearance of the nova, photographs were taken which showed a faint area of luminosity surrounding the star; this nebulous aura was probably first noticed by Dr.Max Wolf at Heidelberg in the spring of 1901. Successive plates revealed that this nebulosity was growing at the astonishing rate of about 2" per day, an expansion which, at the assumed distance of the star, corresponded to a speed about equal to the velocity of light! Evidently the light of the nova was merely illuminating a dark nebula

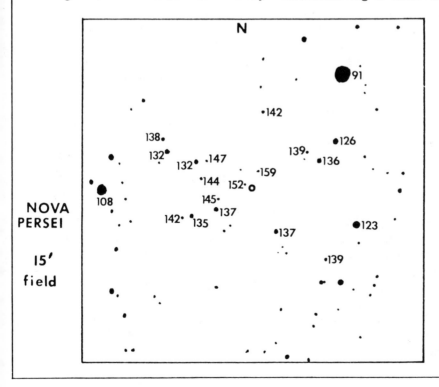

NOVA
PERSEI

15'
field

which had already existed, and the apparent expansion of
the nebula was actually due to the light of the star
traveling outward. From an analysis by P.Couderc (1939) it
seems that the illuminated nebula probably had the form of
a thin sheet of material oriented "flat-side-on" as seen
from the Earth and lying slightly closer to us than the
star itself. Such a "light echo" seems to be unique among
novae, though several diffuse nebulae show rapid changes
which seem to be attributable to a similar process. NGC
2261 in Monoceros (Hubble's Variable Nebula) is probably
the most notable case.

Almost 15 years after the outburst, the actual nebular shell around Nova Persei finally became visible; its
appearance in 1949 is shown in the photograph on page 1431.
This shell is a portion of the outer layers of the star,
blown off into space during the explosion at a velocity of
about 750 miles per second. The existence of this shell was
known from spectroscopic analysis, and it was expected to
eventually become visible in the telescope. Similar nebular
shells were photographed surrounding Nova Aquilae 1918 and
Nova Herculis 1934; these resemble the planetary nebulae
superficially, but are rather short-lived phenomena, usually vanishing after a few years. A peculiar fact about the
shell of Nova Persi was its unsymmetrical appearance, suggesting that the matter had been ejected chiefly from one
hemisphere of the star. Still expanding slowly, the cloud
is now gradually fading; the expansion rate has been fairly
uniform at about 0.4" per year.

GK Persei still shows the largest light variations of
any of the former novae which have been thoroughly studied.
It is usually a 13th magnitude object, but may brighten to
nearly 11.0 or fade to 14.0; there does not appear to be
any regular periodicity. The star has been quite active in
recent years; in August 1966 it rose to 11th magnitude,
with another short outburst in September 1967. After an
erratic period of oscillations it faded to 13th magnitude
but in January 1975 rose again to 12.0 over the short interval of 6 days. The color of the star is equivalent to
class O or early B, with emission features; the star may be
classed as a subdwarf but not a true white dwarf; the absolute magnitude is near +5.

NOVA PERSEI 1901. The nebulous shell is well shown in this photograph made with the 200-inch reflector at Palomar, almost 50 years after the outburst of the star.

DESCRIPTIVE NOTES   (Cont'd)

Observers with adequate telescopes will find it re-
warding to follow any future variations of GK Persei. The
identification field on page 1429 shows a 15' area centered
on the star, with stars to about 16th magnitude shown.
Comparison magnitudes, courtesy of the AAVSO, are given
with decimal points omitted to avoid confusion with star
images; thus "137" indicates a magnitude of 13.7.

The most important discovery about novae in recent
years has been the finding that some of the post-nova stars
are close and rapid binaries. Nova Herculis 1934 was the
first to be so identified;   the period is a mere 4.65
hours. Nova Aquilae 1918 is a similar system with a period
of about 3h 20m. GK Persei now joins the list of former
novae recognized as double stars; the period is 1.904 days
but the components, in this case, do not appear to form an
eclipsing binary system. R.P.Kraft (1963) found the spectrum composite, consisting of a "blue continuum with
emission lines of hydrogen, helium, ionized oxygen, and the
absorption lines of a late-type star"; the derived spectral
types are sdBe and about K2 IV, with computed masses of at
least 1.29 and 0.56 respectively. From its color, luminosity and mass, the explosive blue component would appear to
have a density of several hundred times that of the Sun,
nowhere near the density of a true white dwarf. Refer also
to Nova Aquilae 1918, Nova Herculis 1934, Nova Puppis 1942,
Nova Pictoris 1925, and Nova Cygni 1975. Eruptive "dwarf
novae" are described chiefly in the articles on SS Cygni
and U Geminorum; for recurrent novae see T Corona Borealis,
RS Ophiuchi, and WZ Sagittae)

M34   (NGC 1039)   Postion 02388n4234. A bright open
             star cluster easily located about 5° WNW from
Algol, near the Perseus-Andromeda border. This is one of
Charles Messier's discoveries, found in August 1764, and
entered in his famous catalgue as "a cluster of small stars
a little below the parallel of Gamma Andromedae; in an ordinary telescope of 3 feet one can distinguish the stars...
its position was determined by Beta Persei....diameter 15'."
Bode in 1774 found M34 visible to the naked eye under good
conditions, and T.W.Webb called it "a very grand low power
field; one of the finest objects of its class". To Admiral
Smyth it was "a scattered but elegant group...8--13 mag on

STAR CLUSTER M34 in PERSEUS. This bright group lies about midway between Algol and Gamma Andromedae.   Lowell Observatory photograph with the 13-inch camera.

a dark ground; several of them form coarse pairs". The experienced observer Walter Scott Houston calls this a rather sparse cluster and finds it "not more spectacular in large telescopes, as it does not seem to have the needed fainter stars to buttress the view. Rather I feel that 15 x 65 binoculars give the best impression.........more magnification merely spreads out the few bright stars that the binoculars show perfectly well".

In general appearance, brightness and size, M34 resembles the cluster M36 in Auriga; both clusters appear at their best with fairly low power wide-angle eyepieces. The central knot of bright stars measures about 9' in size and the total diameter of the group may be about 20'. A study by A.Wallenquist, however, has identified probable members out to more than 20' from the cluster center; his estimate for the star density at the center is about 21 stars per cubic parsec. Some 80 stars are presently recognized as true cluster members; the brightest of these are several white giants of spectral type B8 and apparent magnitude 8½; each of these stars has about 60 times the solar

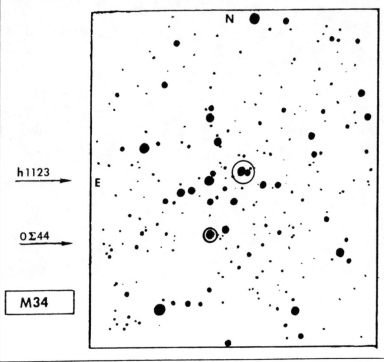

h1123 →

0Σ44 →

M34

DESCRIPTIVE NOTES   (Cont'd)

luminosity. H.L.Johnson (1957) in a photometric study of
M34 derived a true modulus of 8.2 magnitudes, giving the
distance as 1430 light years. A somewhat greater distance
of about 1500 light years was obtained by Wallenquist; the
true diameter of the 9' central mass is then about 4 light
years, and the extreme diameter of the whole group about
18 light years. Color-magnitude studies of the cluster seem
to imply that it is a somewhat older group than the famous
Pleiades or M36; the current estimate of age is slightly
over 100 million years.

The easy double star h1123 lies in the heart of the
cluster, and is identified at the center of the chart on
page 1434. This fixed white pair has a separation of 20"
with magnitudes both close to 8½. A more difficult pair is
0Σ44, about 6' to the SSE; it has a fixed separation of
1.3" with possibly a slight decrease in PA since the early
measurements of 1850. The spectral type is about B9.

M76   (NGC 650) Position 01388n5119. Faint planetary nebula, usually called the "Little Dumb-
bell" or the "Barbell" Nebula. It is located in the extreme
western portion of the constellation, a little less than a
degree NNW from Phi Persei, and about 8° SW of the great
Perseus Double Cluster. Often regarded as the faintest of
the Messier objects, M76 is also one of the more irregular
examples of a planetary nebula and appears as a roughly
rectangular or box-shaped mass measuring about 2' x 1'.
The appearance is fairly similar to the brightest portion
of the "Dumbbell Nebula" M27 in Vulpecula; hence the popular name. In the NGC catalogue it was listed as two objects
and often appears in modern lists as NGC 650 and 651.

Discovery of M76 is credited to P.Mechain in September 1780; Charles Messier found it some 6 weeks later and
thought that it might be comprised of "small stars containing nebulosity....the least light employed to illuminate
the micrometer wires causes it to disappear". T.W.Webb
found it a "pearly-white nebula, double, a curious miniature of M27 and, like it, gaseous....the preceding portion
a little brighter". Lord Rosse thought to find some hint
of a spiral structure with "subordinate nodules and streamers" but Isaac Roberts in 1891 suggested that the appearance was probably that of a broad ring seen edge-wise.

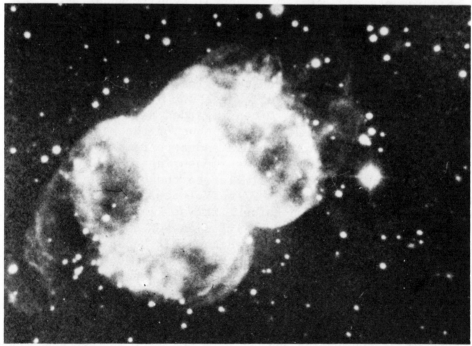

PLANETARY NEBULA M76 in PERSEUS. An example of a planetary which lacks the usual disc shape. The photographs were made at Lowell Observatory (top) and at Palomar (below).

DESCRIPTIVE NOTES   Cont'd)

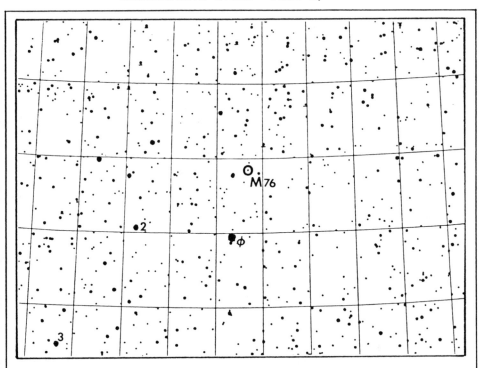

Long exposure photographs show that the nebulous filaments
seen by Lord Rosse, which project outward from the four
corners of the rectangular central mass, are not portions
of a spiral pattern, but instead form parts of great arcs
or loops which encircle the main structure.

According to Lick Publications XIII, the nebula is
"quite irregular, but evidently to be included as one of
the larger members of the planetary class. The central and
brighter portion....is an irregular, patchy oblong 87" X
42" in PA 40° from the ends of which faint, irregular,
ring-like wisps extend; total length 157" in PA 128°......
brightest patch at southern end of central part". Various
observers rate the central star as magnitude 16, 16½, or
17; the Skalnate Pleso Catalogue gives it as 16.6, and the
total integrated light of the nebula is given as 12.2.

As with virtually all the planetary nebulae, the
exact distance is not well determined. C.R.O'Dell (1963)
gives a rather precise figure of 537 parsecs or about 1750
light years; I.S.Shklovsky (1961) obtained 820 parsecs; the
Skalnate Pleso catalogue has 2510, while L.Kohoutek (1961)

has 1070. O'Dell's value for the distance would give the
nebula an actual diameter of about 1 light year and a total
luminosity of 2 to 3 times that of the Sun. The central
star, from these results, must have an absolute magnitude
of about +8, or about 0.07 the solar luminosity. B.A.
Voroncov-Vel'jaminov (1948) derived a much larger distance
for the nebula of about 2000 parsecs; this would increase
the luminosity of the central star to about that of the
Sun. The star is, however, one of the hottest known, with
a computed temperature of about 60,000°K. Radial velocity
measurements indicate a speed of approach of about 15 miles
per second. (A general outline of facts and theories about
the planetary nebulae will be found in the article on M57
in Lyra)

NGC 869   Mean Position 02172n5654. This is the famous
NGC 884   "Double Star Cluster" in Perseus, one of the
truly classic examples of a galactic cluster,
and a wonderfully beautiful object for moderate telescopes.
Among the all-time favorites for amateur observers, it may
be seen without optical aid as a hazy patch of light about
midway between the stars of Perseus and the familiar "W"
figure of Cassiopeia. A small telescope reveals that this
luminous spot consists of two fine open star clusters together in the field, the pair forming one of the most impressive and spectacular objects in the entire heavens. In
the mythological outline of the constellation it marks the
"Sword Handle" of Perseus and is often identified on star
atlases by the inconsistent designation "h - χ" Persei.
The cluster called "h" is NGC 869, and "χ" is NGC 884.
    A low power wide-angle eyepiece with a field of about
1° is needed in order to get both clusters together in the
field of view. A power of 15X is suitable on a 2-inch glass
and 30X on a six-inch. With an 8-inch or 10-inch, powers of
40X or 50X may be used. Short focus telescopes of the RFT
or "rich-field" design are excellent on objects of this
type. W.T.Olcott considered this the finest cluster for a
small telescope in the heavens. "The field is simply sown
with scintillating stars, and the contrasting colors are
very beautiful". The present writer remembers a night in
the autumn of 1963 when he attached a wide-angle Erfle

THE DOUBLE STAR CLUSTER IN PERSEUS. This splendid group
contains some of the most brilliant stars known. Eight-
inch reflector photograph by Kent de Groff.

eyepiece to the 13-inch refractor at Lowell Observatory, in order to show the Double Cluster to a group of friends; the view was absolutely dazzling (with brilliant rainbow flares around every star image, since the 13-inch is actually an astrographic camera and is not corrected for use as a visual objective!)

The existence of the Perseus clusters was noted at least as far back as 150 BC; Hipparchus and Ptolemy both mention the group, but refer to it as a "nebula" or "cloudy spot", one of the half dozen then recognized. The actual nature of such objects remained a mystery until the invention of the telescope. Another mystery, incidentally, lies in the fact that this splendid object was never included by Charles Messier in his famous catalogue, although it was certainly known in his day, and he included other bright clusters such as Praesepe and the Pleiades.

Among the early photographs of the Double Cluster is a fine plate made by Isaac Roberts with his 20-inch reflector in January 1890. In describing the result, Roberts stated that "the photograph presents to the eye the stars in the two clusters and in the surrounding part of the sky with a completeness and accuracy of detail never before seen.....an appearance of grandeur that can only be fully realized by aid of the photographic method. Any written description will convey only a very inadequate idea of it. The stars are shown in their true relative positions and magnitude to about the 16th, and among them are many apparent double, triple, and multiple stars. They also appear to be arranged in clusters, curves, festoons and patterns that are suggestive of some physical connection between the groups..... Again, we see that these clusters......appear to be quite free from nebulosity, and therefore, by this hypothesis, are anterior in time to any of the other nebulae or clusters which have been referred to. We seem to be thus on the way, by the aid of photography, to an intelligible classification of some of the stages in the evolution of the Universe, but of course we must proceed with due caution...."

In a short description of the Double Cluster, published in the *Harvard Annals* in 1908, I.S.Bailey gave the apparent diameter of each cluster as 30'; in a star count reaching to the 12th magnitude, he found some 400 stars in

NGC 869 and about 300 stars in NGC 884. As early as 1913, W.S.Adams and A.van Maanen had measured the radial velocity of the cluster and found it to be about 26 miles per second in approach. The stars chosen for radial velocity measurements were bright B and A stars displaying the so-called "c" characteristic, unusually sharp spectral lines, now known to be one of the identifying features of the supergiants. Astronomers began to realize that the Double Cluster was an object of exceptional interest. During the past 30 years it has been one of the most intensively studied clusters in the heavens.

The distance of the great swarm of stars cannot be measured directly, since it is too remote to show the slightest measurable parallax. In a spectroscopic study the bright members are found to be supergiants of types A, B, and O; stars of exceedingly great luminosity which must be at a vast distance in order to appear as faint as they do. These stars may be used as "standard headlights" but the problem is complicated by the necessity of correcting the observations for some loss of light due to absorption by dust in space. Although there is no visible nebulosity in the immediate vicinity of the cluster, the surrounding region is heavily mottled with irregularly distributed dust clouds. These show plainly on some of Dr.E.Barnard's fine Milky Way photographs. Current evidence suggests that the light of the cluster stars is dimmed by about 1.6 magnitudes by interstellar material.

In a thorough study of the cluster, H.L.Johnson and W.A.Hiltner (1955) found the corrected modulus to be about 11.8 magnitudes; the resulting distance is about 7400 light years. This result is in fairly good agreement with the estimate of about 8000 light years given by Harlow Shapley some 25 years earlier. Supporting this distance determination is another result obtained though a different line of inquiry; the investigation of the structure of our Galaxy. The Sun is located in one of the spiral arms of the system, and the Perseus cluster, the nucleus of a great swarm of giant stars, would seem to mark the next spiral arm going outward, away from the galactic center. The distance of the Perseus cluster is estimated by this method to be some 7000 to 8000 light years. Here we may use our Sun as a comparison object. At a distance of 7400 light years it

THE DOUBLE CLUSTER IN PERSEUS. The two splendid star groups are just half a degree apart. Lowell Observatory photograph made with the 13-inch camera.

would appear as a star of magnitude 16.6; if actually in
the Perseus cluster it would, of course, be further dimmed
by dust in the intervening space, and would appear about
18.2. Star no brighter than our Sun are thus very difficult
to detect at the distance of the cluster, and even fainter
stars are completely out of range. The bright members,
which present such a glittering spectacle in the small telescope, are all great blazing supergiants of almost unimagi-
nable brilliance.

According to studies by Johnson, Morgan and Hiltner
the 10 brightest cluster members have the following magnitudes and spectra:

1. Mag 6.38; Spectrum A1	6. Mag 7.40; Spectrum B2
2. 6.49 B1	7. 7.44 A2
3. 6.55 B3	8. 8.05 B2
4. 6.66 B2	9. 8.40 B1
5. 7.00 B8	10. 8.53 B2

All these stars are A & B type supergiants with absolute magnitudes ranging from -7.3 to -4.9; these figures
have been corrected for a light loss of 1.6 magnitudes by
absorption in space. The brightest stars of the cluster
thus have luminosities approaching 60,000 times that of the
Sun, and are comparable to such supergiants as Rigel in the
Orion Association. There are 17 other stars in the central
portions of the clusters which have absolute magnitudes
brighter than -2. In addition, both clusters are embedded
in a vast swarm of giant stars forming an association about
7° x 5° in size, or several hundred light years in diameter
and containing over 100 early type stars of spectral types
O, B, and A, The absolute magnitudes of these stars range
from -3.3 to about -7.5.

A very interesting feature of the Perseus clusters is
the presence of M-type red supergiant stars, a number of
which may be detected in amateur telescopes. Herschel,
D'Arrest, Smyth, Rosse, and Webb all recorded their presence, and T.E.Espin in 1892 listed nine of them down to
magnitude 10.5. According to Espin, there are three red
stars in NGC 884, none in NGC 869, four lying between the
clusters, and two in the outlying regions. In a list of M-
type supergiants near the Double Cluster, V.M.Blanco (1955)

DESCRIPTIVE NOTES (Cont'd)

published magnitudes and spectra for 17 such stars. These ranged in brightness from magnitude 7.6 to 9.2, with spectral classes from M0 to M5. The brightest star on this list has an absolute magnitude of about -5.7 and a luminosity of some 15,000 suns, comparable to Betelgeuse in Orion. Among other red supergiants lying within a few degrees of the cluster, and believed to be probable members, are the pulsating variables S, T, RS, SU, and YZ Persei, all with semi-regular variations.

To modern astronomers, these are important objects which have much to teach us about stellar evolution. The highly luminous supergiants are "spendthrift stars" which cannot maintain their enormous energy output for very long and must have been formed relatively recently. If we plot the members by magnitude and color on the familiar H-R diagram, we find that some of the most luminous stars, near absolute magnitude -7.5, are still main sequence stars. At the same time we see that some of these objects have already evolved to the red giant stage and now populate the upper right section of the diagram. These are evidently stars

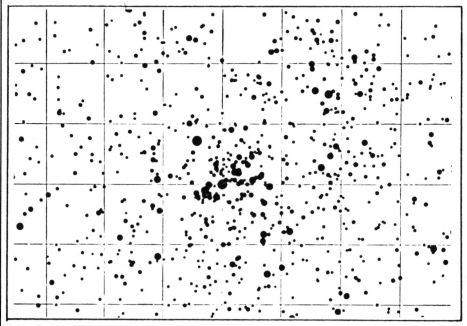

NGC 869. The distribution of the stars to about 16th magnitude, from a study by P.T.Oosterhoff at Leiden Observatory.

DESCRIPTIVE NOTES   (Cont'd)

of exceptionally large mass which were the first cluster
members to begin their evolution. In the beautiful "Jewel
Box" cluster NGC 4755 in Crux we find an identical situation. These two clusters are apparently quite identical in
type and age, and rank among the youngest star groups yet
identified. Although an estimated age of only about one
million years has been quoted frequently in astronomical
texts, current research (1976) suggests that both of these
clusters may be several times that age. There is also some
evidence that the two portions of the Double Cluster are
not exactly the same age, and possibly do not form a truly
physical pair.

This suspicion dates back at least to 1930 when R.J.
Trumpler found a difference of about 0.5 magnitude in the
computed distance moduli of the two groups. W.P.Bidelman
(1943), K.A.Barkhatova (1950), W.Becker (1963) and A.Blaauw
and J.Borgman (1964) have all obtained similar results; the
computed difference in the two moduli range from 0.3 to 0.5
magnitude. A study of all the data has been made by Yerkes
astronomer R.E.Schild and published in the *Astrophysical*

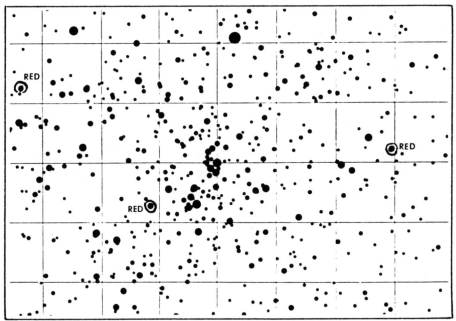

NGC 884. The distribution of the stars to about 16th magnitude, from a study by P.T.Oosterhoff at Leiden Observatory.

DESCRIPTIVE NOTES   (Cont'd)

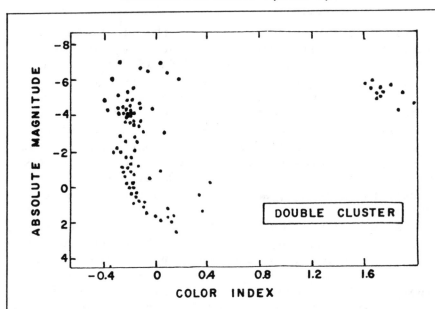

ABOVE: The H-R Diagram for the Perseus Clusters, from
observations by H.L.Johnson and W.A.Hiltner.
BELOW: A comparison of diagrams for various clusters.

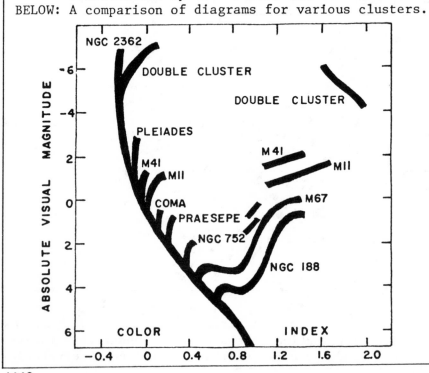

DESCRIPTIVE NOTES   (Cont'd)

Journal in October 1965. His conclusion is that NGC 869 is
a less-evolved cluster than NGC 884, and that 869 is also
somewhat closer than its neighbor cluster. Further studies
having verified these findings, Schild wrote in 1966 that
"NGC 884 is found to be 350 parsecs more distant than 869,
and five million years older. Distances found are, respectively, 2500 and 2150 parsecs, and ages determined are 11.5
and 6.4 million years. Surrounding 884 appears to be an
association of stars 65 parsecs in diameter.....and of the
age and distance of 884. The M-supergiants and extreme Be
stars appear to be associated with the 884 stellar population.". The study also showed that the other cluster, 869,
appears to be associated with another younger stellar population which extends over 200 parsecs; some of the farther
members may lie about midway between the two clusters, but
none appear to be sufficiently distant so as to actually
link the two groups.

Each of the clusters has an actual diameter of about
70 light years; the total mass of both clusters may be near
5000 times the solar mass, and the total light output is
something like 200,000 times that of the Sun. NGC 869, as
we have seen, is now thought to be the younger cluster of
the two, and still among the youngest star groups known;
only a few clusters have been identified as being probably
still younger, as NGC 2362 in Canis Major. At the other
extreme are such objects as the globular star clusters, and
a few very ancient galactic clusters such as NGC 188 in
Cepheus, all believed to be more than 10 billion years old.
The difference in stellar population is strikingly shown
when each of these clusters is portrayed on the H-R diagram.
A pattern resembling that of the Perseus clusters is indi-
cative of extreme youth, while one resembling NGC 188 implies great age. (For a discussion of the use of the H-R
diagram in cluster age-dating, refer to M13 in Hercules.)

The Double Cluster lies in the richest region of the
Perseus Milky Way. "Here again we enter upon one of the
most splendid portions of the Galaxy" says T.W.Webb. "Night
after night the telescope might be employed in sweeping
over its magnificent crowds of stars.." While observing in
the area, note also the smaller cluster NGC 957, about 1½°
to the ENE, described by C.E.Barns as a "field shot with
diamond dust".

NGC 1275   (Perseus A) Position 03164n4120. An unusual
galaxy, a most interesting and controversial
object, listed in the Shapley-Ames catalogue as an E-type
system of magnitude 12.7. It is the brightest member of a
small cluster of galaxies lying about 2° to the east of
Algol, and slightly north. The nearby elliptical galaxy
NGC 1270 is undoubtedly a member of the same cluster; both
systems show a nearly identical red shift of about 3280
miles per second. The distance of the group is currently
thought to be about 300 million light years.

NGC 1275 is a strong source of radio emission, and
bears the number 3C84 in the Third Cambridge Catalogue of
radio sources; it is also known as "Perseus A". Photographs
with large modern telescopes show that the galaxy has a
peculiar distorted appearance; the main body may be either
an S0 galaxy or a very compact spiral, but there are huge
dust and gas clouds all along the entire northern edge.
Some of these patches show a red shift about 1800 miles per
second greater than that of the main body.

W.Baade and R.Minkowski have suggested that this
object actually consists of two galaxies which appear to
be passing through each other. The main galaxy may be a
compact spiral or a large elliptical; the other a loose
later-type spiral whose arms are distorted from tidal
force. According to this picture, the loose spiral is seen
in front of the compact one in the northern part of the
system, and in that region the actual collision is now in
progress. The total duration of such an inter-passage of
galaxies would be on the order of a million years. From
spectral studies it may be surmised that the two galaxies
are moving through each other with a relative velocity of
about 1800 miles per second. Presumably, the southern part
of the loose spiral has already penetrated through its
brighter neighbor, and is now out of sight beyond its central hub. One difficulty with this interpretation is that
the system does not really look like a double galaxy at
all; the fainter galaxy (if such it is) shows neither a
nucleus nor a definite spiral pattern. If actually a galaxy
it may be an irregular system rather than a spiral.

Some other peculiar galaxies, once thought to be
collisional systems, have been discredited for several
reasons, and it now seems unlikely that the hypothesis will

GALAXY NGC 1275 in PERSEUS. This unusual system is a strong source of both radio energy and X-rays. Palomar Observatory photograph with the 200-inch reflector.

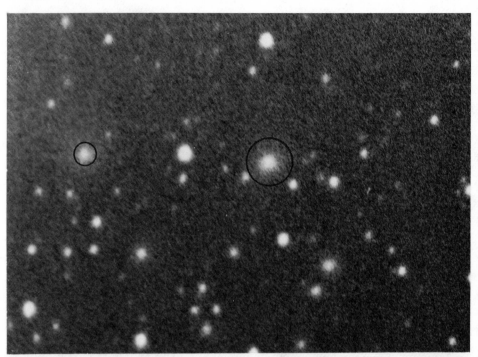

NGC 1275. Top: The field of the eruptive galaxy showing
(large circle) NGC 1275, and (at left) NGC 1270. Below:
Red light photograph showing the extending filaments.

survive in the case of NGC 1275. The collision theory
attributes the strong radio emission to high temperatures
produced in the colliding gas clouds, but it is now certain that a number of "radio galaxies" emit far too much
energy to be explained by this mechanism. Cygnus A, whose
appearance suggests two systems in contact, is among the
most powerful radio sources known, but the energy comes
chiefly from two large invisible "lobes" on either side of
the visible object. These appear to be connected with a
strong magnetic field. In such galaxies it is thought that
enormous outbursts of some sort have occurred in the central nuclei; such an outburst seems to be in progress at
the present time in the strange galaxy M82 in Ursa Major.

NGC 1275, in addition to its radio emission, is now
known to be a source of X-ray energy as well. The X-ray
source, some 35' in apparent size, is centered on the galaxy, but covers an enormously greater area, fully three
million light years in diameter. The tenuous gas pervading
this area might explain the observed X-ray emission, if it
is at a temperature of about 80,000,000°K. According to
another theory proposed in 1974, high energy electrons are
accelerated by the magnetic field of the galaxy, and are
interacting with the low-energy photons that comprise the
3° background radiation attributed to the original high
compression phase or "big bang" stage of the Universe. By
this interaction the energies of the photons are increased
to the X-ray level, a process known as the "inverse Compton
effect".

Red light photographs show an extensive system of
long filaments extending outward on all sides of NGC 1275;
the material is exploding into space at a velocity of over
1500 miles per second. The appearance is remarkably similar to the strange galaxy M82, and shows us the same violent outpouring of material on a titanic scale. Vast outbursts or explosions of some sort have evidently occurred
in the nuclei of such systems, possibly resulting from the
formation of a massive "hyperstar", "collapsar", or "black
hole" in the heart of the galaxy. One outburst, which might
seem trivial by comparison, has been observed in NGC 1275,
the appearance of a 15.5 magnitude supernova in 1968.

(Refer also to 3C273 in Virgo, M82 in Ursa Major,
M87 in Virgo, Cygnus A, and NGC 5128 in Centaurus)

STAR CLUSTERS IN PERSEUS. Top: NGC 1245 is located 3° SW
from Alpha Persei. Below: NGC 1528 is about 7.5° east of
Alpha. Lowell Observatory 13-inch telescope photographs.

PHOENIX

LIST OF DOUBLE AND MULTIPLE STARS

NAME	DIST	PA	YR	MAGS	NOTES	RA & DEC
△250	33.4	85	54	8½- 9	Dist & PA dec, Spect both K2	23244s5033
I 1059	1.1	188	28	7½- 11	PA dec, spect G5	23251s5037
I 23	0.8	334	42	8 - 9	PA slow inc, spect F3	23254s5643
Cor258	1.9	198	47	8½- 9	PA slight dec, spect F8	23307s5718
ι	6.7	270	30	5 - 13	(B603) spect A2	23324s4254
I 25	0.8	42	47	8 - 8	PA dec, spect F2	23326s5746
θ	4.0	275	59	6½- 7	(△251) PA slow inc Spect A3	23368s4655
Hd 303	2.0	65	42	7 - 10	(Hu 1337) relfix, spect G5	23373s4736
Hu1550	0.7	192	59	8 - 8½	PA inc, spect F5	23388s4151
δ28	3.8	74	33	7 - 11	(h5416) relfix,	23404s4635
	45.2	215	14	-11½	spect G5	
δ28c	0.5	171	34	11½-12	(I1607)	
I 305	3.1	123	36	7 - 11	Spect K0	23455s5110
Slr 14	1.0	5	29	8- 8½	Binary, about 188 yr; PA dec, spect G5	23479s5159
I 1477	0.5	292	59	6½ - 7	PA inc, spect G3	23577s4434
h5437	1.9	312	59	6½- 10	PA slow inc, spect G0	23580s5322
Hd 180	5.4	176	28	6- 11½	(L9721) relfix, cpm, spect sgG1	00037s4921
Hd 181	0.4	206	59	6 - 8	(L9740) PA & dist dec, spect sgG4	00065s5416
h3347	25.0	81	16	7 - 13	Spect B8	00066s5027
h3364	39.8	239	55	7½- 10½	Dist & PA inc, spect G0	00223s5415
I 45	0.5	231	59	8 - 8½	(h3376) PA dec,	00312s5536
	7.0	247	30	-10½	spect A2	
ξ	13.2	253	33	6 - 10	Relfix, spect A8, (h3387) cpm	00395s5647
Mld 1	6.2	162	53	7½- 8	relfix, spect G0	00397s5603
h3390	14.2	313	36	7 - 9½	relfix, spect G5	00409s4527
η	19.8	217	33	4½- 11	(h3391) spect sgA0	00411s5745
Hd 183	14.3	307	01	6- 13½	(L207) Spect K0	00435s4750

LIST OF DOUBLE AND MULTIPLE STARS (Cont'd)

NAME	DIST	PA	YR	MAGS	NOTES	RA & DEC
I 47	1.1	11	59	7 - 7½	PA inc, spect F2	00495s4359
I 49	0.7	38	42	8 - 8½	PA dec, spect F0	00585s5251
h3415	1.0	144	47	7½- 7½	PA dec, spect A2	01016s4055
Hd 184	3.7	259	38	7½- 11	Relfix, spect F5	01017s4048
β	1.4	346	54	4 - 4	(Slr 1) PA dec,	01039s4659
	57.5	52	27	- 11½	slow binary, spect G8	
ζ	0.8	40	49	4 - 7	(Rst 1205) PA inc.	01063s5531
	6.4	243	53	- 8	(Rmk 2) AC relfix; all cpm; spect B6; A = ecl.var.	
Slr 2	1.2	191	54	7½- 9	PA dec, spect A5	01065s4656
Hu1342	0.3	314	45	8 - 8	PA dec, spect F8	01072s5652
h3422	14.0	56	54	7½- 11½	Relfix, spect K5	01130s5554
h3430	2.6	228	42	7- 9½	PA dec, spect F8	01185s5737
Hd 185	4.6	240	30	7½-11½	Spect K0	01214s4114
I 447	1.4	309	13	6½-10½	Spect A0	01297s4550
I 51	1.5	11	42	7 - 9½	cpm; spect G5	01322s4557
I 1139	7.3	75	25	7½- 14	Spect F0	01564s4125

LIST OF VARIABLE STARS

NAME	MagVar	PER	NOTES	RA & DEC
ζ	3.9--4.4	1.670	Ecl.bin; also visual triple star; Spect B6+B6	01063s5531
ρ	5.2--5.3	0.12	Delta Scuti type; spect F0	00484s5116
R	7.4--14.0	268	LPV. Spect M3e	23539s5004
S	7.2--9.0	141	Semi-reg; spect M6e	23565s5651
T	8.6--14..	281	LPV. Spect M5e	00280s4641
U	9.5--13..	226	LPV.	00327s5029
V	8.5--14..	257	LPV. Spect M4e	23297s4616
W	7.6--14..	331	LPV. Spect M6e	01179s5611
Z	8.7--12..	255	LPV.	23565s5341

LIST OF VARIABLE STARS (Cont'd)

NAME	MagVar	PER	NOTES	RA & DEC
RR	8.7--13..	427	LPV.	23562s3943
RS	9.8--11..	239	LPV. Spect M2e	01119s5700
RU	9.9--11..	289	LPV. Spect M1e	23254s4744
SV	9.7--13..	265	LPV.	23466s4323
SW	9.9--11.7	2.553	Ecl.bin; spect 08	23579s3954
SX	7.1--7.4	.0550	Dwarf cepheid, spect A2 (*)	23439s4151
SY	8.8--9.3	Irr	RW Aurigae type? spect F8	01283s4258
SZ	9.5--10..	Irr	Spect K5	01319s4330
TT	9.9--10.7	Irr	RW Aurigae type?	01434s4211
AE	7.7--8.0	.3624	Ecl.bin; W U.Maj type; Spect G0	01305s4947

LIST OF STAR CLUSTERS, NEBULAE, AND GALAXIES

NGC	OTH	TYPE	SUMMARY DESCRIPTION	RA & DEC
----	I.5325	⊖	Sb/Sc; 12.5; 1.6' x 1.5' F,S,R, gbM	23260s4136
7689	△347	⊖	Sc; 12.3; 2.6' x 1.8' pF,L,R, gbM	23299s5422
7690		⊖	S0/Sa; 13.0; 1.6' x 0.6' cB,S,1E, psmbM	23302s5158
----	I.5328	⊖	SB? 12.7; 1.4' x 1.3' vF,S,R	23304s4519
7702		⊖	S0; 13.1; 1.3' x 1.0' B,cS,E, gsbM; lenticular with outer ring	23327s5617
7744		⊖	SB; 12.8; 1.1' x 1.0' cB,S,v1E, svmbM	23424s4312
7764		⊖	I/SB; 12.8; 1.1' x 0.8' B,pL,R, gbM	23484s4101
7796		⊖	E2; 12.9; 0.8' x 0.7' pB,cS,R, pgbM	23565s5544
625	△479	⊖	Sa/S pec; 12.3; 2.4' x 0.8' B,L,mE, gpmbM	01329s4141

DESCRIPTIVE NOTES

SX   (GC 32998) (SAO 231773) Variable. Position
23439s4151, about 6½° west of Alpha Phoenicis.
This is one of the most famous of the dwarf cepheids,
originally discovered as a star of large proper motion in
1938, and found to show rapid light variations by Dr.O.J.
Eggen at Canberra, Australia, in 1952. At that time it had
the shortest period known for any type of pulsating star,
about 79 minutes. According to the Moscow "General Catalogue of Variable Stars" (1970) the precise period is 0.054965
day, or 79m 10s. The visual magnitude ranges from 7.1 to
about 7.5; photographically it is 7.1 to about 7.8, owing
to the change in color during the cycle. Some maxima are
higher than others by half a magnitude, indicating that the
star is oscillating in at least two superimposed periods.
The cycle of the amplitude variations is about 4.6 hours,
or very close to 3½ times the main period.

SX Phoenicis appears to be a subdwarf A-type star;
from the computed distance the actual luminosity must be
only 2 or 3 times that of the Sun. The Moscow Catalogue
reports a probable absolute magnitude of +4.1 which places
the star at least two magnitudes below the main sequence.
In a study made in 1975 at the European Southern Observatory, R.Haefner found that the spectral type, usually given
as A2 to A5, actually reached F4 at a low minimum. From
current theories of stellar structure it appears that stars
of this type have an unusually low mass for their spectral
types; the figure for SX Phoenicis is believed to be about
0.24 solar mass, and the diameter probably about 80% that
of the Sun. The star shows an annual proper motion of 0.89"
in PA 163°; the distance may be about 140 light years.

In a study of this odd star made at the observatory
of the University of Chile in 1970, J.Stock and S.Tapia
reported that "an analysis of the radial velocity curves,
from 500 spectra, together with photometric data, show that
the observations cannot be reconciled with the model of a
single pulsating star....In addition, the intensities and
profiles of the absorption lines vary rapidly, often within
minutes....SX Phoenicis is probably a binary, or even a
multiple system". The stars of this type resemble the RR
Lyrae stars, but have smaller masses and lower luminosities.
(Refer also to CY Aquarii, AI Velorum, and RR Lyrae)

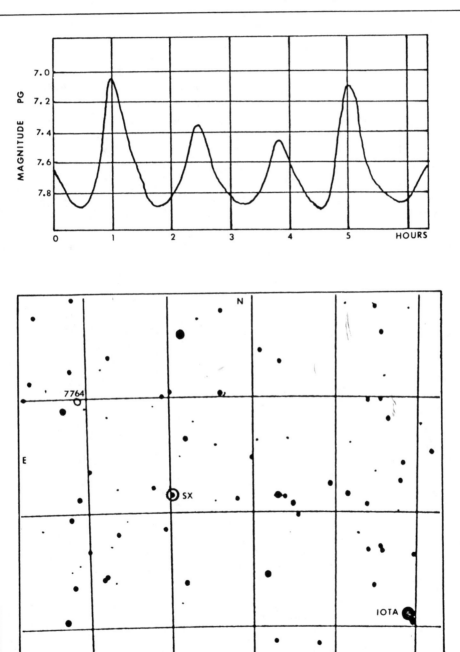

SX PHOENICIS Identification chart. Stars to about 9th magnitude are shown. Grid squares are 1° high.

PICTOR

LIST OF DOUBLE AND MULTIPLE STARS

NAME	DIST	PA	YR	MAGS	NOTES	RA & DEC
I 728	0.5	60	56	8½-8½	Relfix, spect K0	04338s5249
h3681	41.0	254	12	6½- 11	Spect G0	04403s4722
Cor 23	3.6	229	43	7 - 9½	no certain change;	04405s4755
					cpm, spect F0	
I 732	2.1	16	43	8½- 11	spect F2	04488s5148
ι	12.3	58	52	5½- 6½	(△18) relfix, cpm,	04498s5333
					Both spectra F0	
h3715	9.9	112	53	7½- 9	relfix, spect F0	04582s4932
λ52	8.9	339	38	7½-10½	Spect F5	05205s5322
θ	0.2	152	60	7 - 7½	(I345) (△20) PA	05236s5222
	38.2	287	38	- 7	dec, spect A0	
h3777	54.8	345	38	6½- 10	dist inc, spect F5	05328s5456
h3777b	11.3	104	26	10- 12		
h3784	5.0	67	41	8 - 9½	cpm, slow PA inc,	05368s4608
					spect G0	
h3793	12.1	127	56	7 - 11	PA inc, spect A0	05402s4816
I 63	1.1	15	59	7½- 9	slight dist dec,	05470s4856
	31.9	211	01	- 12	spect A0	
β1493	0.2	250	59	6½- 7½	PA dec, spect F5	05494s5247
h3816	22.9	180	13	7- 11½	spect K2	05518s4758
h3822	55.9	305	38	6½- 7½	Spect K2, K0	05562s5326
h3822b	20.1	125	17	7½- 13		
I 3	0.9	5	51	7 - 7½	no certain change,	06120s6128
					spect B9	
△27	40.1	229	50	6½- 8	Optical, dist dec,	06156s5911
					spect G0, F8	
μ	2.4	231	37	6 - 9	(h3874) relfix,	06312s5843
					spect B9	
I 5	2.4	270	54	6½- 8½	cpm, spect G0	06374s6130
I 6	0.9	251	46	8 - 8	no certain change,	06419s6142
					spect F5	

PICTOR

LIST OF VARIABLE STARS

NAME	MagVar	PER	NOTES	RA & DEC
δ	4.7--4.9	1.673	Ecl.bin, lyrid; spect B0	06093s5457
R	6.9--10.0	172	Semi-reg; spect M1e--M4e	04448s4920
S	7.3--14..	427	LPV. Spect M7e--M8e	05096s4834
T	8.0--14..	201	LPV. Spect M6e	05137s4658
V	9.5--11..	180	Semi-reg	06126s5952
Y	8.4--10..	116	Semi-reg; spect M2	05098s4538
RR	1.2--13..	---	NOVA PICTORIS 1925 (*)	06352s6236
RU	9.7--13..	208	LPV.	06297s6011
RV	9.0--11.8	3.972	Ecl.bin; spect B9	04563s5213
RW	9.5--12..	287	LPV.	06089s6007
SS	8.4--9.3	57:	Semi-reg	05462s4516
ST	8.8--9.1	18.75	Cepheid; spect G0	06135s6127

LIST OF STAR CLUSTERS, NEBULAE, AND GALAXIES

NGC	OTH	TYPE	SUMMARY DESCRIPTION	RA & DEC
1705		⊘	S0; 12.9; 0.7' x 0.5' pF,S,R, pmbM, BN	04532s5326

DESCRIPTIVE NOTES

RR   Nova Pictoris 1925. Position 0635z2s6236, about
1½° southwest of Alpha Pictoris. A brilliant
nova, first seen by R.Watson in South Africa on the morning
of May 25, 1925. Watson reported the discovery to the Royal
Observatory at the Cape, and the first scientific observations of the new star were made there that same evening. A
thorough study of the nova was made at the Cape over a number of years, and a complete analysis of the changing spectral features was presented in a monumental report by H.
Spencer Jones in Vol.X of the *Annals* of the Cape Observatory.

     At discovery the nova was magnitude 2.3, but was soon
found to be still brightening, with irregular oscillations,
until it reached its peak brightness of magnitude 1.2 on
June 9. By July 4 the star had faded to 4th magnitude, but

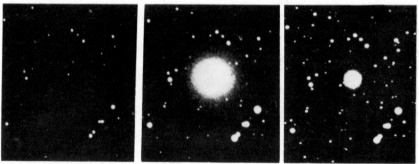

ROYAL   OBSERVATORY   PHOTOGRAPHS

unexpectedly rose again to magnitude 1.9 on August 9. The
maximum was thus a long extended one, classifying the star
as a "slow nova" in contrast to such rapid and sharp maxima
as displayed by Nova Aquilae 1918 and CP Puppis in 1942.
The star had been bright for some time before its discovery,
as it appears as a 3rd magnitude object on a plate made in
mid-April.

     A third, and last, outburst of the star, from magni-
tude 3.7 to about 2.3, took place some 60 days after the
chief maximum; the star thereafter faded with slight fluctuations until it disappeared fron naked-eye view in mid-
December. Ten years later it was a 9th magnitude object and
still fading slowly. The star appears on previous photographs of the region as an object of magnitude 12.7, and

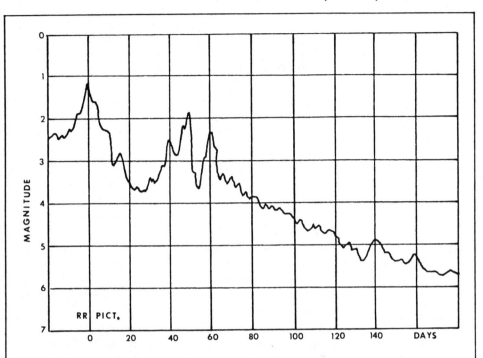

RR PICT.

apparently constant in light. In 1975, the nova was close
to its pre-outburst magnitude of about 12½.

Beginning in 1928, telescopic observations revealed
that the image of the star, under high powers, appeared as
a triple body. Two years later one part had vanished and
the other two had expanded outward. The rate of the expansion was about 0.2" per year, and the separation of the two
principal nuclei had increased to 1.0" in early 1931. These
fragments were apparently knots of nebulous material and
not stellar bodies. A similar appearance was detected
around the image of Nova Herculis 1934 after the outburst.

With an amplitude of about 11½ magnitudes, Nova Pictoris showed a light increase of about 40,000 times during
the outburst. At the peak of its luminosity, spectroscopic
studies showed that the outer layers of the star were being
blasted into space at a velocity of about 700 miles per
second. Still higher velocities, of up to 950 miles per
second, were measured some six months later, when the star
had declined to the limit of naked-eye visibility.

The distance of no nova is known very accurately, as
even the nearest have been too distant for trigonometrical

parallax measurements. Actual luminosities are thus somewhat uncertain, though it seems clear that the "slow novae"
of the RR Pictoris and DQ Herculis type have considerably
lower peak luminosities than the "fast novae" such as CP
Puppis.

According to a summary of the evidence by C.P.Gapos-
chkin, RR Pictoris had a peak absolute magnitude of about
-6.3, equivalent to nearly 30,000 times the light of the
Sun. From a study of the expanding nebulosity, D.B.Mclaugh-
lin (1936) obtained a somewhat different result, giving the
probable absolute magnitude as -7.3. The pre-nova brightness, in any case, was fairly comparable to the luminosity
of the Sun. The computed distance is slightly over 1000
light years; our Sun at that distance would appear as a
star of magnitude 12.4.

Studies of the post-nova star, in recent years, have
brought forth some evidence to support a theory that all
novae are close binary stars. In the case of RR Pictoris,
the short-period variability occupies a cycle of 0.14498
day, or slightly under 3½ hours. (Refer also to Nova Aquila
1918, GK Persei 1901, DQ Herculis 1934, CP Puppis 1942, and
Nova Cygni 1975)

KAPTEYN'S STAR (LFT 395) (GC 6369) (HD 33793) Position
05097s4500. A faint and inconspicuous
star, but remarkable for its unusually large proper motion,
exceeding that of any other star in the heavens with the
single exception of Barnard's Star in Ophiuchus. Kapteyn's
Star was discovered in 1897 by Professor J.C.Kapteyn of the
University of Groningen. The star is sometimes referred to
by its number in the Cordoba Zone catalogue; CZ 5h 243. It
is located in a rather blank region of the southern sky,
about midway between Canopus and 41 Eridani; the field lies
some 8½° NW from Beta Pictoris.

Kapteyn's Star has an annual proper motion of 8.70"
in PA 131°, and thus requires some 414 years to change its
position by 1° in the sky. Like Barnard's Star, it is a red
dwarf; the apparent visual magnitude is 8.8 and the photographic magnitude about 10.0. The spectral type is dM0. A
direct parallax measurement leads to a distance of 12.7
light years, about twice the distance of Barnard's Star.
With an absolute magnitude of about 10.8, Kapteyn's Star

DESCRIPTIVE NOTES   (Cont'd)

is some 250 times fainter than our Sun, but about 10 times more luminous than Barnard's Star.

As in the case of Barnard's Star, the abnormally large apparent motion is the result of the nearness of the star combined with an unusually high true space velocity of about 175 miles per second, one of the highest known. The motion across the line of sight corresponds to about 100 miles per second, and the radial velocity is 145 miles per second in recession.

The proper motion of the star from 1900 to the year 2050 is shown on the charts below. Stars to about the 10th magnitude are shown; grid squares are 1° on a side with north at the top. (Refer also to Barnard's Star, page 1251)

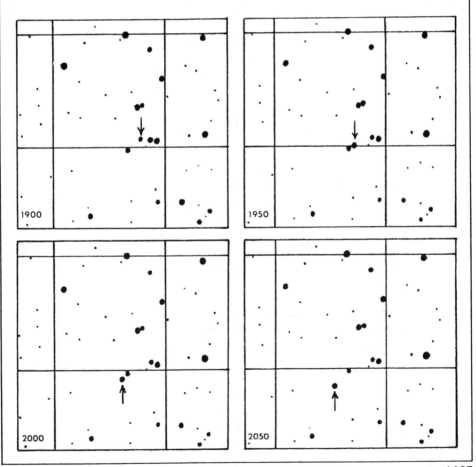

PISCES

LIST OF DOUBLE AND MULTIPLE STARS

NAME	DIST	PA	YR	MAGS	NOTES	RA & DEC
Σ2959	11.0	115	59	6½- 10	PA inc, dist dec,	22545s0331
Σ2959b	11.2	97	59	10- 13	Spect A0	
2	3.8	85	58	6 - 13½	cpm, PA dec, spect gK1	22569n0042
β714	0.4	121	58	7 - 10	PA dec, spect A2	23115s0254
Σ2995	5.0	29	50	8 - 8½	relfix, spect G5	23140s0153
Σ3009	7.1	230	44	7 - 9	relfix, spect K0	23217n0326
A896	0.8	69	44	7½- 10	cpm; spect F8	23262s0106
Σ3019	11.0	185	51	7 - 8	relfix, spect A3	23281n0458
Hu 298	0.2	337	69	7 - 7½	Binary, 31 yrs; PA inc, spect dF6	23297n0649
β723	3.7	164	60	7½-11½	spect K0	23381n0008
Σ3030	2.4	222	39	8½- 8½	relfix, spect F0	23382s0040
Σ3031	14.3	311	23	7½- 8½	relfix, cpm, spect F8	23386n0559
Σ3033	3.1	6	48	8½- 8½	relfix, spect F2	23414n0658
Hu 300	1.0	115	66	8½- 9	slight PA dec, spect G0	23430n0613
Bar 19	1.0	356	58	8½- 8½	PA slow inc, spect F8	23445n0458
A899	3.6	38	30	7½- 14	spect K0	23451s0102
Σ3045	1.5	264	43	8 - 10	relfix, spect A2	23519n0211
A2100	0.4	219	69	7½- 8	PA dec, spect F0	23542n0427
27	1.3	292	58	5 - 10	(β730) PA inc, spect G9	23561s0350
Wei 45	1.8	88	45	8½- 9	relfix, spect G0	23577n0130
β281	1.5	178	66	7½- 11	(h998) PA dec,	00002n0151
	35.0	332	22	-11½	spect G0	
Hdo 2	16.9	127	58	7½- 12	PA & dist inc, spect F0	00062n0745
34	7.7	161	58	6- 10½	(Σ5) relfix, spect B8	00075n1052
Σ8	7.8	291	33	7½- 8½	relfix, spect F8	00090s0321
35	11.8	149	58	6 - 7½	(Σ12) (UU Piscium) cpm, relfix, spect sgA9, dF0; A= var.	00124n0833
Σ15	4.7	199	44	7½- 10	cpm, relfix, spect G5	00133s0553
β---	6.0	142	08	7½- 12	spect A5	00135n0750

LIST OF DOUBLE AND MULTIPLE STARS   (Cont'd)

NAME	DIST	PA	YR	MAGS	NOTES	RA & DEC
Σ23	3.0	256	54	7½- 10	Optical pair, dist	00148n0002
	103	281	25	-11½	dec, spect F8	
Σ20	11.6	232	33	8 - 9	relfix, spect F5	00148n1614
38	0.1	303	58	7½- 7½	(Σ22) (A1803)	00148n0836
	4.3	237	55	- 8	spect F2	
	63.7	142	15	- 12½		
Σ25	1.4	192	58	8½- 8½	relfix, spect F5	00161n1543
β1015	0.2	25	69	8½- 8½	Binary, about 250 yrs, PA inc, spect F5	00180n1202
β1093	0.7	108	69	7 - 8	slow binary, PA inc, spect A0	00183n1042
42	28.5	324	53	6½- 10	optical pair, PA dec, dist dec, spect gK2 (Σ27)	00198n1312
β488	3.4	343	39	7½-10½	PA slight dec, spect A0	00214s0345
	40.5	58	14	-12½		
49	20.9	103	21	7 - 10½	(Σ32) optical, dist inc, spect A0	00282n1546
51	27.7	83	33	5 - 9	(Σ36) relfix, spect A0	00298n0641
55	6.6	193	39	5½- 8	(Σ46) relfix, cpm spect K0, F3; nice colors	00373n2110
OΣ18	1.7	191	70	7½- 9½	Binary, about 545 yrs; PA inc, spect dF6	00398n0354
	42.8	270	13	- 12		
β495	0.3	190	61	7½- 7½	Binary, about 170 yrs; PA dec, spect G0	00461n1825
65	4.5	297	55	6 - 6	(Σ61) neat relfix pair; cpm, spect gF0, gF2	00472n2726
β496	5.4	360	14	7 - 13	cpm, relfix, spect F0	00490n1231
Σ67	2.0	353	66	8½- 9	PA dec, spect F8	00495n1020
A2307	0.3	20	66	7½- 8½	PA dec, spect F0	00508n0349
66	0.4	237	69	6½- 7	(OΣ20) binary, PA dec, about 210 yr; spect A1	00519n1855

LIST OF DOUBLE AND MULTIPLE STARS   (Cont'd)

NAME	DIST	PA	YR	MAGS	NOTES	RA & DEC
Σ74	3.2	300	50	8 - 9	relfix, spect F0	00522n0910
β500	0.4	297	67	8 - 8	PA slow inc, spect A2	00527n3023
β302	0.6	136	58	6½- 8	PA inc, spect A2	00556n2108
A2210	4.2	249	58	7½-13½	PA inc, spect G5	00576n1756
Σ82	1.7	305	42	8½- 9½	relfix, spect F0	00581n0913
Σ87	6.5	198	52	8 - 8½	relfix, cpm, spect K0	01028n1508
ψ1	30.0	160	59	5 - 5	(Σ88) relfix, cpm, spect both B9; AC dist slow dec.	01030n2113
	92.6	123	59	- 11		
77	32.9	83	54	6- 6½	(Σ90) easy cpm pair, spect dF5,dF4	01032n0439
	31.8	313	10	- 14		
0Σ22	8.8	196	40	7- 10	relfix, spect F0	01044n1117
β303	0.5	250	66	7 - 7½	PA slight inc, spect F0	01070n2352
h634	41.1	255	25	6½- 11	optical, dist inc, PA dec, spect G0	01083n0917
Σ98	19.6	249	33	7 - 8	relfix, spect A0	01101n3150
0Σ26	10.8	258	33	6 - 10	relfix, spect K0	01102n2948
	114	342	10	- 12		
φ	7.7	225	36	4½- 10	(Σ99) (85 Piscium) relfix, spect K0; cpm pair	01110n2419
	144	173	12	- 13		
ζ	23.6	63	59	4½- 5½	(Σ100) (β1029), easy relfix pair, spect A5, dF6; BC PA dec. (*)	01111n0719
ζb	1.0	8	58	5 - 12		
h636	20.7	288	11	7½- 10	spect A0	01116n3017
A2102	0.4	149	58	7 - 10	spect F2	01132n0931
β4	0.3	127	68	7½ - 8	Binary, 180 yrs; PA dec, spect F0	01187n1117
	23.3	250	17	- 13½		
A2213	4.6	124	32	7 - 14	spect A0	01233n1009
Σ122	6.0	329	41	7 - 9	relfix, spect B8	01243n0317
95	0.4	165	61	7- 7½	(β1164) Binary, 64 yrs, PA dec, spect dG0	01251n0506
	147	222	25	-11½		
A1910	0.2	150	67	7 - 7	spect A0	01270n2234
	18.2	350	17	- 12½		

PISCES

LIST OF DOUBLE AND MULTIPLE STARS   (Cont'd)

NAME	DIST	PA	YR	MAGS	NOTES	RA & DEC
Σ129	8.5	282	21	8½- 9	relfix, cpm; spect F0	01276n1224
η	1.0	36	58	3½- 11	(99 Pisc) (β506) difficult cpm pair, PA inc, spect G8	01288n1505
Σ132	43.3	348	21	7 - 10	Multiple star; AB PA dec, spect G5	01293n1642
	68.4	231	21	- 11		
	133	114	21	- 10		
	14.2	154	08	- 14		
Σ132d	5.0	291	24	10- 10½		
0Σ31	4.3	80	70	7 - 11	slight PA dec, spect K0	01307n0757
100	15.6	78	38	7 - 8	(Σ136) relfix, spect A3	01322n1218
	75.0	311	11	- 13		
Σ138	1.5	51	66	7½- 7½	Binary, PA slow inc Spect F8	01334n0723
103	1.0	285	57	7 - 9	(β5) relfix, spect G5	01366n1622
	90.5	186	10	- 12		
Σ145	10.8	32	10	6- 10½	relfix, AB cpm; Spect dF3	01385n2530
	82.0	338	52	- 11		
Σ146	24.1	306	54	8½- 8½	relfix, spect F8	01386n0952
Σ155	5.0	326	51	8 - 8½	cpm, PA slight dec, spect F2	01416n0914
α	1.9	292	66	4 - 5	(Σ202) Binary, PA dec, about 720 yrs; Spect A2 (*)	01594n0231
h647	26.1	32	26	9 - 9½	PA slight dec, Spect M8; color contrast	02000n0727

LIST OF VARIABLE STARS

NAME	MagVar	PER	NOTES	RA & DEC
χ	4.91±.04	.5805	(8 Piscium) α Canum type, Spect A2p	23244n0059
19	5.5--6.0	Irr	(TX) Spect N0	23438n0313
35	5.5--5.6	.8417	(UU) Ecl.bin; Spect sgA9, F4; also visual double	00124n0833
47	4.9--5.3	49	(TV) Semi-reg; spect M3	00254n1737
R	7.1--14..	344	LPV. Spect M3e--M6e	01280n0237
S	8.2--15.3	405	LPV. Spect M5e--M7e	01150n0840
T	9.2--12.3	257	Semi-reg; spect M5	00294n1419
X	8.8--14..	353	LPV. Spect M6e	01094n2158
Y	9.0--12.0	3.766	Ecl.bin; Spect A3, K0	23319n0739
Z	7.0--7.9	144	Semi-reg; spect N0	01134n2530
RT	7.5--9..	70:	Semi-reg; spect M3	01111n2652
RU	9.8--10.5	.3903	Cl.Var; Spect A7--F3	01117n2409
RX	8.8--14.5	280	LPV. Spect M1e	01229n2108
SV	8.8--10..	102	Semi-reg; spect M5	01439n1850
SZ	8.5--9.0	3.966	Ecl.bin; Spect K1, F8	23108n0224
TT	9.5--11..	147	Semi-reg; spect M4	01581n0518
TZ	9.6--15..	375	LPV.	01201n2507
UV	8.3--9.2	.8610	Ecl.bin; Spect G5	01143n0633
UZ	6.4--6.6	4.156	α Canum type, spect A2p	01431n0819
VX	5.91±.01	0.16:	δ Scuti type, spect A4	01272n1806
VY	6.57±.02	0.16:	δ Scuti type, spect A7	23252n0435

LIST OF STAR CLUSTERS, NEBULAE AND GALAXIES

NGC	OTH	TYPE	SUMMARY DESCRIPTION	RA & DEC
7541	430^2	⊖	SB/Sc; 12.6; 2,7' x 0.7' B,L,mE, mbM	23122n0415
7619	439^2	⊖	E1; 12.6; 0.8' x 0.6' cB,pS,R,psbM; On Peg-Pisces border; 7' pair with 7626	23178n0755
7626	440^2	⊖	E2; 12.7; 0.9' x 0.7' cB,pS,R,psbM; pair with 7619 Brightest members of small group	23182n0756

LIST OF STAR CLUSTERS, NEBULAE AND GALAXIES   (Cont'd)

NGC	OTH	TYPE	SUMMARY DESCRIPTION	RA & DEC
7679		⊘	S0/SB; 13.1; 1.4' x 0.8' pB,S,R,mbM	23262n0315
7716		⊘	Sb; 12.9; 1.3' x 1.1' F,pL,1E,1bM. 10^m star 2' south	23339n0001
7782	233^3	⊘	Sb; 13.1; 2.1' x 0.9' pF,pL,1E,g1bM; 7781 sp = F,S,R	23514n0742
7785	468^2	⊘	E5; 12.9; 1.1' x 0.6' pB,pS,E,psbM; 8^m star $4\frac{1}{2}$' W	23528n0538
95	257^2	⊘	Sc; 13.1; 1.1' x 0.9' F,pL,R,gbM	00196n1012
128	854^2	⊘	E8 pec; 12.8; 2.2' x 0.4' pB,pS,1E; rectangular main mass (*)	00267n0235
470	250^3	⊘	Sb/Sc; 12.4; 1.7' x 1.1' pB,pL,1E. Pair with 474	01171n0309
474	251^3	⊘	E0/S0; 12.8; 0.4' x 0.4' pB,S,R,smbM	01175n0310
488	252^3	⊘	Sb; 11.2; 3.5' x 3.0' pB,L,R,svmbM; delicate spiral pattern (*)	01191n0500
514	252^2	⊘	Sc; 12.3; 2.2' x 2.0' F,L,1E,vg1bM	01213n1239
520	253^3	⊘	Pec; 12.4; 3.0' x 0.7' F,cL,E, eruptive galaxy (*)	01220n0332
524	151^1	⊘	E1?; 11.8; 2.0' x 1.9' vB,pL,mbM	01221n0916
628	M74	⊘	Sc; 11.0; 9.0' x 9.0' F,vL,R, Fine face-on spiral (*)	01340n1532
718	270^2	⊘	Sb; 12.6; 0.8' x 0.7' pB,S,R,psmbM	01507n0357
741	271^2	⊘	E1; 13.0; 0.9' x 0.8' pF,S,R	01538n0523

DESCRIPTIVE NOTES

ALPHA Name- AL RISCHA, from the Arabic *Al Risha*, "The Cord", possibly derived from the earlier Babylonian *Riksu*, which also means a cord. Another name, OKDA, is from the Arabic phrase *Ukd al H·aitain* which seems to refer to a "flaxen cord" or bond which ties the two fish of Pisces together. Magnitude 3.96; spectrum A2; position 01594n0231. The star is located in the extreme southeastern portion of the constellation, about 7° NW of Mira in Cetus.

Alpha Piscium is a fine binary star for moderate telescopes, first noted by William Herschel in 1779, and gradually becoming more difficult for small instruments as the separation steadily decreases from a measured 3.6" in 1831. During the first century of observation the PA slowly changed from 336° to 307°; in 1966 it was 292° and the distance between the components had diminished to 1.9".

According to the ADS Catalogue, the individual magnitudes are 4.33 and 5.23; the spectral types about A2 and A3; the fainter star has a peculiar spectrum with strong metallic lines. From an orbit calculation by W.Rabe (1961) the period of the pair may be about 720 years with periastron due about 2060 AD. The semi-major axis of the compu-

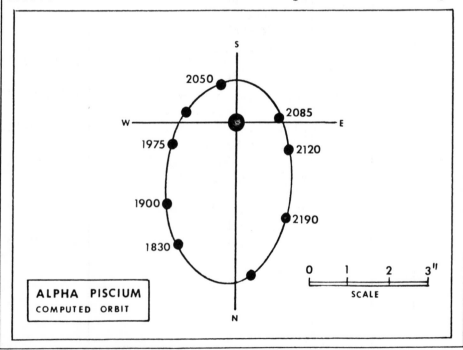

ALPHA PISCIUM
COMPUTED ORBIT

ted orbit is 2.65", with an eccentricity of 0.60 and an
inclination of 142°. Orbital motion is retrograde, and it
appears that both components are spectroscopic binaries of
uncertain period.

A peculiar feature of Alpha Piscium is an apparently
illusionary color contrast described by various observers.
T.W.Webb calls them "greenish white and blue" but then adds
"I found contrast certain, but fainter star troublesome as
to color, usually ruddy or tawny, sometimes blue". In 1855
he thought them "pale yellow and brown yellow" and on other
occasions "pale yellow and fawn-colored". K.McKready in his
Beginner's Star Book (1912) calls the components "pale
green and blue", while C.E.Barns (1929) simply notes *"weird
coloring"* and, perhaps wisely, refuses to commit himself to
any more definite statement.

A direct parallax measurement gives the probable
distance as about 130 light years; the resulting total luminosity of the system is about 35 times that of the Sun.
Alpha Piscium shows an annual proper motion of about 0.03";
the radial velocity (somewhat variable) averages some 5
miles per second in recession.

The suspected variability of the primary star seems
to have first been announced by E.C.Pickering at Harvard in
1879; he found that the magnitude difference between the
components appeared to vary from about 0.7 to 1.1, but in
no definite period. Miss Agnes Clerke (1905) regarded the
variability of at least the primary as certain, but thought
that possibly both stars changed in light. No definite
evidence of light changes has been noted in recent years.

ZETA   Position 01111n0719. A fine easy double star,
located virtually on the Ecliptic, about 13°
WNW from Alpha Piscium. It seems to have been first recorded by William Herschel in 1781, and the components have
shown no certain relative change in nearly two centuries,
although the physical association is proven by the common
proper motion of 0.15" per year. With a separation of 23½"
this is an easy object for a beginner's telescope; Olcott
calls it "a fine object" and Barns refers to it as a colorful pair; the tints having been described by various observers as "yellowish and pale lilac", "white and greyish",
or "pale yellow and rose". The spectral types are A5 and

THE MOON AND VENUS near conjunction in the field of Zeta Piscium (image at top of print). The photograph was made by the author with the Lowell 13-inch telescope on March 11, 1959.

DESCRIPTIVE NOTES   (Cont'd)

dF6; the individual magnitudes, according to the ADS Catalogue, are 4.2 and 5.3, but the discrepancies in reported
measurements suggest that both stars may be variable. The
Harvard photometric magnitudes for the components are 5.57
and 6.49. Both stars show a variable radial velocity, and
are assumed to be spectroscopic binaries.

A trigonometrical parallax indicates a distance of
about 140 light years; the projected separation of the two
stars is then about 1000 AU. The primary star has an actual
luminosity of about 8 to 10 times that of the Sun. The mean
radial velocity of the system is about 6 miles per second
in recession.

A 12th magnitude companion to the fainter star, at a
distance of 1", was discovered by S.W.Burnham in 1888. A
very difficult object, it has been measured on various
occasions (as in 1958) but, according to the Lick Observatory *Index Catalogue of Visual Double Stars* (1961), it was
not visible at all in 1914, 1922, 1936, or 1951. With an
absolute magnitude of about +9.1, this third star must be
some variety of low-luminosity dwarf; it is doubtful that
a spectrum will ever be obtained for this object.

19   (TX Piscium) Variable. Position 23438n0313;
the star is the easternmost member of the so-
called *Circlet of Pisces*, the oval ring of stars about 7° x
5° which represents the western fish of the pair, and which
includes the stars Gamma, Kappa, Lambda, Iota, Theta, 19
and 7 Piscium. 19 Piscium is chiefly of interest for its
strong red color, being one of the few N-type stars within
range of naked-eye sighting, or available to field glasses.
Its spectral class is usually given as N0, but on the new
"carbon star" classification system it would be called $C6_2$.
This is one of the very low temperature giants which show
bands of carbon compounds in the spectrum. As in the case
of the very similar star Y Canum Venaticorum, no parallax
has been detected, proving that the distance must be in the
range of 400 light years or more. From theoretical studies,
the absolute magnitude of the star is believed to be close
to -2.0; the computed distance is then about 1000 light
years. 19 Piscium shows an annual proper motion of about
0.04"; the radial velocity (variable) averages about 6.5
miles per second in approach.

To appreciate the unusual ruddy tint of the star,
observers should compare its light with the nearby bluish
stars 21 and 25 Piscium, of spectral types A2 and A0 respectively. (See chart, page 1475)

As a variable, TX Piscium is not of great interest,
as the visual range is only about half a magnitude, from
about 5.5 to 6.0. When observed photographically the range
increases to about 1 magnitude. Owing to the fairly small
amplitude and the unusual color, the light changes are not
conspicuous visually, and the variability was considered
doubtful for many years. T.E.Espin, however, observed a
well defined maximum of 5.2 in August 1884, and O.C.Wen-
dell at Harvard found a range of 0.36 magnitude during a
one-year period in 1905-1906. There appears to be no regular periodicity, and T.W.Webb refers to the star as one of
a class which "for long intervals are nearly constant in
light, and then for a short time rise to a maximum......In
estimating magnitudes of red stars it is well to put them
out of focus, and thus compare *discs* of light instead of
points."   (Refer also to Y Canum Venaticorum, R Leporis,
V Hydrae, S Cephei, and Omicron Ceti)

VAN MAANEN'S STAR   (Wolf 28) (LFT 76) A white dwarf
star of unusual interest. Position
00465n0509, about 2° south of Delta Piscium. This is one of
the few white dwarf stars which can be seen and identified
easily in amateur telescopes. It was discovered by A. van
Maanen in 1917 through a comparison of photographs made in
1914 and 1917. With the exception of the faint companions
to Sirius and Procyon this is probably the nearest white
dwarf star to the Solar System, at a distance of 13.8 light
years. The star was discovered through its large proper
motion, which amounts to 2.98" yearly in PA 155½°

Van Maanen's Star has an apparent visual magnitude of
12.4, and an absolute magnitude of about 14.2; the actual
luminosity is about 1/5800 that of the Sun. The spectral
type is variously given as either late DF or early DG; the
surface temperature is near 6000°K. This was the first of
the "late type" degenerate stars to be discovered, though
a fair number of DG and DK stars are now known.

Definitely among the smallest stars known, this very
dense star has a computed diameter of about 7800 miles, or

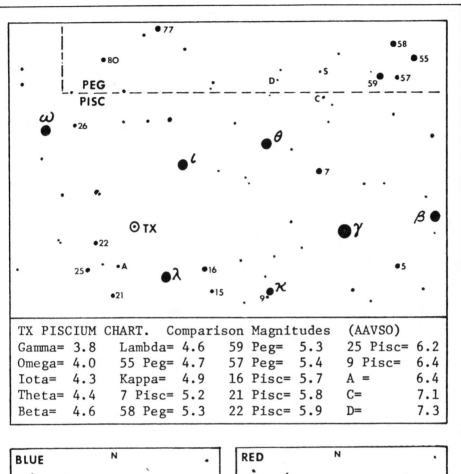

TX PISCIUM CHART. Comparison Magnitudes (AAVSO)

Gamma= 3.8	Lambda= 4.6	59 Peg= 5.3	25 Pisc= 6.2
Omega= 4.0	55 Peg= 4.7	57 Peg= 5.4	9 Pisc= 6.4
Iota= 4.3	Kappa= 4.9	16 Pisc= 5.7	A = 6.4
Theta= 4.4	7 Pisc= 5.2	21 Pisc= 5.8	C= 7.1
Beta= 4.6	58 Peg= 5.3	22 Pisc= 5.9	D= 7.3

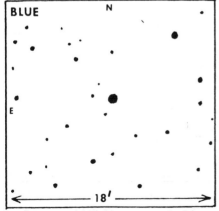

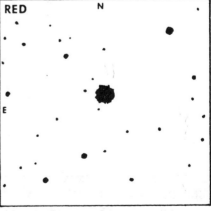

TX PISCIUM. The strong color is illustrated by comparison of red and blue exposures. From Lowell Observatory plates

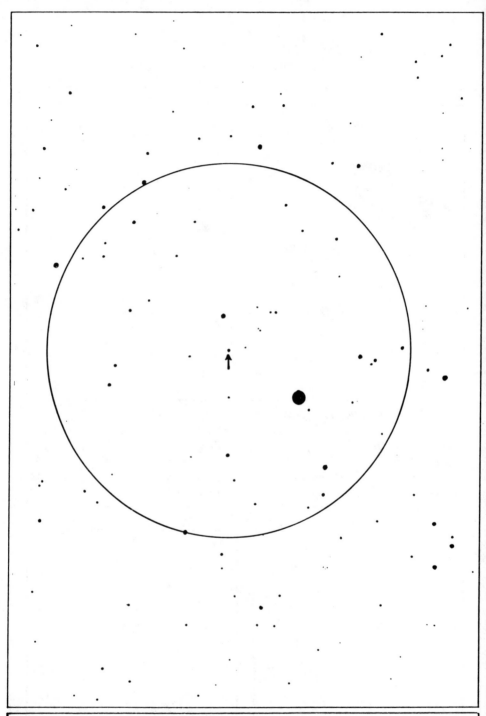

FIELD OF VAN MAANEN'S STAR. The circle is 1° in diameter
with north at the top; stars to about 14th magnitude are
shown.   Lowell Observatory 13-inch telescope plate.

1476

DESCRIPTIVE NOTES   (Cont'd)

very close to the size of our Earth. With a mass comparable to that of the Sun, the density is found to be nearly one million times that of water, or about 20 tons to the cubic inch. This is about 10 times the computed density of the famous Sirius B, most famous of the degenerate stars. The surface conditions on such a body are almost impossible to imagine; the atmospheric pressure on Van Maanen's Star is about 2000 times that present on Earth, and the surface gravity must exceed 50,000 times that of our own feather-weight planet. Ordinarily an object the size of the Earth would be totally beyond the range of any telescope at a distance of a dozen light years. But the Van Maanen object is a *star*, visible in the light of its own radiation, even though actual energy production may have ceased ages ago. Presumably such an object must be among the oldest stars in the Universe, and will now spend many additional aeons of slow cooling as it eventually approaches the "black dwarf" state billions of years from now.

The identification chart (page 1476) was made from a plate obtained with the 13-inch camera at Lowell Observatory in December 1961. Observers of the future must thus make allowance for the large proper motion of the star, about 5' per century. The brightest star in the field is GC 959 or LFT 73, magnitude 6.7; this star also has a large proper motion (1.37" annually) in almost the same direction but from parallax measurements appears to be about twice the distance of Van Maanen's Star. The two objects are evidently not related. (Refer also to Sirius B (page 394), Procyon B (page 449) and Omicron-2 Eridani (page 890)

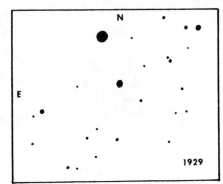
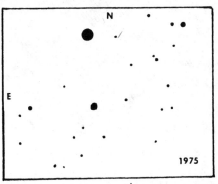

THE PROPER MOTION OF VAN MAANEN'S STAR. EACH FIELD IS 18' WIDE; THE TOTAL DISPLACEMENT IN 46 YEARS IS 137". FROM LOWELL OBSERVATORY PLATES.

SPIRAL GALAXY M74 in PISCES. One of the finest examples of a large face-on spiral. Palomar Observatory photograph made with the 200-inch reflector.

M74 (NGC 628) Position 01340n1532, about $1\frac{1}{2}°$ ENE from Eta Piscium. Large spiral galaxy, first observed by P.Mechain in September 1780, and confirmed by Messier the following month. Mechain described it as a nebula which "contains no star; it is fairly large, very obscure and extremely difficult to observe....One can make it out with more certainty in fine, frosty conditions.." D'Arrest found it "pale and tenuous, very much denser towards the centre; the central part is almost round, diam 40". Resolved; uncertain whether it has a true nucleus.." John Herschel, in his *General Catalogue* of 1864, made the surprising mistake of classifying M74 as a globular star cluster, and this error was carried over into the N.G.C. Lord Rosse, in 1848, was probably the first to detect the spiral form, which was recorded with great clarity by the early astro-photographer Isaac Roberts, in a photograph made in December 1893.

M74 is one of the faintest and most elusive of the Messier objects, and requires a dark clear sky and a suitably low power eyepiece. Under the best conditions it may appear as a circular, quite featureless glow, about 6' in diameter with averted vision, and with a bright, nearly stellar nucleus. Curiously however, M74 was noted and mapped on the charts of the *Bonn Durchmusterung* (about the year 1860) which was compiled from observations with a 3-inch telescope; it bears the number BD+15°238. K.G.Jones (1968) points out that an extended object of low surface brightness "may well be easier to see in small telescopes of wide field than in larger instruments".

On the best long-exposure plates the apparent size increases to about 10'; S.van den Berg (1960) reports a total size of 10.6' x 9.0' and a total integrated magnitude (pg) of 9.74. The integrated spectral type is F5.

M74 is one of the most perfect examples of a face-on Sc-type spiral, resembling the large M101 in Ursa Major but somewhat more symmetrical in form. A.Sandage (1961) in the *Hubble Atlas of Galaxies* states that the chief spiral arms of the system have a thickness of about 1000 parsecs, and are bordered on their inner edges by thin dust lanes which may be traced deep into the nuclear hub. At a distance of close to 30 million light years, the 9' diameter corresponds to about 80,000 light years; the total absolute magnitude is about $-20\frac{1}{2}$, or about 13 billion times the light of the Sun. The galaxy thus seems to be not greatly inferior in size to our own, but has a much lower luminosity, unless the computed distance is considerably underestimated. E.Holmberg (1964) derived a distance modulus of 29.7 magnitudes for M74, and a total mass of about 40 billion suns. A.Sandage (1961) adopted a distance modulus of 30.0 magnitudes; S.van den Bergh (1960) has 29.4. All these figures agree in giving a distance in the range of 25 to 33 million light years. The measured red shift of the system, however, is 426 miles per second (corrected for the solar motion in our own Galaxy), which, according to the presently accepted value for the Hubble constant, implies a somewhat greater distance of about 42 million light years. As in the notable case of the edge-on spiral NGC 4565 in Coma, we have a situation here in which the known red shift gives too large a distance. The greater distance, however, would bring the luminosity and computed mass into closer agreement with the figures obtained for many other large spirals.

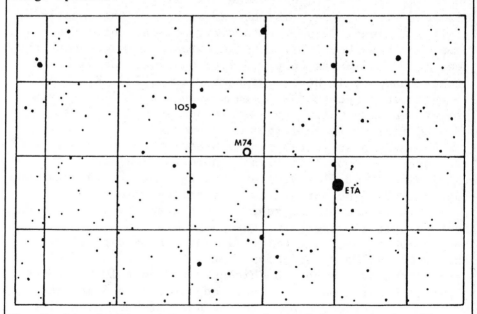

The field of M74, with stars to about 9th magnitude. The grid squares are 1° high with north at the top.

SPIRAL GALAXY NGC 488 in PISCES. A compact spiral with a
fine pattern of delicate spiral arms. Palomar Observatory
photograph made with the 200-inch reflector.

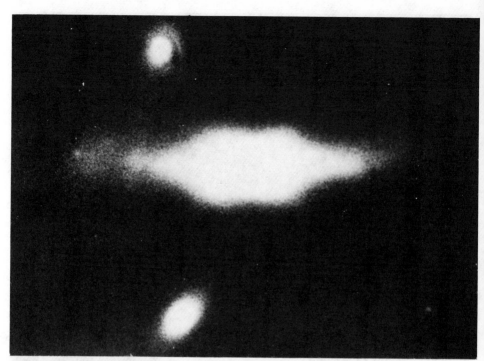

UNUSUAL GALAXIES IN PISCES. Top: NGC 128 has an odd box-shaped central mass. Below: NGC 520, a distorted or erupt-ing system. Palomar Observatory 200-inch telescope plates.

PISCES AUSTRINUS

LIST OF DOUBLE AND MULTIPLE STARS

NAME	DIST	PA	YR	MAGS	NOTES	RA & DEC
6	6.8	59	27	6- 13	(B1009) Spect A0	21292s3410
B1013	0.2	101	39	8½- 9	PA dec, spect F8	21462s3552
h3059	25.2	253	30	7½- 10½	Relfix, spect B9	21479s2810
B1014	2.8	145	32	8 - 12	Spect A0	21537s3138
h5311	40.6	292	19	7 - 10	spect K0	21566s2918
	48.6	227	19	- 10½		
Stn 56	11.3	35	35	7½- 11	relfix, spect F0	21567s2752
η	1.6	116	55	5½- 6½	(β276) relfix,	21580s2842
					spect B8	
B547	2.3	294	35	8 - 14	spect A2	22003s2718
B549	3.6	213	59	7 - 13	spect K0	22017s2608
Daw156	1.3	317	59	7½- 11	PA inc, spect G0	22076s3404
β769	0.9	352	52	7 - 8½	relfix, cpm pair,	22087s3443
					spect F2	
B557	0.3	321	52	8 - 8	PA dec, spect F2	22191s3446
λ474	0.1		55	7 - 8	(h3118) spect K0;	22266s2855
	33.7	296	59	- 11	AB uncertain	
β	30.4	172	52	4½- 7½	(△240) nice cpm	22287s3236
					relfix pair, spect	
					A0	
B568	4.9	310	35	7½- 13	spect A3	22352s3344
h5356	86.6	160	51	6½- 7	(H VI 119) wide cpm	22370s2836
h5356b	3.2	67	59	7 - 8	pair, spect G0,F5;	
					BC PA inc.	
Hd 301	0.2	91	54	7 - 7	Binary, 27 yrs; PA	22472s3304
					dec, spect F2	
h5365	4.8	278	33	7½-12½	spect A3	22490s3608
	55.6	35	19	-11		
γ	4.3	262	57	4½- 8½	(h5367) PA slow	22498s3308
					dec, spect A0	
δ	5.0	244	53	4½- 10	(Hwe 91) cpm pair,	22532s3248
					slight PA inc,	
					spect gG8, G3	
α	30.0	36	00	1 - 14	FOMALHAUT, Spect A3	22549s2953
					Optical (*)	
h5371	9.0	343	50	7½- 9	relfix, spect G0	22551s2622

LIST OF VARIABLE STARS

NAME	MagVar	PER	NOTES	RA & DEC
π	4.7--5.0	7.975	Cepheid, spect F0	23007s3501
R	8.5--12..	293	LPV. Spect M3e	22152s2951
S	8.3--13..	272	LPV. Spect M3e--M5e	22009s2818
V	8.0--9..	148	Semi-reg; spect M5	22526s2953
RV	9.0--12..	361	LPV.	21358s2726
RX	9.0--10..	366	Semi-reg	22103s2731
RY	8.8--13..	223	LPV.	21445s3626
ST	8.5--12..		LPV.	22515s3440

LIST OF STAR CLUSTERS, NEBULAE AND GALAXIES

NGC	OTH	TYPE	SUMMARY DESCRIPTION	RA & DEC
----	I.5135	⊖	S0?/Pec; 13.1; 0.9' x 0.8' F,pL,R	21453s3511
7135		⊖	S0; 13.0; 0.9' x 0.7' pB,pL,1E, vgbM	21468s3507
7154		⊖	I; 13.5; 0.8' x 0.5' B,pL,1E, glbm; central bar	21524s3503
7172		⊖	Sa; 12.9; 1.6'x 0.6' pB,pL,1E,gbM; nearly edge-on with dark equatorial band; Compact trio of E galaxies 7173, 7174, 7176 lies 7' to south.	21591s3207
----	I.5156	⊖	SB; 13.2; 1.0' x 0.5' pF,pS,mE	22004s3402
7314		⊖	Sb/Sc; 11.9; 3.5' x 1.5' cF,L,mE,vlbM, elong N-S	22330s2618
7361		⊖	Sb; 12.8; 3.0' x 1.0' F,pL,mE,vglbM. Nearly edge-on spiral, elong N-S	22395s3019
----	I.5269	⊖	S0; 13.1; 1.3' x 0.5' vF,pS,mE, spindle-shape	22550s3618
----	I.5271	⊖	S0?; 12.6; 2.0' x 0.8' pF,pS,mE,cbM	22553s3401

ALPHA   Name- FOMALHAUT, from the Arabic *Fum al Hut,*
the "Mouth of the Fish". The 18th brightest
star in the sky; magnitude 1.17; Spectrum A3 V. Position
22549s2953. Opposition date (midnight culmination) is about
September 6.

Sometimes called "The Solitary One", Fomalhaut lies
in a rather empty region of the southern Autumn skies, and
to dwellers at the latitude of New York it is the southernmost of the visible 1st magnitude stars. *"On early acquain-
tance,"* said Martha E. Martin (1907) *"the loneliness of the
star, added to the sombre signs of approaching autumn,
sometimes gives one a touch of melancholy....In November
and December, when...the winter stillness has fallen upon
us, a glance toward the southwest will discover Fomalhaut,
still placid and alone, on its way to adorn the warmer
southern skies..."*

Strangely, Fomalhaut is often described in observing
books as "reddish" although it is actually an A-type star
nearly as white as Deneb. R.H.Allen and Mary Proctor both
repeat this curious error. Possibly the effects of the
Earth's atmosphere are responsible for this impression, as
the star is always seen at a rather low altitude by observers in the northern hemisphere.

According to Admiral Smyth, the star appears as *Fom
Alhout Algenubi* in an almanac manuscript dated 1340. The
Alfonsine Tables of 1521 call it *Fomalhant,* while Riccioli
has *Phomaut, Fomauth, Phomelhaut,* and other variations.
Lalande labels it *Fumalhant* and *Phomahant;* Schickard calls
it *Fomalcuti.* An alternate Arabic name, according to R.H.
Allen, was *Al Difdi al Awwal,* "The First Frog". Fomalhaut
is associated in Greek legend with the terrifying monster
Typhon, who, it was believed, now lies imprisoned under the
fuming mass of Mt.Etna in Sicily. In Syrian and Canaanite
lands it was honored as the symbol of the sea god or fish-
god *Dagon,* whose temple at Gaza (possibly the Biblical
Azzah) was so dramatically destroyed by the Hebrew strong-
man Samson, as related in the 16th Chapter of the *Book of
Judges.* The Biblical story, however, does not tell us if
Dagon's temple, as some modern astro-archeologists suspect,
was oriented to the rising of Fomalhaut. Incidentally, the
identification of Dagon as a sea god or fish-god has been
questioned by Isaac Asimov; he suggests that the name does

DESCRIPTIVE NOTES   (Cont'd)

not derive from the Semitic *dag* (fish) but from *dagan* or
"grain". If so, Dagon may have been an agricultural deity
rather than a sea god. Schickard seems to be the origina-
tor of the relatively modern legend that Fomalhaut symbolizes the New Testament story of St.Peter and the coin found
in the mouth of the Fish. Fomalhaut was known to the star-
watchers of ancient Persia as one of the four "Royal Stars"
of Heaven, the others being Regulus, Aldebaran, and Antares.
     Fomalhaut lies at a distance of 23 light years, and
is a main sequence A3 star about twice the diameter of our
Sun and 14 times more luminous. The computed absolute magnitude is about +2.0. The star shows an annual proper motion of 0.37" in PA 117°; the radial velocity is 4 miles
per second in recession.
     A companion star of the 14th magnitude at 30" was
reported by T.J.J.See in 1897; it appears to be merely a
faint field star, having no real connection with Fomalhaut.
Proper motion studies of the region, however, have identified a much more distant companion which does seem to show
the same parallax and motion as the bright star; it is
located just 2° distant toward the south. This is the star

THE DISTANT COMPANION TO FOMALHAUT lies 2° south of the
bright star; the position is 22536s3150.

DESCRIPTIVE NOTES   (Cont'd)

GC 31978 or BS 8721; it appears in W.J.Luyten's Catalogue
of proper motion stars as LTT 9283. The star is a dwarf of
spectral type dK5, apparent magnitude 6.49, about 10% the
luminosity of our Sun. This is an exceptional case of a
wide common motion pair; the true separation (minimum) must
be some 51,000 AU, or nearly one light year! It seems quite
unlikely that the two stars form a truly gravitationally
connected pair today; the attraction over such a vast space
is negligible. We might speculate that Fomalhaut and its
distant companion are the last surviving members of a low
density cluster which gradually dispersed and scattered,
as such clusters as the Coma Berenices group appear to be
doing today.

LACAILLE 9352   (LFT 1758) (Cordoba 31353) Position
23026s3609, in the extreme southeast
corner of the constellation, about 1° SSE of Pi Pisces
Austrinus. This small star is noted for its remarkably
large annual proper motion, the fourth fastest known. This
amounts to 6.90" per year in PA 79°, and the star thus
requires about 520 years to traverse 1° in the heavens. The
only stars known with a larger proper motion are Barnard's
Star in Ophiuchus (page 1251); Kapteyn's Star in Pictor
(page 1462); and Groombridge 1830 in Ursa Major. (Page 1978)
A list of star of large proper motion appears on page 1257.
      Lacaille 9352 is a red dwarf of spectral class dM2e
with an apparent visual magnitude of 7.39, and a photographic magnitude of about 8.6. It is one of the nearest
stars to our Solar System at a distance of 11.9 light years,
just a little more distant than 61 Cygni or Tau Ceti. With
an absolute magnitude of about +9.6, the actual luminosity
is about 1/85 that of our Sun.
      The computed space velocity of the star is about 75
miles per second, and the motion is almost directly "side-
on" as seen from the Earth; the radial velocity is a relatively slight 5.8 miles per second in recession. The large
apparent motion of the star is thus a result of this circumstance, combined with the small distance of the star. It
is interesting to reflect that an equally near star might
go unnoticed for some time, if it was moving directly in
the line of sight, either toward or away from us, as there

would then be no large proper motion to call attention to its nearness. Most of the faint nearby neighbors of the Sun have been detected by proper motion surveys, which have shown us also that low-luminosity dwarfs vastly outnumber giant stars in space.

The chart of the field of Lacaille 9352 is based on plates obtained at Lowell Observatory in 1964. Stars to about 11th magnitude are plotted, and north is at the top. The grid squares are 1° on a side. The arrow indicates the direction of motion of the star, and the distance covered during the interval 1860 - 2060 AD. (Refer also to Barnard's Star in Ophiuchus)

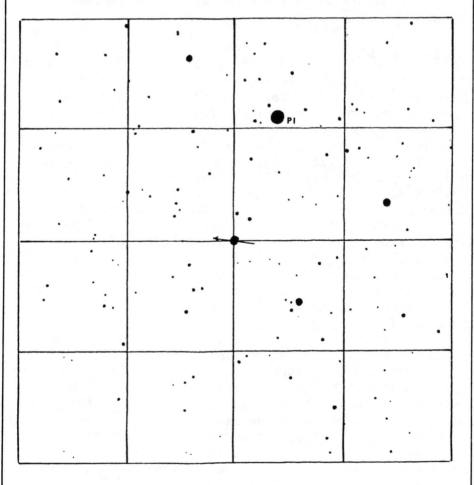

PUPPIS

LIST OF DOUBLE AND MULTIPLE STARS

NAME	DIST	PA	YR	MAGS	NOTES	RA & DEC
h3834	4.8	220	51	6 - 9	Dist inc, PA dec, Spect F5, F8	06032s4505
△23	2.2	83	41	7 - 7½	Binary, about 460 yrs; PA inc, spect G5	06035s4827
h3856	34.5	4	13	6½- 9½	relfix, spect K0	06215s4537
I 282	1.3	306	59	7½- 10	PA dec, spect F0	06229s5030
I 156	1.4	129	46	6 - 8½	cpm; spect B9	06244s4809
R65	0.5	277	26	6 - 6	(△30) Binary, 50	06286s5012
	12.4	314	26	- 8½	yrs; PA dec, spect G5; ABC all cpm.	
R65c	0.5	53	26	9 - 9	binary, 105 yrs; PA inc.	
Cor 39	10.2	204	59	7½- 9	spect K0	06345s4814
△31	12.8	321	32	5 - 7½	relfix, spect gG7, A0	06373s4810
h3882	1.9	25	33	8 - 12	(Hu 1415) spect B8	06373s4501
	18.1	331	33	- 10		
h5443	15.6	107	38	6 - 10	relfix, spect B3	06396s4018
△32	8.2	277	51	6½- 7½	relfix, spect A3	06406s3820
h3889	42.3	266	17	7 - 9	relfix, spect B5, B9	06416s5024
h3895	26.0	64	20	7- 10½	relfix, spect G5	06454s4745
I 158	1.8	186	43	7- 10½	spect G0	06460s4830
I 159	6.6	324	39	6½- 11	spect K0	06485s4523
I 1498	9.9	288	33	6½-12½	spect B9	06513s4355
h3900	2.1	279	59	7 - 9	slight PA dec, spect A0	06524s3409
φ20	1.5	290	36	7½-11½	spect K0	06534s3830
I 65	0.2	274	26	7 - 7	binary, 16.7 yrs; PA inc, spect F5	06555s3526
I 1161	1.9	24	42	7½- 11	spect A5	06562s4158
I 66	1.9	254	42	8 - 9½	(h3905) spect B9	06565s3521
	14.8	269	20	- 9½		
△38	20.6	122	32	6 - 7	relfix, cpm; spect G0, G0	07024s4333
h3928	3.3	140	59	6½- 8	Dist & PA slow dec, spect F0	07037s3442
	37.2	287	20	- 10		
h3931	57.3	313	20	7 - 10	spect B9	07044s4215
	72.5	40	20	- 9	C spect= F5	

LIST OF DOUBLE AND MULTIPLE STARS   (Cont'd)

NAME	DIST	PA	YR	MAGS	NOTES	RA & DEC
β757	2.7	66	43	6 - 9	relfix, cpm, spect B5	07107s3628
I 1505	2.9	196	41	7½- 10	spect A2	07117s3824
I 7	0.7	202	26	8 - 8½	binary, 84 yrs; PA dec, spect K0	07161s4654
Δ45	22.8	157	12	7½- 8½	relfix, spect A2	07200s4826
h3957	7.7	195	35	8 - 8½	relfix, spect F8	07205s3549
h3966	7.0	142	51	7 - 7	relfix, spect A3	07230s3712
h3968	25.9	141	20	7½- 12	spect K0	07244s4201
h3969	17.4	227	38	7 - 7½	relfix, spect F8, F8	07252s3413
h2391	16.7	290	33	6½-11½	spect K5	07260s2644
Δ49	9.0	52	50	6½- 7	relfix, spect B3	07269s3145
Σ1104	2.1	358	54	6½- 8	PA inc, spect F8;	07271s1453
	20.4	188	10	- 11	ABC all cpm	
	42.4	8	16	-11½		
σ	22.4	74	52	3½- 8½	(Δ51) relfix; spect K5, G5	07276s4312
λ80	0.3	86	49	8½- 9	spect A0	07288s2759
h3973	9.7	35	47	7½- 8½	relfix, spect B8	07297s2048
How 18	1.9	201	66	8 - 9	relfix, spect B9	07319s2335
Hh 269	9.6	114	52	6 - 6	(S552) (HN 19) PA slow inc, spect dF4 & dF5	07322s2322
Doo--	12.7	114	58	7 - 12	AB PA inc, dist inc	07322s1346
	17.4	304	58	- 13½	Spect F5; AC PA dec	
	24.4	25	58	- 12	AD PA inc, dist dec	
	40.7	129	58	- 13	AE dist inc	
	65.7	131	58	- 10	AF dist inc	
Doo--F	6.6	108	58	10-10½		
I 1167	1.5	310	43	7½- 9½	spect A0	07327s4158
h3982	38.4	156	20	5 - 9½	(p Puppis) spect B8	07334s2815
h3982b	42.1	130	20	9½- 10		
Σ1120	5.2	176	28	6 - 12	(A3092) in cluster	07338s1423
	20.0	35	28	- 9½	NGC 2422, spect B5	
Σ1121	7.4	304	52	7 - 7½	relfix, spect B9;	07343s1422
	17.5	133	09	- 12	in cluster NGC 2422	
h3983	0.5	159	43	9 - 10	(Rst 3532) Spect	07348s1346
	7.3	299	15	- 9½	A0	

PUPPIS

LIST OF DOUBLE AND MULTIPLE STARS   (Cont'd)

NAME	DIST	PA	YR	MAGS	NOTES	RA & DEC
B730	0.3	67	51	7½- 7½	Spect B8	07350s2719
Daw 90	6.7	129	33	6½- 13	Spect B3	07357s2648
m	0.1	44	26	5 - 5½	(B731) spect B8	07362s2515
β201	2.8	331	35	8 - 8½	(Arg 47) relfix, spect A3	07368s2009
k	9.8	318	51	4½- 4½	Fine pair, relfix, Spect B8,B5	07368s2641
kᴮ	7.2	222	27	4½- 14	(β1061) slight PA dec, spect B5	
h3994	14.8	17	36	7- 9½	spect B8	07379s4856
	22.9	216	35	-10½		
I 160	1.2	147	59	6 - 8½	relfix, cpm; spect B5	07380s3802
λ84	8.3	279	59	6 - 11	PA dec, spect K0	07380s1933
I 353	0.9	36	42	8 - 8½	spect A0	07381s4310
Hn 90	3.0	278	10	9 - 9½	Spect F5	07388s1622
	8.6	141	10	- 11		
	97.8	215	10	- 9		
Hn 91	2.1	220	42	8½- 11	slow PA inc, spect F8	07392s2013
Hu 709	1.8	293	21	9 - 9½	spect A	07401s1809
	24.9	2	42	- 10		
	22.4	334	42	- 10½		
Hu 709c	0.3	208	40	10- 10½		
Hu 710	0.2	124	62	7 - 7½	binary, 87 yrs; PA dec, spect G5	07406s1657
I 486	0.8	277	36	8½- 9	spect B9	07409s2839
I 354	0.9	126	39	7½- 9	spect B8	07410s4227
h3995	6.3	254	30	8½- 10	relfix, spect A0	07413s2200
Δ55	51.8	133	17	7 - 8	spect F5, G0	07429s5020
2	16.9	340	23	6 - 7	(Σ1138) relfix,	07432s1434
	100	228	32	- 10½	AB cpm; spect A0,A0	
h4002	19.5	90	33	7½-11½	spect F2	07438s5010
I 781	3.3	339	44	7½-11½	spect B9	07446s2042
Hu 1428	0.3	16	51	8 - 8½	PA inc, spect B8	07453s4641
Δ56	49.7	177	00	7 - 8	spect B2	07454s4123
Hu 1429	0.5	284	43	8 - 8½	spect B8	07456s4316
5	2.2	5	60	5½- 7½	(Σ1146) dist & PA dec, spect dF5,dG3	07456s1204

LIST OF DOUBLE AND MULTIPLE STARS   (Cont'd)

NAME	DIST	PA	YR	MAGS	NOTES	RA & DEC
I 161	10.6	84	39	5 - 10	Spect B3	07457s3823
ξ	4.8	191	42	4 - 13	(β1063) Spect G3; 5^m field glass star at 286"	07472s2444
I 186	1.3	225	53	8 - 8	PA inc, spect F5	07474s3026
h4009	9.7	322	59	8 - 8½	PA inc, spect G0	07484s3202
B146	0.9	67	59	6½- 10½	PA inc, spect A0	07489s2424
9	0.6	292	66	5½- 6	(β101) Binary; 23 yrs; PA inc, spect G0	07495s1346
He 8	3.0	274	59	5 - 9	(Hwe 65) PA dec, spect F5	07504s3435
h4013	19.0	197	44	7- 10½	spect K5	07519s1812
	67.0	271	02	- 9½		
h4013c	3.0		02	9½-13½		
Hu 54	1.6	10	54	8½- 9	no certain change, cpm, spect F5	07523s1242
h4019	5.4	155	32	7½- 10	relfix, spect G0	07536s4142
Cor 64	5.6	146	31	7½- 11	relfix, spect K0	07540s4339
Hd 200	10.0	37	33	5 - 13	(λ92) Spect B3	07557s4358
	22.7	87	33	- 14		
I 26	0.5	45	51	6½- 7½	PA inc, spect B5	07559s4745
V	7.0	68	33	4½- 10	(h4025) (Hrg 131)	07568s4907
	19.0	48	33	- 11½	Primary= Ecl.bin;	
	39.2	38	33	- 10	spect B1+B3 (*)	
I 30	1.7	349	47	7½- 9	relfix, spect B8	07573s4733
△59	16.9	47	54	6½- 6½	(h4028) relfix, both spectra B3	07578s4950
h4024	10.1	83	37	8 - 9½	relfix, spect G5	07581s2923
I 1070	0.3	353	59	8 - 8½	(h4032) PA inc,	07583s4710
	29.2	351	13	- 9½	spect B9	
β333	1.6	44	42	7½- 10	relfix, spect A2	07591s2212
β202	7.4	163	33	7½- 9½	relfix, spect B9	07599s2704
	29.6	240	16	- 11½		
β203	6.9	244	54	7½- 8½	(h4037) relfix,	08006s2724
	63.6	74	54	- 9½	spect G5	
I 785	6.1	280	27	7½- 11½	spect B3	08008s3104
I 8	2.3	305	47	6½- 9½	PA dec, spect A0	08008s4432
h4038	27.0	346	20	5½- 7½	relfix, spect B9	08010s4110

LIST OF DOUBLE AND MULTIPLE STARS (Cont'd)

NAME	DIST	PA	YR	MAGS	NOTES	RA & DEC
h4035	34.8	134	51	6 - 8	relfix, spect gG8	08011s3219
I 487	1.8	22	59	6½- 10½	spect K2	08024s2533
Hu 623	5.3	64	59	7½- 13	relfix, spect A0	08032s1326
h4046	22.0	88	19	6 - 8	(I 189) relfix,	08038s3326
h4046b	14.1	6	19	8 - 9	spect G5,A	
I 164	0.6	79	45	7½- 9	spect G5	08045s3957
ρ	29.1	15	47	3 - 14	(Rst 5284) spect F6	08054s2410
					Primary var (*)	
β334	3.0	349	55	8 - 9½	slight PA dec,	08060s2159
					spect G5	
B149	4.0	310	26	6½- 14	spect B5	08061s2328
Ho 352	5.2	184	06	6 - 12	spect B3	08072s1606
△63	5.9	82	51	6½- 7½	relfix, cpm, spect A0	08081s4230
19	1.8	245	00	5 - 11	(β1064) spect K0;	08089s1247
	33.3	299	00	- 13	cluster NGC 2539 in	
	60.5	276	00	- 9	field.	
h4051	16.8	206	20	6½-13½	(Daw 43) Spect B0	08092s3709
	18.0	265	20	- 13		
h4057	25.2	299	33	5 - 9	relfix, spect A3	08097s4250
Hd 114	1.4	260	54	8 - 9½	relfix, spect F0	08111s2408
h₂	51.1	341	20	4½- 9	(h4062) spect gK0	08123s4012
λ98	5.5	68	20	6½- 12	spect K2	08131s3532
h4063	17.7	350	20	7½- 9	relfix, spect B8	08136s3713
β454	2.0	11	42	6½- 8½	cpm, slight PA dec,	08139s3046
					spect G5	
β905	3.5	11	33	8 - 10	relfix, spect K0	08142s1610
Cor 70	2.1	144	42	8½- 9	spect F8	08150s3442
How 22	0.2	162	44	8½- 9	(Rst 1386) spect F5	08154s2707
	3.5	116	25	- 9		
h4073	1.9	176	51	7 - 7½	cpm, slight PA dec,	08164s3713
					spect A0	
h4078	11.1	137	47	8½- 11	relfix, spect K0	08185s2356
h4085	6.5	275	33	5 - 13	PA inc? spect B3	08195s3619
h4087	1.5	278	55	7½- 8	PA dec, spect F5	08204s4050
	13.2	241	34	- 13		
	30.5	305	20	- 14		
B767	0.1	233	26	6½- 7	(Ho 353) spect F2	08207s2611
	33.1	223	15	- 13		

PUPPIS

LIST OF DOUBLE AND MULTIPLE STARS   (Cont'd)

NAME	DIST	PA	YR	MAGS	NOTES	RA & DEC
h4090	20.2	13	13	8½- 10	relfix	08217s4239
h4088	26.9	285	20	6½- 12	spect A0	08218s2848
EB-268	42.1	86	17	6 - 9	(S568) Optical, spect K5, K1	08229s2353
h4093	8.2	124	53	7 - 7½	(B1605) relfix cpm	08245s3854
h4093b	0.1	351	56	8 - 8	pair, spect both A0; BC PA dec.	

LIST OF VARIABLE STARS

NAME	MagVar	PER	NOTES	RA & DEC
ρ	2.7---2.8	.1409	Delta Scuti type, spectrum F6 pec; also visual double Rst 5284 (*)	08054s2410
L_2	3.0--6.0	141	Semi-reg; spect M5e--M6e; (*)	07120s4433
U	8.5--14.6	317	LPV. Spect M5e--M8e	07585s1242
V	4.5--5.1	1.454	Ecl.bin; lyrid; spect B1 + B3 (*)	07568s4907
W	7.6--13..	120	LPV. Spect M3e	07443s4204
X	8.0--9.2	25.96	Cepheid; Spect F6--G0	07306s2048
Y	8.0--8.6	110	Semi-reg; spect M7	08107s3500
Z	7.3--14.5	499	LPV. Spect M6e--M9e	07305s2033
RR	9.4--10.7	6.430	Ecl.bin; spect A	07452s4115
RS	7.0--8.1	41.39	Cepheid; spect F8--K5	08112s3426
RT	8.5--9.2	100:	Semi-reg; spect N	08036s3838
RU	8.9--11.1	425	Semi-reg; spect N	08053s2246
RV	8.5--10..	188	LPV. Spect M1e	06409s4219
RW	8.5--13..	332	LPV. Spect M3e--M6e	06079s5012
ST	8.5--9.6	18.89	Cepheid; W Virginis type	06472s3713
SU	8.1--13..	340	LPV.	07546s4400
SV	9.5--11.5	168	LPV. Spect M5e	08149s1339
SW	8.9--10.0	2.747	Ecl.bin; spect F0	08171s4236

PUPPIS

LIST OF VARIABLE STARS   (Cont'd)

NAME	MagVar	PER	NOTES	RA & DEC
TU	9.5--14.0	238	LPV.	08152s3427
TY	9.0--9.6	.5807	Ecl.bin; W Ursa Major type	07306s2041
			spect A9	
UZ	9.6--10.5	.7949	Ecl.bin; lyrid, spect A6	07394s1316
VX	8.2--8.9	3.012	Cepheid; spect F5--F8	07305s2149
VZ	9.1--10.2	23.169	Cepheid; spect F5--G1	07366s2823
WX	9.0--9.8	8.9378	Cepheid; spect F6--G1	07399s2545
WZ	9.9--10.5	5.0274	Cepheid; spect F5--G5	07582s2334
XY	9.2--11.4	13.778	Ecl.bin; spect A3e	08072s1150
XZ	8.0--10.7	2.1924	Ecl.bin; spect A0	08114s2348
YY	9.7--11.0	27.955	Ecl.bin; spect A0	07337s1917
ZZ	9.4--11.3	6.3381	Ecl.bin; spect A2	07462s1910
AC	8.9--10..	Irr	Spect N	08204s1545
AD	9.4--10.2	13.594	Cepheid; spect F5--G2	07460s2527
AO	9.0--14..	390	LPV.	07537s3926
AP	7.2--8.0	5.0843	Cepheid; spect F5--G0	07560s3959
AQ	8.1--9.6	29.857	Cepheid; spect F2--G2	07563s2859
AR	8.4--10.1	75:	RV Tauri type; spect F0-F8	08012s3627
AS	7.7--11..	328	LPV. Spect M7e	08078s3801
AT	7.7--9.1	6.6648	Cepheid; spect F8--G1	08105s3648
AU	8.6--9.5	1.1264	Ecl.bin; spect A0, A1	08159s4133
AZ	9.3--9.8	.86737	Ecl.bin.	08164s3446
BG	9.5--10.2	Irr	Spect M2	08055s2032
BH	8.4--9.1	1.9158	Ecl.bin.	08066s4153
BN	9.4--10.8	13.673	Cepheid; spect G4	08043s2957
CF	9.4--12.6	7.6456	Ecl.bin.	06040s4907
CH	8.0--12..	530:	LPV.	06435s3629
CP	0.2--18..	---	NOVA PUPPIS 1942 (*)	08099s3512
CQ	9.0--10..	Irr	Spect M5	07437s1903
DY	7.0--16..	---	Nova 1902	08117s2625
EI	9.0--10..	Irr	Spect M6	07363s1651
ET	9.2--14..	330	LPV.	07525s2757
HS	8.0--20..	---	Nova 1963	07515s3131
HU	9.0--11..	240:	Semi-reg; spect M3	07536s2831
HZ	7.7--18..	---	Nova 1963	08013s2820
KQ	4.9--5.2		Ecl.bin? Spect M2e+B2	07315s1425
KV	9.3--10.1	3.6677	Ecl.bin; spect A0	07459s4825
LS	9.4--10.3	14.146	Cepheid	07570s2910
LW	9.6--10.2	59.349	Ecl.bin.	08034s2632

LIST OF STAR CLUSTERS, NEBULAE AND GALAXIES

NGC	OTH	TYPE	SUMMARY DESCRIPTION	RA & DEC
2298	△578	⊕	Mag 10; diam 3'; B,pL,R,vRi; stars eF; class VI	06472s3557
2310		⊖	S0; 12.8; 2.0' x 0.5' pB,pL,vmE; spindle-shaped edge-on lenticular.	06524s4048
2421	67[7]	⠩	L,cRi; diam 8'; about 60 stars mags 11...13; class F	07341s2030
2422	M47	⠩	B,vL,pRi, Mag 5; diam 20'; about 25 stars mags 6.... class D; includes double star Σ1121 (*)	07343s1422
2423	28[7]	⠩	vL,Ri,pC, diam 20'; about 60 stars mags 12.... class D; 37' north of NGC 2422	07348s1345
2427		⊖	SB; 12.4; 5.0' x 3.0' F,L,pmE	07350s4730
----	Mel 71	⠩	pL,F,mC; diam 8'; about 65 stars mags 10... class G	07353s1156
2432	36[6]	⠩	pL,pC; diam 7'; about 20 faint stars	07387s1858
2439		⠩	B,pL,pRi; diam 9'; about 50 stars mags 9... class G; Includes star R Puppis, suspected variable, not confirmed.	07389s3132
2437	M46	⠩	! vB,vRi,vL; Mag 8; diam 25' 150 stars mags 10... class F; Planetary nebula NGC 2438 on north edge (*)	07396s1442
2438	39[4]	◎	pB,pS,vlE, Mag 11; diam 65"; 17^m central star; on north edge of cluster M46.	07396s1436
2440	64[4]	◎	S,B, Mag 11½; Diam 50" x 20" with 16^m central star	07399s1805
2447	M93	⠩	L,pRi,1C; Diam 18'; Mag 7; about 50 stars mags 8.....13 class G (*)	07424s2345
2451		⠩	vvL, scattered group of bright stars incl 4^m c Puppis	07436s3751

LIST OF STAR CLUSTERS, NEBULAE AND GALAXIES (Cont'd)

NGC	OTH	TYPE	SUMMARY DESCRIPTION	RA & DEC
2452		◎	F,S,1E; Mag 13; Diam 20" x 15" with 19^m central star	07456s2713
2455			cL,pRi; Diam 5'; about 20 stars mags 12..... class D	07468s2110
2477	△535		! B,vRi,L, Diam 25'; mag 7; about 300 stars mags 11..... class G　(*)	07505s3825
2467	22^4	□	pB,vL,R; Diam 4' with several stars 8½– 12^m in center;　F streamers to 15' radius	07513s2616
2482	10^7		L,cRi,vlC; Diam 18'; about 50 faint stars, class E; 1.5° ENE from double star Xi Puppis	07528s2410
----	H2		pL,F, Diam 9'; about 20 stars mags 11...... class D	07531s2547
2479	58^7		pL,Ri, Diam 8'; about 40 faint stars;　class F	07547s1735
2489	23^7		pL,cRi,pC; Diam 7'; about 35 stars mags 11... class G	07562s2956
2509	1^8		B,pRi,1C; Diam 4'; about 40 stars mags 10.... class G	07585s1856
2525	877^3	⊘	SBc; 12.3; 2.2' x 1.4' cF,pL,1E,vglbM	08033s1117
2527	30^8		vL,pRi; Diam 20'; about 40 stars mags 11....	08033s2801
2533			pL,Ri,C; Diam 8'; about 40 stars mags 12....	08050s2945
2539	11^7		vL,Ri,1C; Diam 20'; about 100 stars mags 11..... class F; Double star 19 Puppis in field	08084s1241
2546	△563		L,1C, Diam 25'; about 50 stars mags 9....	08101s3732
2568			S, mC; diam 2'; about 15 faint stars.	08164s3658
2567	64^7		pL,pRi,1C; Diam 10'; 50 stars mags 11..14;class F	08166s3029
2571	39^6		vL,1C; Diam 8'; about 25 stars mags 9..... class C	08169s2935

LIST OF STAR CLUSTERS, NEBULAE AND GALAXIES (Cont'd

NGC	OTH	TYPE	SUMMARY DESCRIPTION	RA & DEC
2580			cL,pRi,pC; Diam 9'; about 30 stars mags 12... class C	08194s3009
2587			pL, Irr; Diam 6'; about 30 stars mags 9.... class C	08213s2920

DESCRIPTIVE NOTES

ZETA Magnitude 2.25; Spectrum O5; position 08018s 3952. Zeta Puppis is a supergiant star, one of the most luminous stars known in our Galaxy; at a computed distance of about 2400 light years, the actual luminosity is close to 60,000 times that of the Sun, and the derived absolute magnitude is -7.1. The star is comparable to Rigel in energy output; if such a star were as near to us as Vega, it would appear to us about 12 times brighter than Venus at her best! Zeta Puppis shows an annual proper motion of about 0.03"; the radial velocity is 14½ miles per second in approach.

The star is located in a rich region of the winter Milky Way (to Northern Hemisphere observers) and the owner of a good wide-field telescope will find many stunning fields in the surrounding region. Just 2½° to the northwest is the splendid rich star cluster NGC 2477, while 5° NNE is the spot where the brilliant nova CP Puppis flared up in November 1942. The star AP Puppis, shown on the Skalnate Pleso Atlas, about 1° west, is a cepheid with a 5-day period of oscillation.

NU   Magnitude 3.18; Spectrum B7 or B8 III;
position 06362s4309, about 10° NNE from the
brilliant star Canopus. The computed distance of the star
is about 600 light years, giving the true luminosity as
1600 times that of the Sun (absolute magnitude -3.2). The
annual proper motion is 0.01"; the radial velocity is 17
miles per second in recession.

XI   Magnitude 3.34; Spectrum G3 Ib; position
07472s2444. Xi Puppis is a yellow supergiant
star with a computed actual luminosity of about 5800 suns
(absolute magnitude -4.6); the derived distance is about
1200 light years. The annual proper motion is less than
0.01"; the radial velocity is 1.6 miles per second in
recession.
    The 13th magnitude companion at 4.8" was first noted
by S.W.Burnham with the 36-inch refractor at Lick Observatory in 1889. The projected separation of the pair is 1770
AU, but it is not certain that the two stars are physically
related; there has been no definite relative change in the
pair since discovery. The faint star, if it is at the same
distance as the primary, has an actual luminosity about
equal to our Sun.
    Xi Puppis lies in a fine region of the Puppis Milky
Way, rich in myriads of faint stars. The difficult double
star B146 lies 0.5° to the NE, and the attractive cluster
M93 is just 1.6° to the NW. Note also the curious diffuse
nebula NGC 2467, about 1.7° to the SSE.

PI   Magnitude 2.81; Spectrum K4 or K5 III.
Position 07154s3700, about 8° below the stars
Eta and Epsilon in Canis Major, which mark the feet of
Orion's Great Dog. Pi Puppis is a moderate sized giant
star, probably somewhat over 100 times the solar luminosity
(computed absolute magnitude -0.3). It is some 140 light
years distant. Slight variability has been suspected; the
range of catalogue magnitudes is from 2.70 to about 2.85.
The annual proper motion is 0.01"; the radial velocity is
9½ miles per second in recession.
    Pi Puppis forms a charming color contrast group with
the two components of Upsilon Puppis, which lies 26' to
the NNE. The two stars are magnitudes 4.7 and 5.1, with a

DESCRIPTIVE NOTES   (Cont'd)

separation of about 4'; both spectra are B3 V. These two
bluish stars possibly form a true pair, as the measured
proper motions, although only 0.01", agree closely, as do
the radial velocities of 12 and 14 miles per second in
recession.   The western star has an emission line spectrum
while the eastern component shows a variable radial velocity and is probably a spectroscopic binary. The bluish
pair contrasts beautifully with the bright orange tint of
Pi itself.   From the lack of any measurable parallax, the
two B-stars appear to be at least three times more remote
than Pi itself; their distance is probably about 500 light
years, and each star has about 100 times the luminosity of
the Sun.

RHO　　Magnitude 2.80 (slightly variable). Spectrum
F6 IIp; Position 08054s2410. Rho Puppis is one
of the brightest and best known examples of a "Delta Scuti"
type of variable star, a small class of pulsating stars
which resemble the RR Lyrae (cluster variables) stars, but
have shorter periods and smaller amplitudes. The variability of Rho Puppis was discovered by O.J.Eggen at Mt.Stromlo
in Australia, and was independently detected at the Cape
Observatory in South Africa. Visually, the star is not a
very exciting variable, as the amplitude is only 0.15 magnitude; the range is about 2.72 to 2.87. This star, however,
has one of the shortest periods known for any pulsating
variable, only 3h 23m, or 0.14088143 day. The star Delta
Scuti, often regarded as the standard star of the type, has
a slightly longer period (4.65 hours) and a slightly
earlier spectral class (F3). The two stars appear to be
quite comparable in actual luminosity; the absolute magnitude in each case is about +0.3. In actual luminosity these
are among the brightest of the Delta Scuti stars; other
examples range from about +1.0 to +2.2.   Stars of the class
often seem to be somewhat underluminous for their spectral
types; a normal F6 II star, for example, should have an
absolute magnitude of about −2.0.

　　　Rho Puppis is approximately 100 light years distant,
and shows an annual proper motion of 0.10"; the radial
velocity is 28 miles per second in recession. (Refer also
to Delta Scuti, Epsilon Cephei, and RR Lyrae)

SIGMA   Magnitude 3.28; Spectrum K5 III; Position 07276s4312, about 7½° NW from Gamma Velorum. Sigma Puppis is a common proper motion double star with a fixed separation of 22.4" in PA 74°; magnitudes 3.3 and 8½. E.J.Hartung (1968) refers to it as "a brilliant orange star with white (by contrast) companion....a fine sight in the star sprinkled field....7.5 cm shows it well." At a computed distance of 180 light years the actual luminosity of the faint star is about equal to our own Sun; the primary is some 120 times more luminous. The two stars share the proper motion of 0.20" annually in PA 336°; the spectral types are K5 III and G5 V; and the projected separation of the two stars is about 1200 AU.

The third magnitude primary is itself a spectroscopic binary with a period of 257.8 days and an eccentricity of 0.17; the mean radial velocity is about 54 miles per second in recession.

TAU   Magnitude 2.95; Spectrum K0 III; Position 06487s5033, at the southern edge of the constellation, about 4.3° NE of Canopus. The computed distance of the star is about 125 light years; the resulting actual luminosity is 75 times that of the Sun (absolute magnitude +0.1). Tau Puppis shows an annual proper motion of 0.08"; the radial velocity is 22 miles per second in recession. The star is a spectroscopic binary with a period of 1066 days and an eccentricity of 0.088.

L₂   Magnitude 3.0 (variable); Spectrum M5e; Position 07120s4433, about 2.7° SW of Sigma Puppis. The star is also designated GC 9604 and h3943. One of the brightest of the red variable stars, usually classed as a semi-regular type, occasionally as a member of the Omicron Ceti or LPV type. It was discovered by B.A. Gould in 1872, and the average period of about 141 days was first determined by Isaac Roberts. The star is a naked-eye object throughout much of its cycle, falling to 6th magnitude at minimum, and rising sometimes to brighter than 3rd at maximum. A rather symmetrical light curve is one of the unusual features of the star; the time required to rise from minimum to maximum is almost exactly half of the total period. At maximum the spectral class is gM5e.

L₂ Puppis is also the visual double star h3943, the companion being a 9½ magnitude star about 1' distant in PA 214°; the last measurement reported in the Lick Observatory *Index Catalogue of Visual Double Stars* was made in 1913, when the separation was 62". The fairly large proper motion of the primary, however, is carrying it away from the faint star, and the separation is increasing steadily. Hence the star no longer appears in most modern lists of double stars.

From the spectral features, the assumed luminosity of the star is about -3.1, or over 1400 times the light of the Sun; this implies a distance in the range of 600 - 650 light years. Direct parallax measurements, however, lead to a much smaller distance of about 200 light years, and the moderately large proper motion supports the smaller distance estimate. The annual proper motion is 0.34" in PA 18°; at 650 light years this corresponds to a transverse velocity of about 320 miles per second, an improbably high value. Also, of course, a star at 650 light years would scarcely show a measurable parallax. Evidently this is one star in which the spectroscopic features are not a reliable indicator of luminosity. Tentatively accepting the smaller distance of about 200 light years, the actual luminosity at maximum appears to be about 200 times the Sun, and the absolute magnitude about -1.0. This suggests a luminosity class of II or III. The star shows a radial velocity of 32 miles per second in recession, which varies somewhat during the course of the star's pulsations. (Refer also to Omicron Ceti)

V   Position 07568s4907. A bright eclipsing variable star of the Beta Lyrae type, discovered by A.S.Williams in 1886. It is located in the extreme SE corner of the constellation, a little more than 8° southeast of L₂ Puppis, and 2½° southwest of Gamma Velorum. The star consists of two brilliant B-type giants revolving nearly in contact in a period of 1.4544877 day, or 1d 10h 54m 28s. As the orbit is oriented only 15° from the edge-on position, both components alternately eclipse each other in the course of each revolution. The star never drops below naked-eye visibility during its cycle; the photographic range, according to the Moscow *General Catalogue of Variable*

DESCRIPTIVE NOTES   (Cont'd)

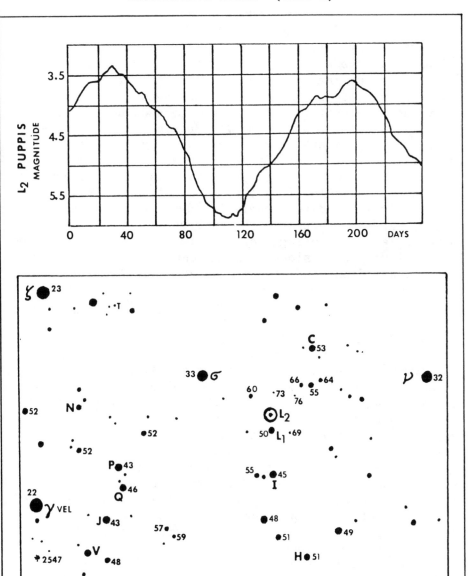

L$_2$ PUPPIS and V PUPPIS FIELD. The chart is about 12°
high. Numbers indicate comparison magnitudes according
to the AAVSO, with decimal points omitted.

Stars (1970) is 4.74 to 5.25. A binary of this type shows a continuously varying magnitude; the light curve is a sinusoidal wave with minima of two different depths midway between the maxima. In this case, the primary minimum is only a few hundredths of a magnitude deeper than the secondary one, indicating that the two stars do not differ very greatly in size or brightness. V Puppis is one of those stars which presents us with the "Beta Lyrae problem"; the distortion of the radial velocity measurements by moving gas streams between the stars, making the derived orbital elements unreliable. Thus the masses of about 19 suns each, derived from the direct analysis of the radial velocity curves, are now thought to be in error. In his comprehensive table of eclipsing binary stars, S.Gaposchkin adopted the following dimensions , masses and luminosities for the system, based on studies by H.van Gent and D.M.Popper:

	Spect.	Diam.	Mass	Abs.Mag.	Lum.
A	B1 V	6.1	16.5	-2.80	1100
B	B3 IV?	5.3	9.7	-2.15	580

The computed separation of the two stars, center to center, is about 5.3 million miles, which implies that their surfaces are nearly in contact. Both stars are ovoid in shape as a result of rapid rotation and tidal distortion; the amplitude of the radial velocity curve is close to 360 miles per second. No eccentricity can be detected in this system; the orbits seem perfectly circular. Spectroscopic studies show that the gas stream moves from the fainter star to the brighter one, gradually altering the masses of the components, and changing the orbital elements. Like Beta Lyrae, this system is being observed at a critical point in its evolution.

From the derived luminosities, the distance appears to be about 1300 light years. The annual proper motion is 0.02"; the mean radial velocity is about 12 miles per second in recession.

V Puppis is also the visual double star h4025, but it is not certain that the companions are physically associated with the bright star. The closer star is a 10th magnitude object at 7.0"; there are two more distant stars at 19.0" and 39.2". (Refer also to Beta Lyrae)

CP   Nova Puppis 1942. Position 08009s3512, about 5° NNE from Zeta Puppis. One of the brightest novae of modern times, probably first noticed by Bernhard Dawson at the University Observatory, La Plata, Argentina, on the night of November 8, and soon found independently by other observers. The first report from the United States was made by the amateur observer and AAVSO member F.Hartmann of St.Albans, Long Island, on the morning of November 9. The following morning the star was independently discovered by Dr.Edison Pettit at Mt.Wilson; the brightness was then about 1st magnitude, and the maximum of about 0.3 was reached the following day.

The nova faded rapidly at a rate of about 0.5 magnitude per day, and became invisible to the naked eye by the end of November. The star was thus one of the "fast novae" characterized by a single sharp maximum and a rapid decline. A "slow nova" such as DQ Herculis or RR Pictoris may remain near peak brightness for several weeks.

The pre-nova brightness of CP Puppis is not definitely known, as it does not appear on previous photographs of

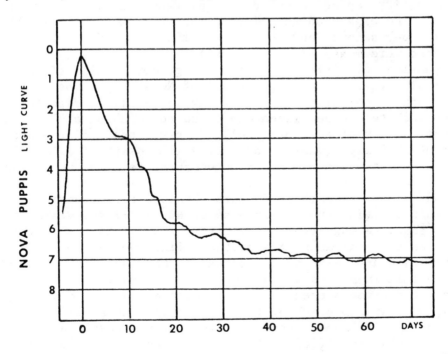

the region which show stars to nearly 18th magnitude. On
Lowell Observatory 13-inch camera plates, no image at all
appears at the position; these plates reach to about magnitude 17.3. The total light range was thus at least 17½ or
18 magnitudes; at the time this was the greatest range observed for any of the normal or "classical" novae. Nova
Cygni 1975, however, surpassed this record when it flared
up with a 19-magnitude increase in the summer of that year.
The abnormally large ranges have suggested to some astronomers that these two stars were actually supernovae or
possibly intermediate types. The light curves, spectra, and
expansion velocities were all reasonably normal, however,
and it is now thought that CP Puppis and Nova Cygni were
possibly "virgin novae", stars which have not previously
exploded. On this theory it is assumed that probably all
novae are recurrent, though the intervals between outbursts
may be many centuries or millenia.
      There is some reason to believe, however, that CP
Puppis did have an unusually high absolute magnitude. Since
the distance of the star is not accurately known, the true
luminosity can be computed only by rather indirect methods.
      The expansion velocity at the maximum was about 600
miles per second, and velocities of up to 1200 miles per
second were measured a few days later. These are typical
expansion velocities for the outer gaseous shells of the
normal novae. The nebulous shell of Nova Puppis was detected photographically by Dr.F.Zwicky in 1956; its diameter on
200-inch telescope plates was found to be about 5.5". From
observations of the nova-shell, Zwicky finds a distance of
about 5200 light years, and a true luminosity of 1.3 million suns at maximum, the computed absolute magnitude being
-10.5 without correction for absorption. In a spectroscopic
study at Mt.Wilson, R.F.Sanford obtained rather similar
results, giving the distance as about 4800 light years and
the absolute magnitude as slightly brighter than -10. This
places the star among the most brilliant normal novae, and
there appears to be no doubt that it had one of the highest
actual luminosities on record. Dr.Pettit estimated that it
radiated as much energy in two months as the Sun does in
20,000 years. D.B.McLaughlin calculated that should our Sun
favor us with such a dazzling display, the resultant radiation would be sufficient to melt the crust of the Earth to

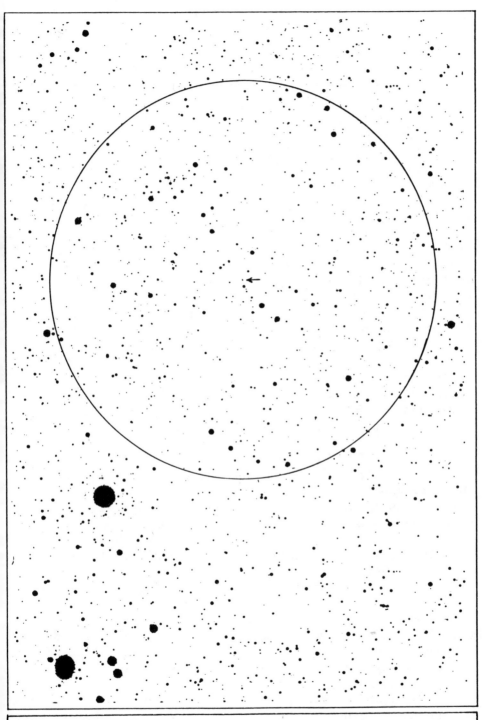

FIELD OF NOVA PUPPIS 1942. The circle is 1° in diameter
with north at the top; stars to about 16th mag are shown.
From a Lowell Observatory 13-inch camera plate.

a depth of 12 miles. Some hint of the violence of such an
outburst may be realized from the fact that an 18-magnitude
rise corresponds to an increase of about 15 million times
in luminosity. C.P.Gaposchkin in her book *The Galactic
Novae* (1957) describes CP Puppis as "a noteworthy star on
account of its large range, rapid development, high termin-
al excitation, and the presence of [Fe II]". Whether or
not the star is a close binary, as are so many other novae,
is not definitely known. R.P.Kraft, however, in 1964, found
that the star's spectrum shows "enormous radial velocity
variations in the course of a few hours" which suggests
that it is probably a very close and rapidly rotating pair
of stars.

The identification chart on page 1507 was made from
a Lowell Observatory 13-inch camera plate, obtained by the
author in January 1965. The star at that time was close
to magnitude 15.0, of a strong bluish color, and still
fading slowly. (Refer also to Nova Aquilae 1918, Nova DQ
Herculis 1934, Nova Persei 1901, Nova Pictoris 1925, and
Nova Cygni 1975. Recurrent novae are described chiefly
under T Corona Borealis and RS Ophiuchi)

M46   (NGC 2437) Position 07396s1442. A rich open
             star cluster, beautifully situated in the glow-
ing stream of the Puppis Milky Way, some 14½° east of the
brilliant star Sirius in Canis Major. M46 appears in small
telescopes as a fine circular cloud of small stars, nearly
half a degree in diameter. Charles Messier, who discovered
the cluster in 1771, achieved only partial resolution and
thought the group to contain a little nebulosity. "The
stars cannot be distinguished except in a good telescope"
hr wrote. John Herschel called it "superb", while Admiral
Smyth recorded it as "a noble though loose assemblage of
stars 8 to 13 mag. Most compressed trending S.f. & N.p."
T.W.Webb describes it as "a beautiful circular cloud of
small stars"; D'Arrest found it well sprinkled with many
multiple stars in groups.

At least 150 stars between magnitudes 10 and 13 are
recognized as cluster members; the total population is very
probably in excess of 500 stars. The brightest members are
blue giants near spectral class A0, each having about 100
times the solar luminosity. According to a study by J.Cuffey

NOVA PUPPIS 1942. The region is shown five years before
the outburst (top) and about a month after maximum (below).
Lowell Observatory 13-inch telescope photographs.

(1941) the distance of the group is about 4700 light years; a more recent figure, based on a study of the cluster H-R diagram, is about 5400 light years. The true diameter of the group is then about 30 light years. A.Wallenquist, however, adopts a considerably smaller distance of about 3200 light years; he identifies 197 probable members, and finds a star density of about 9 stars per cubic parsec in the cluster center. The total integrated photographic magnitude of M46 is about 6.6; the apparent size is given by various observers as 25', 27' or "about 30' on long exposure photographs". For the visual observer it is at its best in a good 10-inch or 12-inch reflector with a 1° ocular, and evokes the impression of a celestial meadow strewn with fireflies.

A curious feature of M46 is the presence of the small planetary nebula NGC 2438, located well within the apparent borders of the cluster, some 7' north of the center. This was first noticed by Sir William Herschel, and was described by John Herschel in 1827 as "exactly round, of a fairly equable light....has a very minute star a little north of center....it is not brighter in the middle or fading away, but a little velvety at the edges..." Lalande and Lord Rosse found it annular; the ring is about 65" in apparent size, with a faint central star.

Controversy over the possibility of true membership in the cluster is now ended by radial velocity measurements which show that the two objects have different motions in space. The cluster is receding at about 25 miles per second but the nebula is moving at about 47 miles per second, also in recession. C.R.O'Dell (1963) derives a distance of some 3300 light years for the nebula, and an actual diameter of 70,000 AU, or just a little over one light year. It appears that the nebula is probably closer to us than the cluster, but the distance of neither object is known with sufficient precision to make this definite. The central star of the nebula appears about 16th magnitude visually, but is a relatively easy object to photograph, owing to its strong radiation in the blue and ultraviolet; the computed surface temperature is about 75,000°K, one of the hottest stars known.

The bright cluster M47 lies about 1½° distant, toward the west. (See page 1512)

STAR CLUSTER M46 in PUPPIS. Note the small planetary
nebula on the north edge of the group. Lowell Observatory
photograph with the 13-inch telescope.

M47   (NGC 2422) Position 07343s1422. A bright open
star cluster located a little more than 1½°
to the west of M46, and discovered by Messier at the same
time, in February 1771. Owing to an erroneous position, M47
has long been regarded as one of the "missing" Messier objects, but the identification with NGC 2422 now seems
definite. M47 is a brighter and sparser group than M46, and
is described by Admiral Smyth as "a very splendid field of
large and small stars dispersed somewhat in a lozenge shape
and preceded by a 7 mag. with a companion about 20" N.f.
of it..." T.W.Webb found it a "grand broad group visible
to the naked eye: too large even for 64X. Some brilliant
5 or 6 mag. stars including Σ1121". Walter Scott Houston
(1973) gave a brief description in his *Deep-Sky Wonders*
column in *Sky & Telescope* magazine: "a coarse scattering
of bright stars lies on a dim sheen of fainter ones, so a
10-inch is really needed to enjoy its splendor fully".

M47 appears to be from one half to one third the
distance of M46, and must be in reality a somewhat smaller
group than its more distant neighbor. T.Schmidt-Kaler in
his table of star clusters (1976) gives the probable distance as 0.48 kiloparsec, or about 1540 light years;

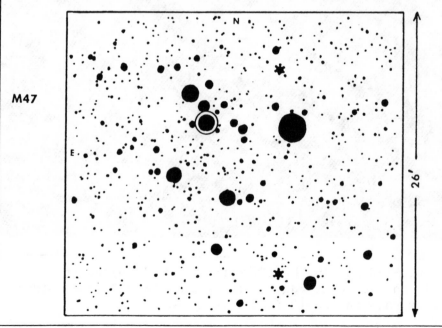

M47

26'

STAR CLUSTERS IN PUPPIS. The rich group M46 is at the top;
the bright M47 is near the lower edge of the print. Lowell
Observatory photograph with the 13-inch telescope.

A.Wallenquist (1959) had reported a probable distance of
about 1780 light years, and a diameter of 17 light years.
Some 45 stars are identified as true cluster members; the
brightest of these is a B2 giant of magnitude 5.7. All the
other bright stars of the cluster are B and A-type stars.
Wallenquist derives a central density of 16 stars per cubic
parsec. The stellar population is fairly comparable to the
Pleiades, suggesting that this is a rather young star
cluster. Two orange stars are known in the group, marked
by the asterisk symbol on the chart on page 1512; these are
magnitudes 7.83 and 7.93, and have the color indices of
early K-type giants. If true members of the cluster, the
actual luminosity of each star is about 200 suns.

Near the center of the cluster is the easy double
star Σ1121, a neat 7½" pair of fixed separation and PA;
both spectra are B9, and the apparent magnitudes about 7
and 7½. The projected separation at the accepted distance
is about 4100 AU, a gap which would nicely hold 50 orbits
of Pluto, lined up edge-to-edge. From such considerations
we can comprehend something of the scale involved in a
typical galactic star cluster. (Σ1121 is indicated by the
encircled image on the chart, page 1512). Another pair, on
the west edge of the cluster, and the brightest object on
the chart, is Σ1120; magnitudes 6 and 9½, separation about
20", spectrum B5.

M93   (NGC 2447) Position 07424s2345. A bright open
star cluster in the Puppis Milky Way, just 9°
south from M46. This is another of Messier's discoveries,
found in March 1781, and described as "a cluster of small
stars without nebulosity...between Canis Major and the prow
of [Argo] Navis". Messier gave the apparent size as 8'.
T.W.Webb refers to it as "a bright cluster in a rich neighborhood" and Smyth calls it "a neat group of a star fish
shape...S.p. portion being brightest, with individuals of
7 - 12 mag... Mistaken for a comet by Chevalier d'Angos of
the Grand Master's Observatory in Malta.." Walter Scott
Houston finds M93 visible to the naked eye under excellent
sky conditions.

M93 is a smaller but brighter group than M46; the
central mass being distinctly triangular or wedge-shaped
with outer branches and scattered sprays of stars to a

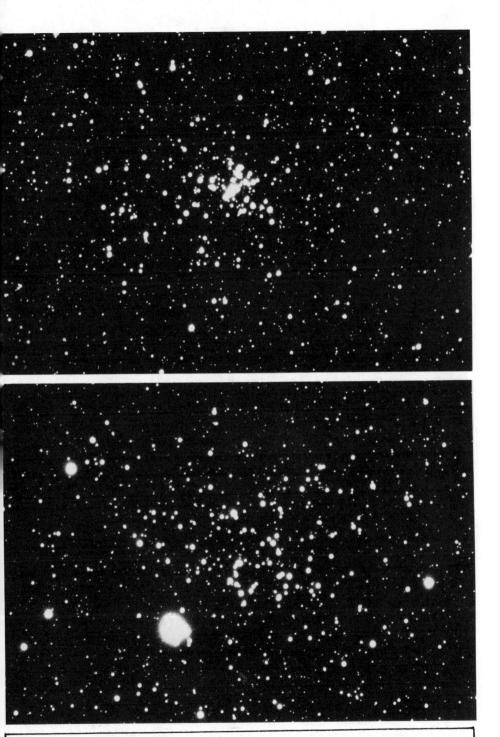

STAR CLUSTERS in PUPPIS. Top: The compact group M93.
Below: NGC 2539 and the bright star 19 Puppis. Lowell
Observatory photographs made with the 13-inch telescope.

diameter of about 24'; to K.G.Jones (1968) the over-all
pattern suggests a butterfly with open wings. A.Wallenquist
(1959) adopts a total diameter of about 18' and finds 63
probable members within this area; his value for the distance is about 3400 light years. At this distance, each
minute of arc corresponds to 1 light year; the true diameter of the group, then, is about 18 light years. Several
blue giants of type B9 are the brightest members of M93.

While sweeping the starry fields of the Milky Way
near M93, the observer may find, about 3° to the SE, the
curious nebulosity NGC 2467, appearing in small telescopes
as a hazy glow some 2' in radius, surrounding an 8th magnitude B-type star. Palomar Observatory 48-inch Schmidt
camera plates show a complex field of irregular streamers
extending to the NE, filling a 30' field. This little-known
nebula lies at a computed distance of about 3000 light
years, comparable to the distance of M93.

NGC 2477   (Dunlop 535) Position 07505s3825, about 2½°
NW from Zeta Puppis. Probably the finest of
the galactic clusters in Puppis, but not noted by Messier
and omitted also from many observing guides because of its
position low in the southern winter sky. It is a striking
group, somewhat smaller than M46, but richer and more compact, containing about 300 stars crowded into a 20' field.
All the observers of NGC 2477 have commented on the unusual
richness of this cluster; P.Doig in 1925 called it "almost
globular", while Harlow Shapley in his book *Star Clusters*
(1930) found it "in superficial appearance...the richest of
galactic clusters; or perhaps it is the loosest of globular
clusters" According to A.Becvar's *Atlas Coeli* catalogue
(1951) the over-all diameter is about 25'; H.B.S.Hogg in
her list in the *Handbuch der Physik* gives the integrated
magnitude as 5.7, suggesting naked-eye visibility under
good conditions.

Published values for the distance of NGC 2477 are
very discrepant; R.J.Trumpler, from a study of the spectral
types, found about 860 parsecs; Shapley has 1200, Hogg has
700, D.H.Menzel gives 1000, and Becvar about 1900. From
Shapley's comparative studies, it seems that NGC 2477 is
probably about 1.7 times more remote than M46. (Refer also
to M46, page 1508)

STAR CLUSTER NGC 2477 in PUPPIS. An unusually rich galactic
star cluster in the Puppis Milky Way. Lowell Observatory
photograph made with the 13-inch telescope.

PYXIS

LIST OF DOUBLE AND MULTIPLE STARS

NAME	DIST	PA	YR	MAGS	NOTES	RA & DEC
Gls 96	0.1	96	59	6½- 6½	(φ314) PA inc,	08261s3457
	25.5	143	19	- 9½	spect B3	
9	0.3	66	60	6 - 6½	(I 489) PA dec,	08293s1924
					spect A0	
I 805	1.9	45	56	7½- 9½	spect A0	08293s2531
h4106	1.0	208	55	7½- 9½	PA inc, dist dec,	08295s3630
	47.9	218	43	- 10½	spect K0	
β205	0.5	187	26	7 - 7	Binary, about 128	08309s2426
	26.4	354	00	- 14	yrs; PA dec, spect A7	
β206	1.8	279	45	8 - 9	relfix, spect G0	08333s2456
I 355	0.7	46	42	8 - 8½	PA inc, spect K2	08340s2053
h4115	22.4	158	19	6½- 11	spect A5	08355s3334
	45.0	20	19	-12½		
	29.9	197	19	-13		
β207	4.3	103	38	6½- 10	relfix, spect K5	08364s1934
β208	0.6	245	38	6 - 9	Binary, about 215	08370s2230
	85.0	202	18	- 10½	yrs; PA inc, spect dG6	
I 314	0.8	233	13	6½- 8	binary, 54 yrs; PA dec, spect F3	08375s3626
ζ	52.3	61	05	5½- 10	(h4120) spect G4, G0	08376s2923
I 813	3.3	300	60	7½- 13	relfix, spect K0	08421s2005
λ106	17.9	238	55	7 - 12	PA inc, spect K0	08425s2336
λ106b	3.5	333	55	12- 13		
Hd 204	10.0	310	00	7 - 12	spect F2	08431s3204
	13.0	275	00	- 12		
Ho 356	1.6	276	54	8 - 8½	PA inc, spect A0	08471s2614
h4144	2.4	315	35	7 - 9	spect B9	08485s3545
φ296	0.1	74	59	7 - 7	PA dec, spect F2	08507s3622
Cor 77	10.2	200	30	8 - 9	spect F5	08550s2939
h4166	13.7	153	19	6½- 7½	(Rst 3619) all cpm	09012s3324
h4166b	0.8	77	47	7½-11½	Spect A0	
Hwe 23	3.1	306	40	8½- 9	relfix, spect A0	09042s3124
𝒦	2.1	263	11	5 - 10	(I 491) spect gM0	09059s2539
β410	1.8	158	25	7 - 9	slight PA dec, spect A0	09076s2536
ε	17.8	147	20	5½- 9½	cpm, spect A3.	09078s3010
ε b	0.3	88	51	10- 10	BC PA dec	

| PYXIS |

LIST OF DOUBLE AND MULTIPLE STARS   (Cont'd)

NAME	DIST	PA	YR	MAGS	NOTES	RA & DEC
Daw 131	0.9	124	39	8- 9½	spect A0	09130s3010
h4200	3.0	73	54	7 - 8	relfix, spect A0	09186s3133
h4199	11.7	111	20	8 - 9½	spect A0	09186s2734
	8.4	271	00			
I 170	1.4	35	43	7½- 10	spect F2	09192s3508
Jac 5	0.6	264	47	8 - 8½	PA inc, spect B8	09246s2834

LIST OF VARIABLE STARS

NAME	MagVar	PER	NOTES	RA & DEC
R	9.0--13..	364	LPV.	08434s2801
S	8.1--13..	207	LPV. Spect M3e	09029s2453
T	7.0--14.5	---	Recurrent nova; 1890, 1902	09026s3211
			1920, 1944, 1966 (*)	
U	8.0--8.6	345:	Semi-reg; spect K5	08278s3009
V	8.5--11..	70:	Semi-reg; spect K2	08514s3437
RZ	8.6--9.3	.6563	Ecl.bin; spect B7	08499s2718
TT	8.8--9.4	1.516	Ecl.bin; spect B9	08464s2558
TU	9.0--10..	88	Semi-reg; spect M5	09080s1943
TX	9.4--9.8	1.124	Ecl.bin; spect A3	08405s3210
TY	6.4--7.0	1.599	Ecl.bin; spect G0	08576s2737
TZ	9.0--9.4	.6973	Ecl.bin; W U.Maj type	08391s3201

LIST OF STAR CLUSTERS, NEBLULAE AND GALAXIES

NGC	OTH	TYPE	SUMMARY DESCRIPTION	RA & DEC
2613	266[2]	⊖	Sb; 10.9; 6.4' x 1.5' cB,L,vmE, nearly edge-on spiral; 1.2° W from double star β208.	08312s2248
2627	63[7]	⟨cluster⟩	Diam 8' with 70 stars mags 11....13; class F	08352s2946
2635		⟨cluster⟩	pL,pRi,C,1E vS,pmC, Irr △ ; Diam 3'; about 20 stars mags 12.... class D; 0.7° NNW from Beta Pyxis.	08365s3435
2658	△609	⟨cluster⟩	pS,Ri,C, Irr R; diam 9'; about 60 stars mags 12.... 0.7° N from Alpha Pyxis.	08414s3229
2818	△564	⟨cluster⟩	Diam 9' with 30 faint stars mags 12.... class E; Contains planetary nebula 13^m, diam 40", "barbell" shape, on W edge of cluster	09140s3624
2888		⊖	E1/E2; 13.1; 0.5' x 0.4' vF,S,R, gmbM. 13^m star on south edge.	09242s2748

T
      Recurrent nova. Position 09026s3211, about 4°
ENE from Alpha Pyxidis. One of the few known
repeating novae, and the present record-holder for the
greatest number of observed outbursts. The star is normally
an object of magnitude 14, but during an outburst it may
rise to 6½ or 7. Five maxima have been recorded for the
star; in 1890, 1902, 1920, 1944, and 1966.

      The first outburst to be detected was that of May
1902, "discovered" some time later on Harvard photographic
plates. The star attained a peak brightness of magnitude
7.3 on May 2, and had faded only to magnitude 8.7 by July
3. Plates made in January 1903 showed it once again in its
normal 14th magnitude state. The discovery of the nova
stimulated a search of the Harvard plate collection, and a
previous maximum in the spring of 1890 was discovered. The
interval of about 12 years suggested a possible reappearance of the nova about 1914, but no outburst was detected
although a close watch was kept. In March 1920, however,
the star was found to be magnitude 7.7 and still growing
in brightness. On April 6th the maximum was reached at magnitude 6.4; the star faded slowly and was still magnitude
8.5 in early June.

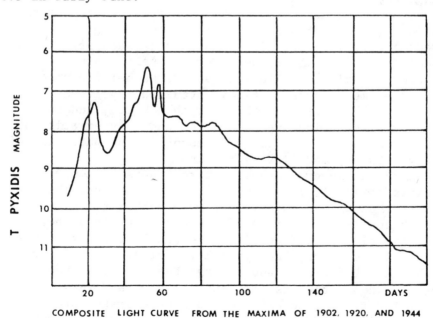

COMPOSITE LIGHT CURVE FROM THE MAXIMA OF 1902, 1920, AND 1944

Spectra obtained at the 1920 outburst left no doubt that T Pyxidis should be classed with the novae, although the remarkable circumstance of three separate outbursts appeared to place it in a class by itself. In 1938 a spectrum of the star at minimum was obtained by M.L.Humason at Mt.Wilson; the spectrum appeared to be a virtually continuous one, tentatively classed as type O, but with superimposed bright lines. The range of about 7 magnitudes corresponds to a light increase of about 600 times; much less than a typical "classical" nova.

The fourth explosion of T Pyxidis, in 1944, was not actually detected until April 1945, more than four months after peak brightness. A check of earlier plates, made at Bloemfontein, South Africa, showed that the star had reached a maximum magnitude of about 7.1 about November 21, 1944. Spectra obtained by A.H.Joy at Mt.Wilson revealed expansion velocities of up to 1250 miles per second, fully comparable to those observed in typical classical novae.

The date of the fifth outburst is given by different authors as either 1966 or 1967; the star began to brighten in December 1966, but did not reach maximum until the following January. The variable star observer Albert Jones of Nelson, New Zealand, was probably the first to report the eruption of the star when he found it to be magnitude 12.9 on December 7, 1966. Within two nights it had increased by an additional four magnitudes. For the next month the nova brightened slowly, oscillations sometimes being noticed in a period of mere hours, until it attained magnitude 6.3 on January 11. The fading of the star then began, and the irregular fluctuations became larger and more erratic. As the star began to dim, the color changed from white to definite yellowish, and finally reddish.

Although several other recurrent novae are now known, T Pyxidis still displays several unique features. The light curve is that of a slow nova; at each outburst the star faded at the leisurely rate of about one magnitude per month, and also showed considerable oscillations when near maximum. In contrast, the typical recurrent novae T Coronae, RS Ophiuchi, U Scorpii, and WZ Sagittae have all shown extremely sudden outbursts with sharp maxima and very rapid fading. T Pyxidis also has the shortest average interval between maxima, and furthermore appears to show a gradual

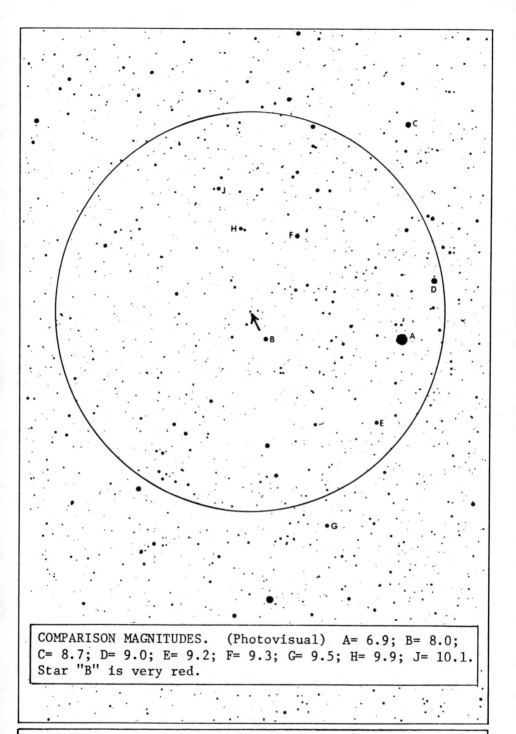

COMPARISON MAGNITUDES. (Photovisual) A= 6.9; B= 8.0;
C= 8.7; D= 9.0; E= 9.2; F= 9.3; G= 9.5; H= 9.9; J= 10.1.
Star "B" is very red.

NOVA T PYXIDIS. Identification field, from a plate made
with the 13-inch telescope at Lowell Observatory. North is
at the top; the circle is 1° in diameter.

progressive increase in the interval between explosions;
from 12 years to 18 to 24; the fourth interval (22 years)
was nearly as long as the third, and much longer than the
first two. The explanation for this feature, if real, is
unknown.

The distance and true luminosity of this star is not
accurately determined, and any estimate depending on an
assumed absolute magnitude is made doubtful by the fact
that the observed recurrent novae appear to be classifiable
into two very different types. T Coronae and RS Ophiuchi
reach very high luminosities at maximum (about -8.5 absolute) but WZ Sagittae is a dwarf star which rises to only
about 30 times the luminosity of the Sun. The status of U
Scorpii is uncertain, but is suspected of being a "dwarf
nova" like WZ Sagittae. To which class should T Pyxidis be
assigned? If a member of the high-luminosity type, the
distance must be well over 10,000 light years; the absolute
magnitude at maximum possibly about -6.5. This is the value
deduced from a study of the light curve, and appears to be
supported by W.J.Luyten's findings that the proper motion
of the star is not greater than about 0.005" per year. The
figure is in fact close to the probable error of the measurement, so it seems safe to say that the star effectively
has no certain proper motion; this implies that the true
distance is unlikely to be less than several thousand light
years. From the strength of the interstellar calcium lines
in its spectrum, a distance of at least 3400 light years
has been derived; the peak absolute magnitude (uncorrected
for absorption) is then about -3.5. Despite the expected
discrepancies in these results, it seems clear that the
star is not a dwarf nova of the WZ Sagittae type, but has
a luminosity of at least -4 or -5 at maximum.

At the 1966 outburst, T Pyxidis was intensively studied by O.J.Eggen in Australia, A.U.Landolt at Cerro Tololo
in Chile, R.M.Catchpole at Radcliffe Observatory, as well
as at more northern stations. Continuing observations, up
to the unusually faint minimum of 15.1 in early 1969, have
shown that the color of the star is bluest near maximum;
rapid "flickerings" of nearly 0.1 magnitude were also detected, possibly indicating that this star, like many other
novae, is a close and rapid binary. (Refer also to T Corona
Borealis, RS Ophiuchi, WZ Sagittae, and U Scorpii)

LIST OF DOUBLE AND MULTIPLE STARS

NAME	DIST	PA	YR	MAGS	NOTES	RA & DEC
Δ12	19.1	103	46	7 - 9	cpm; spect F5	03144s6438
Δ12b	0.4	38	45	9½- 9½	(Rst 67) BC PA dec	
ζ	130	222	60	5 - 5½	wide cpm pair; spect G1, G2	03171s6242
Δ14	57.5	271	16	7 - 8	cpm, relfix, spect F2, F5	03371s5956
h3592	5.2	15	51	6 - 9	cpm; spect K0	03433s5425
Δ17	62.9	142	16	7½- 8	(I 269) spect A2	03596s5431
Δ17b	3.4	79	33	8 - 11		
	27.7	196	00	- 13		
h3641	10.2	230	55	5½- 11	optical, dist inc, PA dec, spect gK1	04142s6219
L 1430	0.4	42	59	6½- 7½	(Gale 1) PA inc, spect A0	04155s6104
θ	4.1	4	43	6 - 8	(Rmk 3) cpm; dist dec, spect B9	04171s6323
h3670	31.9	99	17	6 - 8½	relfix, cpm, spect K0, G5	04330s6256

LIST OF VARIABLE STARS

NAME	MagVar	PER	NOTES	RA & DEC
R	7.0--13..	278	LPV. Spect M4e	04330s6308
RX	8.4--10.0	---	Irr or Semi-reg; spect K0	03472s6651
RY	9.5--11.0	237	LPV. Spect M3e	03490s5730

LIST OF STAR CLUSTERS, NEBULAE AND GALAXIES

NGC	OTH	TYPE	SUMMARY DESCRIPTION	RA & DEC
1313	Δ206	⊖	SB; 10.8; 5.0' x 3.2' pB,L,E,vgbM, nuclear bar	03176s6640
1536		⊖	SBc; 13.2; 1.5' x 0.9' vF,R,pL,vlbM	04100s5636
1543		⊖	E0/S0; 12.0; 3.0' x 1.5' B,pL,E, smbMN	04117s5752
----	I.2056	⊖	E0p; 12.3; 0.6' x 0.6' F,pL,R,bM	04156s6020
1559		⊖	SB; 11.1; 3.0' x 1.5' vB,vL,vmE, mbM	04170s6255

SAGITTA

LIST OF DOUBLE AND MULTIPLE STARS

NAME	DIST	PA	YR	MAGS	NOTES	RA & DEC
Σ2437	0.6	36	59	8 - 8	PA & dist dec; spect G5	18597n1906
h2851	15.3	131	26	7½- 12	PA slow inc, spect	18598n1903
	44.2	298	26	- 10	G5	
β139	0.6	138	66	6½- 8	relfix, spect B8	19103n1646
	113	285	19	- 7½		
	28.6	103	58	- 9½		
	27.7	103	11	- 13		
Σ2484	2.6	234	59	7½- 9	PA inc, spect F8	19121n1859
Σ2504	9.0	285	59	6½- 8	relfix, spect F5	19188n1903
0 Σ375	0.7	171	69	7½- 8½	PA inc, spect G5	19324n1801
ε	89.2	81	49	5½- 7½	(HVI 26) optical,	19350n1621
	99.4	280	12	- 12½	spect gG8, B8	
J 138	21.2	138	10	7- 13½	Spect A0	19359n1708
	38.2	129	10	- 11½		
J 139	12.5	333	14	8 - 11½	Spect B3	19362n1708
HN 84	28.4	301	31	6½- 8½	(Hh 630) cpm; good colors, spect gM0	19372n1627
ζ	0.2	180	62	5½- 6½	(Σ2585) (AGC 11)	19468n1901
	8.4	311	62	- 9	AB binary, 22.8 yr;	
	75.7	247	62	- 11	PA dec, spect A3; ABC all cpm.	
β149	128	279	00	6½- 10	Spect B8	19560n1621
β149b	8.4	200	01	10-12½		
13341	6.3	190	16	8 - 9	(ADS 13341) Spect A0	20029n1632
Σ2631	4.0	344	54	8 - 9½	relfix, spect G0	20050n2057
Σ2634	5.1	14	45	8 - 9½	cpm, slight dist dec, spect K0, K5	20073n1639
θ	11.9	325	52	6½- 8½	(Σ2637) AB cpm,	20077n2046
	83.9	223	49	- 7	Spect dF1, dG5, gK2 AC optical, dist inc.	

SAGITTA

LIST OF VARIABLE STARS

NAME	MagVar	PER	NOTES	RA & DEC
R	8.5--10.4	71	RV Tauri type; spect G0--G8	20118n1634
S	5.5--6.2	8.382	Cepheid; spect F6--G5	19537n1630
T	8.5--10.0	165	Semi-reg; spect M6	19195n1734
U	6.4--9.0	3.381	Ecl.bin; spect B9+G2 (*)	19166n1931
V	9.5--13.8	---	Erratic (*)	20180n2057
W	8.9--13.2	279	LPV. Spect M4e	19173n1707
X	8.7--9.7	196	Semi-reg; spect N3	20029n2030
VZ	6.0--6.2	Irr?	Spect M4	19578n1723
WW	9.8--11.7	221	Semi-reg; spect M6	20183n1848
WY	5.5--19..	---	Nova 1783	19305n1738
WZ	6.3--16..	---	Recurrent nova; 1913, 1946, 1978 (*)	20053n1733
BF	8.5--10..	Irr	Spect N3	20002n2057
CV	9.8--11.2	Irr	Spect gM4	19407n1807
FG	10...13	---	Spect B9; in nebulous disc diam 30" (*)	20097n2011
	7.2--19..	---	Nova 1977	19371n1801

LIST OF NEBULAE, STAR CLUSTERS, AND GALAXIES

NGC	OTH	TYPE	SUMMARY DESCRIPTION	RA & DEC
----	H20	⊙	pL,1C; Diam 10'; about 20 stars mags 11... class D	19509n1813
6838	M71	⊕	Mag 9; Diam 6'; vL,vRi,vC; stars mags 12.... class not certain (*)	19519n1839
6879		◎	Mag 11; Diam 5" with 15^m central star. Nearly stellar	20081n1646
6886		◎	Mag 11; Diam 9" x 6" with $16\frac{1}{2}^m$ central star.	20105n1950
----	I.4997	◎	Mag 11; Diam 2" with 14^m central star; appearance stellar.	20179n1635

DESCRIPTIVE NOTES

U   Position 19166n1931. One of the finest of the Algol-type eclipsing variable stars, discovered by F.Schwab in 1901, and announced in the *Astronomische Nachrichten*, Vol.157, in November 1901. U Sagittae is a system in which the bright primary star is totally eclipsed by a larger but fainter companion revolving about it; the system closely resembles the eclipsing binary U Cephei. The star is easily located about 1.7° west of the coarse star group called "Collinder 399" in Vulpecula, the snail-shaped asterism which includes the stars 4, 5, and 7 Vulpeculae, and which appears on the left edge of the identification chart on page 1531.

The variability of U Sagittae is easily observed in a small telescope or even in binoculars, and is best followed by comparing the star with the nearby Σ2504. At normal light U Sagittae is somewhat brighter than this star, while at minimum it is nearly 10 times fainter. The exact period between eclipses is 3d 9h 08m 05s. Primary eclipse is total and the star at this time remains at a constant minimum of magnitude 9.2 for approximately 1h 40m.

The brighter component is a bluish main sequence star of type B8 or B9; the faint companion, whose spectrum can be studied only when the primary is in eclipse, has a subgiant spectrum of about type G2; its luminosity class may

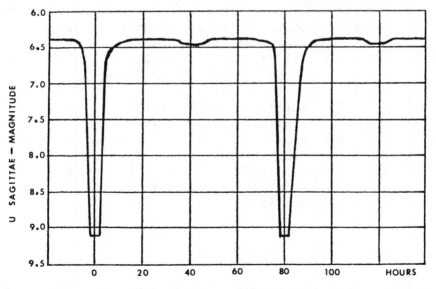

ECLIPSE OF U SAGITTAE. The star is shown in its normal state (top) on Sept.3, 1964, and during a primary minimum (below) two nights later. Lowell Observatory photographs.

be either III or IV. N.G.Roman in 1956, however, assigned types of B7 III and K1 III on the basis of spectra obtained with the Yerkes 40-inch refractor. The Moscow *General Catalogue of Variable Stars* (1970) gives the photographic range of the system as 6.31 to 9.92; the visual range is about 6.4 to 9.2. The fading of the star at total eclipse is much more striking than that of Algol, whose eclipses, of course are only partial.

From the derived orbital elements, the two stars appear to be about 8½ million miles apart, center to center and the orbit has an eccentricity of 0.035 and an inclination of close to 90°. As with many close binaries, there is some doubt concerning the precise orbital elements, owing to the distortion of the radial velocity measurements by moving streams of gas between the stars. Studies in 1951 revealed that the hydrogen emission lines give systematically different velocities than other lines; in addition to a gas stream between the components, there also seems to be a flattened cloud or disc of gas surrounding the brighter star.

In one of the first comprehensive studies of this system, made at Mt.Wilson by A.H.Joy in 1930, the following results were obtained:

	Spect:	Diam.	Mass	Density	Lum.
A	B9	3.4	6.8	0.17	120
B	G2	5.7	2.0	0.01	10

These figues may be compared with more recent results obtained by D.H.McNamara and K.A.Feltz, Jr., at Brigham Young University in 1976:

	Spect:	Diam.	Mass	Density	Abs.Mg.
A	B8 V	2.52	3.5	0.22	-0.4
B	G2 III	3.32	1.4	0.04	+1.8

Slight changes in the period probably result from the exchange of material between the stars. The Moscow catalogue gives the precise period as 3.3806184 days; but in 1975 the best determination was 3.3806205 days. The very similar star U Cephei has shown the same type of period changes.

DESCRIPTIVE NOTES (Cont'd)

From the total expected luminosity of about 130 suns, the distance of U Sagittae would appear to be in the range of 750 - 800 light years, making no correction for loss of light through absorption in space. Obscuration of half a magnitude would reduce the true distance to about 630 light years. U Sagittae shows an annual proper motion of 0.017"; the radial velocity is about 10½ miles per second in approach.

THE IDENTIFICATION FIELD (below) shows stars to about 11th magnitude. North is at the top; the grid squares are 1° on a side. The chart is based on plates made with the 13-inch telescope at Lowell Observatory.

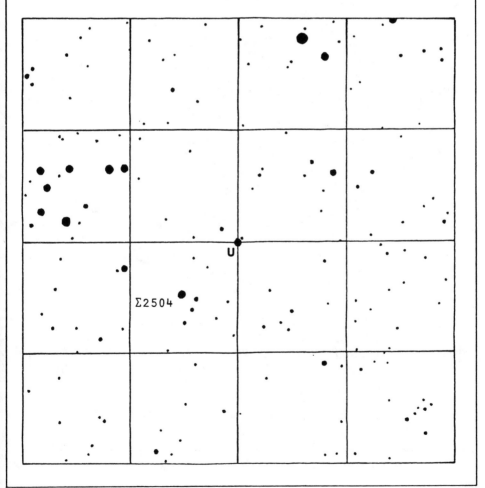

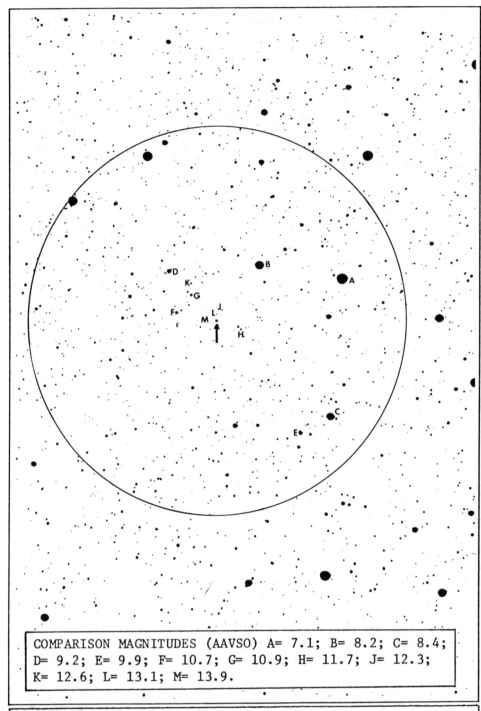

COMPARISON MAGNITUDES (AAVSO) A= 7.1; B= 8.2; C= 8.4;
D= 9.2; E= 9.9; F= 10.7; G= 10.9; H= 11.7; J= 12.3;
K= 12.6; L= 13.1; M= 13.9.

V SAGITTAE Identification chart made from a 13-inch telescope plate at Lowell Observatory. The circle is 1° in
diameter with north at the top; limiting magnitude about 15

DESCRIPTIVE NOTES   (Cont'd)

V   Position 20180n2057, in the extreme northeast portion of the constellation, about 1° ESE from 18 Sagittae. V Sagittae is a curious erratic variable star discovered by L.Ceraski at Moscow in 1902. Its exact classification is uncertain, though it may be related to the cataclysmic variables of the SS Cygni type or to the novae; C.P.Gapschkin refers to it as a "potential nova". It is a hot blue star with a type O spectrum and strong radiation in the ultraviolet. E.Hartwig, in 1903, found an estimated period of about 96 days, but observations at Harvard (1903-1907) showed the variability to be often rapid and irregular, with a range of about 9.5 to 12 or 13.

There appear to be three discernible periods involved in the variations of this star. This was probably first realized by P.M.Ryves who made an intensive series of observations during 1910-12 and 1923-31, at the Cambridge and Leyden Observatories, and at Madrid and Zaragoza in Spain; the star was examined on a total of 1243 nights. He found increases as rapid as 0.9 magnitude in one day and 2.1 magnitudes in five days, as well as decreases of 2 magnitudes in three days. Ryves also noted that the most rapid fluctuations tended to occur just after a high maximum. "The smaller fluctuations sometimes show a marked periodicity, which is different in different seasons. In 1910 a period of 12± days was in evidence, in 1927 one of

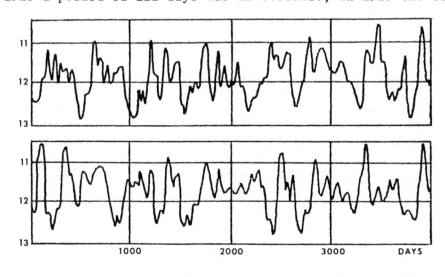

DESCRIPTIVE NOTES   (Cont'd)

17 to 18 days, and in 1928 a very constant period of 16
days was in operation......An interesting point in connection with all these short term variations is the fact that,
however violent they may be and however large the range,
they generally do not alter the general trend of the long-
term variation..."

From the work of Ryves it was shown that the long
oscillation had an amplitude of about 2 magnitudes and a
period of about 530 days. A secondary variation occupied a
cycle of about 130 days, and the third cycle about 2 weeks.
Some of the more rapid variations are shown on an enlarged
portion of the light curve, below.

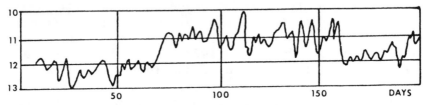

A very thorough study of V Sagittae was made at the Lick
Observatory in 1964 by G.H.Herbig, G.W.Preston, J.Smak, and
B.Paczynski; both the 120-inch and the Crossley reflectors
were used. The complex light variations were shown to be
the result of at least three separate activities:

    I. The star is a short -period eclipsing binary which
resembles many of the post novae; the period is 0.514195
day.
    II. The star shows occasional sudden brightenings by
as much as three magnitudes.
    III. There are constant minor fluctuations with a
time-scale of several days; additional "flickerings" with
a time-scale of about an hour are evident when the star is
faint.
    The odd spectrum of the star, as A.H.Joy pointed out
at Mt.Wilson in 1930, resembles that of a Wolf-Rayet star
of about type WN5; there are broad and hazy emission lines
of hydrogen, ionized helium, oxygen, and neon. The sharpest
features of the spectrum are the rare emission "O III"
features at λ3132 and λ3444 which are often observed in the
spectra of planetary nebulae and novae in their late stages.
It was at Lick in 1962 that these O III lines were found
to become alternately double and single in a cycle of about

DESCRIPTIVE NOTES   (Cont'd)

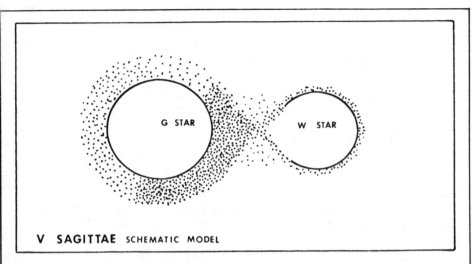

V SAGITTAE SCHEMATIC MODEL

12.3 hours; thus was the binary nature of V Sagittae first
revealed. According to the Lick astronomers, the two stars
form one of those strange systems which we meet again and
again in the cataclysmic variables:

Spect.	Mass	Diam.	Abs.Mag.
WN5	0.74	1.07	-0.5
dG ?	2.8	1.40	+0.1

The smaller and brighter star is the one eclipsed at the
primary minimum; it moves in an orbit about 1.75 million
miles in radius about the gravitational center of the pair,
a point actually beneath the surface of the cooler star.
At times of major eruptions, the hot star emits a semi-
opaque shell of gases which expands at 250- 300 miles per
second and rapidly engulfs the entire system. Much of the
material goes into orbit around the larger and cooler star,
but there is also evidence of continuous mass loss from the
system in the form of an expanding spiral stream of gas.
From the derived absolute magnitudes, the distance of the
star appears to be about 9000 light years.
     V Sagittae is of special interest from its similarity
to the known post-nova stars. It was very possibly a nova
at some time in the past, or may be preparing for such an
outburst in the future. As a binary system, it strongly
resembles the former novae DQ Herculis and V603 Aquilae.

WZ   Recurrent nova. Position 20053n1733, about
2.8° SE from Gamma Sagittae. This is one of
the few known examples of a repeating nova. Three maxima
have been recorded, in 1913, 1946, and 1978, but it is
possible that some outbursts have been missed. The star is
normally beyond the range of amateur telescopes, but rises
to nearly naked-eye visibility at maximum.

The first recorded outburst occurred on November 22,
1913, and the star reached magnitude 7.0 at maximum. The
fading was steady, but slower than for some recurrent novae
and the star declined by only 3½ magnitudes during the
first month. A year later it had faded to 14th magnitude,
but with irregular variations of about 0.8 magnitude total
amplitude.

Between 1935 and 1940 the star continued to oscillate
slightly between magnitudes 15 and 16; the total range of
the nova outburst was thus about 9 magnitudes, or about
4000 times in actual light increase. The observed range is
less than in the normal or "classical" novae, but this
appears to be a property of recurrent novae in general.

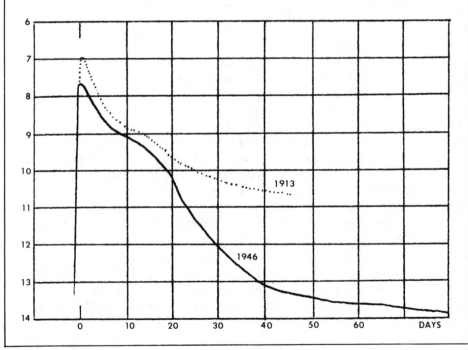

The second explosion of WZ Sagittae occurred on June
28, 1946, when the star was discovered to be brightening
at the rate of about ½ magnitude per hour. On the following
night the maximum was reached at magnitude 7.7 and the star
then began to fade, declining by 3 magnitudes in the first
21 days, and reaching 13th magnitude 40 days after the outburst. A third maximum occurred in December 1978.

A comparison of WZ Sagittae with the two brightest
recurrent novae - T Coronae and RS Ophiuchi - brings to
light some interesting and seemingly contradictory facts.
The relatively small ranges of the recurrent novae suggested originally that such "repeating" stars might be comparatively low-luminosity objects. Studies of T Coronae and RS
Ophiuchi, however, brought forth surprising evidence that
these stars attain high luminosity at maximum, and apparently equal the normal or "classical" novae. For WZ Sagittae, on the other hand, the evidence indicates that this
one star, at least, is a true "dwarf nova". An annual proper motion of about 0.08" first suggested that the star must
be relatively nearby and therefore intrinsically faint. The
presently accepted distance is about 300 light years, but
the figure must be regarded as somewhat uncertain.

J.L.Greenstein, from spectroscopic studies, finds the
star at minimum to have the characteristics of a true white
dwarf with an absolute magnitude of about +10.5, and a
luminosity of about 1/175 that of the Sun. During a nova
outburst, the luminosity rises to about 25 or 30 times that
of the Sun, whereas the normal novae usually exceed 100,000
suns. WZ Sagittae, then, seems to be the first authentic
"dwarf nova" to be discovered; this raises the question of
whether the star should still be classed as a true nova,
or reclassified as an extreme sort of SS Cygni star. The
enigmatic U Scorpii may be another object of this class;
its apparent faintness and high galactic latitude make it
a suspicious object. The four observed outbursts have all
been very brief, and the first spectra were obtained only
in 1979. The star at minimum is an 18th magnitude object
and has never been intensively studied.

The known duplicity of T Coronae, RS Ophiuchi, and
objects of the SS Cygni type brings out a point of special
interest. Stars of these types all seem to be close and
rapid binaries, suggesting that the repeating outbursts may

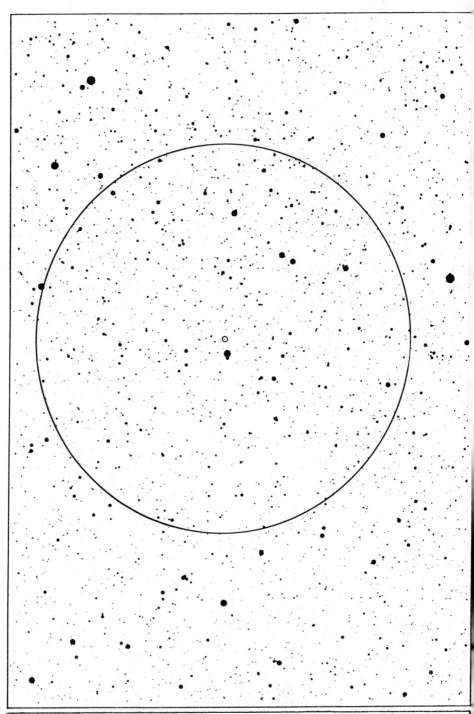

WZ SAGITTAE FIELD, from a Lowell Observatory 13-inch telescope plate. The circle is 1° in diameter with north at the
top; limiting magnitude about 16. The central portion of
the field is shown enlarged in the chart on page 1540.

be attributed to some process of interaction between two close stars, one of which is a partially degenerate dwarf or subdwarf. The theory is strengthened by the discovery (1961) that WZ Sagittae is not only a rapid spectroscopic double, but also an eclipsing pair with the extremely short period of 81.6 minutes. As of 1976, this is still the most rapid binary known. Primary eclipse has a depth of about 0.4 magnitude, and the secondary minimum about 0.2.   The accompanying light curve is based upon observations made by W.Krzeminski at the Lick Observatory in the summer of 1961. The extremely short period implies unusually small diameters; evidently both stars are dwarfs or subdwarfs revolving nearly in contact. DQ Herculis (Nova 1934) and the SS Cygni stars appear to be very similar systems.   WZ Sagittae, however, shows one peculiarity which is quite rare among the white dwarfs, the presence of emission lines of hydrogen. These appear to originate in a gas cloud which forms a rotating ring around the star, and which is supp- lied by material streaming from the companion star at a velocity of about 500 miles per second. At principal eclipse, it is not only the white dwarf which is hidden, but also the encircling ring. This situation explains the peculiar fact that the light curve is "out of phase" with the radial velocity curve; the maximum recession velocity is measured near the time of primary eclipse.

From studies made in 1964, the separation of the two dwarf components appears to be something like 230,000 miles

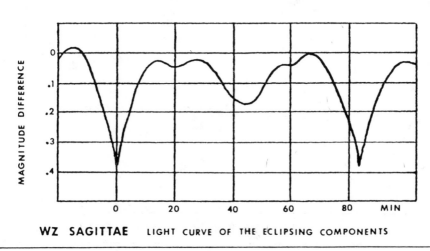

WZ SAGITTAE   LIGHT CURVE OF THE ECLIPSING COMPONENTS

or about the distance between the Earth and the Moon! In a
study by R.P.Kraft and W.Krzeminski, the following data
were obtained, which appears to be our best present picture
of this very odd system:

	Spect.	Mass	Diam.	Lum.	Abs.Mag.
A	DAe	0.59	0.013	0.0056	+10.5
B	dM?	0.04	0.100	?	?

The derived diameter of the white dwarf primary is about
11,000 miles, not even twice the size of the Earth; and the
computed density is 270,000 times that of the Sun, or some
7 tons to the cubic inch. It is also interesting to note
that the companion, about equal in size to Jupiter, has a
computed mass of 0.04 sun, which places it among the smallest stellar masses known. The famous star UV Ceti, usually
considered the record-holder, also has a mass of close to
0.04.   (Refer also to T Coronae Borealis, RS Ophiuchi, T
Pyxidis, U Scorpii, DQ Herculis, and Nova Aquilae 1918)

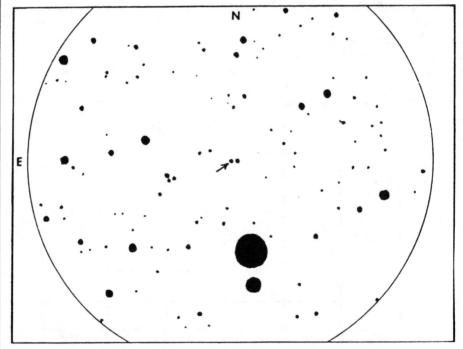

WZ SAGITTAE- 10' field. The bright star is BD+17°4225.

FG   Position 20097n2011, about 0.7° SE of Theta
Sagittae. An unusual variable star, possibly
of a unique type, discovered in 1943 by C.Hoffmeister and
his colleagues at Sonneberg Observatory in Germany. The
star is sometimes referred to by its original provisional
designation "377-1943". At first considered an irregular
type, FG was found to be of special interest when a study
of Heidelberg plates and Harvard plates revealed a slow and
apparently linear increase in brightness over the last several decades; the star was about 13.7 magnitude in 1890,
10.3 in 1959, and about 9.4 in 1967. Simultaneously with
the light increase, a gradual change in spectral type has
occurred. The star was first classified as type Be p by K.
Henize in 1955; it was about B9 in 1960, A3 in 1965, and
about A5 in 1967. At all times, the spectral features were
those of a supergiant of luminosity class Ia, though it is
virtually certain that the star has a much lower luminosity
than this type would imply.

The slightly fuzzy appearance of the star on Palomar
48-inch Schmidt plates prompted an investigation with larger telescopes. In June 1960, photographs were obtained at
Lick Observatory with the new 120-inch reflector; these
revealed a round nebulous disc centered on the star, about
30" in diameter. Spectra showed the usual features common
to planetary nebulae, including the usual emission lines of

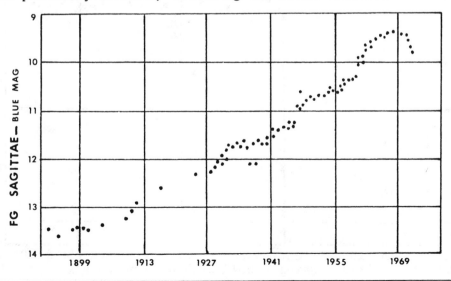

DESCRIPTIVE NOTES   (Cont'd)

ionized oxygen and nitrogen. In this odd star it seems that
we are witnessing the formation of a planetary nebula. In
a study by G.Herbig and A.A.Boyarchuk (1968) a probable
distance of about 2.5 kiloparsecs (8150 light years) has
been derived; the diameter of the nebulous shell is then
about 74,000 AU, or about 1.2 light years. From the known
ejection velocity of 70 km/sec, the age of the visible
shell must be some 3000 years. Since the spectroscopic
studies show that mass ejection is underway at the present
time, it appears that a second gaseous shell is now being
produced.

An independent estimate of distance has been made by
D.J.Faulkner and M.S.Bessell at the Mt.Stromlo and Siding
Springs Observatories in Australia. They find that in 1955
the star had an effective temperature of about 12,000° K,
a diameter of about 6 suns, and a luminosity of 670 suns;
by 1965 the temperature had decreased to 8600°K; the diameter was 13.5 suns, and the luminosity about 880. From
these results they find a somewhat smaller distance (about
4900 light years) than the Lick astronomers; the diameter
of the cloud would then be about 45,000 AU. The formation
of a planetary nebula, on their model, proceeds in stages,

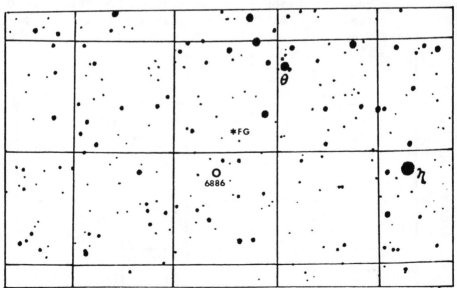

FG SAGITTAE Identification chart, with stars to about 10th
magnitude. Grid squares are 1° high with north at the top.

DEEP SKY OBJECTS IN SAGITTA. Top: The star FG Sagittae and its nebula, photographed at Lowell and Lick Observatories. Below: The cluster M71, photographed at Mt.Wilson.

with an interval of several thousand years between major
outbursts. Each stage of emission is thought to last about
a century. Possibly the star has now passed the peak of
its current cycle of activity; the apparent magnitude has
declined slightly since 1970. Small irregular variations
continue, however; in 1971 the star was varying by about
0.45 magnitude (measured in yellow light) and a rough periodicity of about 60 days seemed to be in evidence.

In a summary of the research on FG Sagittae published
in 1974, R.P.Kraft reported that the decrease in temperature of the star had proceeded at the rate of about 250° K
per year; in 1972 the temperature closely matched that of
Alpha Persei (about 6500°K) which is type F5 Ib. In actual
luminosity, however, FG Sagittae remains at least three
magnitude below that of a normal supergiant of equivalent
type. The star's atmosphere seems to be no longer expanding
and the earlier "P-Cgyni" type spectral features are no
longer in evidence. Kraft called attention to the growing
intensity of the spectral lines of barium, strontium, zir-
conium, and some of the rare-earths elements, which began
about 1967. "Over the past decade, the atmosphere of FG
Sagittae has been contaminated by these heavy elements to
about 25 times their normal solar abundances...on the other
hand...the abundances of iron, titanium, and chromium
have remained approximately constant and similar to the
solar values..."

"Perhaps", Kraft speculated, "before our very eyes
this star is making the transition from a normal giant to
a barium star or possibly an S-type star". If so, the odd
nebula surrounding the star is probably not a "normal"
planetary, since the evidence seems reasonably conclusive
that the majority of planetaries are produced by very old
stars which are approaching the white dwarf state. (Refer
also to M57 in Lyra for a general survey of theories regarding the planetary nebulae)

M71   (NGC 6838) Position 19519n1839, about midway
                between Delta and Gamma Sagittae, but somewhat
south of a line joining them. M71 is a rich and compact
cluster of faint stars, of uncertain type, lying in the
Milky Way about 10° north of Altair. It was probably first
observed by J.G.Koehler at Dresden about the year 1775, but

DESCRIPTIVE NOTES   (Cont'd)

may have been noted by de Cheseaux as early as 1746.   In
June 1780 it was rediscovered by P.Mechain, which prompted
Messier to search for it later that year; he found it "very
faint...it contains no star...The least light extinguishes
it...diam 3½'...it was reported on the chart of the comet
of 1779.."

   John Herschel found M71 to be "very large....a rich
cluster, with stars 11...16 mag". T.W.Webb, with more modest telescopic equipment, thought it "large and dim, hazy
to low power with 3.7-inch; yielding to a cloud of faint
stars to higher powers...Interesting specimen of the process of nebular resolution. About 1°S.p. M71 is a beautiful
L.P.field containing pair and triple group, all about 8 or
9 mag..." This last comment may refer to the scattered
cluster called H20 which is centered about 30' SSW of M71.

   Isaac Roberts, on a photograph made in the 1890's
with his 20-inch reflector, thought to find some evidence
for a spiral arrangement in the stars: "...the surrounding
region densely crowded with stars down to about 17 mag....
arranged in remarkable curves and lines which are very suggestive of having been produced by the effects of spiral
movements..." Six red M-type giants are known in this

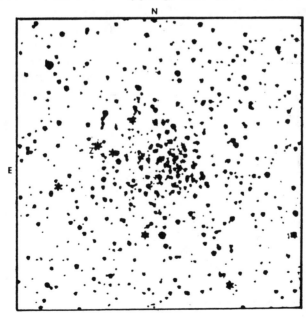

M71

10' FIELD

cluster, and are indicated by asterisk symbols on the chart
on page 1545. Probably true members of the cluster, one of
them is the irregular variable Z Sagittae, with a magnitude
range (photographic) of about 13.5 to 14.9.
        M71 is one of those unusual clusters whose precise
classification is somewhat in doubt. R.J.Trumpler and H.
Shapley both considered it as a galactic cluster, though
an unusually rich one. In most modern lists it is included
as a globular, though it lacks the dense central compression found in most globulars. A study by J.Cuffey (1959) of
Indiana University has not totally resolved the uncertainty. Using the McDonald 82-inch reflector, he found an H-R
diagram very different from a typical open cluster; the
usual main sequence is not present; the diagram reveals a
red giant sequence resembling that of a globular, but with
some unique differences. The red giant sequence shows an
unusually large scatter and a steeper slope than normal;
the usual horizontal branch - if it exists at all - shows
only a rough sprinkle of plotted points with large scatter,
and there appear to be no RR Lyrae stars at all in the
group. Spectra show that the cluster stars are metal-rich,
which does not support the classification as a globular.
According to a study of various clusters by T.Kinman (1957)
the distance is probably about 18,000 light years; the full
diameter of about 6' then corresponds to about 30 light
years. The radial velocity is 50± miles per second in
approach, consistent with either type of cluster.

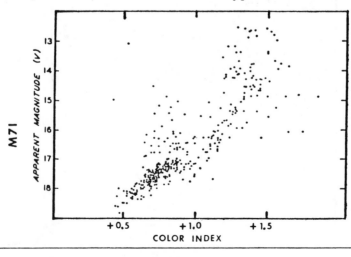

SAGITTARIUS

LIST OF DOUBLE AND MULTIPLE STARS

NAME	DIST	PA	YR	MAGS	NOTES	RA & DEC
Hu 186	0.9	331	52	7½- 11½	PA dec, spect F0	17445s1805
△219	50.1	259	35	5½- 8½	PA dec, spect gG5, F0	17555s3651
h5000	7.5	105	33	7 - 9½	PA slow dec, spect A0; Primary= V1647 (Ecl.bin)	17558s3656
h5003	5.5	106	52	5 - 7	relfix, spect M2, G8	17559s3015
	26.2	239	19	- 13		
β283	8.2	236	33	5½- 12	relfix, spect B0	17589s2247
HN 40	5.4	22	34	7 - 10½	relfix, spect O7;	17593s2302
	10.6	212	56	- 8	In Trifid Nebula	
HN 40c	2.3	281	34	8- 10½	M20 (*)	
L7542	12.5	290	00	6 - 11	(Hd 283) spect K0	18015s3554
Hwe 88	3.2	4	34	7½- 8½	spect B8	18023s3635
Hwe 88b	0.4	15	33	8½-10½		
β244	2.1	263	42	7½- 8½	PA slow inc, spect G5	18055s2752
B379	4.7	339	27	7½- 13	spect G0	18060s2607
β245	3.9	352	51	5½- 8	relfix, spect K0	18068s3045
β132	1.2	198	68	7 - 7½	PA dec, spect A2	18083s1952
11	42.1	287	07	5½- 11	(h5030) spect gK0	18086s2343
B1353	0.1	35	54	7½- 7½	PA dec, spect B8; NGC 6569 in field	18103s3159
μ	17.1	270	53	4 - 10	(h2822) (β292)	18108s2105
	25.8	119	01	- 13	Primary Ecl.bin;	
	47.4	312	25	- 9½	Spect B8; AC PA	
	50.6	115	25	- 9½	inc.	
B384	4.6	330	27	7½- 14	Spect B9	18110s2732
16	6.2	218	33	6 - 13	(β286) relfix, spect O9	18123s2024
η	3.6	104	59	3 - 9	(β760) cpm; spect	18142s3647
	33.3	276	00	- 13	M3 (*)	
	93.2	303	00	- 10		
RS	38.9	86	19	6 - 9½	(h5036) Primary =	18143s3408
	94.1	41	00	- 10	Ecl.bin; spect B5	
RSc	16.5	347	19	10- 11		
β299	54.4	13	16	6½- 9½	(Sh 263) spect B5;	18149s1850
	22.2	22	00	- 13	Rich field in	
	29.4	66	00	- 13	Small Sagittarius	
	22.0	328	00	- 13	Star Cloud	

SAGITTARIUS

LIST OF DOUBLE AND MULTIPLE STARS   (Cont'd)

NAME	DIST	PA	YR	MAGS	NOTES	RA & DEC
β299c	7.1	305	00	12½-12½		18149s1850
β639	0.5	135	58	7 - 7½	PA dec, spect 08;	18157s1839
	17.6	52	21	- 8	rich star field in	
					Small Sagittarius	
					Star Cloud	
δ	25.8	276	00	3 - 14	(λ350) spect K2	18178s2951
	40.1	165	00	- 15	(*)	
	58.1	221	00	- 13		
21	1.8	289	59	5 - 8	(Jac 6) Slow PA	18224s2034
					dec, spect K2	
β133	1.3	251	59	7 - 7	slow PA dec, spect	18246s2640
					A5	
β1128	3.2	193	42	5½- 10	(Hwe 43) PA dec,	18278s3301
	2.9	338	00	- 13	spect A3	
U	66.5	253	18	6½- 9½	(β966) Cepheid var;	18289s1910
					in cluster M25 (*)	
Stn 62	2.1	135	55	7½- 8	PA slow dec,	18311s3451
					spect F0	
Ho 567	0.9	163	43	7 - 10½	spect B5	18337s2022
φ3	4.7	268	31	7 - 14	spect F5	18350s3244
I 1381	1.0	35	44	7½- 11	spect B9	18403s2228
Stn 44	1.5	302	59	8½- 8½	PA inc, spect G5	18410s1956
01 20	0.5	336	59	8 - 8½	PA dec, spect G0	18416s3326
λ358	1.6	46	53	7½- 8	PA inc, spect F8	18417s2551
Kui 88	0.3	182	59	7 - 7	PA dec, spect A0	18458s1839
29	17.6	1	00	6 - 14	(λ362) spect gK4	18467s2023
B409	5.2	258	24	7 - 14	spect A2	18475s2650
h5070	9.5	50	47	8½- 8½	relfix, spect K0	18483s2205
I 1031	0.5	302	47	7½- 9	PA inc, spect A5	18500s2749
Ho 569	18.6	41	05	6½-11½	spect K0	18502s1842
ν¹	2.2	98	30	5 - 11	(β1033) (h5072)	18511s2249
	28.2	59	30	- 10½	spect gK2	
B413	5.4	252	27	7½- 14	spect G0	18530s2507
Kui 89	0.2	231	67	6 - 6½	PA inc, spect B5	18566s1254
I 252	0.9	26	59	8 - 8½	PA inc, spect F2	18581s3435
h5080	5.5	247	51	7½- 8½	relfix, spect A0	18593s3610
ζ	0.4	82	62	3½- 3½	Binary, 21 yrs; PA	18594s2957
	75.0	302	05	- 10	dec, spect A2 (*)	
Hld 33	2.4	57	51	9 - 9½	cpm; spect G0	18594s2843

LIST OF DOUBLE AND MULTIPLE STARS   (Cont'd)

NAME	DIST	PA	YR	MAGS	NOTES	RA & DEC
h5082	7.6	89	58	6 - 9	relfix, spect G6	19002s1919
	20.1	113	58	- 11		
	25.9	260	58	- 13½		
h5082b	13.7	126	33	9½-10½		
HN 129	7.8	308	51	7½- 8½	relfix, spect A0	19012s2258
Hu 261	1.1	219	61	7½- 7½	(HN 126) Binary;	19013s2136
					about 665 yrs; PA	
					dec; spect G0	
S 710	6.0	359	58	6 - 9½	spect B8	19040s1618
S 711	45.4	124	33	7 - 8½	relfix, spect G5,	19050s2655
					A3	
h5091	9.0	210	47	6½- 9½	cpm; slight PA inc	19053s3103
					spect G5	
Cor 233	11.8	255	54	8 - 8½	cpm; spect both F5;	19054s3352
" b	5.6	322	59	8½-10½	BC dist & PA inc.	
π	0.1	152	39	4 - 4	(φ311) difficult	19068s2106
	0.4	121	39	- 6	triple; spect F2	
B427	0.1	8	53	7 - 7½	Binary; 2.68 yrs;	19069s1953
					PA inc, spect K0	
Ho 100	4.8	327	44	8 - 11	spect A0	19080s1213
h5094	23.6	191	53	7½- 7½	(B428) Dist inc, PA	19095s3357
h5094b	2.6	12	50	- 13	dec; spect A0,A0;	
					BC PA dec.	
ψ	0.2	105	39	5½- 6	(B430) binary, 18½	19125s2521
					yrs; PA inc. spect	
					dF5	
β246	0.5	113	52	8 - 8	PA inc, spect B2	19147s1941
S 715	8.2	15	51	7½- 8	relfix, spect A3	19148s1604
S 716	5.0	196	51	8 - 8½	relfix, spect A0	19152s1602
I 253	0.6	136	26	8 - 8	binary; 60 yrs;	19157s3322
					nearly edge-on	
					orbit; spect G0	
B431	1.3	148	59	8½- 9	PA inc, spect G0	19158s3315
I 646	4.4	213	35	7 - 12	spect F5	19161s3645
h5107	13.6	127	52	7½- 9	cpm; spect G0	19177s3309
Ho 272	12.3	52	59	7½-11½	PA & dist inc;	19188s1721
	4.9	301	00	-12½	spect F8	
β¹	28.4	77	53	4 - 8	(Δ226) relfix,	19190s4433
					cpm; spect B8, A3	

LIST OF DOUBLE AND MULTIPLE STARS (Cont'd)

NAME	DIST	PA	YR	MAGS	NOTES	RA & DEC
h5113	12.2	160	59	6 - 10	Optical, dist dec, slow PA dec, spect K0	19219s2924
λ375	12.6	167	39	7½- 12	spect A2	19237s2625
h5117	6.0	261	51	8 - 9½	slow PA dec, spect F8	19248s4359
β142	0.9	37	61	8 - 8	(Scj 22) Binary, about 143 yrs; PA inc; spect dG7	19254s1215
Hu 75	0.7	257	60	7½- 8	PA inc, spect F8	19268s1245
Hh 619	7.5	142	51	5½- 8½	(HN 119) cpm, relfix, spect K3	19268s2705
I 651	0.6	136	59	8 - 8½	PA inc, spect A3	19326s3655
52	2.5	170	59	4½- 10	(β654) cpm; PA slow inc, spect B9	19337s2500
β761	2.5	193	59	7½- 10	(λ388) slow PA dec, spect G5	19363s3951
S 722	10.0	236	40	8 - 8½	spect A2	19363s1702
53	0.2	323	59	7 - 7	(λ389) no certain change; spect A0	19368s2333
I 656	0.1	140	37	8 - 8	spect A0	19369s2211
54	45.6	42	32	5½- 8½	(h599) cpm, spect	19379s1625
	38.0	274	03	- 12	K2, F8	
Σ2565	5.6	37	44	8½- 8½	relfix, spect G5	19425s1321
λ394	0.9	287	59	8 - 9	dist inc; spect F0	19434s2500
β467	3.1	134	33	7½- 10	relfix, spect A5	19435s2139
I 1039	2.0	290	13	6 - 11½	spect dF5	19461s2855
I 122	5.1	339	38	7½- 10½	spect A2	19472s4159
B454	0.1	11	59	8½- 8½	PA inc, spect F8	19501s2535
I 1406	0.4	267	43	8½- 9	(h5152) PA inc,	19503s3024
	6.1	151	54	- 8½	spect F8	
h2904	28.9	49	59	6 - 10	optical, dist inc, PA dec, spect dK5	19513s2404
L8293	23.0	209	01	6- 13½	(Hd 293) spect K0	19571s3750
λ400	1.4	29	59	8½- 10	spect F0	19574s2406
Hd 294	0.3	310	26	8 - 8½	binary; about 340 yrs; PA inc, spect F2	19578s3844
I 1490	0.6	149	59	8 - 9	PA inc, spect F2	19598s4304

LIST OF DOUBLE AND MULTIPLE STARS   (Cont'd)

NAME	DIST	PA	YR	MAGS	NOTES	RA & DEC
λ402	0.8	206	38	7 - 9½	(B463) Spect A2	20003s3644
	22.3	309	29	- 12		
λ404	0.5	86	45	7 - 9	PA inc, spect B8	20024s3309
	21.4	185	49	- 12½		
S2625	12.7	8	59	7½- 11	spect K0	20040s1305
h5168	18.7	80	19	6½- 10½	spect K0	20043s2952
B987	5.1	25	47	7½- 11	spect A0	20077s3228
h5173	7.4	123	49	5½- 12	cpm, slow dist dec, spect dK4	20079s3613
h5178	2.7	10	55	7 - 8½	relfix, spect G0	20105s3416
h5183	38.0	229	19	6½- 12	spect M4	20131s3636
	46.0	180	19	- 12½		
Stn 64	2.2	299	59	8 - 8½	slight PA dec? spect G5	20138s3246
I 1416	0.3	300	54	7 - 7½	binary, 18½ yrs; PA dec, spect G0	20170s3445
h5188	4.4	50	59	6½- 9	PA dec, spect A2	20174s2921
	27.4	322	57	- 8		
	106	332	51	- 10		
h5188d	4.7	186	51	10- 10		
$\varkappa^2$	0.9	234	52	6 - 7½	(β763) PA inc, spect A3	20205s4235
R 321	0.5	20	37	6½- 8	binary, 135 yrs; PA dec, spect K2	20236s3734

LIST OF VARIABLE STARS

NAME	MagVar	PER	NOTES	RA & DEC
μ	3.8--3.9	180.45	Ecl.bin; spect B8; also visual double star	18108s2105
υ	4.3--4.4	137.94	Ecl.bin; lyrid; spect B8 + F2p (*)	19189s1603
R	6.7--12.8	269	LPV. Spect M4e--M6e	19138s1924
S	9.3--15..	231	LPV. Spect M4e	19165s1907
T	7.6--12.7	392	LPV. Spect S6e	19133s1704

LIST OF VARIABLE STARS   (Cont'd)

NAME	MagVar	PER	NOTES	RA & DEC
U	6.3--7.1	6.7449	Cepheid; spect F5--G1; In Cluster M25 (*)	18289s1910
W	4.3--5.0	7.5947	Cepheid; spect F2--G6	18018s2935
X	4.3--4.9	7.0122	Cepheid; spect F5--G9	17444s2749
Y	5.3--6.0	5.7734	Cepheid; spect F6--G5	18184s1853
Z	8.4--15..	451	LPV. Spect M4e--M5e	19167s2101
RR	5.7--14.0	334	LPV. Spect M5e--M6e	19528s2919
RS	6.0--6.5	2.4157	Ecl.bin; spect B5; also visual double star	18143s3408
RT	6.2--14..	305	LPV. Spect M5e--M7e	20144s3916
RU	6.3--14.0	240	LPV. Spect M3e--M6e	19553s4159
RV	7.2--14..	318	LPV. Spect M5e	18246s3321
RW	9.0--11.6	190	LPV. Spect M4--M5	19110s1857
RX	9.4--14.0	334	LPV. Spect M5e	19116s1854
RY	6.5--14.0	---	Irregular; R Corona Bor type; spect G0ep (*)	19133s3337
RZ	8.5--11..	223	Semi-reg; spect Se	20120s4434
SS	9.0--10..	Irr	Spect N	18275s1656
ST	7.6--15..	395	LPV. Spect S4e	18587s1250
SU	8.0--8.9	88:	Semi-reg; spect M6	19007s2247
SW	9.2--13..	290	LPV. Spect M5e	19166s3149
SX	9.3--10.8	4.1540	Ecl.bin; spect A2	18429s3033
SZ	9.0--10..	100:	Semi-reg; spect N	17420s1838
TT	9.3--15..	333	LPV. Spect M4e--M6e	19226s2013
TV	8.8--12..	266	LPV. Spect M7e	19408s4159
TW	9.1--13..	221	LPV. Spect M2e--M3e	19105s2139
TX	9.9--14..	247	LPV. Spect M3e	19111s1731
TY	9.1--14..	325	LPV. Spect M3e	19147s2402
UU	8.8--13..	268	LPV. Spect M6e	19214s3921
UX	7.6--8.4	100:	Semi-reg; spect M5	18520s1635
VX	7.7--11..	732	LPV. Spect M4e	18050s2214
WX	9.6--11.0	2.1291	Ecl.bin; spect A1	17565s1724
WZ	7.8--9.1	21.850	Cepheid; spect G3--K6	18140s1906
XX	8.5--9.2	6.4243	Cepheid; spect F8--G8	18219s1650
XZ	9.0--11.1	3.2755	Ecl.bin; spect A3 + G	18190s2516
YY	9.8--10.5	2.6285	Ecl.bin; spect A0	18416s1927
YZ	7.1--7.9	9.5535	Cepheid; spect G0--G7	18466s1647
AG	8.6--12..	359	LPV. Spect M5e--M7e	19044s2857
AK	9.5--14..	420	LPV. Spect M7e	18252s1648

SAGITTARIUS

LIST OF VARIABLE STARS   (Cont'd)

NAME	MagVar	PER	NOTES	RA & DEC
AL	9.3--13..	77	LPV.	19148s1734
AM	9.5--13..	95	Semi-reg.	19195s3214
AN	8.6--12..	337	LPV. Spect M5e	19241s1837
AO	9.2--10.4	262	Semi-reg; spect M3	18090s2952
AP	6.8--7.5	5.0578	Cepheid; Spect F6--G5	18100s2308
AQ	6.6--7.7	200	Semi-reg; spect N3	19314s1629
AR	8.5--12.0	87.87	RV Tauri type; spect F5e--G6	18567s2347
AX	8.4--9.5	350:	Semi-reg; spect G8	18055s1834
BB	6.8--7.4	6.6370	Cepheid; spect F8--G5	18480s2021
BI	9.4--14..	335	LPV.	19475s1607
BM	9.3--12..	403	LPV. Spect M5e	19352s3630
BN	9.0--10.0	2.5197	Ecl.bin; spect F6	17439s2808
BP	9.2--12..	241	LPV.	19438s4436
BQ	9.6--12.3	8.0195	Ecl.bin; spect A2	19109s3620
BS	9.2--16..	---	Nova 1917	18237s2709
CD	9.5--15..	265	LPV.	18501s3611
DW	9.6--12..	175	LPV.	18593s1717
FL	8.3--14..	---	Nova 1924	17572s3436
FM	8.2--16..	---	Nova 1926	18143s2340
FN	9.0--14..	Irr	Erratic; probably type of Z Andromedae	18509s1903
GR	7.5--16.6	---	Nova 1924	18199s2536
HS	10--16.5	---	Nova 1900	18251s2136
KY	10--17..	---	Nova 1926	17583s2624
LN	9.0--16..	194	LPV.	18204s3326
LQ	10--17..	---	Nova 1897	18253s2757
V350	7.1--7.9	5.1542	Cepheid; spect F5--G4	18423s2042
V356	6.8--7.9	8.8961	Ecl.bin; spect B3 + A2	18449s2020
V363	8.8--16..	---	Nova 1927	19082s2957
V440	9.2--10.5	.4775	Cl.Var; Spect A7--F5	19293s2358
V441	8.2--16..	---	Nova 1930	18190s2530
V505	6.4--7.4	1.1829	Ecl.bin; spect A1 + F6	19503s1444
V525	8.1--9.0	.7051	Ecl.bin; lyrid; spect A2	19040s3014
V526	9.9--10.7	1.9194	Ecl.bin; spect A0	19050s3126
V540	8.5--10.0	Irr	Spect M2	17567s3554
V630	4.0--14..	---	Nova 1936	18055s3421
V675	9.9--10.9	.64229	Cl.Var; Spect A5--F6	18103s3420
V726	9.0--16..	---	Nova 1936	18164s2655

LIST OF VARIABLE STARS   (Cont'd)

NAME	MagVar	PER	NOTES	RA & DEC
V732	6.4--16..	---	Nova 1936	17529s2721
V737	9.5--14..	---	Nova 1933	18040s2846
V771	9.0--9.5	Irr	Spect B0e	17504s2446
V776	9.5--10..	Irr	Spect M0	17521s2800
V777	9.0--9.4	936	Ecl.bin; Spect K5+A	17433s2611
V787	7.2--16..	---	Nova 1937	17568s3030
V909	6.8--16..	---	Nova 1941	18225s3503
V910	9.5--13..	376	LPV. Spect Me	18287s3445
V927	7.5--16..	---	Nova 1944	18044s3322
V928	8.5--16..	---	Nova 1947	18159s2807
V933	9.2--15..	275	LPV. Spect M8e--M9e	18277s3111
V935	9.0--15..	235	LPV. Spect M5e	18288s3114
V963	9.5--16..	256	LPV.	18466s3137
V969	8.5--15..	262	LPV.	18593s3136
V971	8.8--15..	269	LPV.	19047s3244
V983	8.5--15..	209	LPV.	19305s3221
V999	7.5--16..	---	Nova 1910	17569s2733
V1012	8.0--17..	---	Nova 1914	18030s3145
V1014	10--16..	---	Nova 1901	18036s2726
V1015	7.1--12..	---	Nova 1905	18058s3229
V1016	7.0--15..	---	Nova 1899	18169s2512
V1017	6.2--14..	---	Recurrent nova; 1901, 1919, 1973	18289s2926
V1059	4.5--16..	---	Nova 1898	18590s1314
V1148	8.0--16..	---	Nova 1943	18060s2600
V1149	8.5--16.5	---	Nova 1945	18153s2818
V1151	10--16..	---	Nova 1947	18224s2014
V1159	9.0--11..	---	LPV.	19186s4404
V1163	9.0--13..	276	LPV.	19368s2225
V1172	9---	---	Nova 1951	17474s2040
V1175	7.0--13..	---	Nova 1952	18110s3108
V1274	9.5--13..	---	Nova 1954	17459s1751
V1275	7.5--13..	---	Nova 1954	17557s3619
V1583	8.9--16..	---	Nova 1928	18124s2355
V1647	7.0--7.2	3.283	(h5000) Ecl.bin; spect A0; also visual double	17558s3656
V1905	9.1--16..	---	Nova 1932	18306s2523
V1942	6.7--7.1	Irr	Spect N	19163s1600
V1943	7.8--10..	Irr	Spect M8	20039s2722
V1944	7.0--14..	---	Nova 1960	17568s2717

LIST OF STAR CLUSTERS, NEBULAE AND GALAXIES

NGC	OTH	TYPE	SUMMARY DESCRIPTION	RA & DEC
6439		◎	Mag 13, diam 5"; almost stellar; 18^m central star	17454s1628
6440	150[1]	⊕	Mag 10; diam 1'; pB,S,R; stars eF; Class V	17459s2021
6445	586[2]	◎	Mag 13; diam 35" x 30"; pB, pS,1E; 19^m central star; in field with NGC 6440	17463s2000
6469		⊙	Diam 12'; pRi; about 40 stars in Milky Way field; class E	17499s2220
6494	M23	⊙	B,L,pRi; Mag 7; diam 25'; class E; about 100 stars mags 9.... (*)	17540s1901
6514	M20	□	!! vB,vL,Irr; diam 25' with 3 dark lanes; incl double star HN 40; "Trifid Nebula" (*)	17589s2302
----	B86	■	! Very distinct dark nebula 4.5' x 3' in bright region; borders cluster NGC 6520	18000s2750
6520	7[7]	⊙	pS,Ri,1C; Mag 9; diam 5'; class G; about 25 stars mags 9....12; Dark nebula B86 on west edge (*)	18003s2754
6522	49[1]	⊕	B,pL,R; Mag 10½; diam 2'; class VI; stars mags 16.... in field with Gamma Sgtr	18004s3002
6523	M8	□	!!! vB,eL,Irr; Mag 5; Diam 80' x 40'; contains cluster NGC 6530. "Lagoon Nebula" (*)	18016s2420
6530		⊙	B,pL,pRi; diam 10', about 25 stars mags 7.... Class E; In nebula M8	18016s2420
6528	200[2]	⊕	pF,cS,R; Mag 11; diam 1'; Class V; stars mags 16..... in field with Gamma and NGC 6522	18016s3004

LIST OF STAR CLUSTERS, NEBULAE, AND GALAXIES (Cont'd)

NGC	OTH	TYPE	SUMMARY DESCRIPTION	RA & DEC
6531	M21	⸬	pRi,pS,1C; Diam 10'; mag 7; class D; about 50 stars mags 9....12; in field with M20 (Trifid Neb complex) (*)	18018s2230
6537		◎	B,S, nearly stellar; mag 12, diam 5"	18022s1951
6540	198^2	⸬	pF,pS; diam 1'; about 12 faint stars; class G	18031s2750
6544	197^2	⊕	cF,S,Irr R; Mag 9; diam 1'; in rich field 50' SE of M8	18042s2501
6546		⸬	vL,Ri, Diam 13'; mags 11... scattered group in Milky Way field	18042s2319
----	I.4681	□	vF; Diam 8' with three B-stars mags 9....10; portion of NGC 6559 complex	18052s2327
6553	12^4	⊕	F,L,1E; Mag 10; Diam 2'; Class XI; stars mags 20.... 1.6° SE from M8; double star B379 in field.	18063s2556
6559		□	L,vF, diam 5', surrounds 10^m B-type star; connected with I.4681	18068s2408
6558		⊕	vRi,mC,pB;R; Mag 11; diam 2'; type uncertain, may be compact galactic cluster; 1½° SSE from Gamma	18070s3146
6565		◎	Mag 13; diam 10" x 8" appearance nearly stellar	18087s2811
6563		◎	F,L,E; Mag 13; diam 50" x 35" with 18^m central star	18088s3353
6568	30^7	⸬	vL,1C; Diam 15'; about 30 stars mags 11... 0.5° SW from Mu Sgtr.	18098s2136
6569	201^2	⊕	cB,L,R; Mag 10; Diam 2'; Class VIII; stars mags 15.. about 2° SE from Gamma Sgtr; double star B1353 in field	18104s3150
6567		◎	vS; mag 11½; Diam 11" x 7" with 15^m central star	18108s1905

LIST OF STAR CLUSTERS, NEBULAE AND GALAXIES   (Cont'd)

NGC	OTH	TYPE	SUMMARY DESCRIPTION	RA & DEC
6578		◎	vS,F, Mag 13½; Diam 8"; with 16ᵐ central star; in field of 16 Sgtr.	18119s2018
----	B92	▣	Diam 15'; on NW edge of the Small Sagittarius Star Cloud	18127s1820
6583	31⁷	⠿	pRi,pC,E; Diam 2'; class G; about 35 stars mags 12.... 0.5° SE from 14 Sgtr.	18128s2209
6603	M24	⠿	vRi,vmC; Diam 4'; about 50 stars mags 14.... Class G. In brilliant Milky Way field; the Small Sagittarius Star Cloud (*)	18155s1827
6613	M18	⠿	vlC, Diam 7', Mag 8; Class D; about 12 stars mags 9....10; 1° S from nebula M17 (*)	18170s1709
6618	M17	□	!! B,vL,E,Irr; Mag 6; diam 45' x 35'; "Swan Nebula" or "Omega Nebula" (*)	18180s1612
6620		◎	vS,F; Diam 5"; mag 14 with 16ᵐ central star; appearance stellar	18187s2652
6624	50¹	⊕	vB,pL,R; Diam 3'; Mag 8.5; class VI; stars mags 16..... = X-Ray source. 0.7° SE from Delta Sgtr.	18205s3023
6626	M28	⊕	! vB,L,R,eCM; Mag 8; diam 6' Class IV; stars mags 14.... 0.8° NW from Lambda Sgtr. (*)	18215s2454
6629	204²	◎	pB,S,R; Mag 10½; Diam 15"; with 13½m central star; 1.8° N from M28	18227s2314
6638	51¹	⊕	B,S,R; Mag 9½; Diam 2'; Class VI; 0.6° E from Lambda Sgtr.	18279s2532
6637	M69	⊕	B,L,R; Diam 4'; Mag 7½; Class V; stars mags 14.... (*)	18281s3223
----	M25	⠿	(I.4725) B,L,1C; Mag 6; Diam 35'; Class D; about 50 stars mags 6...10; incl U Sgtr (*)	18288s1917

LIST OF STAR CLUSTERS, NEBULAE AND GALAXIES   (Cont'd)

NGC	OTH	TYPE	SUMMARY DESCRIPTION	RA & DEC
6642	205²	⊕	pB,pL,R; Ri; Diam 2'; Mag 8; Class V? stars mags 15.... 1° WNW from cluster M22	18288s2331
6644		◎	Mag 12; diam 2". eS,B; appearance stellar	18295s2511
6645	23⁶	⊙	pL,vRi,pC; Mag 9; Diam 10'; Class G; about 75 stars mags 11...15; 2½° N from M25 (*)	18298s1656
----	I.4732	◎	Mag 13' Diam 2"; vS,F; appearance stellar	18309s2242
6652	△607	⊕	B,S,1E; Mag 8½; Diam 2'; Class VI; stars mags 15....	18325s3302
6656	M22	⊕	!! vB,vL,vRi,vmC; Mag 6; Diam 18'; Class VII; stars mags 11..... One of finest globulars (*)	18333s2358
6681	M70	⊕	B,pL,R,Ri; Mag 8; Diam 4'; Class V; stars mags 14..... 4° NE from Epsilon Sgtr (*)	18400s3221
----	I.4776	◎	vS,F; Diam 8" x 6"; Mag 12½; with 16 central star	18426s3323
6715	M54	⊕	vB,L,R; Diam 6'; Mag 9; Class III; stars mags 15.... 1.5° WSW from Zeta Sgtr (*)	18520s3032
6723	△573	⊕	vB,L,v1E; Mag 6; Diam 7'; Class VII; stars mags 14.... Near Sgtr-Cor Austr border	18562s3642
6774		⊙	vL,1C; scattered group diam 25'; mags 12.... possibly not a true cluster; 1.5° WSW from Upsilon Sgtr.	19139s1621
6809	M55	⊕	vL,pB,R,Ri,pmCM; Mag 7; diam 15'; Class XI; stars mags 15..... (*)	19369s3103
6818	51⁴	◎	B,vS,R; Mag 10; Diam 22" x 15"; greenish disc with 15ᵐ central star; Galaxy NGC 6822 in field, 0.7° to SSE.	19411s1417

LIST OF STAR CLUSTERS, NEBULAE AND GALAXIES   (Cont'd)

NGC	OTH	TYPE	SUMMARY DESCRIPTION	RA & DEC
6822		⊖	I; 11.0; 20' x 10' vF,vL,mE; Irregular member of Local Group of Galaxies (*) Plan.Neb.NGC 6818 in field; 0.7° to NNW	19421s1453
6835		⊖	E8; 13.0; 2.0' x 0.3' F,pL,mE	19518s1242
6864	M75	⊕	B,pL,R,bM,BN; Mag 8; Diam 3'; Class I; stars mags 17.. Near Sgtr-Caprc border (*)	20032s2204
6878		⊖	Sb; 13.1; 1.2' x 0.9' vF,pL,glbM	20103s4441
6890		⊖	E2; 12.7; 1.2' x 1.0' pF,S,R,vglbM	20148s4458
----	New 5	⊖	S0; 12.4; 1.4' x 0.5' S,F,mE	20206s4410
6902		⊖	SBa; 12.4; 0.9' x 0.7' F,cS,R	20212s4350

DESCRIPTIVE NOTES

GAMMA   Name- EL NASL. Magnitude 2.97; Spectrum K0 III. Position 18026s3026. The name is from the Arabic *Al Nasl*, "The Point"; an alternate name is *Al Wazl*, the "Junction". The star marks the tip of the Arrow aimed by the centaur Χειρων, *the Chiron* of the Greeks, although the constellation Centaurus has also been identified with the figure of Chiron. In one Greek tradition, Sagittarius was placed in the sky by Chiron to guide the Argonauts in their voyage, and was entitled Τοξευτης , "the Archer"; Aratus calls it Ρύτωρ τόξου, "the Bow-stretcher", and Ovid speaks of it as *Thessalicae Sagitta*, from the legend that Thessaly was the original home of the centaurs. According to R.H. Allen, the ancient Hebrews saw in it a tribal symbol of Manasseh and Ephraim, from Jacob's last words to their

DESCRIPTIVE NOTES   (Cont'd)

father Joseph: *"His bow abode in strength"*. In later Christian times, Sagittarius was sometimes identified with the
apostle Matthew. Babylonian records refer to the star pattern as "the Archer God of War" or the "Giant King of War";
these titles, however, also appear in inscriptions honoring
the god Nergal, who was identified with the planet Mars.
Akkadian inscriptions refer to the group as the "Stars of
the Bow". In early India the constellation appears to have
been regarded as a horse or horse's head, occasionally as
a horse and rider. The later Hindu name *Taukshika* was evidently derived from the Greek word for "archer".

Gamma Sagittarii is a K-type giant star, about 125
light years distant, and about 85 times more luminous than
our Sun (absolute magnitude +0.1). The star shows an annual
proper motion of 0.20" in PA 195°; the radial velocity is
somewhat variable, but averages about 13 miles per second
in recession. The star is probably a close binary.

One of the finest portions of the Sagittarius Milky
Way lies immediately north and west of Gamma; the bright
star clouds in this region mark the general direction of
the central hub of our Galaxy, and are streaked by many
curious dark winding lanes of dust and nebulosity. From
radio studies, the actual Galactic Nucleus appears to be
located in this region, about 4° WNW from Gamma. (Refer to
page 1619)

Several interesting objects will be found in the
vicinity of Gamma. Some 50' north is the bright cepheid
variable W Sagittarii, visible to the naked eye, and going
through its cycle in a period of 7.5947 days. About half a
degree from Gamma, to the north and west, will be found two
remote globular star clusters, NGC 6522 and 6528, both discovered by William Herschel in 1784. Another cluster of
similar appearance is NGC 6558, about 1.6° to the SSE; it
is listed in various catalogues as either a globular or a
very small and dense galactic type.

DELTA Name- MEDIA, or KAUS MERIDIANALIS, the "Middle
of the Bow". Magnitude 2.71; Spectrum K2 III;
Position 18178s2951. A star of similar type to Gamma, but
possibly slightly less luminous and somewhat closer to us.
The computed distance is about 85 light years; the actual
luminosity about 60 times that of the Sun (absolute magni-

REGION OF GAMMA SAGITTARII. Gamma is the bright star just below center; the Sagittarius Star Cloud occupies the top half of the print.   Lowell Observatory photograph.

tude about +0.7). The annual proper motion is 0.05"; the
radial velocity is about 12 miles per second in approach
Several faint companions to the star are listed in standard
catalogues, but it is not certain that any of them are more
than independent field stars. The 14th magnitude star at
25.8" was first reported by T.J.J.See in 1896.

    In the low power field of Delta Sagittarii, about 45'
to the SE, the telescope will reveal the faint globular
cluster NGC 6624. This object, though not very impressive
visually, is one of the few globulars identified as sources
of X-ray energy; a better known example is the bright M15
in Pegasus. The light of the cluster is dimmed by an estimated two magnitudes in this very dusty region of the galaxy; the total integrated magnitude might be about 6.5 if
the cluster could be seen completely clear of obscuration.
NGC 6624 is another of William Herschel's discoveries,
found in June 1784, and now estimated to be about 45,000
light years distant; the radial velocity is 42 miles per
second in recession. E.J.Hartung finds the cluster resolvable with an aperture of 30 cm.

EPSILON   Name- KAUS AUSTRALIS, "The Southern Bow" or
              the "Southern Segment of the Bow". Magnitude
1.81; Spectrum B9 IV; Position 18209s3425. This is the
brightest star in the constellation, located about 4.6° to
the south of Delta. Epsilon Sagittarii appears to be at
about the same distance as Gamma, roughly 125 light years;
the actual luminosity is then about 250 times that of the
Sun, and the absolute magnitude about -1.1. With a small
telescope the bluish-white tint of the star forms an interesting contrast with the bright yellow tint of both Gamma
and Delta. Epsilon shows an annual proper motion of 0.13";
the radial velocity is 6.5 miles per second in approach.

    A 7th magnitude field-glass companion lies 3.3' to
the north and slightly west; this is GC 25096, magnitude
6.8, spectrum B9. The two stars are not physically associated. T.J.J.See also mentions a 14th magnitude companion to
the bright star at 32" in PA 295° (1896).

    About 35' SE is the spot where Nova V909 Sagittarii
flared up in 1941, reaching a maximum magnitude of about
6.8; it is now fainter than 16.0.

DESCRIPTIVE NOTES   (Cont'd)

ZETA   Name- ASCELLA, from the Latin *Axilla*, the Arm or Armpit of the Centaur. Magnitude 2.61; spectrum A2 III + A2 V; Position 18594s2957. Zeta is a very close and difficult binary star, discovered by W.C.Winlock at Washington in 1867; the components have completed five full revolutions since discovery and in 1972 were in precisely the same alignment as in the year of discovery. From an orbit computation by van den Bos (1960) the period is 21.14 years with retrograde motion and periastron in 1963. Other orbital elements are: Semi-major axis = 0.532"; eccentricity = 0.205; and inclination 110.6°. The greatest apparent separation occurs in 1954, 1975, etc.

Both stars seem to be of spectral type A2, but the brighter star is a giant of luminosity class III whereas the other appears to be a main sequence object. The two stars differ only slightly in apparent magnitude; the published figures are 3.2 and 3.4. At a distance of about 140 light years, the mean separation works out to about 23 AU, roughly the separation of Uranus and the Sun; the total luminosity must be about 145 suns, and the combined absolute magnitude about +0.1. Zeta Sagittarii shows an annual proper motion of 0.02"; the radial velocity is 13 miles per second in recession.

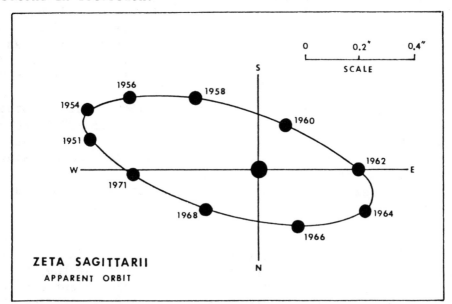

ZETA SAGITTARII
APPARENT ORBIT

There is also a 10th magnitude star at 75", apparently not a true member of the system.

Zeta Sagittarii is the southernmost star of the figure called the "Milk Dipper", consisting of Zeta, Sigma, Tau, and Phi; the handle, we may imagine, extends to Lambda and then on to the bright spot of nebula M8. The small but bright globular star cluster M54 lies about 1.5° from Zeta, to the WSW; another Messier globular, M70, will be found about midway between Zeta and Epsilon.

ETA   Name- ARKAB. Magnitude 3.15; Spectrum M3 II; Position 18142s3647, about 2½° SSW from the bright star Epsilon. Eta Sagittarii has a computed distance of about 90 light years; the actual luminosity is some 40 times that of the Sun, and the absolute magnitude about +1.1. The star shows an annual proper motion of 0.22" in PA 220°; the radial velocity is very slight, about 0.3 mile per second in recession.

The star forms a common proper motion pair with the 9th magnitude companion at 3.6"; the small star was first noted by S.W.Burnham with a 6-inch refractor in 1879. There is only slight evidence for orbital motion in this pair, which has a true separation of about 100 AU; the PA has possibly increased by about 4° since discovery, and the separation of the stars has widened from 2.8" to 3.6". "This brilliant orange star," says E.J.Hartung, "dominates a field sown with scattered stars on a profuse faint ground and the white companion close Sf. is steadily visible with 7.5 cm."

More distant companions include a 13th magnitude star at 33" in PA 276°, and a 10th magnitude star at 93" in PA 303°. These are not physically associated with the bright star, and the separation in both cases is slowly increasing from the proper motion of Eta itself.

The very fine naked-eye star cluster M7 in Scorpius lies some 4½° to the WNW.

LAMBDA   Name- KAUS BOREALIS, the "Northern Part of the Bow." Magnitude 2.80; Spectrum K2 III; Position 18249s2527. The star marks the handle of the "Milk Dipper" of Sagittarius. This is another yellow giant star closely resembling Delta in type and luminosity. It lies

about 70 light years distant; the actual luminosity may be
about 35 times that of the Sun, and the computed absolute
magnitude about +1.1. The star shows an annual proper
motion of 0.19"; the radial velocity is 26 miles per second
in approach.

Lambda Sagittarii lies in a fine Milky Way field,
thoroughly peppered with countless faint star images, and
streaked with dark lanes of cosmic dust. Just 0.8° to the
NW will be found the bright globular star cluster M28;
slightly closer to the bright star, toward the east, lies
another, fainter globular, NGC 6638. Absorption in this
part of the Galaxy averages 2 to 2½ magnitudes, making the
resolution of these clusters more difficult than usual.
After observing these two clusters, the observer should
move the telescope to a spot about 2.3° NE of Lambda, for
a refreshing view of one of the most stunning globular
clusters, the great M22. (Refer to page 1596)

XI^2 Magnitude 3.51; Spectrum K1 III; position
          18548s2110, just 5° north of the Milk Dipper.
Another K-type giant star, probably about 160 light years
distant. With a computed absolute magnitude of 0.0 the
actual luminosity is equivalent to 90 suns. The star shows
an annual proper motion of 0.04"; the radial velocity is
about 12 miles per second in approach.

The star ξ^1 is 0.5° to the north, forming a wide pair
with interesting color contrast; the fainter star is magnitude 5.06, spectrum A0. The two stars, however, do not
form a true physically associated pair.

PI   Magnitude 2.89; Spectrum F2 II or III.
          Position 19068s2106. This star is a very close
and difficult triple, suitable only for large telescopes.
According to measurements by W.S.Finsen in 1939, the A-B
pair has a separation of about 0.1"; magnitudes both about
4.0; the third fainter star, magnitude 6, lies at 0.4".
The last observations reported in the Lick Observatory
Index Catalogue of Visual Double Stars are those of 1939;
it would be of great interest to obtain more modern measurements of this very tight triple; the close pair, at least,
might be expected to show fairly rapid orbital motion.

The computed distance of Pi Sagittarii is about 250 light years, which gives a total luminosity of some 360 suns, and a total absolute magnitude of -0.7. The star shows an annual proper motion of 0.04"; the radial velocity is about 6 miles per second in approach.

SIGMA   Name- NUNKI, from the Babylonian *Tablet of the Thirty Stars;* the name is said to signify "The Star of the Proclamation of the Sea", probably from the fact that the appearance of the star heralds the approach of the "watery" constellations Capricornus, Aquarius, and Pisces, as well as Cetus and Pisces Austrinus. Another name, *Sadira,* appears on some modern charts. Magnitude 2.12; Spectrum B2 V; Position 18522s2622. The star is the brightest and the northernmost of the stars that outline the so-called "Milk Dipper" of Sagittarius; the bowl of the Dipper consisting of Sigma, Tau, Zeta and Phi. This pattern is referred to in some ancient Chinese writings as "The Ladle" but on various Oriental star charts it also appears as a "Sacred Shrine" or a temple.

The computed distance of the star is about 300 light years, which gives an actual luminosity of about 1100 suns and an absolute magnitude of about -2.7. The annual proper motion is 0.06"; the radial velocity is 6½ miles per second in approach.

TAU   Magnitude 3.30; Spectrum K1 II or III. The position is 19038s2745. This is the easternmost of the four stars that comprise the "Milk Dipper", lying about 2.6° ESE from Sigma. The computed distance is about 85 light years; the actual luminosity about 30 times that of our Sun. Tau Sagittarii shows an annual proper motion of 0.26" in PA 192°; the radial velocity is 27 miles per second in recession.

UPSILON   (46 Sagittarii) Position 19189s1603. Magnitude 4.35 (variable); Spectrum about B8p. Upsilon Sagittarii is an unusual binary system and eclipsing variable of small amplitude, possibly one of the largest and most luminous systems known, though the evidence is still not entirely conclusive. The variability of the star was first announced by S.Gaposchkin in 1939; an examination of

about 600 plates in the Harvard collection showed that the
light curve appeared to be that of an eclipsing binary of
the Beta Lyrae type with a period of 137.939 days and two
minima alternating between depths of 0.15 and 0.08 magnitudes. According to the Moscow *General Catalogue of Variable Stars* (1970) the photographic range is from 4.34 to
4.44, and the secondary minimum is about 4.40. R.E.Wilson
at Lick Observatory had previously identified the star as
a spectroscopic binary, with a period identical to that of
the light variations. Gaposchkin in 1944 demonstrated that
the two minima coincide exactly with the points in the
velocity curve that would represent eclipses of one star
by the other, but that the comparison of the two graphs
(below) revealed the strange situation that the supposedly
brighter star is actually in front of the other at primary
eclipse! Only one star is detected spectroscopically; if
the unseen star is actually the more luminous of the two
it is necessary to explain why its light does not appear
in the spectrum. The most probable explanation is that the

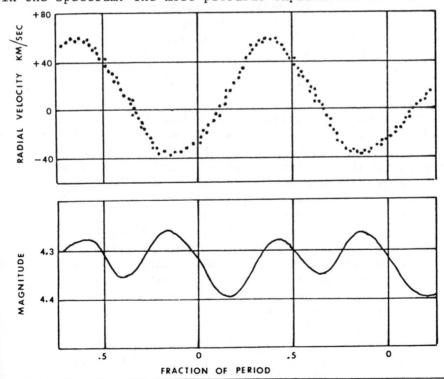

DESCRIPTIVE NOTES   (Cont'd)

lines of the second star are greatly broadened, apparently
to the point of obliteration, by turbulent motions in the
atmosphere of the star and in a gaseous "shell" surrounding
it. In the binary system of Capella we find a similar circumstance.

From the small amplitude of the light curve, it would
appear that the eclipses of this system are nearly grazing
occultations; Wilson's computed orbit is inclined about 43°
from the line of sight. A Beta Lyrae type light curve is
distinctly unusual for a system of such long period, however, as it suggests that the stars must be of very great
size. In his study of 1944, Gaposchkin derived probable
diameters of about 160 suns for both stars, and a total
mass of about 32 suns; the individual masses appear to be
roughly equal. From the spectral class of type B or early
A, and the adopted radius, the computed luminosity of the
visible star is close to absolute magnitude -6.7; the total
light of the system is possibly about -7.5, or over 80,000
times the light of the Sun. If these results are even close
to the truth, Upsilon Sagittarii must be one of the most
luminous objects known in the Galaxy.

The spectrum of this strange star is difficult to
classify. J.L.Greenstein noted in 1940 that the hydrogen
lines are unusually weak, and that the lines of the metals
are comparatively strong, when compared to normal spectra
of B-stars. Wilson called them B3 and about A2; the Moscow
General Catalogue has B8p and F2p, but reports also that W.
Bidelman in 1954 classed the system as *"Ape"* which may seem
a little odd until one realizes that the symbols stand for
"Type A, peculiar, with emission lines". The spectral type
of the second component, of course, is derived from its
relative surface brightness, as deduced from a study of the
light curve. Its very turbulent atmosphere may indicate
that the star is an eruptive giant resembling P Cygni, and
is continually surrounded by an expanding gaseous envelope.
Gas streams between the stars undoubtedly affect the radial
velocity measurements, so it is difficult to say whether
the derived masses and radii are entirely accurate. Beta
Lyrae itself presents the same constellation of problems.

In a study made in 1967, T.A.Lee and K.Nariai pointed
out that the spectrum of Upsilon Sagittarii was unique, in
that "the lines of both neutral helium and numerous metals

are seen at the same time; moreover, the hydrogen lines
are no stronger than those of the metals.... All the lines
show the same radial velocity variations, and, therefore,
they must originate in the same star..." In a study made
in 1962, M.Hack and L.E.Pasinetti found that the star is
deficient in hydrogen by a factor of about 200, and that
the peculiarities of the spectrum make it impossible to
assign a normal spectral type. The star's effective temperature, however, is about 13,000°K, equivalent to a type
near B6. Possibly this is an evolving star which has lost
its outer hydrogen-rich layer, and is producing heavier
elements; as in other very close binaries the exchange of
material may be affecting the normal pattern of evolution
of both components.

   If the derived luminosity of Upsilon Sagittarii is
reasonably correct, the distance must be close to 8000
light years. The star shows only a slight proper motion of
about 0.003"; the radial velocity (mean value) is about 5½
miles per second in recession. (Refer also to Beta Lyrae)

PHI   Magnitude 3.20; Spectrum B8 III; Position
             18425s2703. This is the westernmost star in
the "Milk Dipper" and marks the junction of the bowl with
the handle; it lies about 2° WSW from Sigma Sagittarii.
Phi is another giant star, with a computed distance of
about 590 light years, and an actual luminosity of some
1600 times that of the Sun (absolute magnitude about -3.1).
The star shows an annual proper motion of 0.05"; the radial
velocity is slightly variable, but averages about 13 miles
per second in recession. The star is possibly a spectroscopic binary.

RY   Unusual variable. Position 19133s3337, in the
             south central part of the constellation, some
4½° SE of Zeta in the Milk Dipper. RY Sagittarii was discovered by the British observer Col.E.E.Markwick in 1896;
it was found independently by J.C.Kapteyn in October of the
same year, and by Mrs.W.Fleming at Harvard, from its peculiar spectrum. Kapteyn noted that the star appeared about
magnitude 10.2 on a Cape photographic plate made in October
1888, but was found to be magnitude 7.1 in November 1890.
RY Sagittarii is an intriguing variable resembling the odd

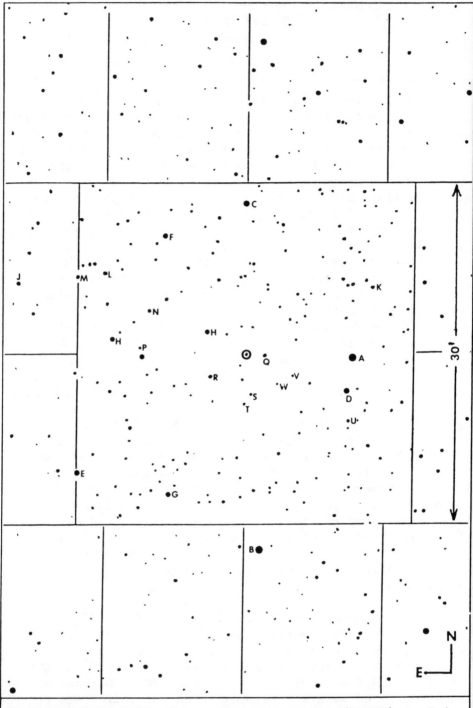

RY SAGITTARII FIELD. Comparison Magnitudes (AAVSO) A= 7.4;
B= 7.5; C= 8.3; D= 8.6; E= 9.5; F= 9.7; G= 9.9; H= 10.2;
J= 10.6; K= 10.8; L= 10.9; M= 11.3; N= 11.6; P= 12.0; Q=
12.2; R= 12.4; S= 12.5; T= 12.7; U= 13.2; V= 13.4; W= 13.9

star R Coronae Borealis, and is the brightest star of its
type in the southern sky. It is normally about magnitude
6.5, and may remain at that brightness for a year or two
at a time, then will suddenly begin to fade, falling within
a few weeks to magnitudes as faint as 14. The light curve
of this unpredictable star thus resembles a "nova in
reverse", and the return to normal is virtually always much
slowerthan the decline, with large oscillations and frequent relapses. No definite periodicity can be detected in
these intervals of fading, though the star does appear to
show a secondary semi-regular cycle of about half a magnitude with an average periodicity of 39 days; this has been
confirmed by both L.Jacchia and L.Campbell (1941) and is
mentioned by J.S.Glasby (1969) as a "peculiar pseudo-
Cepheid-like variation".

Owing to its position in a zodiacal constellation
and south of the Ecliptic as well, the star cannot be followed throughout the year; it is completely unobservable
during December and January while the Sun is passing
through Sagittarius. The records of the star are therefore
not as complete as those of R CorBor. A typical decline of
RY Sagittarii, from the observations of Campbell in 1930,
is shown below at left. One of the star's erratic periods,
shown below at the right, occurred in 1933, when the star
seemed to be rising from one of its faint periods, but then
fell rapidly to an even lower minimum. As these lines were
being written in October 1977, RY Sagittarii was well on
its way to another minimum, having declined to magnitude
9.5 in mid-September, and to about 10.9 by the end of the
month. The total range of this star is about 8 magnitudes,

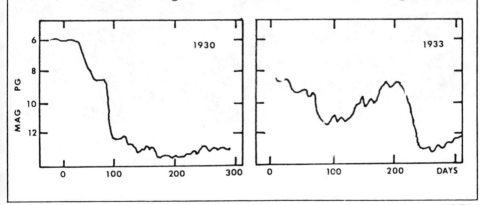

DEEP MINIMUM OF RY SAGITTARII. This is the minimum of
June 1931, as photographed with the 13-inch telescope at
Lowell Observatory.

a range which is exceeded by only a few of the long-period
variables and the novae, but the light curve goes in the
opposite direction! A somewhat idealized or "smoothed-out"
light curve for this star is shown below, based on a good
series of observations compiled by the AAVSO; the curve
covers the period from August 1938 to October 1939. In this
case the star dimmed to about 1% of its normal brightness,
and required approximately one year to return to normal.

Spectra of the RY Sagittarii stars resemble those of
supergiants, but show several peculiarities; unusually
weak hydrogen features and abnormally strong absorption
bands caused by an overabundance of carbon. The star has a
spectral type near G0, or possibly late F, and if actually
a supergiant may have an absolute magnitude of about -4 or
-5. O.J.Eggen in 1970 derived a probable figure of about
-3.5 for the star; M.W.Feast at Radcliffe adopted about -4
in his study in 1969. There is some evidence that at least
a few of these stars are not as luminous as their spectral
types would seem to imply. R CorBor, for example, shows
both a measurable parallax and proper motion, which seems
to indicate that the star cannot be at the great distance
derived from the assumed supergiant luminosity. On the
other hand, one star of the type, W Mensae, appears to be
a member of the Large Magellanic Cloud, from which its
true absolute magnitude is rather quickly found to be about
-5.0. Its spectral type is close to F8 Ip, a little earlier
perhaps than RY Sagittarii; if the two stars are really

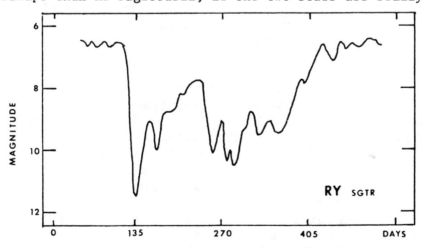

very similar in type then an absolute magnitude of about
-4 seems quite reasonable for RY Sagittarii. This would
imply a distance of about 4000 light years, making no ad-
justment for interstellar absorption; obscuration of 0.7
magnitude would reduce the true distance to 3000 light
years. The star shows only a slight proper motion of 0.02"
annually; no parallax data appears in any of the standard
catalogues.

Stars of the type appear to be going through a very
unusual phase of stellar evolution. From spectroscopic
studies it appears that most of the star's hydrogen has
been consumed or ejected into space, leaving chiefly a
helium "core" in which the generation of heavier elements
is proceeding. In a study at Mt.Stromlo Observatory in
1964, I.J.Danziger found that the carbon-to-iron ratio in
RY Sagittarii is about 35 times the normal value. "In most
respects RY Sagittarii is very similar to R CorBor, and
the continuous opacity source is probably the same for
both - photoionization of neutral carbon". The occasional
fading of the star is thus attributed to the formation of
what might be literally called a "soot cloud" around the
star. Studies by M.W.Feast in 1969 revealed a large infrared excess for this star, attributed to a cloud of solid
particles, presumably graphite grains; the cloud has an
apparent angular size of 0.02" or about 20 AU at the probable distance of about 3000 light years; its temperature
is about 900°K and its expansion velocity some 25 km/sec.
The secondary cycle of this star, with its period of some
39 days, may be due to a semi-regular pulsation in the
outer layers of the star.   (Refer also to R Coronae Borealis, page 702)

M8   (NGC 6523) Position 18016s2420. This is the
             "Lagoon Nebula" in Sagittarius, one of the
finest of the diffuse nebulae, located about 4.7° west and
slightly north from Lambda Sagittarii in the handle of the
"Milk Dipper", and plainly visible to the naked eye as a
glowing comet-like patch just off the main stream of the
Sagittarius Milky Way. In the small telescope it is seen
to consist of a fine irregular nebulosity enveloping the
scattered open cluster NGC 6530. Discovery of M8 is often
credited to LeGentil in 1747, though it seems that it was

THE SAGITTARIUS MILKY WAY. The bright spot above center is the nebulous cluster M8; the Small Sagittarius Star Cloud appears at the top of this print. Photographed by David Healy with a 50 mm f/1.8 lens.

recorded by Flamsteed as a "nebulosum" preceding the "Bow
of Sagittarius" as early as 1680; de Cheseaux in 1746 also
refers to a "cluster in Sagittarius' bow" which is probably
M8. LeGentil described it as "a small nebulosity like the
tail of a comet with numerous stars....like the more transparent and whitish localities of the Milky Way". Messier
in 1764 saw it as "a cluster which looks like a nebula in
an ordinary telescope of three feet but in a good instrument one observes only a large number of small stars... A
fairly bright star nearby is surrounded with a very faint
glow; this is 9 Sagittarii, 7 mag. The cluster appears
elongated NE-SW. Diam 30'."

William Herschel, with his great reflector, described
"an extensive milky nebulosity divided into two parts; the
north part being the strongest. Its extent exceeds 15';
the southern part is followed by a parcel of stars...."
To which John Herschel added the observation that M8 consisted of "a collection of nebulous folds and matter surrounding and including a number of dark, oval vacancies and
in one place coming to so great a degree of brightness as
to offer the appearance of an elongated nucleus. Superimposed upon this nebula and extending in one direction beyond its area, is a fine and rich cluster of scattered
stars which seems to have no connection with it as the
nebula does not, as in the region of Orion, show any tendency to congregate about the stars.." On this latter
point, John Herschel was definitely in error, as modern
studies leave no doubt that the cluster, which occupies the
east half of the nebula, is intrinsically connected with
the nebulosity.

T.W.Webb referred to M8 as "a splendid galaxy object;
visible to naked eye. In a large field we find a bright
coarse triple star, followed by a resolvable luminous mass,
including two stars, or starry centres, and then by a loose
and bright cluster enclosed by several stars: a very fine
combination..." C.E.Barns saw here "myriads of low-mag
stars and a few brighter units resembling somewhat the
Pleiades, involved in wide wastes of incandescent hydrogen
and helium, overflung with dark absorbing patches.....A
naked-eye wonder."

With a diameter of over $\frac{1}{2}°$, a wide-field ocular is
needed to observe the full extent of this object. On long

REGION OF NEBULA M8 in SAGITTARIUS. The larger object in
this low-power view is M8; the Trifid Nebula M20 appears
at the top of the print. Lowell Observatory photograph with
the 13-inch camera.

NEBULA M8 in SAGITTARIUS. The Lagoon Nebula, one of the
finest objects of its type. Lowell Observatory photograph
made with the 42-inch reflector.

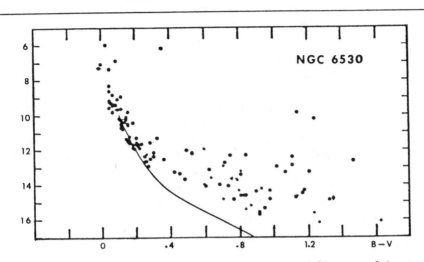

NGC 6530

exposure photographs the nebula is a magnificent object,
showing a wealth of intricate detail in the mixed bright
and dark nebulosity. The most prominent detail is the wide
curving dark channel that cuts nearly through the middle of
the cloud, dividing it in half; its course is from NE to SW
and its width averages about 2' along much of its path. The
dark lane is not completely "dark" but shows some bright
filaments which run approximately parallel to the sides of
the channel, particularly on the west side. This central
dusky area is the source of the name "Lagoon Nebula" which,
according to John C.Duncan, was probably first used by Miss
Agnes M.Clerke in the first edition of her book *The System
of the Stars* in 1890. The name does not seem entirely
appropriate, however, as the central dark feature resembles
a channel rather than a lagoon. In addition to the major
dark lane, there are numerous other smaller dust clouds
scattered here and there against the bright background. A
few of these appear in Prof.E.E.Barnard's *Catalogue of 182
Dark Markings in the Sky* (1919), and may be seen on the
photograph on page 1581 of this book. The peculiar comet-
shaped dark cloud at left center is B88; the very irregular
dark marking just above the bright star at the bottom is
B89.

The west half of M8 is dominated by two bright stars
just 3' apart; the southern star is 9 Sagittarii, spectrum
05, magnitude 5.97. This would appear to be the chief illuminating star of the nebula, though W.Baade has suggested

DESCRIPTIVE NOTES   (Cont'd)

that several additional extremely hot stars probably exist
in the cloud, but are totally obscured by masses of dark
nebulosity. Just 3' WSW from 9 Sagittarii lies the brightest segment of the nebulosity, a "figure 8" shaped knot
about 30" in size and often called from its shape "The
Hourglass"; this is the first detail to appear on photographic plates, and shows clearly on a 1-minute exposure
with the Lowell Observatory 42-inch reflector. The Hourglass, incidentally, has been identified as a source of
radio emission, first detected with certainty in 1973 at
the National Radio Astronomy Observatory at Green Bank,
West Virginia. Immediately to the west of the Hourglass
lies a small dark cloud in which is centered the illuminating star called Herschel 36, magnitude 9½, spectral type
about 07.

The eastern half of M8 contains as its most prominent
feature the loose star cluster NGC 6530, about 10' in diameter; the brightest members are subgiants of type B0 IV.
Using the Lowell 42-inch reflector in 1919, C.O.Lampland
identified 18 erratic variable stars in the cluster and in
the surrounding nebulosity; these appear to be flare stars
and irregular variables of the T Tauri type, which have not
yet reached a stable main sequence state. An H-R diagram
for the cluster stars, and the other bright members of the
M8 complex, is shown on page 1579, based on studies by M.F.
Walker in 1957. This diagram shows the expected features of
a very young star group; the main sequence extends only
from spectral type O5 to about A0, and all fainter members
are displaced well above the normal position, by up to 4
or 5 magnitudes. (The solid line in the diagram represents
the normal main sequence) These are thought to be stars
which are still in the process of gravitational contraction
and are probably not yet producing thermonuclear energy.
M8 has an H-R diagram resembling that of the NGC 2264 group
in Monoceros; both groups are among the youngest clusters
known, and are thought to be not more than a few million
years old. M8 is one of the great nebulous aggregates now
called "H II" regions by astronomers; it is in such clouds
of dust and gas that we find star formation in progress.

One of the remarkable features of the Lagoon Nebula
is the presence of very tiny circular dark nebulae known
as "globules", appearing as so many little black dots on

DETAILS IN NEBULA M8. The cluster NGC 6530 appears just below the center of the print; the dark comet-shaped marking at the left edge is Barnard's Nebula B88. Lowell Observatory photograph made with the 42-inch reflector.

DESCRIPTIVE NOTES   (Cont'd)

the luminous background; a number of them can be detected
in the photograph on page 1581. These small round clouds
have actual diameters of 7000 to about 10,000 AU, and are
thought to be "protostars" or new stars in their earliest
stages of formation. The great Rosette Nebula complex in
Monoceros (NGC 2237-2244) also contains numerous globules.
     The exact distance of M8 is still somewhat uncertain,
owing to strong but variable obscuration throughout the
entire region, but is almost certainly greater than the 770
parsec figure given in the Skalnate Pleso *Atlas Coeli* Catalogue, or the approximate "round figure" of 3000 light
years quoted in a number of modern observing lists. From
studies at Lick Observatory, C.R.O'Dell has found a probable distance modulus of 11.0 magnitudes; at Yerkes Observatory W.A.Hiltner, W.W.Morgan, and J.S.Neff (1964) have
obtained approximately the same results, giving a distance
of about 5150 light years for M8. The apparent dimensions
of the nebula as it appears on most photographs (0.7°x 0.5°)
then corresponds to about 60 x 44 light years; fainter outlying portions increase the total size to about 115 light
years. The cluster shows a radial velocity of about 5½
miles per second in recession.
     The surrounding region contains many fine views for
a good 8-inch or 10-inch telescope. Just 1½° to the NNW
will be found another prominent diffuse nebula, M20, the
"Trifid", in the same low-power field with the compact star
cluster M21. Sweeping an area 1° SE of M8, the diligent
observer may be able to detect the very remote globular
cluster NGC 6544, about 1' in diameter; another 1° to the
SE will bring a second globular, NGC 6553, into view; this
is one of the most difficult globulars to resolve, as the
obscuration in the region appears to exceed 6 magnitudes!
About 1½° to the ENE of M8 is a field containing a number
of nebulous bits, including NGC 6559 and IC 4681; each part
illuminated by one or more faint B-type stars, and possibly
connected to the M8 complex by a continuous faint nebulous
haze. A tenuous gas evidently pervades this entire region,
becoming visible in the presence of any high-temperature
star. The Trifid Nebula and M8 may actually be portions of
the same vast nebulous aggregation; the derived distances
are fairly comparable. (Refer also to M20)

DETAILS IN M8. Top: The "Hourglass" Region, photographed with the Lowell Observatory 42-inch reflector. Below: The field of 7 Sagittarii on the west side of M8. U.S.Naval Observatory photograph.

DESCRIPTIVE NOTES   (Cont'd)

M17   (NGC 6618) Position 18180s1612. The "Swan" Nebula, also called the "Omega" Nebula or the Horseshoe Nebula; one of the most prominent of the diffuse nebulae. It is located near the northern border of Sagittarius, about 2° NNE of the center of the Small Sagittarius Star Cloud, and some 2° south from M16 in Serpens. M17 was probably first noted by the Swiss astronomer de Cheseaux in the spring of 1764, and was found independently by C. Messier in June of the same year. de Cheseaux referred to it as "A nebula which has never been discovered; it has a shape quite different from the others; it has the perfect form of a ray or the tail of a comet, 7' long and 2' wide. Its sides are exactly parallel and well terminated. The centre is whiter than the edges....It makes an angle of 30° with the meridian.."

Messier, a few weeks later, saw M17 as "A train of light without stars, 5' or 6' in extent, in the shape of a spindle, a little like that in Andromeda's belt [M31] but the light is very faint. In a good sky, seen very well with a 3½ foot telescope..." Sir William Herschel, in 1785, spoke of it as "A wonderful extensive Nebulosity of the milky kind...There are several stars visible in it but they can have no connection with that nebulosity and are, doubtless, belonging to our own system scattered before it...." Admiral Smyth described "a magnificent arched and irresolvable luminosity occupying more than a third of the area in a splendid group of stars, principally from 9 to 11 magnitude, reaching more or less all over the field..." Sir William Huggins, in 1866, was the first to study the light of M17 with the spectroscope; he announced that the cloud was truly a mass of glowing gas, and not merely an unresolved cluster of stars. Isaac Roberts, in 1893, obtained the earliest known photograph of M17, with his 20-inch reflector and an exposure of two hours. In Lick Publications XIII the nebulosity is described as filling an area about 26' x 20'; the full dimensions of the outer fainter portions are about 45' x 35', one of the largest nebulae in the sky.

For the visual observer, the main feature of M17 is the long bright comet-like streak across the north edge; on the west end a curved "hook" gives the whole nebula a resemblance to a ghostly figure "2" with the bright streak forming the base. It requires only the slightest use of the

FIELD OF NEBULA M17 in SAGITTARIUS. A low power view showing the Swan Nebula and (top) the star cluster M18. South
is at the top in this print. Lowell Observatory photograph
with the 13-inch camera.

THE SWAN NEBULA. M17 in Sagittarius is one of the brightest of the diffuse nebulae. Lowell Observatory photograph
made with the 42-inch reflector.

imagination to transform this pattern into the graceful
figure of a celestial swan floating in a pool of stars,
perhaps that legendary swan of Finnish mythology, who, as
Jean Sibelius tells us in his evocative tone poem, floats
eternally on the black lake surrounding the mysterious
kingdom of Tuonela. Herschel saw the swan's head as the
curving top of a Greek "omega" (Ω) while other observers
have seen it as a horseshoe or a "2-hooked" bar of nebulosity. The space enclosed by the neck of the swan is evidently an obscuring cloud of some sort, and looks quite
black when compared to the star-strewn sky beyond. Fainter
luminous masses extend to the east and north, forming a
large irregular loop around the whole structure. With long
exposure photographs much patchy detail is seen in the
nebula; filaments and shreds of nebulosity, and numerous
dark slender streaks which seem to radiate out from a point
near the base of the swan's neck.
     This nebula, unlike M8, contains no conspicuous star
cluster, though the entire field is sprinkled with multitudes of glittering star-points, from magnitude 9 or so
down to the limit of visibility. Perhaps the chief illuminating stars of M17 are hidden in the thick masses of
nebulosity. From studies made in 1965, however, it seems
that at least 35 stars in the vicinity are actual members
of the M17 complex, and that the total mass of the visible
nebulosity is sufficient for the formation of about 800
stars of the solar type. This nebula, like many others, is
a moderately strong source of radio emission.
     Published distances for M17 are rather discordant,
but the best modern estimate is probably about 1.7 or 1.8
kiloparsecs, close to 5700 light years. The bright mass
which forms the body of the swan is about 12 light years
in length, and the full extent of the fainter outer portions
is about 40 light years.

M18　　(NGC 6613) Position 18170s1709, about 1°
          south of the Swan Nebula and slightly west.
This is a small galactic star cluster which must be classed
among the minor objects in Sagittarius. It was discovered
by Messier in June 1764, and described as "a cluster of
small stars, a little below M17; surrounded by a slight
nebulosity. Easier to see than M16...Appears nebulous in a

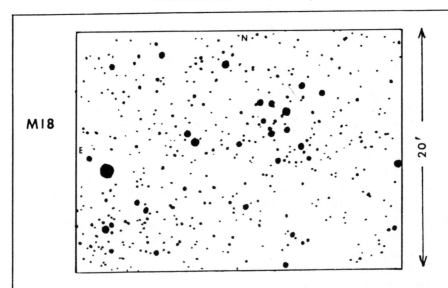

M18

3½ foot telescope; with a better telescope the stars can
be seen....Diam 5'." The whole object is little more than
a loose grouping of about a dozen stars, the brighter members arranged in several coarse pairs, on a background
flecked with countless faint star-points. Smyth speaks of
the field as containing "a long and straggling assemblage
of stars...The whole vicinity is very rich and there are
several splendid fields about a degree to the south". Webb
also mentions a "region of surpassing splendor" to the
south of M18; both descriptions refer to the star-crowded
mass of the Small Sagittarius Star Cloud, listed by Messier
under the designation M24.

M18 is one of the most neglected of the Messier
objects, and is often omitted from lists of galactic star
clusters. H.S.Hogg gives it an apparent diameter of about
7' and an integrated magnitude of 8.0; the distance is not
accurately known, but is probably close to 1.5 kiloparsecs
or about 4900 light years. The impression of nebulosity
in the group, reported by Messier, was doubtless due to
the unresolved background of faint stars. Modern photographs, made with the 48-inch Schmidt telescope at Palomar,
do show a faint nebulosity enveloping the cluster, but this
cannot be detected visually, and was far beyond the range
of Messier's modest instruments. (See photograph on page
1585.)

A FIELD OF STAR CLUSTERS IN SAGITTARIUS. M24 is the large group at center. M25 is at the left edge and M23 at the right edge; M16 in Serpens is near the top of the print with M17 below. Bright nebula M8 is at lower right.

REGION OF THE TRIFID NEBULA. M20 is one of the major show objects in Sagittarius. The star cluster in the field is M21. Lowell Observatory 42-inch telescope photograph.

DESCRIPTIVE NOTES   (Cont'd)

M20   (NGC 6514) Position 17589s2302. The Trifid
Nebula, one of the major deep-sky objects in
Sagittarius, located about 1½° NNW from the Lagoon Nebula
M8, and possibly a part of the same vast complex of cosmic
nebulosity. It was probably first observed by LeGentil in
1747, during his examination of M8, and was rediscovered by
Messier in June 1764. Messier, however, seems to have seen
it only as a cluster of faint stars, the nebulosity probably lying below the limit of detection in his telescope.
Admiral Smyth also found its presence "indicated only by a
peculiar glow" which surrounded "the delicate triple star
in the center of its opening, the nebulous matter resisted
light of my telescope..." Sir William Herschel found the
nebulosity conspicuously divided by a curious pattern of
dark lanes, and catalogued the brightest portions as four
separate objects. John Herschel was probably the first to
call it the "Trifid" Nebula, and described it as "consisting of 3 bright and irregularly formed nebulous masses,
graduating away insensibly externally, but coming up to a
great intensity of light at their interior edges where they
enclose and surround a sort of 3-forked rift or vacant area,
abruptly and uncouthly crooked and quite void of nebulous
light... A beautiful triple star is situated precisely on
the edge of one of these nebulous masses just where the
interior vacancy forks into two channels".
        The nebula, on Lowell Observatory 13-inch telescope
plates, measures about 20' x 15' but faint outer portions
increase the total size to about 25'. The "Trifid" pattern
will be found in the bright southern portion of the nebula,
where the three dark rifts radiate out from the central
triple star HN 40 or GC 24537, an O7 type giant with a computed absolute magnitude of about -5.2. This star appears
as a double in small telescopes and as a multiple system in
larger ones; the chief components are magnitudes 6.9, 8.0,
and about 10.5, with distances of 10.6" and 5.4". Using the
great 36-inch refractor at Lick Observatory, S.W.Burnham
found a total of six stars in this system; details are
given in the short table on page 1593. The primary of HN 40
appears to be the chief source of illumination of the nebula
though it is possible that other very hot stars exist in
M20, totally hidden by the dark obscuring masses.

THE TRIFID NEBULA. Fine detail in the dark rifts appears in this photograph made at Lick Observatory with the 120-inch reflector.

DESCRIPTIVE NOTES   (Cont'd)

Modern observers will find the "trifid" structure
fairly easy to detect in a good 8-inch or 10-inch telescope
with moderate powers; the author of this book has always
found it quite noticeable with a 10-inch and good sky
conditions; in the U.S. Naval Observatory 40-inch reflector
at Flagstaff, Arizona, it looks for all the world like a
huge illuminated transparency, the dark rifts appearing
just as they do on photographs. C.E.Barns described M20
as "a dark night revelation, even in modest apertures....
Bulbous image trisected with dark rifts of interposing
opaque cosmic dust-clouds..." Each of the three dark lanes
measures about 45" in width; the east and west lanes merge
into the surrounding blackness of the sky on the edges of
the nebula, but the path of the southern lane is blocked
by one of those "bright rim" features which we find also in
the outer environs of M8.

The Multiple Star HN 40 in the Trifid			
	Magnitudes	Separation	P.A.
A-B	7 - 10.6	5.4 "	23°
A-C	- 8.8	10.6	212°
A-F	- 13.8	22.1	106°
C-D	8.8- 10.5	2.2	282°
C-E	12.4	6.2	191°

The claim that large-scale changes have occurred in
the Trifid since Herschel's time appears to rest on some
150-year-old drawings which show the central star located
precisely in the middle of one of the dark channels, rather
than inside the tip of one of the three luminous lobes as
it appears today. No evidence of any real change has been
detected in the approximately 80 years since the Trifid was
first photographed.
      As with most of the diffuse nebulae, published dis-
tances for M20 show considerable discrepancies, ranging
from the Skalnate Pleso *Atlas Coeli* Catalogue value of 670
parsecs up to C.R.O'Dell's result of 2340 parsecs obtained
from a study at Lick Observatory in 1963. A "compromise"
figure of about 1600 parsecs is quoted in a number of modern lists, and matches reasonably well the probable distance of the Lagoon Nebula M8. Possibly the two objects are

portions of one huge H II association. The Lick studies, however, suggest that the difference in the two distance moduli may be as great as 0.8 magnitude, which would place M20 about 1500 light years more distant than M8. The most that can be said, then, is that the possible connection of the two is not proven, and apparently not supported by the best current distance estimates. Both objects are sources of fairly strong radio emission.

The small compact star cluster M21 is located in the field with M20, about 0.7° to the NE.

M21   (NGC 6531) Position 18018s2230. Galactic star cluster, easily found in the same low-power field with the Trifid Nebula, about 0.7° to the NE. M21 is one of Messier's discoveries, found in June 1764 during observations of the Trifid. Messier identified the brightest star of the cluster as 11 Sagittarii, which is something of a puzzle, as the star which now bears that designation is more than 2° distant, toward the SE. He also thought to find some hint of nebulosity in the group although there is none within the range of small telescopes. The cluster is a fairly compact group highlighted by about six 8th-mag stars in a tight knot, surrounded by several dozen more scattered members. Admiral Smyth described it as "A coarse cluster of telescopic stars in a rich gathering galaxy region... About the middle is a conspicuous pair, A, 9 mag.yellowish; B, 10 mag.ash-coloured; PA 317°, d. 30.9" (1875). Messier included some outliers in his description and what he mentions as nebulosity must have been the grouping of minute stars in view..."   T.W.Webb gives no real description of the cluster, but merely states that it is located in "a lucid region".

In a field about 12' in diameter some 3 dozen stars of 8th to 12th magnitude populate this cluster. From a study by S.N.Svolopoulos in 1953, the actual membership of 57 stars in the area seems clearly demonstrated; the brightest stars are giants of type B0. The derived distance of M21 is 680 parsecs or about 2200 light years; evidently it is a much closer object than the Trifid itself. A.Wallenquist reports an actual diameter of about 17 light years, and a central density of 9.36 stars per cubic parsec.
(Refer to photograph on page 1590)

STAR CLUSTERS IN SAGITTARIUS. Top: The galactic cluster
NGC 6645. Below: The great globular cluster M22. Lowell
Observatory photographs made with the 13-inch telescope.

M22   (NGC 6656) Position 18333s2358, just above
the Sagittarius Milk Dipper, about 2.3° NE
from Lambda Sagittarii. M22 is a wonderful globular star
cluster, ranking among the six finest in the entire heavens
and is equalled, in the northern sky, only by the great M13
in Hercules. Discovery of M22 is usually credited to the
obscure German astronomer Abraham Ihle in 1665; virtually
nothing appears to be known about Ihle, however, and it has
even been suggested that the name is a misprint for "Hill".
Admiral Smyth mentions the fact that an Abraham Hill was a
member of the first council of the British Royal Society,
"and was wont to dabble with astronomy....Hevelius, however, appears to have noticed it [M22] previous to 1665."
Halley mentions it in 1716, and LeGentil observed it in
1747 in a telescope of 18-foot focal length, wherein it was
seen as "very irregular, long-haired, and spreading some
kind of rays of light all around its diameter". Lacaille in
1752 compared it to the nucleus of a comet; Messier in 1764
described it simply as a "round nebula without stars, near
25 Sagittarii".
        William Herschel was apparently the first observer to
recognize M22 as a dense cluster of faint stars; his son
John Herschel found it to be "a magnificent globular star
cluster; gradually brighter in the middle but not to a
nucleus. All the stars of two sizes: 10 and 11 mag., and 15
mag., as if one shell over another...larger ones ruddy."
Admiral Smyth described M22 as "a fine globular cluster:
consists of very minute and thickly condensed particles of
light with a group of small stars preceding by 3 min. somewhat in a crucial form". T.W.Webb pointed out that the
clear resolution of the cluster gave a valuable clue to the
structure of "many more distant or difficult nebulae".
        It has always seemed to the author of this book that
J.R.R.Tolkien, in his delightful fantasy *The Hobbit*, un-
wittingly created an exquisite description of M22 when he
spoke of the fabulous jewel called the "Arkenstone of
Thrain": *"It was as if a globe had been filled with moonlight and hung before them in a net woven of the glint of
frosty stars...."*
        In total light, M22 probably ranks in third place
among all the known globulars in the heavens; it is exceeded only by Omega Centauri and 47 Tucanae, while the great

GLOBULAR STAR CLUSTER M22 in SAGITTARIUS. One of the finest
of the star clusters, a great globe of several hundred
thousand suns.   Mt.Wilson Observatory photograph

M13 holds fourth place. It is also one of the easiest of
the globulars to resolve; the brightest members are 11th
magnitude, and hundreds of its stars are within range of a
good 8-inch telescope. About 75,000 star images were actually counted on a plate obtained by John C.Duncan with the
Mt.Wilson 60-inch reflector in 1918; the total population
of the cluster is thought to be at least half a million
stars. M22 is distinctly elliptical in outline; Harlow
Shapley found about 30% more stars along the major axis
than along the minor; the longer dimension is tilted toward
PA 25°. S.I.Bailey, in 1902, reported 16 variable stars in
the cluster; another 16 have been detected up to 1973. The
majority of these are short-period pulsating stars of the
RR Lyrae type. One long period variable star, noted by
Bailey, appears to be a Mira-type variable with a cycle of
about 200 days and a variation of about magnitude 14.5 to
17.5. This star, and four other long-period variables in
the field, do not appear to be true members of the cluster.
      According to H.B.S.Hogg's *Bibliography of Individual
Globular Clusters* (First Supplement, 1963) the total integrated magnitude of M22 is 6.48, the total apparent size
about 17', the integrated spectral type is F6, the apparent
modulus about 14.15 magnitudes, and the radial velocity is
89 miles per second in approach.
      There can be do doubt that M22 is one of the nearest
of all the globulars, and is definitely closer than any
globular in the northern sky. Shapley in 1919 estimated the
distance as about 27,000 light years, but later revised
this to 22,000. Both figures are now thought to be serious
over-estimates since the full extent of stellar obscuration
in the region was not recognized at the time. From the H-R
diagram of the cluster, the apparent modulus is seen to be
about 14.1 magnitudes, but a correction of about 1.8 mags
is required due to the light loss from absorption in space.
The final result is then (m-M=12.3) or about 9600 light
years. The true diameter of the cluster is not easy to determine; not only is there much dimming by cosmic dust, but
the field is crowded with faint stars of the Milky Way. The
bright central mass, however, measures about 50 light years
in diameter. This is one of the closest globulars to the
galactic plane. It also lies less than a degree from the
Ecliptic, so astrophotographers will occasionally have an

DESCRIPTIVE NOTES   (Cont'd)

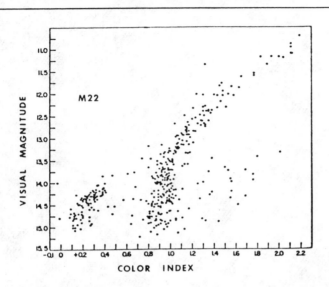

opportunity to record a bright planet in the field with the
cluster; the planet Mercury, for example, passed through
the field on December 12, 1977; Venus was in nearly the
same position in early January, 1978. Such events are not
always mentioned in astronomical publications, so the observer must make his own predictions by periodically check-
ing planetary positions in the *Nautical Almanac*.

About a degree to the WNW will be found the small and
faint cluster NGC 6642, looking like a fuzzy spot less than
1' in diameter, and first noted by Herschel in 1789. This
object has appeared in catalogue lists as either a remote
globular cluster or a very compact galactic type. On Lowell
Observatory plates it resembles a distant globular, and was
identified as such by W.Baade at Palomar in 1948. There are
a number of very faint and heavily obscured clusters in
this region whose exact classification is uncertain.

M23   (NGC 6494) Position 17540s1901. An attractive
galactic star cluster located in the NW part
of Sagittarius, approximately on a line between ξ Serpentis
and μ Sagittarii, about 4.7° NW from the latter. M23 was
probably first observed by Messier in June 1764, and was
described as a cluster about 15' in diameter with stars
"very close to one another". John Herschel found about 100
stars in the group, from 9th magnitude to about 13th; these

STAR CLUSTER M23 in SAGITTARIUS. One of the many attractive objects in Sagittarius for the small telescope. Lowell Observatory photograph made with the 13-inch telescope.

DESCRIPTIVE NOTES   (Cont'd)

are scattered more or less evenly across the whole field,
with very little concentration toward the center. A number
of the stars, however, appear to be arranged in curving
arcs and chains, which to the imaginative observer may suggest the outline of a Chinese temple, or perhaps some bit
of oriental calligraphy. Flammarion saw near the center a
circle of six stars, with "nine stars in an arc" in the NE
portion of the cluster. K.G.Jones (1968) points out that
"three or more" of the star chains appear to be portions of
concentric arcs which are seemingly focussed toward the
bright star (mag 8.23) on the NE side of the cluster. That
tireless observer Admiral Smyth described M23 as "A loose
cluster, an elegant sprinkling of telescopic stars over the
whole field under moderate magnification. The most clustering portion is oblique in a direction S.p. and N.f.......
Precedes a rich outcropping of the Milky Way". A chain of
faint stars runs out to the bright (mag 6.52) white star on
the NW, some 18' from the cluster center. The author of
this book finds the most pleasing view of M23 to be obtained with about 45X on a 10-inch f/6 reflector; the field of
a wide-angle ocular just takes in the cluster with a rim of
dark sky all around. C.E.Barns (1929), with his usual
exuberance, calls the whole group "a blazing wilderness of
starry jewels".

From color and magnitude studies, the majority of the
members appear as somewhat reddened main sequence stars;
the most luminous members are of type B9. Several G-type
giants exist in the group, however; the brightest example
in the main mass of the cluster will be found in the small
elliptical formation of 6 stars at the cluster center; it
is the westernmost star of this flattened ring. In a field
slightly over half a degree in diameter, A.Wallenquist has
identified 129 probable cluster members, and derived a central density of about 31 stars per cubic parsec. In another
study made in 1953 at the Norman Lockyer Observatory in
Devon, about 150 stars were identified as probable members
within a radius of 13.6' from the cluster center. From
these and other studies, the best value for the distance of
M23 appears to be about 660 parsecs, or somewhat over 2000
light years; the extreme diameter of about 25' then corresponds to about 15 light years. The Trifid Nebula M20 will
be found about 4° to the SSE (Refer to page 1591).

THE SMALL SAGITTARIUS STAR CLOUD. This appears to be the
Messier object M24. Mt.Wilson Observatory photograph made
with the 10-inch Bruce telescope.

M24   Milky Way star cloud, containing the galactic cluster NGC 6603. Position 18155s1827. This is the bright star cloud between the Lagoon Nebula M8 and the Swan Nebula M17 complex; it is located approximately 6° NE from M8, and is the brightest portion of the Milky Way to be found in the stretch between the Scutum Star Cloud and the great Sagittarius Cloud which marks the direction to the Galactic Center, just above Gamma Sagittarii. M24 is often called the "Small Sagittarius Star Cloud" to distinguish it from the larger cloud near Gamma. Although visible to the naked eye, it does not appear to have been mentioned by any observer until catalogued by Messier as a large cluster, in June 1764. Roughly rectangular in shape, the cloud measures about 2° x 1° with the longer dimension oriented more or less NE and SW; it may be seen at the top of the photograph on page 1575. Several very distinct dark nebulae border the cloud along its northwest rim.

There has long been confusion regarding the exact identification of M24; some catalogues identify it with the large star cloud itself, and others with the much smaller cluster NGC 6603 located in the northern portion. It seems certain that the small cluster could not have been detected with Messier's modest telescopes, and that his description refers to the entire Milky Way cloud: "Cluster on the parallel of the preceding and near the tip of Sagittarius' bow, in the Milky Way; a large nebulosity in which there are many stars of different magnitudes; the light which is spread throughout this cluster is divided into several parts....Diam 1° 30'." Admiral Smyth described the whole complex as "a beautiful field of stars; the whole is faintly resolvable, though there is a gathering spot with much star-dust. A double star, H&S 264, follows in S.f.quadrant and a wider one, H&S 263, S.p." "A magnificent region..... visible to the naked eye as a kind of protuberance in the Galaxy" says T.W.Webb. According to Miss Agnes Clerke in 1905, the star cloud was named *Delle Caustiche* by Father Secchi from "the peculiar arrangement of its stars in rays, arches, caustic curves and intertwined spirals". The field is indeed one of the finest and richest in the heavens for wide-angle telescopes; even the Great Cloud near Gamma does not produce so fine a spectacle since it appears to be much more remote and the individual stars are much dimmer.

DESCRIPTIVE NOTES   (Cont'd)

The enclosed cluster, NGC 6603, is located in the northern portion of the star cloud, just 15' east and a little north from the star GC 24950, magnitude 6.4; this is the brightest star centered in the narrowest portion of the cloud. Evidently the "gathering spot" mentioned by Admiral Smyth, NGC 6603 is a faint but very rich group about 5' in size, but not easy to detect in any aperture smaller than 8 inches. H.B.S.Hogg gives the total integrated magnitude as 11.4; the brightest stars are about 14th, however, so that the cluster usually appears as Flammarion described it: "a little nebulous patch". A distinct ruddy-hued star will be noticed immediately to the south. A distance of about 5 kiloparsecs, or 16,000 light years has been quoted for the cluster in Becvar's *Atlas Coeli* Catalogue; K.G. Jones also gives the same figure, but there appears to be some uncertainty as to whether the star cloud itself is at this same distance. If so, the true diameter of the small cluster is about 20 light years, while the cloud itself measures about 560 light years across the longer dimension.

The prominent dark nebula B92 may be seen on the photograph (pg 1602) just above the center of the cloud; it measures about 15' x 10' and contains a single 12th magnitude star apparently floating in a black vacancy. This particular example is one of the easier ones to detect visually, and is shown in a "close-up" view on page 1605. Once regarded as true "holes" or tunnels between masses of stars, such dark patches are now recognized as clouds of non-luminous dust blocking off the light from stars beyond. Another interesting example, B86, will be found in the heart of the great Sagittarius Star Cloud near Gamma; it is shown in the photograph on page 1643.

M25 (I.C. 4725) Position 18288s1917, about 3.5° east of the Small Sagittarius Cloud. M25 is a scattered galactic star cluster, a bright but not especially rich aggregation containing about 50 stars brighter than 12th magnitude, and perhaps several dozen fainter members. It was discovered by de Cheseaux in 1746 and was reobserved by Messier in 1764 as "a cluster of small stars in the neighborhood of the two previous clusters...the nearest star is 21 Sagittarii, 6 mag...the stars seen with difficulty in a 3½-foot telescope. No nebulosity seen. Position

DETAILS IN THE SMALL SAGITTARIUS STAR CLOUD. Top: The
cluster NGC 6603, photographed at Lowell Observatory with
the 13-inch telescope. Below: Dark nebula B92.

determined from Mu Sagittarii....Diam 10'." J.E.Bode
thought the cluster somewhat nebulous; T.W.Webb called it
coarse and brilliant. The most detailed description is that
of Admiral Smyth who found here "a loose cluster of large
and small stars; the gathering portion of the group attains
an arched form and is thickly strewn in the South where a
pretty knot of minute glimmers occupies the centre, with
much star-dust around". M25, through some error, was not
mentioned in John Herschel's *General Catalogue* of 1864, and
as a result was also passed over during the compilation of
the great N.G.C. The designation IC 4725 is its number in
the *Index Catalogue* of 1908, a supplement to the NGC.
        According to studies by A.Sandage in 1960 and A.U.
Landolt in 1963, the cluster has a corrected distance modulus of close to 8.9 magnitudes, giving the probable distance as slightly over 2000 light years. The apparent size
of 35' then corresponds to about 20 light years. The most
luminous main sequence stars of M25 are type B4; at least
two G-type giants are also known to be members. From a
study by A.Wallenquist (1959) 86 members have been identified in a 34' diameter field; the computed over-all density
is then 0.7 stars per cubic parsec, rising to about 18.4
stars per cubic parsec in the cluster center. M25 shows a
radial velocity of about 2.5 miles per second in recession.
        This cluster is unusual in containing, as one of its
members, the bright classical cepheid variable star called
U Sagittarii, first noticed by J.Schmidt in 1866; the probability of membership in the cluster was discussed by P.
Doig in 1925. From studies by J.B.Irwin in 1955, it seems
that the star is almost certainly a true cluster member.
U Sagittarii is a normal cepheid with a visual range of
magnitude 6.3 to 7.1 and a period of 6.744925 days; the
spectral range is from type F5 to about G1. The presence
of this star is unusual since cepheids are extremely rare
in galactic clusters, and only a few such cases are known.
Another example is S Normae, which appears to be a member
of NGC 6087. The presence of cepheids in galactic clusters
is interesting not only from its connection with problems
of stellar evolution, but is also important for the purpose
of accurately calibrating the cepheid period-luminosity
relation. Allowing for considerable absorption in the area,
the absolute magnitude of U Sagittarii at mean brightness

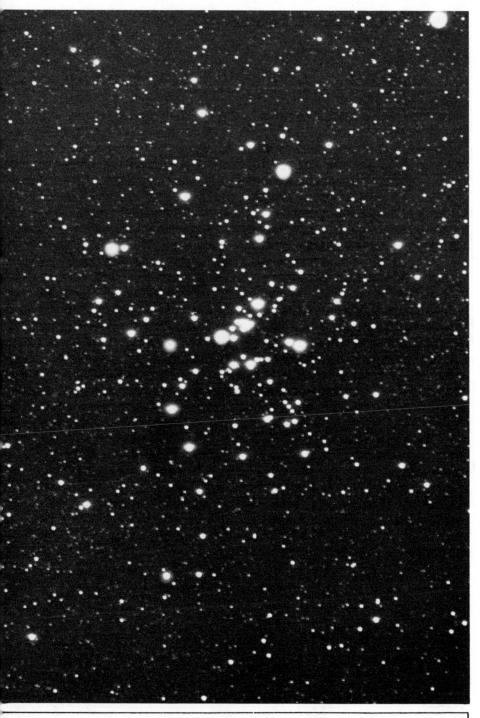

STAR CLUSTER M25 in SAGITTARIUS. A bright group for small telescopes, this cluster contains the cepheid variable star U Sagittarii. Lowell Observatory photograph.

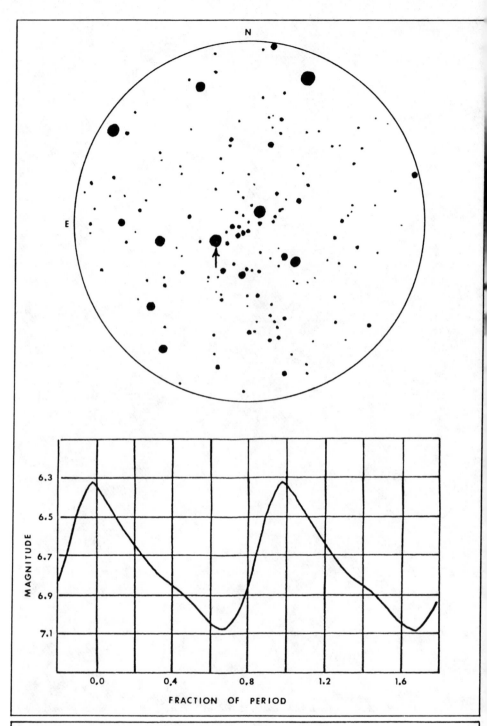

U SAGITTARII FIELD. Top: The cluster M25, showing the location of the star. The field is 30' in diameter. Below: The visual light curve of U Sagittarii.

appears to be about -3.6 visual or close to -3.0 photographic. As with all the cepheids, the color of the star
becomes somewhat yellower as the star fades to minimum; the
photographic range is thus greater than the visual.   U
Sagittarii is also a visual double star, apparently first
measured by S.W.Burnham in 1879; the 9½ mag companion is
66.5" distant in PA 253°. Since both stars are thought to
be true members of the cluster, this can be regarded as a
physical pair, with a projected separation of about 40,000
AU. It appears in double star catalogues under the number
β966, or ADS 11433. According to measurements reported in
the ADS Catalogue, the fainter star is itself a close pair
with a separation of about 0.7" in PA 122°; no certain
change has occurred in this pair since the early measurements of Burnham in 1880. The Lick Observatory *Index Catalogue of Visual Double Stars* gives the individual magnitudes of the BC pair as 9.6 and 10.1; their projected separation is about 430 AU.

M28   (NGC 6626) Position 18215s2454. Globular star
              cluster, easily found in the low-power field
of Lambda Sagittarii, about 0.8° to the NW. It was first
noted by Messier in July 1764, and described briefly as a
"nebula containing no star....round, seen with difficulty
in 3½-foot telescope; Diam 2'." William Herschel was the
first observer, apparently, to resolve M28; John Herschel
found it to be "very bright, round, very much compressed,
resolved into stars 14...15 mag. A fine object". Owing to
an estimated 2½ magnitudes of obscuration in this dusty
region of the Milky Way, M28 is not one of the more striking globulars for moderate telescopes; it appears as a
round fuzzy spot requiring considerable aperture for real
resolution. Also, of course, it suffers by comparison with
the great M22 which is only 3° away. M28 is, however, one
of the more compact and denser clusters; J.H.Mallas (1970)
calls the center "intense" and finds it "oddly shaped" in
small telescopes. H.Shapley had previously measured the
somewhat elliptical outline of the cluster; the longer axis
is oriented toward PA 50°. E.J.Hartung (1968) finds it
well resolved with an aperture of 30 cm.
     According to comprehensive lists published by H.B.S.
Hogg (1963) M28 has a total apparent diameter of about 15'

and a total integrated photographic magnitude of 8.48; the
integrated spectral type is F9. The average magnitude of
the 25 brightest stars is given as 14.73. M28 shows only
a slight radial velocity, less than 1 mile per second in
recession. Allowing for the strong absorption in this
region, the estimated distance appears to be about 5 kiloparsecs; T.D.Kinman in 1959 derived a figure of 4.8 kpsc,
or slightly over 15,000 light years. The 15' apparent size
then corresponds to about 65 light years; the corrected
absolute magnitude of the cluster is close to -7, or about
50,000 times the light of the Sun. Eighteen variable stars
have been detected in M28 up to 1973; most of these are RR
Lyrae type stars, but A.H.Joy in 1949 detected one star
which appears to be a Population II cepheid, resembling W
Virginis, with a period of 12.937 days. Another odd star
in this cluster is Variable #17, with a period of about 90
days; it appears to be an object of the RV Tauri type, with
a 2-magnitude light range.

M54   (NGC 6715) Position 18520s3032. A bright
               but rather small globular star cluster, found
without difficulty about 1.5° WSW from Zeta Sagittarii,
which is the southernmost star of the bowl of the "Milk
Dipper". M54 is one of Messier's discoveries, found in July
1778 and described as a "very bright nebula, discovered in
Sagittarius...It is bright in the centre and contains no
star, seen with an achromatic telescope of 3½-feet (F.L.).
Its position has been determined from Zeta Sagittarii, 3rd
magnitude". John Herschel found the cluster "at first
gradually, then suddenly much brighter in the middle; well
resolved, clearly seen to consist of stars which are chiefly of 15 mag. with a few outliers of 14 mag. 2½' diam."
On the best photographic plates the extreme diameter increases to about 6'. H.B.S.Hogg gives for this cluster a
total integrated magnitude of 8.74 and an integrated spectral type of F7. M54 seems perfectly round, showing no
evident ellipticity, and has a rather high surface brightness which permits fairly high magnification. Although
quite bright and strongly compressed, this is not an easy
cluster to resolve; under good conditions it may show some
sign of granularity with a 10-inch to 12-inch aperture;
smaller telescopes show only a round fuzzy spot.

STAR CLUSTERS IN SAGITTARIUS. Top: M28 as photographed at
Lowell Observatory with the 5-inch and 42-inch telescopes.
Below: M55 as photographed at Mt.Wilson Observatory.

DESCRIPTIVE NOTES   (Cont'd)

Hogg gives the apparent distance modulus of M54 as
17.4 magnitudes, but a correction of about 1.4 mags must be
made for obscuration; the true distance is then found to
be 15.1 kiloparsecs, or close to 50,000 light years. The
extreme diameter of the cluster is about 70 light years.
Up to 1973, the large total of 82 variables had been identified in this cluster; the majority are again RR Lyrae
type stars, but two semi-regular red variables are known
in the group, with periods of about 77 and 101 days. The
cluster shows the large radial velocity of about 76 miles
per second in recession.

M55   (NGC 6809)   Position 19369s3103. A large but
loose-structured globular star cluster located
about 7° to the east and slightly south from Zeta Sagittari
in the "Milk Dipper". It was discovered by Lacaille in 1752
while observing at the Cape of Good Hope, and described as
resembling the "shadowy nucleus of a large comet". Messier,
in the summer of 1778, confirmed its existence and found
it resembling "a nebula which is a whitish spot; extending
for 6' around the light is even and does not appear to contain a star.." Messier had apparently searched for it as
early as 1764, but without success; the Herschels with
their great reflectors resolved it easily into a circular
swarm of faint stars. J.E.Gore, observing with a 3-inch
refractor in India, thought to see "glimpses of stars in
it with power 40X; it will not bear higher powers with this
aperture". Flammarion found M55 "plainly stellar....a huge
agglomeration of stars uniformly distributed and immersed
in a pale nebulosity. Diam.about 6' but a little elongated
N-S. This cluster should be admirable in the southern hemisphere; for us it is a little pale".
The modern observer will find M55 easily visible in
a good pair of field glasses as a hazy "star" of about 7th
magnitude; in the small telescope it appears as a circular
glow about 10' wide and the apparent size increases to 15'
on photographic plates. Early observers commented on the
unusual "openness" of this cluster, and found the center
to be so little compressed that individual stars could be
counted easily, with apparently blank sky between. This
impression is due to the fact that only a relatively small
percent of the members exceed a brightness of 13th- 14th

magnitude, and the cluster does not begin to "fill in"
until one reaches about 17th where a vast swarm of stars
quite suddenly appears. On 48-inch Schmidt camera plates
made at Palomar, the center is a solid mass of light.

      H.B.S.Hogg (1963) gives the total apparent diameter
of M55 as 14.8' and the integrated photographic magnitude
as 7.08. Six short-period variables are known in the group,
all of which appear to be RR Lyrae type stars. The measured
radial velocity of the cluster is about 105 miles per second in recession; the average magnitude of the 25 brightest
stars is 13.58. M55 is believed to be among the nearer
globular clusters, and is estimated to be a little less
than 20,000 light years distant; the derived diameter is
then about 80 light years, and the actual luminosity close
to 100,000 times the Sun. (Photograph on page 1611)

M69   (NGC 6637) Position 18281s3223, about 2½° NE
               from Epsilon Sagittarii. Small globular star
cluster, discovered by Lacaille at the Cape of Good Hope in
1752; he found it resembling "the small nucleus of a comet"
and detected no sign of resolution. Messier observed it in
August 1780, and described it as a "nebula without star in
Sagittarius...Near to it is a 9 mag.star; the light is very
faint; can be seen only in a good sky, and the least illumination of the micrometer wires extinguishes it.......The
position was determined from Epsilon Sagittarii. This nebula has been observed by M.de LaCaille and reported in his
catalogue. It resembles the nucleus of a little comet....
Diam 2'." John Herschel, with his 18-inch reflector, found
it "all clearly resolved into stars, 14--15 mag. A blaze
of stars". In moderate telescopes it is a mere hazy spot,
however, becoming truly impressive only in large instruments. The apparent diameter is given by various observers
as 2' or 3'; it increases to about 4' on long exposure
photographs.

      H.B.S.Hogg, in the First Supplement to her detailed
Bibliography (1963) gives the integrated photographic magnitude as 8.94, the integrated spectral type as G5, and the
apparent size as 3.8'. The radial velocity is 59 miles per
second in recession. The 9th magnitude star in the field,
mentioned by Messier, lies 4.3' distant to the NNW, toward
PA 325°. Eight variable stars are known in M69, including

DESCRIPTIVE NOTES   (Cont'd)

two long period variables (196 days ±), one of which rises
to magnitude 13.0 at maximum.

A.Sandage and F.D.A.Hartwick (1968) have made a
study of M69 with the 200-inch reflector at Palomar; 467
stars in and around the cluster were measured. The resulting H-R diagram resembles that of the great cluster 47 Tucanae. M69 was also found to be one of the most metal-rich
globular clusters known. From this study, a modulus of 15.2
magnitudes was derived, giving the probable distance as
about 36,000 light years. The absolute magnitude is about
-8.1 and the actual diameter about 70 light years. T.D.
Kinman, in 1959, had derived a somewhat smaller distance
of about 7.5 kiloparsecs; the *Atlas Coeli* Catalogue (1960
edition) gives a value of 7.2, whereas H.B.S.Hogg has 7.0.

The similar globular cluster M70 lies about 2° away
toward the east, and a smaller cluster, NGC 6652, is 1°
distant, toward the SE. The double star β1128 lies about
1° to the south.

M70   (NGC 6681) Position 18400s3221. A small
globular star cluster located about midway
between Zeta and Epsilon Sagittarii, some 2° east of M69.
It was discovered by Messier in August 1780, and described
as "a nebula without star, near the preceding [M69] and on
the same parallel. Near to it is a 9 mag.star and four
small telescopic stars, almost in the same straight line,
close to one another and situated below the nebula as seen
in a reversing telescope. The [position of] the nebula
determined from the star Epsilon Sagittarii . Diam 2'. "
The Herschels found M70 to be a compact globular, with
stars 14--17 magnitude, and a bright condensed center. The
appearance is very similar to that of M69; on Lowell Observatory 13-inch camera plates M70 seems slightly the fainter
of the two, and also somewhat more irregular in outline,
with a more granular or "clumpy" structure around the outer
edges. In total light the two clusters are very nearly
equal.

H.B.S.Hogg gives the apparent size of M70 as 4.1' and
the total integrated magnitude as 8.95; the integrated
spectral type is G2, and the measured radial velocity is
123 miles per second in recession. Only two variable stars
have been discovered in this cluster, both apparently RR

DESCRIPTIVE NOTES   (Cont'd)

Lyrae type variables. The 9th magnitude star mentioned by
Messier is 14' to the west and a little south. On the NE
edge of the cluster there is also a small extending arc of
stars described by K.G.Jones as "a little slightly curved
'tail' of small stars, shooting off like sparks to the NNE.
These may be the stars mentioned by Messier". This group
consists of a small clump connected with the cluster, plus
two brighter field stars. E.J.Hartung finds an aperture of
15 cm necessary to demonstrate the resolution of M70; the
nucleus appears to be more compressed than that of M69,
however, requiring higher powers. Although the two clusters
seem so similar in size and brightness, M70 is thought to
be a more remote group than M69. T.D.Kinman finds a probable distance modulus of about 16.5 magnitudes, equivalent
to about 20 kiloparsecs or 65,000 light years; the true
diameter is then about 80 light years.

Flammarion mentions that the cluster is "decorated
with a pretty double star to the NE"; this may be the pair
of stars in the "tail" described above; there is also a
wider pair some 200" due north of the cluster, both stars
12th magnitude, with a separation of 75".

M 75   (NGC 6864) Position 20032s2204. A small but
very rich globular star cluster located in a
rather blank region in eastern Sagittarius, less than half
a degree from the Capricornus border, about 8° SW of Beta
Capricorni. M75 was probably first seen by P.Mechain in
August 1780 and was confirmed by Messier within two months;
Messier listed it as a "Nebula without stars", but thought
it possibly "composed of very small stars and to contain
nebulosity". William Herschel, in 1784, described it as
"a miniature of M3 and pale to the gaze"; he also made an
attempt to estimate its distance on the assumption that
the average bright star in the cluster was equal in actual
luminosity to Sirius. His results, in modern units, imply
a distance of a little more than 6000 light years, which
is much too small, but which remains a remarkably intelli-
gent guess, considering that at the time the distance of
not a single star was known.

M75 is one of the more compact globular clusters,
and resolvable only in fairly large telescopes. In standard
catalogues it is assigned to class I, the highest degree

of concentration, and is claimed by some observers to be
the equal of the better known globular M80 in Scorpius,
listed by Herschel as the most compressed cluster he had
ever observed. M75 is both a fainter and more distant object, however, and is only clearly resolved in large telescopes; the average magnitude of the 25 brightest stars is
given by Hogg as 17.06. Other data from her catalogue are
presented here: Total apparent diameter = 4.6', total (pg)
magnitude = 9.50, integrated spectral type = G1, radial
velocity about 123 miles per second in approach. About a
dozen variables have been identified in the cluster, but
periods and light ranges are not yet determined.

Admiral Smyth described the cluster as "a lucid white
mass among some glimpse stars" and thought that Messier's
claim of suspected resolution was distinctly "bold". There
are, however, four or five 12½ - 14 magnitude stars in the
field, surrounding the cluster in a neat semi-circle on
the N,E, and S; the brightest star of this arc is 1.5' out
from the cluster on the SE side. Very possibly these stars
were glimpsed by Messier and gave the impression of partial resolution.

M75 is very probably the most remote globular in the
Messier Catalogue; its distance is currently thought to be
about 95,000 light years, and the full diameter about 125
light years. Interstellar absorption is about 0.5 magnitude in the region; after allowing for this degree of dimming, the true absolute magnitude (pg) is found to be close
to -8.3, or approximately 160,000 times the solar luminosity.

NGC 6822   ("Barnard's Galaxy") Position 19421s1453, a
little less than 2° NE from the wide double
star 54 Sagittarii, in the NE portion of the constellation.
This is an irregular dwarf galaxy generally considered to
be a member of the Local Group of Galaxies, and was discovered by the sharp-eyed E.E.Barnard with a 5-inch refractor in 1884. For the small telescope it is not a particularly easy object, though its visibility depends chiefly
upon the darkness of the sky and the type of telescope
used. Hubble found it "fairly conspicuous" in a short focus
4-inch finder with a low-power ocular, but "barely discernible at the primary focus of the 100-inch". Low powers are

IRREGULAR GALAXY NGC 6822 in SAGITTARIUS. A system which resembles a smaller version of the Magellanic Clouds. Mt. Wilson Observatory photograph with the 100-inch telescope.

DESCRIPTIVE NOTES   (Cont'd)

essential on objects of this nature. The author of this book has always found the galaxy not particularly difficult on 6 to 10-inch telescopes with wide-angle oculars; it is actually somewhat easier to detect than the Veil Nebula in Cygnus, and is one of the few dwarf members of the Local Group within range of small instruments. W.T.Olcott (1929) referred to it as "one of the most remarkable objects in the heavens" but called it "a marvellous star cluster" and gave the calculated distance as 700,000 light years; this passage, of course, was written several decades before the major revision of the extra-galactic distance scale which occurred in the 1950's.

The maximum dimensions of NGC 6822 are about 20' x 11' but the brightest portion is an elongate central core or bar measuring about 8' x 3' and oriented almost due N-S. With large telescopes the galaxy is well resolved into an irregular mass of hundreds of thousands of stars, gathered here and there into irregular clumps and clusters; the brightest individual stars being about 15th magnitude. This is one of the closest objects of its kind, and is estimated to be about 75% the distance of the great Andromeda spiral M31, or about 1.7 million light years. At this distance the actual dimensions are found to be about 10,000 X 5,000 light years, and the total luminosity perhaps equal to 50 million suns. Compared with the Large Magellanic Cloud, however, this is a relatively minor system, and has been usually passed over in astronomical guidebooks. In a list published in 1975, S.van den Bergh gave the total integrated magnitude of the system as 9.2 (photographic); the object is difficult visually only because of its large area and low surface brightness.

E.P.Hubble found the stellar population of NGC 6822 to be similar to that of the Magellanic Clouds. The most luminous members are blue giants with absolute magnitudes of up to -7; there are 5 patches of diffuse nebulosity in the system, and a number of cepheid variables have been identified with periods ranging from 12 to 64 days. Hubble in 1925 measured a corrected red shift of about 60 miles per second for one of the brighter emission patches; the mean value for the entire system is possibly somewhat less, about 45 miles per second according to a study by Sandage and Humason.

DESCRIPTIVE NOTES   (Cont'd)

Having found this galaxy, the observer may find it
interesting to attempt to find the small planetary nebula
NGC 6818, located in the same field, about 36' to the north
and slightly west. This appears as a small pale bluish or
grey-green disc about 20" in diameter, with a slightly
elliptical outline. The central star, about 15th magnitude,
is visible only in large instruments. This nebula, of
course, has no connection with NGC 6822, but is a much
closer object, lying perhaps about 3000 light years away.

**THE SAGITTARIUS
MILKY WAY**
For observers in the Northern Hemisphere, the finest portions of the
Milky Way lie in the constellations
Cygnus, Scutum, and Sagittarius. The brightest part of all
lies just north of the star Gamma Sagittarii, and is known
as the Great Sagittarius Star Cloud; here we find a vast
milky swarm of millions of stars, marking the direction to
the center of the Galaxy, some 30,000 light years distant.
As the entire system rotates in space, the Sun moves around
the Center at a velocity of about 250 km/sec., requiring
roughly 200 million years to make one revolution, a period
which has been called the *"cosmic year"*. Interstellar dust
and dark absorbing material is so thick in the region of
the Great Cloud as to hide completely whatever lies beyond.
Therefore, the actual nucleus of the Galaxy will probably
never be seen, although the Sagittarius Cloud is considered
to be a portion of the actual central hub.

Standing in some clear, open spot, late on a moonless
night in June or July, we can trace the path of the Milky
Way across the entire heavens, as a wide glowing band, from
the Perseus-Cassiopeia region, now low in the northeast,
up through the bright star clouds of Cygnus, nearly over-
head, and thence southward through Aquila and Scutum, to
the culminating splendor of the Sagittarius Cloud above the
southern horizon. Brilliant Vega sparkles nearly in the
zenith at this hour, and white Altair approaches its culmination high in the south; between the two bright stars we
can follow the course of the "Great Rift" which now begins
to divide the Milky Way into two parallel streams all the
way to the southern horizon. The Rift, of course, is nothing more or less than the dark band of dust clouds that
appears so prominently in many of the edgewise spiral gal-

DESCRIPTIVE NOTES   (Cont'd)

axies such as NGC 891 in Andromeda and NGC 4565 in Coma
Berenices. From the Earth, we are seeing such a galaxy -
our own - close up, and our view toward the center is
blocked by the equatorial dust-lane. Wide-angle photographs
of the Milky Way strikingly resemble such systems as NGC
891; this technique also demonstrates that we are nowhere
near the center of our Galaxy, but are observing from a
point more than halfway out, on one of the outer spiral
arms.

* * * * * * * * *

        In the mythology of many ancient cultures the Milky
Way is a heavenly River, a Sky-road, a Great Path to the
world beyond the land of Men, a cosmic bridge linking
Heaven and Earth. In these concepts we find a deep though
obvious symbolism involving the metaphor of life itself as
a journey or voyage, and Man as both the traveler and the
bridge between the worlds. Rivers, roads, and bridges, of
course, are all symbols of the journey of life, and the
endless journey that the Universe itself appears to be
making, toward an unknown destination. *"Life is a bridge,"*
writes an ancient Zen master, *"Build no house upon it. But
walk on!"* Ovid, in the first book of the *Metamorphoses*,
introduces the Milky Way as the Road of the Gods:

    *"There is a way on high, conspicuous in the clear
heavens, called the Milky Way, brilliant with its own
brightness. By it the gods go to the dwelling of the great
Thunderer and his royal abode. Right and left of it the
halls of the illustrious gods are thronged through open
doors; the humbler deities dwell further away, but here the
famous and mighty inhabitants of heaven have their homes..
This is the region which I might make bold to call the
Palatine of the Great Sky..."*

        Anaxagoras and Eratosthenes refer to it as the *Circle
of the Galaxy*; Aratus titles it *"that Shining Wheel which
men call the Galaxios"*; the name is from the Greek *gala* or
galactos which simply means "milk". Hence the Latin term
Via Galactica or "Milky Way". Near-Eastern peoples seem to
have regarded it as a heavenly River; according to R.H.
Allen the Akkadian name seems to mean *The River of the*

DESCRIPTIVE NOTES   (Cont'd)

Divine Lady, though in other inscriptions it seems to be connected with the concept of a huge *River Serpent* or *Great Serpent* encircling the world. The ancient Arabian title was simply *Al Nahr*, "The River", transformed by later Semite tribes into *Al Nahr di Nur*, "The River of Light". The Hebrew word *Aroch* or *Arocea* is thought to be derived from *Aruhah*, a "long bandage" wound around the sky.

In an anonymous Pre-Confucian poem in the *Shih Ching* or Book of Songs, occur the following lines in which the vast and impersonal majesty of the Heavens is contrasted with the futile overbusyness of the world of man:

> *"The men of the East, working endlessly,*
> *But gaining no comfort;*
> *The men of the West, splendid in their*
> *   fine garments;*
> *The sons of the boatmen,*
> *Rudely clad in the skins of the bear;*
> *The sons of the slave,*
> *Taking whatever may be found;*
> *If they have rich wine*
> *They find no virtue in simple fare;*
> *Their jade pendants are long*
> *Yet they wish them longer...*
> *While above*
> *The Milky Way in Heaven*
> *Shines on all brightly.*
> *The Weaving-Lady labors there, beside the River,*
> *Asking no rewards of Earth.*
> *The Ox-Star glitters and shines,*
> *Yoked to no cart of man..*
> *In the East shines the Opener of the Dawn,*
> *In the West gleams the Star of Evening.*
> *Curved are the wide nets of Heaven;*
> *From the ancient days, in their appointed places.."*

In various Chinese writings we find the Milky Way referred to as the *River of Heaven, the Great Path, the Celestial River*, or the *Han River;* in one legend it is regarded as the ultimate source of the earthly Yellow River of central China. The legend of the *"Weaving-Girl"* or the *"Lady of the Han River"*, identified with the star Vega, is

"Look now upon the River of Heaven,
Sky-Road of the Immortals,
White with the star-frost of a billion years..."

DESCRIPTIVE NOTES   (Cont'd)

one of the most popular tales of Old China. It is mentioned again in a poem of the Han Dynasty, of unknown authorship, written about the time of Christ:

"Infinitely apart lie the Herd-boy star
And the streaming whiteness
Of the Lady of the Han River,
Working endlessly at her loom...
....How vast a distance separates them!
Always the immeasurable River yawns before them;
Forever gazing....never able to speak...."

Except on one night of the year, we are told, when a bridge of birds spans the Celestial River and allows the Heavenly Lovers to meet, on the "seventh night of the seventh moon". A thousand years after the story was first told, that sensitive and solitary young T'ang Dynasty poet Li Ho, meditated on the Milky Way from his small garden at Ch'un Ku, and evoked for us the slow fading of the year in his *Song of the Twelve Moons:*

"The bamboo mat tonight is cool as Autumn jade;
Near the Milky Way the stars grow cold;
Bubbles of dew form on the stairs;
A last flower trembles on its branch....
Grasses fade in the empty garden;
The night sky is paved with jade;
How swiftly the moving wind sweeps past!
The northern constellations glitter
And curve down the sky;
In the summer palace
The fireflies have lost their way...."

Li Ho strikes off images "like sparks from flint" says Robert Payne, one of the most knowledgeable students of Chinese literature, "and the sparks turn into fireflies and will-o'-the-wisps". Gazing on the endless pageant of the stars, Li Ho views the cycle of the dying year with a motionless serenity, a calm detachment in which we can see something of the cardinal viewpoint of Zen: *"If you want to see into it, see into it directly. If you begin to* think *about it, it is altogether missed."* Or, as our own sage, Walt Whitman would certainly have phrased it:

Just wander off by yourself
And look up from time to time
In perfect silence
At the stars....

Using the metaphor of the *Great Path*, the monk Ekai, in the
13th Century, put it into words as the essence of his
teaching, in four lines as simple as they are profound:

" *The Great Path has no gates;*
Thousands of roads enter it.
When one passes through this gateless gate
He walks freely between heaven and earth. "

The much-admired 4th Century hermit-poet Tao Yuan Ming,
born some 900 years before, shared much of Ekai's vision.
Finding rather early in life that society attached no particular value to his talents, and that he could no longer
"kow-tow for five pecks of rice a day", Tao Yuan Ming re-
turned to his own private world where he devoted himself
to the things which he loved; the world of nature, flowers,
children, poetry and wine, and his tiny garden with its
three paths and *five willows,* where, on summer evenings, he
might watch the rising of the Milky Way *over the eastern*
fence. For over 1500 years no Chinese scholar has been able
to read any reference to "five willows" or the "eastern
fence" without thinking immediately of Tao Yuan Ming's very
special little world, and the infinite serenity with which
the old hermit-sage contemplated both the earth and the
heavens:
"Somewhere there lies a deeper meaning...
I would like to say it,
But have forgotten the word..."

The 11th Century Sung Dynasty poet Su T'ung-Po, living perhaps in more troubled times, finds it difficult to achieve
the freedom and the serenity of the ancient sages; in his
moody *self*-awareness he is much closer to the doubt-ridden
and restless anxiety which plagues the modern world. Even
while contemplating the silent circling of the stars and
the majestic panorama of the Celestial River, he finds himself tormented by doubts concerning his own destiny:

DESCRIPTIVE NOTES (Cont'd)

"It is nightfall; the clouds have vanished;
The sky is clear,
Pure and cold...
Silently I watch the River of Stars,
Turning in the Jade Vault...
Tonight I must enjoy life to the full,
For if I do not,
Next month, next year,
Who can know where I shall be? "

One of the delights of Chinese poetry is the skillful use
of what one might call "multi-dimensional ideographs",
words and phrases which can express two or three different
thoughts simultaneously; each line of such a poem may be
read and understood at several different levels of meaning.
The Milky Way, with its deep symbolical undercurrents, is
often used in this way, creating effects which tend to re-
mind one of a surrealistic canvas by Salvador Dali. Consid-
er a simple poem by the 8th Century painter-poet Wang Wei,
one of the founders of the great Chinese tradition of landscape painting, and justly praised for his unrivalled ability to "paint pictures with words":

"The cold mountain fades into green twilight;
The autumn stream flows murmuring on....
Leaning on my gnarled staff beside the rustic gate,
In the rushing wind I hear the cry of the aged cicada. "

A simple evocation of autumn nostalgia, we may say, without
realizing that the *"autumn stream"* is almost certainly a
symbolic reference to the Milky Way, as well as to the
stream of *time,* whose image reappears again in the *"rushing*
wind", as the poet contemplates both the autumn of the year
and the autumn of his own life, both fading now into the
"green twilight". In the *"cold mountain"*, the *"gnarled*
staff" and the cry of the cicada, the poet creates visions
of rich symbolic significance to the Chinese reader, while
the image of the *"rustic gate"* evokes that indefinable mood
which the Japanese call *"sabi"*, for which there is no exact
English equivalent, but which might be roughly defined as
that direct inward perception by which we find deep significance or great artistic quality in some outwardly simple

DESCRIPTIVE NOTES   (Cont'd)

and unpretentious object. A diamond bracelet from Tiffany's
for example, might contain no *sabi*, while a woodcarving by
a simple uneducated fisherman might be packed to the (ahem)
gills with it.
    The image of the Milky Way, as both the River of
Heaven and the Great Path of life, appears frequently in
the works of the unrivalled Tu Fu of the T'ang Dynasty,
whose poems, more than those of any other writer, evoke for
us the very heart and mind of eternal China. Waiting for an
imperial audience on a spring night, Tu Fu watches the
River of Heaven moving over the *"ten thousand households"*
of the city, and wonders *"how long the night will last"*.
Visiting an old friend, he catches a glimpse of the heavens
over the roof and contemplates the fact that *"our lives
have moved on - as do the stars in clusters"*. Watching the
new moon suddenly engulfed in storm clouds, he raises his
eyes to the Milky Way, *"shining unchanging, over the freez-
ing mountains of the border"*, then sees that the same white
star-frost has covered his garden... In times of war he
writes of the drums and bugles of battle and the bitter
years of endless strife, but over it all he finds that the
stars still shine, remote and serene in the heavens, and
*"Over the Triple Gorge the Milky Way pulsates between the
stars.."* Spending a restless night, far from home, he
watches the night dew turn to thick mist, until *"one by one
the stars go out... only the fireflies are left...."*
    Tu Fu, as one of the most intelligent, perceptive,
and sensitive men of his age, naturally found it difficult
to conform to the demands of his society and, like Tao Yuan
Ming several centuries before, spent much of his life in
chronic insecurity or in official or unofficial retirement.
In his later years he became something of the eternal wan-
derer or traveling hermit, sailing up and down the great
rivers of China in the little boat which had become his
final home; the Milky Way appears in his poems of those
years as both the celestial river and the earthly river;
he himself as a traveler on both this earth and the world
beyond. One night in the early summer of 768 AD, Tu Fu
watched the eternal panorama of the stars from the deck of
his little craft on the vast Yangtze River, and wrote his
Night Thoughts While Traveling, a poem which has preserved
for us, after 1200 years, the most perfect picture of his

DESCRIPTIVE NOTES   (Cont'd)

last days, and remains possibly the loveliest farewell
gift that any poet has given to the world:

"A light breeze rustles the reeds
Along the river banks. The
Mast of my lonely boat soars
Into the night. Stars blossom
Over the vast desert of
Waters. Moonlight flows on the
Surging river. My poems have
Made me famous but I grow
Old, ill and tired, blown hither
And yon; I am like a gull
Lost between heaven and earth. "

TRANSLATED BY KENNETH REXROTH

The Milky Way, in American Indian legend, is the Path to
the land of the hereafter; the Algonquins saw the camp-
fires of the departed warriors in the bright stars along
the way.   Longfellow, in the *Song of Hiawatha,* tells us
that the Iroquois had a similar tradition:

"Showed the broad white road in heaven,
Pathway of the ghosts, the shadows,
Running straight across the heavens,
Crowded with the ghosts, the shadows,
To the Kingdom of Ponemah,
To the land of the hereafter...."

The mysterious *Kingdom* is evidently the subject of the
ancient chant or dirge attributed to the Chippewas:

"To the golden lodge of evening...
To the land of the departed..
Who knows their pathway?
Their lodge of evening?
All the Old Men have not seen it...
All the Wise Men know nothing of it..."

In these lines we find a humble and pleasing agnosticism
which many modern scientists and theologians might do well
to emulate. The concept of the Milky Way as the Road to
Heaven appears again in Milton as the *"Broad and ample road*
whose dust is gold, and pavement stars......The Way to

God's eternal house.."   The Norsemen had a very similar
tradition, and saw the Milky Way as the Path of the slain
warriors on their way to Valhalla; hence the name *Wuotanes
Weg* or "Wotan's Way", sometimes rendered *Vetrarbraut,* the
"Winter Path", or simply *"The Path of the Spirits".* In some
versions of the legend, the Milky Way is the *Asgard Bridge*
or Bridge to Valhalla, though in later tradition it is the
Rainbow *Bifrost* which becomes the celestial bridge, as in
the majestic closing scene of Wagner's myth-drama *Das
Rheingold.* In Anglo-Saxon lore the Milky Way is *Irenges Weg*
or Iringe's Way, honoring one of the descendants of the
legendary King Waetla or Ivalde, whose giant sons were said
to have constructed the ancient Roman Road from London to
Dover, still called *Watling Street* or *Walsyngham Way.* From
this legend is derived the name *Vatlant Street* or *Watlinga-
strete,* the "Path of the Waetlings".
    A similar legend, possibly of quite modern origin,
identifies the Milky Way with *King Arthur's Causeway,* the
road which in olden times ran from the mysterious ruins of
Glastonbury Abbey to the traditional site of Camelot at
South Cadbury. Somewhere in this legend-haunted country-
side, it is said, King Arthur rests in enchanted sleep, to
awaken in some future day of great national crisis, and
lead the Britons once again to victory over their enemies.
But this will only happen, it is prophecied, in the reign
of "King George, son of King George".
    According to R.H.Allen, the Milky Way was called the
Hulde Strasse or "Saint Hilda's Street" by the Midland
Dutch; it was the *Arianrod* or "Silver Street" to the Celts,
and the *Wiar Strasse* or "Weather Street" to the people of
Westphalia. It was the *Milch Strasse* in Germany, the *Melk-
path* in East Friesland, the *Strada di Roma* in medieval
Italy, and the *Linnunrata* or "Bird's Way" in Finland. In
Sweden it is the "Winter Street", and in the well known
lines by Miss Edith Thomas is depicted as the path of the
celestial pilgrims on their way to heaven:

> *"Silent with star-dust, yonder it lies -
> The Winter Street, so fair and so white;
> Winding along through the boundless skies,
> Down heavenly vale, up heavenly height...
> Faintly it gleams, like a summer road,*

DESCRIPTIVE NOTES (Cont'd)

When the light in the west is sinking low,
Silent with star-dust! By whose abode
Does the Winter Street in its windings go?
And who are they, all unheard and unseen -
O, who are they, whose blessed feet
Pass over that highway smooth and sheen?
What pilgrims travel the Winter Street? "

The Turks had a similar tradition with their title *Hagjiler*
Yuli, "The Pilgrim's Road"; various Arabic names such as
Al Majarrah and *Tarik al Laban* all mean something close to
"The Milky Path" or "Milky Track", while another Arabian
title, *Umm al Sama*, is translated "The Mother of the Sky".
R.H.Allen states that both the Eskimos and the Bushmen of
Africa called the Milky Way *The Path of Ashes*.

 In Old England it was the *Way of Saint James*, the
equivalent of the Spanish *El Camino de Santiago;* the name
originating, it is said, from a popular legend that Theo-
domir, Bishop of Idria, was guided by a miraculous star to
find the burial place of St.James in 835 AD. From the Field
of the Star or *"Campus Stella"* where the discovery was made
evidently comes the title *St.James of Compostella.* In still
another tradition, popular in the Middle Ages, the Milky
Way represented the Biblical *Jacob's Ladder* upon which the
angels descend to Earth. In Welsh legend the angels are
replaced by the powerful enchanters and spirits of Celtic
myth, and the Milky Way becomes the *Caer Gwydyon*, the celestial stronghold of Gwydyon, that curious figure of Welsh
lore who appears to be a combination hero, wizard, fairy,
and demigod. The Biblical *Mazzaroth* of the *Book of Job* is
possibly a reference to the Milky Way, though the identification with the Circle of the Zodiac now appears to be
more likely.

 The 19th century Finnish dramatist and poet Zachris
Topelius gives us a Milky Way legend which seems to echo
the ancient Chinese tale of the Heavenly lovers and the
River of Stars:

"They toiled and built a thousand years
In love's all powerful might;
And so the Milky Way was made -
A starry bridge of light...."

DESCRIPTIVE NOTES (Cont'd)

Al Biruni, in the 11th Century, gives the Sanskrit title of the Milky Way as *Akash Ganga,* "The Bed of the Ganges"; another name in use in North India was *Bhagwan ki Kachahri,* translated "The Court of God". In Roman tradition it was the *Coeli Cingulum* or "Heavenly Girdle"; sometimes it is referred to as the *Milky Circle,* and in one legend is identified with the blazing path scorched across the sky by the rash Phaeton when he attempted to drive the Chariot of the Sun. This legend is immortalized in the Roman name *Vestigium Solis.* Manilius, in the days of Julius Caesar, sees it as the path of the illustrious heroes to the celestial realms; hence the classical title *Heroum Sedes.* Still another Roman name, *Circulus Junonius,* honors the Queen of Heaven, and may possibly be traced back to the Egyptian legend of the goddess Isis, who is said to have created the Galaxy by scattering grains of corn across the sky. In one of the Homeric hymns we find a reference to *Galaxure,* "The Lovely One", possibly a personification of the Milky Way. In ancient Mexico she was the "Sister of the Rainbow".

Western poets have found the Milky Way an inexhaustible source of rich imagery, explaining Swift's satirical *Edict* of 1720 in which hopeful authors and poets are specifically forbidden to make any mention of it. To William Wordsworth, whose vision perhaps comes closest to that of the great Chinese poets, it was *"heaven's broad causeway paved with stars.."* The earlier English poet Edward Young spoke of it as *"this gorgeous arch with golden worlds inlaid.."* W.H.Hayne saw there

"Pure leagues of stars from garish light withdrawn
Behind celestial lace-work pale as foam...."

John Milton, in *Paradise Lost,* refers to

"................that Milky Way
Which nightly as a circling zone thou seest
Powder'd with stars....."

And Dante, in the *Paradiso,* finds it

"Distinct with less and greater lights
Glimmering between the two Poles of the world..."

Longfellow, in *The Galaxy,* sees the Milky Way as a

"...torrent of light and river of the air,
Along whose bed the glimmering stars are seen
Like gold and silver sands in some ravine....
..........The Spaniard sees in thee the pathway
Where his patron saint descended in the sheen
Of his celestial armor, on serene
And quiet nights
When all the heavens were fair..."

Matthew Arnold, in *A Summer Night*, contemplates the vast
and ageless world of the stars, and finds there

"Plainness and clearness without shadow of stain!
Clearness divine!
Ye heavens, whose pure dark regions have no sign
Of languor, though so calm, and, though so great,
Are yet untroubled and unpassionate;
Who, though so noble, share in the world's toil,
And, though so task'd, keep free from dust and soil!
I will not say that your mild deeps retain
A tinge, it may be, of their silent pain
Who have long'd deeply once, and long'd in vain -
But I will rather say that you remain
A world above man's head, to let him see
How boundless might his soul's horizons be,
How vast, yet of what clear transparency!
How it were good to abide there, and breathe free;
How fair a lot to fill
Is left to each man still!

The almost legendary Chinese sage Lao Tzu, speaking to us
across a gulf of 2500 years, finds the ultimate source of
wisdom in the contemplation of the infinite; in lines which
foreshadow some of the speculations of modern cosmologists
he tells us, in the *Tao Teh Ching* or "Book of Tao":

"Before the heavens and Earth existed
There was something nebulous,
Silent, infinite, unfathomable,
Without beginning or end,
Standing alone, changing not,
Ever present and revolving eternally,
The source of the ten-thousand things...

HEART OF THE GALAXY. The Great Sagittarius Star Cloud is a portion of the Hub of the Galaxy. The bright spot at the top of the print is the nebula M8. Lowell Observatory photo made with the 5-inch camera.

I do not know its name
And call it Tao;
If a name must be given,
I shall call it the Ultimate...

. . . .

The Way of heaven achieves all without striving;
Answers all without speaking;
Accomplishes all without deliberate design;
The net of Heaven is broad, with wide meshes,
Yet nothing slips through..

. . . .

He who walks in harmony with the eternal Way
Shall be the master of existence;
Knowing the ancient beginning
Is the essence of enlightenment.
This is the mystic unity of the Way..."

* * * * * * * * * * * * *

The probable composition of the Milky Way appears to have
first been suspected by the Greek philosopher Democritus
in the days of Socrates, though it seems that Pythagoras
also held the view that the Circle of the Galaxy was composed of vast numbers of faint stars. Manilius, at about
the beginning of the Christian era, refers to this theory
when he writes:

"Is the spacious band serenely bright
From little stars, which there their beams unite,
And make one solid and continued light? "

Diodorus of Sicily, a few decades before, had noticed that
the Milky Way formed a great circle around the sky, and
theorized that the Galaxy marked the line where the two
starry hemispheres had been joined. Aristotle is credited
with the view that the Milky Way was a "gathering of celestial vapors", but various other theories are also attributed to the great philosopher-scientist down through the
Middle Ages. Francis Bacon, in the time of Shakespeare,
wrote that

"The way of fortune is like the Milky Way in the sky,
which is a meeting or knot of a number of small stars, not
seen asunder, but giving light together...."

It was Galileo, of course, who set the matter at rest for all time when he made the first telescopic observations of the *Via Galactica* in 1609- 1610:

"I have observed the nature and material of the Milky Way. With the aid of the telescope this has been scrutinized so directly and with such ocular certainty that all the disputes which have vexed philosophers through so many ages have been resolved, and we are at last freed from wordy debates about it. The Galaxy is, in fact, nothing but a congeries of innumerable stars grouped together in clusters.. "

Galileo's conclusions were soon confirmed, with the exception of that naive and wistful expectation that the philosophers would henceforth be freed from "wordy debates" about it.

The idea that the Milky Way formed the "framework of the sidereal universe" was probably first suggested by Thomas Wright of Durham, England, in 1750; he pictured the galactic system as a vast disc-shaped aggregation in which the Sun was embedded among myriads of other stars. Several decades later, Sir William Herschel began to compile strong observational evidence in support of this view; by making star counts in various parts of the sky he found that the number of faint stars increased enormously as his survey approached the Milky Way, and that the structure of the galactic system must be something like a great disc or "grindstone". At the time, the distance of not a single star was known, so Herschel could only make sagacious conjectures about the probable dimensions of the whole stellar system. A century later, the studies of J.C.Kapteyn proved the essential correctness of Herschel's model of the Galaxy which was now estimated to contain at least 45 billion stars.

Harlow Shapley, in 1918, summarized the results of his attempts to measure the scale of the Galaxy by a study of its star clusters, and reported that "the method succeeds in the gross, but not in detail. We may confidently expect that analyses of star motions, star clouds, and individual stellar distances will in time reveal with more than present clarity the significance of our galactic system in the total material universe". Shapley's studies

made it clear that the approximately 100 known globular
star clusters were distributed in space in the form of a
roughly spherical system, and that the center of this vast
system was undoubtedly the center of the Galaxy. Since the
full extent of stellar obscuration was not realized in the
1920's, the over-all size of the Galaxy was much over-
estimated; Shapley's figure of about 300,000 light years,
however, is still a reasonably accurate value for the full
diameter of the Milky Way's family of globular clusters.
The actual galactic disc is now believed to be approximately 100,000 light years in diameter, and possibly 10,000
light years thick at the central hub; the Sun is located
some 30,000 light years out from the center. Probably at
least 300 billion stars populate the Galaxy, which is now
definitely known, from radio studies, to be a large spiral
resembling the Andromeda Galaxy M31. The exact type is not
quite certain, but is probably Sb or an intermediate Sb-Sc.
From a recent analysis of probable models by M.Schmidt at
Palomar, the total mass of the system appears to be about
180 billion solar masses; a rounded figure of "about 200
billion" often appears in modern texts.

THE GALACTIC CENTER. Looking toward the Sagittarius Star
Cloud, then, we are looking across some 300 light centuries
of space, toward the center of the Galaxy. For a fair-size
telescope, especially the rich-field variety, the region
is a truly fascinating area for exploration. "*These magnificent star clouds are the finest in the sky*" wrote Prof.
E.E.Barnard in 1913. "*They are full of splendid details;
one necessarily fails in an attempt to describe this wonderful region of star masses. They are like the billowy
clouds of a summer afternoon; strong on the side toward the
Sun, and melting away...on the other side. Forming abruptly
at their western edge against a thinly star strewn space,
these star clouds roll backwards toward the east in a
broadening mass to fade away into the general sky...In the
dimmer regions, below the great bright clouds, the dark
details become more interesting and delicate.....*" It was
Barnard who, in the summer of 1889, made the first really
successful wide-angle photographs of the Milky Way at Lick
Observatory, using the 6-inch f/7 Willard portrait lens
which had been purchased second-hand from a photographer in

DESCRIPTIVE NOTES   (Cont'd)

San Francisco. With this lens, and later with the 10-inch
Bruce doublet, Barnard and E.B.Frost compiled a superb
collection of Milky Way photographs, eventually published
as *A Photographic Atlas of Selected Regions of the Milky
Way*, in two volumes, by the Carnegie Institution in 1927.
In describing Plate 26 of the *Atlas*, which depicts the huge
Sagittarius Star Cloud, Barnard stated that *"These great
clouds...are the most magnificent of the galactic clouds
visible from this latitude. For many years I had been familiar with their remarkable structures, while comet-seeking
with a 5-inch refractor at Nashville. Previous efforts to
photograph them with several different lenses had been a
failure, mainly because the lenses were too slow......The
Willard lens, with its short focus and large aperture,
brought out for the first time the remarkable forms and
structural details of these immense star clouds...... The
whole star cloud is broken up with many rich structures
which it would be difficult to describe. Nebulous matter
seems to be spread through parts of this great cloud; but
perhaps this appearance may be due to masses of extremely
distant small stars..."*
Since Barnard's time, special extremely wide-angle
optical systems have been designed for sky photography. In
the remarkable photograph opposite, the entire sweep of
the central Milky Way is shown as portrayed with the extra
wide-field Henyey-Greenstein camera. The three dark spikes
are the supports for the camera which photographs the reflection of the sky in a bowl-shaped mirror. The Great
Sagittarius Cloud is just to the upper left of the inter-
section of the spikes, and the Great Rift is remarkably
well shown. The photograph should be compared with one of
a typical edge-on galaxy, as NGC 891 in Andromeda, shown on
page 155.
If we now approach the Sagittarius Cloud itself, we
will find a view resembling the photograph on page 1632;
the central portion is shown on a larger scale on page 1639.
Here lie untold millions of stars, clustering about the
very hub of our Galaxy; here lie stars beyond stars until
vision fails, and the fainter stars merely blend into an
unresolvable background. Here too, the telescope picks up
great obscuring masses and ill-defined dark clouds, some of
which appear to wind like black lanes through the bright

VIEW OF THE SOUTHERN MILKY WAY, looking toward the Center of the Galaxy, as photographed with the Henyey-Greenstein camera. The brightest portion is the Sagittarius Cloud.

starry clouds of the Galaxy. Somewhat north of the center
of the Cloud is one of the most prominent dark nebulae in
the sky, known as Barnard 86. It rims the west edge of the
compact little open cluster NGC 6520, and appears as a
distinct inky spot against the surrounding star-shimmer.
This is one of the few dark nebulae that may be appreciated
in amateur telescopes, and lies 2.7° north and slightly
west from Gamma Sagittarii. It may be seen near the center
of the photograph on page 1639, and is shown in detail on
page 1643. A few degrees to the west, the star clouds are
obscured by the dust masses of the Great Rift, which here
closely delineates the central line of the Galactic plane.
Beyond this obscured region, about 4° WNW from Gamma, lies
the actual nucleus of the Galaxy, at position 17425s2859;
the nearest bright star to the spot is the cepheid variable
X Sagittarii (4th mag) located 1.2° to the NNE. The chart
below shows star to about 8th magnitude; the grid squares
are 1° high with north at the top.
        In the search for the actual Galactic Nucleus, for-
ever obscured by the cosmic dust of the Great Rift, various
techniques have been developed, based on the fact that many

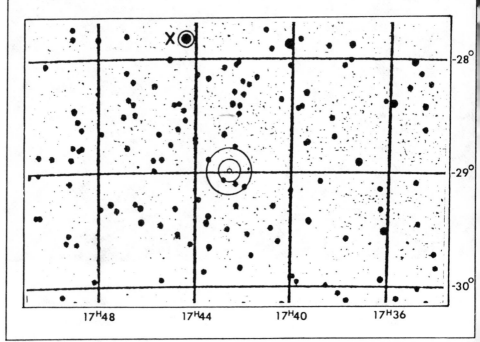

THE SAGITTARIUS STAR CLOUD. This is the richest section of the Milky Way. The small dark nebula B86 appears above the center of the print. Lowell Observatory photograph.

DESCRIPTIVE NOTES (Cont'd)

other types of radiation possess a much greater penetrating
power, as infrared, radio radiation, and X-rays. In 1951
the Soviet astronomer A.A.Kaliniak and his associates were
the first to detect infrared energy coming from the region
of the Galactic Center; later studies by J.Stebbins and
A.E.Whitford at Mt.Wilson revealed an elongated region of
infrared radiation extending some 8° along the central line
of the Rift, and centered at about -29° declination. From
studies by E.E.Becklin and G.Neugebauer at Mt.Wilson in the
summer of 1966, the actual Galactic Nucleus has been ident-
tified as a sharp "core" less than 15" in diameter, lying
in the center of a larger feature some 5' across. These
observations were made at wavelengths of 2.0 to 2.4 microns
or about 22,000 angstroms, far beyond the region of visible
light.

The infrared position coincides with the source of
microwave radiation received by radio telescopes as a faint
"cosmic static". The reception of this radiation was first
announced by K.G.Jansky of the Bell Telephone Laboratories
in 1932; further research has shown that it comes from a
number of areas of the sky, but is strongest along the line
of the Milky Way, particularly in the star clouds of Cygnus
and Sagittarius and in the neighboring constellations of
Ophiuchus and Scorpius. The greatest intensity comes from
the direction of the Galactic Hub, producing the strong
radio source called "Sagittarius A", now identified as the
actual Galactic Nucleus itself.

In using the term "nucleus" astronomers are not referring merely to the great starry hub of the Milky Way,
but to the much smaller nucleus believed to exist within
it. In observing the Great Andromeda Galaxy M31, for example, we find an extremely sharp increase in brightness in
the center, producing the appearance of a nearly stellar
nucleus. This object remains unresolvable even in large
telescopes and measures about 2.5" x 1.5"; the actual size
must be about 50 light years. Such an object may be something like a super-globular star cluster, containing perhaps more than 10 million stars. Observations with the Lick
Observatory 120-inch reflector indicate that the density of
such a nucleus is about 1700 solar masses per cubic parsec;
the average separation between stars would be only a few
hundred AU. In some unusual galaxies, we see very violent

DESCRIPTIVE NOTES   (Cont'd)

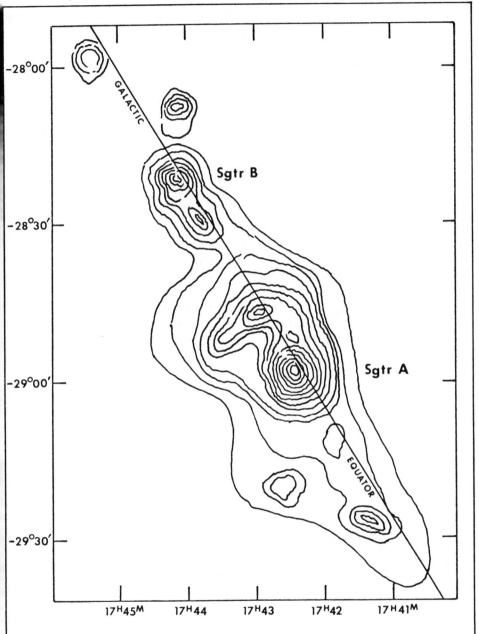

RADIO MAP OF THE REGION OF THE GALACTIC CENTER, from
observations at 3.75 centimeters. The strongest feature,
Sagittarius A, is the Galactic Nucleus.

outbursts of some sort occurring in the nuclei; the strange "Seyfert galaxies" (as M77 in Cetus) are mild examples of this eruptive activity, while really violent specimens (as M82 in Ursa Major) appear to be undergoing vast explosions involving masses of several hundred million suns. In such phenomena, according to various theorists, we may be seeing whole chains of supernovae, explosions of "hyperstars", or matter collapsing with titanic violence into black holes. Evidently, the nucleus of a galaxy is more than merely a super-cluster of stars.

From radio observations and infrared studies, we have learned of the existence of a similar nucleus in our own Galaxy, but all attempts to record it photographically have failed. In his book *Evolution of Stars and Galaxies* (1963), W.Baade described his work on the problem, using both the 48-inch Schmidt telescope and the 200-inch reflector at Palomar. Exposures of up to 7 hours were made on red sensitive plates, but revealed nothing except a number of exceedingly reddened globular star clusters. The computed position of the Galactic Nucleus appeared quite blank except for a faint sprinkling of distant stars. Baade originally estimated that the absorption in front of the nucleus must be some 9 or 10 magnitudes; a great under-estimate

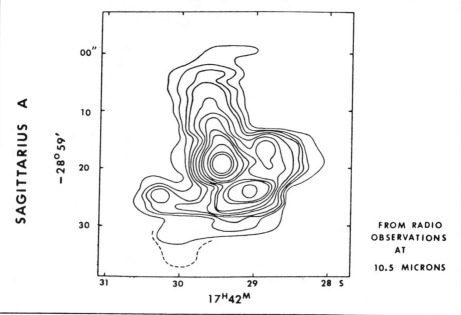

FROM RADIO
OBSERVATIONS
AT
10.5 MICRONS

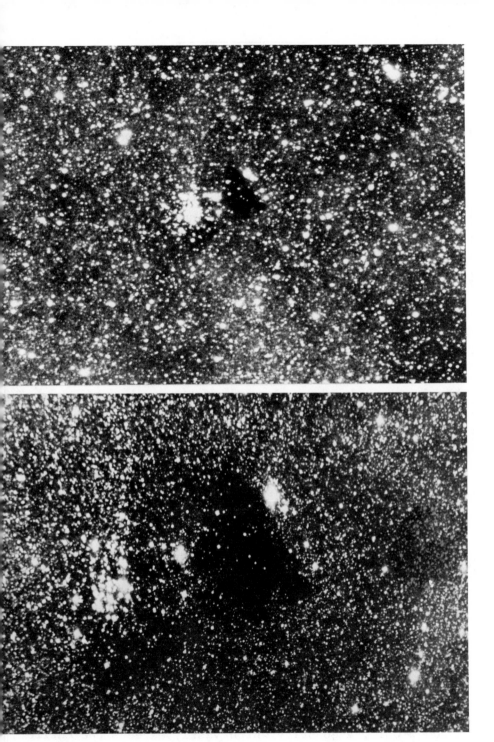

DARK NEBULA BARNARD 86, photographed (top) with the 13-
inch camera at Lowell Observatory, and (below) with the
Mt.Wilson 100-inch reflector.

as it turned out, since the figure is currently believed to be about 25 to 30 magnitudes! But although the mysterious Nucleus may remain forever unseen, its position has been very accurately determined by radio studies; the precise figures are: 17h 42m 29.5s -28° 59' 20". This corresponds to a galactic longitude of 327.8° on the older coordinate system; on the newer system the Nucleus is the zero point for computing galactic longitude, and is called 0°. The position is less than ½° from the Ophiuchus border.

At least three processes appear to contribute to the radio energy from the region: (1) Simple emission resulting from the combination of free electrons with hydrogen nuclei; (2) Synchrotron radiation from electrons accelerated to high velocities in strong magnetic fields; (3) The 21-centimeter radiation of neutral hydrogen, by which the spiral pattern of the Galaxy has been traced. It was through this latter radiation, also, that an expanding cloud or ring of gas has been detected in the region of the Center, apparently indicating that a violent outburst of some sort occurred there about 10 million years ago. The ring, some 300 parsecs in radius, is expanding at about 100 km/sec., and involves a total mass of about 100 million suns. The actual Galactic Nucleus is found to be about 10 parsecs in diameter with a 1.5 parsec "core"; the average density of matter in the core is some 10 million times that in our own familiar region of the Galaxy.

The region near the Galactic Center has been a "happy hunting ground", as Bart J.Bok phrases it, for astronomers looking for new and exotic molecular lines. Formaldehyde, carbon monoxide, and the OH radical have been identified, while the presence of methyl alcohol was detected in 1970. Writing in the March 1975 issue of the *Astrophysical Journal*, B.Zuckerman and B.E.Turner stated that a "truly astronomical source of *ethyl alcohol*" had been located in the cloud called Sagittarius B2; "preliminary estimates indicate that the alcoholic content....if purged of all impurities and condensed, would yield approximately 10^{28} fifths at 200 proof", a quantity which vastly exceeds the entire mass of the Earth (6 X 10^{27} grams).

(Refer also to the sections on the Milky Way in Aquila, Cygnus, Scutum, and Scorpius)

SCORPIUS

LIST OF DOUBLE AND MULTIPLE STARS

NAME	DIST	PA	YR	MAGS	NOTES	RA & DEC
2	2.5	274	46	5 - 7½	(β36) relfix, cpm Spect B3	15506s2511
I977	0.5	104	59	7½- 8	PA inc, spect F8	15527s2636
β38	4.4	348	45	8- 10½	slow PA dec, spect A0	15599s2453
ξ	0.9	358	66	5 - 5½	(Σ1998) AB binary	16016s1114
	7.4	53	59	- 7	46 yrs; PA inc, spect both F5; AC PA dec (*)	
Σ1999	11.4	100	50	7½- 8	4' S from Xi;	16017s1118
	80.7	83	59		Spect G8, K3	
β	0.5	132	59	2½- 9½	(β947) Dist dec,	16025s1940
	13.7	23	58	- 5	PA inc; spect B0; AC reflix (*)	
11	3.3	253	58	6 - 10	(β39) cpm; slight PA dec, spect A0	16048s1237
Δ199	16.2	298	20	6½- 13	(λ265) Spect A0,	16052s3857
	44.3	185	54	- 7	AC cpm.	
β949	0.4	191	25	7½- 7½	binary, 55 yrs; PA inc, spect F7	16057s0958
Hu 155	0.8	70	49	9 - 9	PA inc, spect F8	16058s1237
L6706	7.8	85	34	7 - 7½	(Brs 11) cpm, spect both dG2	16063s3232
I557	0.4	218	59	7½- 8	PA inc, spect F0	16063s3055
I1082	0.1		35	8 - 8½	(Daw 145) spect B5	16068s4000
	2.5	145	34	- 12		
ν	1.2	2	67	4 - 6½	(β120) all cpm;	16091s1921
	41.4	336	55	- 6½	multiple system;	
ν c	2.3	51	67	7 - 8	Spect B2, A0	
12	4.2	75	54	6½- 8½	(h4839) PA dec, spect B9	16092s2817
B307	1.5	228	52	6½- 10	spect B7	16107s2418
λ268	1.7	174	47	8½- 8½	PA inc, spect F0	16114s3901
Σ2019	0.2	265	58	8 - 8	(Rst 3936) PA dec	16115s1017
	22.2	154	38	- 9½	AC relfix, spect F8	
λ270	8.8	139	00	7½- 13	spect B8	16137s2938
λ272	0.2	274	60	7½- 7½	PA slow dec, spect F2	16161s3522

LIST OF DOUBLE AND MULTIPLE STARS   (Cont'd)

NAME	DIST	PA	YR	MAGS	NOTES	RA & DEC
55G	23.0	319	00	5½- 6½	(Brs 12) easy cpm pair; spect F5,F8	16164s3047
I 91	15.1	298	59	6½- 9	Dist inc, PA dec, Spect A0	16172s3918
Sh 225	47.1	333	16	7 - 8½	spect B9	16172s1956
I 562	0.8	202	52	7½-10½	PA inc, spect A0	16174s2924
Sh 226	0.2	195	59	7½- 8	(Hh504) (B1808)	16175s2000
	12.7	22	40	- 8	Spect A0	
σ	20.0	273	59	3 - 9	(Sh 224) Spect B1 Primary var (*)	16181s2529
h4843	12.4	267	33	7½- 12	Spect F5	16182s3311
β624	1.1	320	59	8 -9½	relfix, spect A2	16199s2300
h4845	1.9	130	42	8 - 8½	PA slow dec; Spect F0	16204s4108
h4848	6.3	154	52	7 - 7½	relfix, cpm, spect A0	16207s3305
	92.0	357	00	- 9		
B868	0.1	130	47	6 - 6	spect B8	16211s3727
h4850	5.4	354	57	6 - 7	cpm, dist dec, spect both G0	16215s2935
B310	3.8	206	27	7½-13½	spect B8	16225s2918
λ277	0.9	194	46	8 - 9½	(I 94) PA dec, spect G0	16226s2949
α	3.0	275	59	1 - 6½	ANTARES. Relfix; Spect gM1 (*)	16264s2619
δ146	8.5	131	34	5½- 12	(Δ202) spect B1	16282s4143
	58.0	180	21	- 10		
I 95	1.6	359	52	7½- 9½	relfix, spect A0	16298s3325
h4867	16.4	294	34	6 - 9	cpm; spect B3	16349s4319
h4878	0.1	60	43	9 - 9	spect G0	16371s2754
	8.3	359	38	- 8½		
R283	0.2	12	60	7 - 7½	PA dec, spect G5	16392s3659
β1116	2.1	8	43	7 - 10½	cpm, PA inc, spect A0	16412s2722
	24.2	197	33	- 14		
Δ209	23.4	141	59	7½- 8½	PA slow dec, spect A5, A	16449s3648
I 99	1.1	72	42	8 - 8½	PA slight dec, spect A0	16460s4352
λ288	19.3	136	00	7 - 13	spect A2	16461s3356
	20.0	162	00	- 13½		

SCORPIUS

LIST OF DOUBLE AND MULTIPLE STARS   (Cont'd)

NAME	DIST	PA	YR	MAGS	NOTES	RA & DEC
I 993	3.6	99	36	7 - 11	spect K0	16463s3134
h4889	6.8	6	51	6 - 8	cpm, relfix; spect B9	16476s3725
λ291	2.7	2	33	7 - 11	slow PA dec, spect A0	16491s2531
B1833	0.4	71	60	5½- 7	(λ293) Spect B0;	16505s4143
	56.6	21	00	- 7½	in cluster NGC 6231	
B1833c	6.5	47	20	7½- 13	(λ294) (h4892)	
	8.6	300	33	-10½		
λ296	20.0	242	00	7 - 13	in cluster NGC 6231	16506s4139
λ297	13.4	129	00	7 - 12	" " " "	16506s4145
	24.0	120	00	- 12		
B1834	4.3	312	31	6½- 13	Spect WC6 + 08; In cluster NGC 6231	16508s4144
h4893	7.3	52	34	8 - 10	In NGC 6231	16509s4145
I 576	5.0	269	34	6 - 12	In cluster H12; spect 07e	16515s4104
I 577	0.8	14	44	8 - 10	spect G5	16519s3108
Rst 1913	0.8	140	44	7½-10½	spect K0	16531s3114
Rst 5421	7.2	129	45	7 - 13	In cluster H12;	16534s4026
	7.6	252	45	- 9½	spect B3	
	15.7	238	03	- 10½		
λ313	2.4	230	42	8 - 9½		16536s3818
Jsp 700	9.1	130	45	7½-11½	spect B5	16545s3755
RV	6.0	325	25	7 - 13	(I 1304) Primary cepheid variable	16551s3332
λ315	0.2	92	57	7 - 7	(B885) binary, 40 yrs; PA inc, spect A3	16555s3732
Ho 410	9.6	347	32	7- 12½	spect A2	16585s3318
WNO 5	17.6	296	59	8 - 8½	optical, PA & dist inc; spect G0	17021s3342
Hd266	6.5	84	31	6 - 10	(L7123) spect A2	17030s3709
	43.3	187	00	- 13		
B894	2.8	356	27	6½-12½	spect B0	17031s3523
Co 208	0.5	139	34	7 - 9	spect A0	17038s4422
I 407	0.3	179	52	7½- 8	PA dec, spect A3	17046s4133
Hwe 86	2.8	143	55	7 - 8½	relfix, spect B5	17105s3814
h4926	14.4	334	33	6½-10½	(L7171) spect G5	17110s3943
	16.9	210	33	-11½		

SCORPIUS

LIST OF DOUBLE AND MULTIPLE STARS   (Cont'd)

NAME	DIST	PA	YR	MAGS	NOTES	RA & DEC
λ322	0.2	307	54	7½- 7½	PA dec, spect B3	17126s3341
I 408	1.7	168	43	7 - 9	spect B5	17128s4217
β1119	0.3	297	59	7½- 8	PA dec, spect G5; on Ophiuchus-Scorpius border	17141s3007
Hd268	19.5	174	34	6½- 12	spect B6e	17151s3230
β416	1.5	181	68	6 - 7½	(Mel 4) (h4935) AB	17154s3457
	30.8	136	48	- 10	binary, 42 yrs; PA dec, spect dK5; ABC all cpm	
L7211	0.5	20	47	7½- 7½	(Hd 269) slight PA inc? spect A0	17158s4410
B908	2.7	114	27	6½- 12	spect B4	17193s3745
	12.0	208	27	- 13½		
I 1317	0.3	190	47	8½- 9	PA inc, spect A0	17193s3809
H1d 28	3.6	233	38	8½- 9	spect K0	17213s3030
I 595	0.8	150	43	7½- 9½	spect A0	17215s3431
Δ217	13.5	169	52	6½- 8½	relfix, cpm, spect B9	17254s4356
B342	0.4	85	59	7 - 7½	PA inc, spect A2	17260s3829
Hwe 39	4.4	321	34	7 - 9½	(Ho 646) spect B5	17268s3340
	14.6	314	34	-11½		
	58.7	29	35	- 9½		
Hwe 87	3.1	232	51	7½- 8½	relfix, spect F8	17278s3859
H1d 136	1.1	109	37	8 - 8½	spect B9	17282s4100
I 603	1.2	80	39	7½- 8½	spect B9	17296s4529
h4962	5.5	102	33	5½-10½	(Ho 647) relfix; in cluster NGC 6383	17315s3233
	13.3	83	07	-10½		
B915	0.2	175	59	8½- 8½	(Co 216) PA inc, spect F0	17342s3745
	14.0	194	40	- 9		
δ148	1.2	138	43	8 - 9½	spect F5	17373s3537
φ341	0.1	27	59	7 - 7	PA inc, spect A2	17411s4243
Co 220	13.5	235	20	7½- 9½	spect B8	17433s4329
I 1336	0.2	208	60	7 - 7	binary, 34 yrs; PA inc, spect B9	17437s3804
B1868	1.2	218	44	8½- 12	spect A0	17451s4521
RY	2.2	102	38	8 -11½	Primary cepheid; spect G0- G7	17476s3342
	13.3	221	38	-10½		
Stn 37	9.9	190	52	6½- 7½	relfix, spect A0; NGC 6451 in field	17480s3033

LIST OF DOUBLE AND MULTIPLE STARS (Cont'd)

NAME	DIST	PA	YR	MAGS	NOTES	RA & DEC
β1123	0.1	341	59	7 - 7	PA inc, spect B9; in cluster M7	17500s3443
λ342	0.4	242	59	6½- 6½	PA dec, spect gG8; in cluster M7	17500s3453
B1871	0.1	111	57	6½- 7½	PA inc, spect B9; in cluster M7	17501s3444
V453	13.5	168	44	6½-12½	(Jsp 748) Ecl.bin; spect B0	17530s3228
Co 222	3.5	124	52	7½- 9	relfix, spect F5	17534s3956
I 1013	0.7	151	59	6½- 8½	PA dec, spect A0	17545s3908
R306	3.6	17	47	7 - 9½	(λ343) spect A0	17546s3601
	11.8	99	31	- 13		

LIST OF VARIABLE STARS

NAME	MagVar	PER	NOTES	RA & DEC
α	0.88- 1..	1730:	ANTARES. Semi-reg; Spect M1 (*)	16264s2619
μ¹	3.0--3.3	1.4403	Ecl.bin; lyrid; spect B2 + B7 (*)	16485s3758
σ	2.8--2.9	.24684	β Canis Major type; Spect B1; also visual double star (*)	16181s2529
R	9.7--15.5	223	LPV. Spect M3e; in M80 field.	16147s2249
S	9.8--15..	177	LPV. Spect M3e; in M80 field.	16147s2246
T	7.0----	---	Nova 1860 in globular cluster M80 (*)	16141s2251
U	8.8--18..	---	Recurrent nova; 1863, 1906, 1936, 1979 (*)	16196s1746
X	9.9--14.3	200	LPV. Spect M2e	16056s2124
Z	8.7--13.4	352	LPV. Spect M6e--M7e	16030s2136
RR	5.1--12.3	279	LPV. Spect M6e--M8e (*)	16534s3030
RS	6.1--12.9	320	LPV. Spect M5e--M8e	16520s4501
RT	7.2--15.6	449	LPV. Spect M6e--M7e	17001s3651

NAME	MagVar	PER	NOTES	RA & DEC
RU	7.8--13.7	368	LPV. Spect M7e	17388s4344
RV	6.7--7.4	6.0613	Cepheid; Spect F5--G5; Also visual double star	16551s3332
RW	8.6--15..	389	LPV. Spect M5e	17115s3323
RX	9.9--13..	280:	LPV.	16089s2446
RY	7.5--8.4	20.315	Cepheid; Spect G0--G7; Also visual double star	17476s3342
RZ	8.2--12.8	160	LPV. Spect M3e--M4e	16016s2358
SS	8.5--9..	Irr	Spect K2	16520s3233
ST	9.0--13.0	194	Semi-reg; spect S4	16334s3108
SU	8.0--9.4	414:	Semi-reg; Spect N0	16374s3217
SV	8.7--14.9	256	LPV. Spect M3e	17450s3541
SW	9.5--13..	261	LPV. Spect M5e	17217s4347
SX	8.5--9.5	Irr	Spect N3	17441s3541
SY	8.8--12..	235	LPV. Spect M4e	17505s3424
SZ	9.9--12.0	321	LPV. Spect M6e	16531s3934
TU	8.6--12..	373	LPV. Spect M7e	17043s3146
WW	9.2--13.0	431	LPV. Spect M6e	16242s3112
YY	9.3--14..	327	LPV.	16351s2828
AH	7.1--11..	714	Semi-reg; spect M3e	17080s3216
AI	9.0--12.5	71	RV Tauri type; Spect G0--K2	17530s3348
AK	8.7--10.2	Irr	RW Aurigae type; Spect F5 pec	16514s3649
AX	9.0--11..	138	Semi-reg; spect M6	16388s2701
BM	6.0--8.1	850	Semi-reg; spect K0; in M6	17377s3211
FV	7.9--8.6	5.7279	Ecl.bin; spect B9	17105s3248
GH	9.5--12..	277	LPV. Spect Me	17494s4338
KP	9.4--16..	---	Nova 1928	17409s3542
V380	9.5--10.5	187	Semi-reg; spect M0	16529s3014
V382	9.0--16..	---	Nova 1901	17486s3525
V393	7.8--8.6	7.7125	Ecl.bin; Spect B9	17455s3502
V449	7.0--7.6	38.8:	Ecl.bin? Spect A2	17337s3206
V453	6.5--7.0	12.004	Ecl.bin; lyrid, spect B0; Also visual double star	17530s3228
V482	8.1--9.1	4.5278	Cepheid; spect F5	17275s3334
V499	8.8--9.4	2.3333	Ecl.bin; lyrid, spect B5	17258s3258
V500	9.0--10.0	9.3166	Cepheid; spect K0	17454s3028
V635	8.5--10..	Irr	Spect S7	17189s4142
V636	6.3--7.0	6.7966	Cepheid; spect G5	17191s4534

LIST OF VARIABLE STARS (Cont'd)

NAME	MagVar.	PER	NOTES	RA & DEC
V696	7.5--16..	---	Nova 1944	17498s3549
V697	8.0--16..	---	Nova 1941	17479s3724
V701	8.2--8.9	.7619	Ecl.bin; lyrid, spect B5	17311s3228
V703	7.6--8.0	.1152	Cl.var; Spect F0	17390s3230
V707	9.9--15..	---	Nova 1922	17450s3636
V711	9.7--16..	---	Nova 1906	17508s3421
V718	9.0--10.3	200?	Ecl.bin?	16102s2221
V719	9.8--18..	---	Nova 1950	17422s3359
V720	7.5--18..	---	Nova 1950	17486s3523
V721	9.5--18..	---	Nova 1950	17391s3439
V722	9.4--13..	---	Nova 1952	17453s3457
V723	9.5--21..	---	Nova 1952	17467s3523
V727	8.0--9.0	Irr	Spect M1	17123s3303
V728	5.0--13..	---	Nova 1862	17355s4527
V760	7.3--7.7	1.7309	Ecl.bin; spect B8	16215s3446
V764	8.6--9.1	6.8084	Ecl.bin; spect A2	17524s4509
V818	11---14	Irr	Scorpius X-1; strong X-ray source (*)	16171s1531
V825	8.0--13..	---	Nova 1963	17466s3332

LIST OF STAR CLUSTERS, NEBULAE AND GALAXIES

NGC	OTH	TYPE	SUMMARY DESCRIPTION	RA & DEC
----	I.4592	□	vvL,eF, Irr; 170' x 45' Surrounds Nu Scorpii	16091s1920
6072		◎	pF,R,1bM; Mag 14, Diam 50" x 30" with 17½^m center star	16097s3607
6093	M80	⊕	Mag 8; Diam 7'; Class II; vB,pL,R,eC,eRi; stars mags 14.... Incl. Nova T Scorpii (1860). 4.2° NNW from Alpha Scorpii (Antares) (*)	16141s2252
6121	M4	⊕	Mag 7.4; Diam 20'; Class IX; B,vL,iR, stars mags 11..... 1.2° W from Antares (*)	16206s2624

LIST OF STAR CLUSTERS, NEBULAE, AND GALAXIES   (Cont'd)

NGC	OTH	TYPE	SUMMARY DESCRIPTION	RA & DEC
6124	△514	☼	B,pL,R, Mag 8; Diam 25'; Class E; about 100 stars mags 9....12 (*)	16222s4035
6144	10⁶	⊕	Mag 10; Diam 3'; Class II; cL,mC,gbM; stars mags 14.... ½° NW from Antares	16242s2556
6139	△536	⊕	Mag 10; Diam 2'; Class II; B,pL,R,bM; stars mags 16....	16243s3844
----	I.4605	□	eF; Irr; Diam 20'; surrounds 5ᵐ star 22 Scorpii	16272s2500
6153		◎	vS,F; Diam 20"; Mag 11½; eF central star	16280s4008
6178		☼	S,B,Irr; Diam 4'; Class F; About 12 stars mags 8.....	16321s4531
6192	△483	☼	pL,pRi, Diam 7'; Class F; About 75 stars mags 11...14	16368s4317
6222		☼	L,Ri, 1bM; Diam 3'; Class D; About 25 stars mags 12...13	16471s4439
6227		⸙	Rich Milky Way field N.p. NGC 6231. Not a true cluster	16481s4049
6231	△499	☼	! vB,L,pRi; Mag 6; Diam 15'; Class E; With 8 stars mag 7 + 100 stars mags 10....13; Contains Wolf-Rayet and P Cygni type stars (*)	16507s4143
6242	△520	☼	B,L,Ri; Diam 10'; Class F; About 45 stars mags 8....11	16522s3925
----	H12	☼	vvL,B,E; Irr; Diam 40'; Class C; about 200 stars in rich Milky Way field 1° NNE from cluster NGC 6231; Faint neby IC 4628 inv. (*)	16527s4038
6259	△456	☼	B,vL,vRi; Daim 15'; Class E; About 100 stars mags 11....	16571s4436
6266	M62	⊕	Mag 6.5; Diam 6'; Class IV; vB,L,pRi,1E, gbM; stars mags 11....14. Unusual irregular outline. On Scorpius- Ophiuchus border (*)	16581s3003

LIST OF STAR CLUSTERS, NEBULAE, AND GALAXIES   (Cont'd)

NGC	OTH	TYPE	SUMMARY DESCRIPTION	RA & DEC
6268	Δ 521		B,pL,cRi; Diam 10'; Class F; About 30 stars mags 10....	16586s3939
6281	Δ 556		L,pRi,1C; Diam 9'; Class D; Triangular outline, about 25 stars mags 9....11. 2.2° E from Mu Scorpii	17014s3749
----	I.4637		vS,F, appearance nearly stellar; Diam 10"; Mag 13½	17016s4048
6302		□	pB,pS,mE; Diam 2' x 1' with 10^m B6 central star; much flattened Fig-8 shape; poss unusual planetary neb.	17105s3703
6318	Δ 522		pL,R,Ri; Diam 5'; Class G; About 60 stars mags 12...14	17143s3924
6322			vL,1C,pRi; Diam 12'; Class E, about 25 faint stars	17152s4250
6334		□	vL,cF,Irr; complex field of neby 30' diam, with 8^m A0-type star; 2½° NW from Lambda Scorpii (*)	17172s3601
6337		◎	S,F, bluish ring-shaped neb 38" x 28" with eF central star	17189s3825
___	H14		S,F, Diam 6'; Class F; about 40 stars mags 15.... 2.4° SW from Lambda Scorpii	17212s3859
6357		□	F,L,E, Diam 4' x 1' with filaments extending to 50'; incl. 10^m B-type star	17213s3407
----	H16		L,F, Diam 15'; Class E; About 20 stars mags 10.... In field of Lambda Scorpii	17274s3649
6383			pL,F, Diam 6'; Class E; about 12 faint stars sur-rounding 6^m double star h4962	17314s3233
6380		⊕	vS,eF, Diam 2'; stars mags 16.... with 8½^m star on S edge; 2° South from Lambda Scorpii	17319s3902

LIST OF STAR CLUSTERS, NEBULAE, AND GALAXIES   (Cont'd)

NGC	OTH	TYPE	SUMMARY DESCRIPTION	RA & DEC
6388	△457	⊕	vB,pL,R; Mag 7; Diam 4'; Class III; stars mags 17...	17326s4443
6400	△568	⊙	pL,pRi,Irr; Diam 7'; Class D; About 25 stars mags 9...10	17361s3655
6404		⊙	F,L,pRi, Diam 4'; Class G; About 25 stars mags 13...	17363s3313
6405	M6	⊙	!! L,Irr;1C; Mag 6, Diam 25'; Class E; about 50 stars mags 7...10; incl BM Scorpii (*)	17368s3211
----	H17	⊙	pB,pL,pRi; Diam 10'; star 8^m + 20 stars mags 10.. Class D	17372s4003
6416	△612	⊙	vL,Ri,1C; Diam 20'; Class E; 25 faint stars. Near M6.	17410s3220
----	I.4663	◎	vF,S; Mag 13, Diam 15"; appearance nearly stellar	17417s4453
6425		⊙	pS,1Ri,1C; Diam 6'; about 15 stars mags 11....	17438s3125
6441	△557	⊕	Mag 8, Diam 3'; Class III; B,pL,R,vRi; stars mags 17... $3\frac{1}{2}^m$ star G Scorpii prec $4\frac{1}{2}'$	17468s3702
6451	13^6	⊙	pL,pRi,mC; Diam 6'; Class E; about 50 stars mags 10...13	17474s3011
6453		⊕	F,S,R; bM; Mag 11, Diam 1'; Class IV; in field with M7	17480s3437
6475	M7	⊙	!! vvL,vB,pRi; Mag 5; Diam 60'; Class E; about 50 stars mags 7...11. Fine naked-eye cluster (*)	17507s3448
----	H18	⊙	F,pL,Irr; Diam 15'; Class D; about 80 stars mags 10.... In field with cluster M7	17530s3517
6496		⊕	Mag 10; Diam 3'; Class XII; pL,1E,gbM; on border of Cor. Australis	17555s4415

SCORPIUS

DESCRIPTIVE NOTES

ALPHA Name- ANTARES, from the Greek αντί Αρης, the "Rival of Mars", so called from its conspicuous red color. The Roman name *Cor Scorpionis*, the "Heart of the Scorpion", is the equivalent of the Arabic title *Kalb al 'Akrab*, and the French *Le Coeur du Scorpion*. Antares is the 15th brightest star in the sky. Magnitude 0.92 with slight irregular variations; Spectrum M1 Ib; Position 16264s2619. Opposition date (midnight culmination) occurs about May 30. There are only two M-type supergiants among the 1st magnitude stars; the other example is Betelgeuse in Orion.

In Greek and Roman tradition the constellation represents the Scorpion whose sting caused the death of Orion; in another legend it appears as the monster which frightened the horses of Phaeton when that bold youth attempted to drive the Chariot of the Sun. Scorpius is one of the most prominent and most striking of the summer constellations, and has been known by its present name throughout recorded history. According to an ancient Persian tablet from Susa, Scorpius was known as *Akrabu* and was associated with the 8th month of the year, *Arah shamna*. Another name in Persian tradition was *Kazhdum*, which refers to a scorpion or a scorpion-monster; the Turkish name *Uzun Koirughi* seems to mean "The Long-tailed One". The ancient Akkadian name *Girtab* has been translated "The Stinger".

The Biblical *Chambers of the South* in the *Book of Job* may be a reference to Scorpius, since the Pleiades, almost directly opposite Scorpius on the celestial sphere, are mentioned in the same passage. The Jewish writer Aben Ezra, in the 12th Century, identifies Scorpius with the Hebrew *Kesil* or *Akrabh*, the tribal emblem of *Dan*, usually depicted as a basilisk or crowned serpent. The constellation, in very early times, occupied the area which is now divided into Scorpius and Libra; the names for α *and* β Librae, the "Southern Claw" and the "Northern Claw", still preserve the ancient tradition. Aratus calls the entire group Τέρας μέγα, the "Great Sign"; Riccioli labels it with the phrase *Acrobo Chaldaeis*, apparently honoring the priestly stargazers of ancient Babylonia. In Mesopotamian lands, and on Phoenician cylinder seals, we find the constellation portrayed as a half-human "scorpion-man"; amulets engraved with the figure of this mysterious divinity were probably

1655

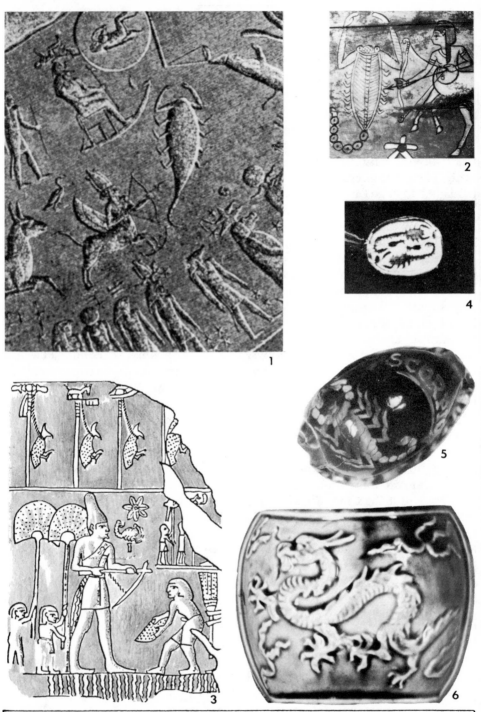

SCORPIUS IN ANCIENT TRADITION. (1) On the Egyptian zodiac at Denderah. (2) On a Ptolemaic age sarcophagus. (3) On the mace-head of the Pharaoh Ip. (4) Scarab of the 18th Dynasty (5) Carved cowrie shell. (6) The Chinese "Azure Dragon".

intended to protect the wearer from scorpion stings. R.H.
Allen states that other Babylonian titles for Scorpius were
Bilu-sha-ziri, "The Lord of the Seed"; *Lugal Tudda*, the
"King of Lightning" or the "Lusty King", and *Kakkab Bir*,
"The Vermillion Star", the last name referring more speci-
fically to Antares itself.

On the famous Egyptian Zodiac at the Temple of Denderah, Scorpius is shown (Fig.1) much as it appears on the
charts of modern astrologers; another portrayal of the
constellation is seen on a sarcophagus of late style, probably dating from the Ptolemaic period (Fig.2) and now in
the British Museum. In a very ancient Egyptian legend we
are told that Horus, son of Osiris, was once slain by a
scorpion's sting, but was restored to life by his mother
Isis, with the aid of certain magical formulae supplied by
Thoth, god of wisdom, science, literature, and astronomy.
From this legend grew the custom of protecting buildings,
homes, and temples from the attacks of evil spirits by the
presence of engraved tablets or stelae called "Cippi of
Horus". The finest known example is the *Metternich Stele*
found during excavations in Alexandria in 1828, and apparently erected to protect the city temple from the attacks
of scorpions. According to the text, it was made for the
prophet and scribe Ankh-Psemtik, about 370 BC.

The symbol of the Scorpion was one of the oldest of
the Egyptian hieroglyphics, and appears in inscriptions
pre-dating the beginning of the 3rd millenium BC. According
to Prof. E.A.Wallis Budge, it was pronounced *"serq"*, and
was the personal symbol or signature of the very early
Pharaoh called *"Ip"* or *"The Scorpion King"* who ruled in
Upper Egypt shortly before the unification of the "Two
Lands" under Narmer or Menes, founder of the 1st Dynasty,
about 3100 BC. The mace-head of the Scorpion King, with his
hieroglyphic symbol, was found in the ruins of ancient
Hierakonpolis, and is now in the Ashmolean Museum at Oxford
(Fig.3). The figure of the scorpion appears also on scarabs
and Egyptian jewelry items, undoubtedly for use as a talisman; the scarab shown in Fig.4 is thought to date from the
time of the 18th Dynasty, about 1400 BC. A similar carving
on a cowrie shell, the work of a modern Indian lapidary, is
shown in Fig.5. Evidently, the ancient traditions have
lost none of their power.

DESCRIPTIVE NOTES   (Cont'd)

The ancient Egyptian priesthood of sorcerers and healers called the "Charmers of Selket" was dedicated to the service of the Scorpion goddess *Selket* or *Selkis*. The goddess herself appears to have been one of the many manifestations of Isis; hence Antares became a star sacred to Isis. The English archeologist F.C.Penrose believed that a number of early Greek temples were oriented to the rising of Antares, including the temple at Corinth (about 750 BC) and the first Temple of Apollo at Delphi (about 630 BC). Antares was one of the four *Royal Stars* of Persia, the other three being Regulus, Aldebaran, and Fomalhaut. But the reverence for the great star was tempered by a certain degree of fear; the name *Insidiata*, given by Philip Caesius in his *Coelum Astronomico-Poeticum* of 1662, signifies "The Lurking One". Pliny's term, *rutilans*, appears in the lists of the *Rudolphine Tables* and simply means "with a red glow".

It is interesting to find that the Pre-Columbian cultures of ancient America also saw a scorpion in the pattern of the constellation; according to V.W.von Hagen, the Mayan title *"Zinaan ek"* is translated "The Stars of the Scorpion".

George F.Kunz, in *The Curious Lore of Precious Stones* (1913) states that the figure of Scorpius was one of the most popular subjects for ancient and medieval amulets:

"....in the case of the bezoar stone, a generally recognized antidote for all sorts of poisons, it was held that the scorpion's bite could be most effectively healed by a bezoar upon which the creature's figure had been cut during the time when the constellation Scorpio was in the ascendancy".

The mysterious *bezoar stone*, incidentally, is not any recognized variety of mineral or gem; according to medieval writers it is a type of hard concretion occasionally found in the digestive organs of certain ruminant animals, particularly the wild goat of Persia! Pliny gives his opinion that "red agate" or carnelian was an especially effective talisman against scorpions; the medieval Arabic writer Ahmad ibn Yusuf, some 750 years ago, further specified that the image of the scorpion should be engraved on the gem *"when the Moon is in that sign"*.

Scorpius, in the Middle Ages, became something of a symbol of unyielding fortitude, from the legend that the

creature, if surrounded by enemies, would sting itself to death rather than submit to capture. Lord Byron refers to this tradition when he writes
"The mind that broods o'er guilty woes
Is like the Scorpion girt by fire......"

In ancient China, however, the constellation was not seen as a scorpion; it was the major portion of the large and regal figure of the *Azure Dragon* or *Dragon of the East* (Fig.6) while Antares itself was titled *"The Fire Star"*. On some oriental charts the region of Antares - and the bright stars nearby - is labeled *Ming T'ang*, the "Hall of Light" or the "Emperor's Council-Hall". The term *Fang*, a celestial hall or "room" was applied to the pattern formed by Beta, Delta, Pi, and Rho, which we now identify as the Scorpion's head. It was in this region that a bright nova is said to have appeared in the summer of 134 BC, causing Hipparchus to begin the compiling of a comprehensive catalogue of stars. The authenticity of this story, however, has been questioned by some modern astronomers, who suspect that the account of this object may be confused with the bright comet of 136 BC, observed by the Romans. Another archaic Chinese text describes an occasion when *"the Sun and the Moon have not met harmoniously in Fang"*; this may be a reference to a solar eclipse.

A curious type of artifact called a *"ju-i"* is found occasionally in ancient Chinese tombs; it takes the form of a curved and twisted rod or sceptre, sometimes made entirely of jade but more often of precious wood, lacquer, or animal horn, frequently inlaid with jade stones. It appears to have been some variety of ceremonial sceptre, used by the Emperor and the court nobles during important rituals. The name literally means *"as you wish"* or perhaps *"may you obtain your desires"*, suggesting that the *ju-i* was possibly used in some type of prayer to Heaven, or in supplication to the Emperor. Dame Una Pope-Hennessy, in her book *Early Chinese Jade*, has suggested that the form of the *ju-i* was derived from the star pattern of the celestial Azure Dragon which we now call Scorpius. On this hypothesis, the flattened flower-like head of the *ju-i* symbolizes the portion of the constellation called *Fang*, though it may also represent, as some Chinese scholars claim, the sacred *ling chih*

fungus which became one of the Taoist symbols of immortality. There are references in old Chinese texts to various
types of wands or staffs used by wizards and diviners who
could "read the secrets of the stars". The Chinese undoubtedly also saw additional symbolic significance in the fact
that the Azure Dragon is seen in the sky with his tail
still immersed in the Milky Way; he is emerging from his
home in the Celestial River as he prepares to descend to
Earth, to confer great blessings on the sages who do him
reverence. The Chinese dragon, of course, is not the hideous maiden-devouring monster of medieval Christian myth;
he is the wise and majestic incarnation of the awesome
power and infinite splendor of Nature. When portrayed with
5 claws, as in Figure 6, photographed from a Chinese dragon
vase, he becomes the *Imperial Dragon*, a symbol of the King
or Emperor, and the *Mandate of Heaven* under which he rules.

Jade carvings and amulets of the Azure Dragon are
known from a very early period in Chinese history; two fine
specimens from the late Chou Dynasty (about 350 BC) are
shown above in Figures 8 and 9.   These bear a very curious
resemblance to the traditional portrayals of the great god
Quetzalcoatl of the Toltecs, the *Kukulcan* of the Maya, and
worshiped throughout ancient America as the *Feathered Serpent*.   Quetzalcoatl is shown in Fig.10 as he appears on a
carved stone lintel, dating from about 800 AD, at the ruins
of Yaxchilan in Southern Mexico, near the Guatemala border.

The strange similarity in these heavily stylized representations becomes even more intriguing when we realize that
at least a thousand years separates the Chinese carvings
from the Mayan one. We might speculate that the pattern
of the constellation Scorpius provided the original inspiration for the traditional portrayals of Quetzalcoatl, but
there is no direct evidence that this is true; on at least
one of the few surviving Pre-Columbian *codices* the strange
Feathered Serpent seems to be identified with the stars of
Draco.

To observers in the Earth's northern hemisphere,
Antares becomes visible each year in the southeastern sky
about mid-Spring, its ruddy glow heralding the approach of
Summer. *"I am not sure that the color of Antares is any
reason why it should be associated with the blooming of red
flowers,"* wrote Martha E.Martin in *The Friendly Stars, "but
I find in my journal of out-of-door things that I have un-
consciously made the association. One entry, on June 30,
says: 'Antares is shining splendidly tonight and rivals in
color the wild red lilies that were blooming today in the
coppice down beyond the spring'..... its redness is still
further intensified by the white sweep of the Milky Way,
which is now very bright, and which lies within a few
degrees of Antares.."* Although Antares is often referred
to in star books as the reddest of the bright stars, actual
photoelectric measurements have shown that Betelgeuse in
Orion has a slightly deeper tint; a color index (B-V) of
+1.87 has been measured for Betelgeuse, and +1.80 for the
Heart of the Scorpion. Since both stars are somewhat variable however, neither figure is an absolutely permanent
value. Admiral Smyth called Antares "fiery red", but T.W.
Webb thought the tint not uniform: "The disc appears yellow
with flashes of deep crimson alternating with a less prop-
ortion of fine green, the latter mixture perhaps accounted
for by the 7 mg [companion] star near enough to be usually
involved in the flaming rays of the principal..."   E.J.
Hartung (1968) calls it "a brilliant orange red star with
a fine spectrum of well-distributed dark lines and bands".
Antares is a supergiant star of exceptionally great
dimensions; among the bright stars it is probably exceeded
in size only by Betelgeuse. Actual interfermometer obser-

COR SCORPIONIS. The red giant Antares is at center; the globular cluster in the field is M4. Lowell Observatory photograph made with the 5-inch Cogshall camera.

DESCRIPTIVE NOTES   (Cont'd)

---

vations have shown that the star has an easily measurable
apparent disc of about 0.041" diameter; to obtain the true
diameter, in miles, this figure is used as the quantity (d)
in the formula:

$$D = \frac{d \times 93,000,000}{\pi}$$

where π is the annual parallax in seconds of arc. The solution of the equation gives the diameter of the star as
about 600 million miles, or about 700 times the size of the
Sun. This result is about double the value which has been
given in older texts; the difference is due to a recent
revision of the distance, now believed to be about 520
light years. The new distance estimate is based upon the
spectral characteristics rather than upon a direct parallax
which, in any case, is almost immeasurably small. Antares,
from this result, appears to have an absolute magnitude of
about -5.1, and an actual luminosity of 9000 times that of
the Sun. The measurement of the star's diameter is somewhat complicated by the fact that Antares may not even be
truly spherical in shape. There are irregularities in its
light curve during times of occultation by the Moon which
suggest that the disc has the form of a flattened oval,
about 0.041" across the equator and 0.026" through the
poles. A curious fact is that the center of brightness is
not precisely at the center of the ellipsoid; possibly the
star is somewhat egg-shaped.
        The huge size of Antares is deceptive, for the mass
is probably not over 10 or 15 times that of the Sun, and
the resulting density is thus much less than a millionth
of the density of the Sun. The surface temperature is a
comparatively cool 3100°K. Recent studies have shown that
the star is slightly variable in light; the range appears
to be about magnitude 0.86 to 1.06, though the Moscow *Catalogue of Variable Stars* (1970) gives the extreme minimum
as 1.8, with a semi-regular periodicity of about 1733 days,
or 4.75 years. The variability might explain the often-
mentioned fact that Antares is rated as fainter than Beta
Librae in the catalogue of Ptolemy, but as the latter star
is classed as 1st magnitude, the variability of both stars
would be required to fully explain this discrepancy.

Betelgeuse and Antares both show slow and erratic pulsations, during which, as the light fluctuates, the diameter varies also. The radial velocity of Antares ranges from 0.0 to about 1.8 miles per second in approach; the slight differences are presumably due to the pulsation of the star. According to a note in the A.D.S.Catalogue, a periodicity of about 5.8 years in the radial velocity has been detected, but it is not certain that this variation, even if confirmed, implies binary motion. It is of some interest to note that the main period of Betelgeuse also appears to be about six years. Antares shows an annual proper motion of about 0.03", shared by a number of other bright stars in this region of the sky. (See Page 1668)

Antares was one of the first stars to be detected with radio telescopes. Using a three-element tracking interferometer at Green Bank, West Virginia in March 1970, C.M.Wade and R.M.Hjellming of the National Radio Astronomy Observatory detected weak radio emission from the star at a wavelength of 11.1 cm.; the observations were confirmed by additional studies made in June and November of the same year. The radio spectrum, in this case, indicates that the radio energy does not originate in the tenuous gas cloud which surrounds the star, but in the much denser atmosphere of the star itself. Betelgeuse has also been identified as a source of radio radiation.

THE COMPANION TO ANTARES. Alpha Scorpii has a small companion star, usually described as green in color, possibly an optical effect due to contrast with the deep "saffron-rose" of the primary. It was probably first observed by Prof. Burg at Vienna on April 13, 1819, during an occultation of Antares by the Moon; the reappearance occurring at the dark limb: "At 12h 03m 17.1s I observed the emersion of a star 6.7 mag, which, about 5 seconds after suddenly appeared to me like a star of the first magnitude, and it is from this transition that I have dated the time of emersion. Perhaps Antares is a double star, and the first observed small one is so near the principal star that both, viewed even with a good telescope, do not appear separated..." The small star was seen again by Grant in India in 1844, and independently by Prof. O.M.Mitchel with the 11.7-inch refractor at the Cincinnati Observatory in 1845-1846.

ANTARES. North is at the left in this photograph made with the Lowell Obseryatory 13-inch camera. The bright globular cluster is M4; the fainter one at left edge is NGC 6144.

The companion is 3.0" from the primary in PA 275°, and about magnitude 6.5, not a particularly easy object for small telescopes, though, as in the case of Sirius, the visibility depends critically upon the steadiness of the air. When seeing conditions are good, it appears quite plainly in a 6-inch telescope as a little spark of glittering emerald, almost drowned out in the blazing ruddy light of the giant Antares. W.Dawes, in the 1850's, observed the star repeatedly with an aperture of 6 inches; E.J.Hartung (1968) finds it visible "in good conditions with 7.5 cm". D.W.Edgecomb used Antares as a test object for his new 15½" mirror in July 1887, and reported that "the companion.... stands out clear and bold with 140 [X]. Images very fine." The author of this book has always found it fairly easy in the Lowell Observatory 7-inch refractor when the air is calm. As to the color of the elusive companion, there are the usual differences of opinion; Hartung finds it "pale green which I have seen well with 30 cm in bright sunshine against the blue sky". W.T.Olcott calls it "vivid green", while K.McKready (1912) refers to it as blue; Mary Proctor in *Evenings With the Stars* (1924)speaks of it as "the wily companion of verdant hue", while E.Crossley, J.Gledhill, and J.M.Wilson, in their *Handbook of Double Stars* (1879) quote the color estimates of various observers: "blue", "purple", "very blue", and "green". From the effect of contrast alone, it is not surprising that the small star should appear greenish. Dawes, however, observed the star emerging from the dark limb of the Moon during an occultation in 1856, and noted the tint as greenish, before the red primary itself had appeared. Hartung points out that even in such circumstances only a few seconds are available for judgment; he thinks that the true color is a pale yellow. Interested observers should not fail to take advantage of occultations of Antares, particularly those which occur between full moon and new, when the reappearance will take place at the moon's dark edge, and the companion will emerge first, remaining visible for 5 or 6 seconds before the primary appears. Occultations of Antares occur in "seasons" as in the long series beginning September 10, 1967, and ending September 14, 1972; the star was occulted at every passage of the Moon during this interval, for a total of 64 occultations!

The companion star has a spectral type of about dB4, and an absolute magnitude of about +0.5. With a luminosity of about 50 times that of our Sun, it is definitely under-luminous for its type. The exact classification of this odd star is uncertain; the spectral lines are much broadened by rapid rotation and there are peculiar features, such as emission lines of doubly ionized iron, apparently produced in a turbulent cloud surrounding the star. This "shell" is evidently some 4" or 5" in diameter, but has not been seen visually in any telescope. Antares definitely forms a true physical pair with the green companion, as both stars share a common proper motion of about 0.03" per year, but there has been no evidence for orbital motion since discovery. The projected separation of the components is about 500 AU, or about 6 times the diameter of the orbit of Pluto. The orbit for Antares published by J.Hopmann in 1957 must be classed as completely fictional; though impossible to dis-prove it cannot be verified either! It gives the system a computed period of 853 years with periastron in 1888, and a semi-major axis of 3.27" with an eccentricity of 0.0. The orbit is assumed to be oriented nearly edge-on as seen from the Earth, and the companion is assumed to be moving, at this time, almost directly toward or away from us, so there is no evident orbital motion or change in PA. On this hypothesis, the companion should, in the next few cen-turies, move closer in toward the primary, eventually van-ishing into its light and then reappearing on the other side. At closest apparent separation, possibly around the year 2115 AD, the pair should be irresolvable in even the greatest telescopes. Greatest apparent separation of about 3.0", on the opposite side of Antares (PA about 91°) is predicted for 2300 AD. Another set of possible orbital elements, by W.D.Heintz (1955) lead to a rather similar result; he finds a probable period of about 900 years, and an inclination of about 86°.

The Antares system is enveloped in a vast reddish nebula some 5 light years in diameter, very elusive visual-ly but covering an area more than 1° across on red-sensitive photographs. This cloud evidently shines from reflected light of the star and appears to be composed of fine solid particles of some sort rather than gaseous material; polar-ization studies suggest that the composition is some sort

of metallic dust. Other red giant stars are known to be
surrounded by similar clouds, though usually on a much
smaller scale. The Antares nebula appears on the Skalnate
Pleso *Atlas of the Heavens* with the designation IC 4606;
its apparent dimensions are given as 80' x 85'.

Antares lies in a curious and heavily obscured area
of the southern sky; immediately to the north of the star
a huge murky cloud of some sort begins to blot out the
light of distant stars, extending some 4° to the north and
merging with several tentacle-like dark lanes which flow
out for more than 10° to the east and north east, toward
the star clouds of the Ophiuchus Milky Way. Throughout this
region, considerable nebulosity appears in the vicinity of
many bright stars which illuminate the surrounding murk
like street lamps on a foggy night. About 3° NNW from
Antares we find the brightest portion of this nebulosity,
surrounding the bright double star Rho Ophiuchi; this is
IC 4604, and is the bright object near the top of the print
on page 1669. About 1.3° almost due west from Antares will
be found the large globular cluster M4, easily found even
in good binoculars, and undoubtedly one of the nearest of
all the globulars. In the same field, 0.7° NW of Antares,
will be found a much smaller globular, NGC 6144, a faint
blur only in a 6-inch glass. The dark area just to the
right of the Rho Ophiuchi nebula is presumably the "hole
in the heavens" which William Herschel reported during his
comprehensive survey of this part of the sky in 1785.

Antares is the brightest member of the huge Scorpio-
Centaurus Association, described below.

THE SCORPIO - CENTAURUS ASSOCIATION

A large scattered aggregation of early-type stars extends
over about 90° of the southern sky, and includes many of
the bright stars in the constellations of Crux, Centaurus,
Lupus, and Scorpius. This aggregation is called the Scorpio
Centaurus Association. It is a part of a general population
of bright B-type stars in the region defining a larger
structure often called the Local System or the Local Star
Cloud, a sub-unit in one of the spiral arms of our Galaxy.
The Group is the nearest aggregation of early B-type stars,
and its study is important for the accurate calibration of

NEBULOSITY IN THE FIELD OF ANTARES. The red giant is at lower left; the cluster to the right is M4. The Rho Ophiuchi nebulosity appears at top. Lowell Observatory photo.

the luminosities of stars of this type. The existence of
the Scorpio-Centaurus Association was established by J.C.
Kapteyn in 1914, from a study of the proper motions of
bright southern B-type stars. In 1921 N.H.Rasmusen found
that the motions of the stars all converge toward a point
in the eastern portion of the constellation Columba. Further studies of the stars and their radial velocities were
undertaken by J.S.Plaskett and J.A.Pearce between 1928 and
1936; in 1946 a thorough analysis and summary of the known
facts was presented by A.Blaauw in the *Publications* of the
Astronomical Laboratory at Groningen.

The members of the Scorpio-Centaurus Association show
a common space motion toward the southwest, and the computed convergent point is near RA 6 hours and Declination -35°
in the vicinity of the star Beta Columbae. The proper motions are all rather small, averaging about 0.03" annually,
but the fact of common motion is clearly evident. The entire group forms a flattened and tilted structure crossing
the central plane of the Milky Way in the region of the
Southern Cross, and reaching a galactic latitude of about
+20° at its eastern end near Beta and Delta Scorpii.

At the present time about 100 stars are recognized
members of the Association. The number of possible fainter
members is unknown. In a preliminary study by F.C.Bertiau
(1958) some 35 fainter stars in the area near Pi and Delta
Scorpii were identified as probable members; these are
stars of 6th to 8th magnitude, with spectral types ranging
from B0 to A0. A thorough study of this group would possibly identify dozens of new members. In a photometric investigation of the group made at Cerro Tololo in Chile in 1962
and 1963, A.Gutierrez-Moreno and H.Moreno confirmed 102
stars as certain or very probable members; their plot of
the Association has an elongated shape with the major plane
of symmetry lying at an angle of about 18° from the plane
of the Galaxy.

The true space motion of the Association with respect
to the Sun is about 15 miles per second; the member stars
in Scorpius show slight approach radial velocities (for
example Antares itself) but as one proceeds westward toward
the convergent point the radial velocities are nearly all
found to be positive. The highest velocities, such as that
of Delta Crucis, are about 15.5 miles per second, away from

the Sun. The distance to the center of the group is close to 550 light years; the apparent center lies near Alpha Lupi and Zeta Centauri. Some of the members appear to be as close as 450 light years.

Antares itself is the brightest member of the huge Association; virtually all the other members, however, are early B-type stars. Prominent among these are Beta Crucis, Sigma Scorpii, Epsilon Centauri, Alpha Lupi, and Delta Scorpii. The peculiar "shell star" 48 Librae is also a member. Several spectral variables of the β Canis Majoris type are included in the group, as Theta Ophiuchi, Beta Crucis, and Sigma Scorpii. Delta and Mu Centauri are emission type B stars with slight variability in light. Double and multiple stars, both visual and spectroscopic, are common in the Association. Typical examples are Beta and Nu Scorpii, Antares, Gamma and Lambda Lupi, and Eta Centauri.

In addition to the general drift of the Association, an accurate study of the stellar motions has shown that the group is gradually expanding. From such findings, the age of the group appears to be no more than about 20 million years. One star, Zeta Ophiuchi, if accepted as a member, would be considerably younger than this figure, but its discrepant proper motion seems to make its membership unlikely. Antares itself, which has already reached the red giant stage, may be the most massive star in the group as well as the most evolved. Red supergiants are found also in some other bright clusters of early type stars, as in the great Orion Association, the "Jewel Box" NGC 4755 in Crux, and the magnificent Double Cluster in Perseus.

BETA   Name- GRAFFIAS, sometimes called AKRAB. from Riccioli's *Aakrab schemali*. R.H.Allen suggests that "Graffias" is derived from the Greek Γραψαιος which signifies "crab" since "the ideas and words for crab and scorpion were almost interchangeable in the early days, from the belief that the latter creature was generated from the former....This was held even by the learned Saints Augustine and Basil of the 4th Century". Beta and Delta are the brightest stars of the Chinese asterism *Fang*, a room, chamber, or house; the star itself is labeled *Tien Tze*, a "Chariot of Heaven" on some oriental charts. Magnitude 2.55; Spectrum B0 V; Position 16025s1940. The computed

DESCRIPTIVE NOTES   (Cont'd)

distance of the star is about 600 light years; the actual
luminosity must then be about 2700 times that of the Sun
(absolute magnitude -3.7 or -3.8). Beta Scorpii shows an
annual proper motion of 0.027"; the radial velocity is
about 4 miles per second in approach. The star is one of
the bright members of the Scorpio-Centaurus Association,
and is located about 8½° NW from Antares.

Beta Scorpii is one of the finest bright double stars
in the sky for small telescopes, and will be found not at
all difficult in a good 2-inch refractor. Individual magnitudes for the components have been measured as 2.63 and
4.92; separation 13.7" in PA 23°. There has been no certain
change in distance or PA since the early measures of the
younger Herschel and J.South in 1823; the stars, however,
show the same proper motion. There is very little color
contrast in this pair, though C.E.Barns calls them "intense
and colorful". To T.W.Webb they were "pale yellow and
greenish", W.T.Olcott has them "white and bluish", Mary
Proctor thought them "white and lilac", while E.J.Hartung
speaks of them as a "splendid pale yellow pair".   Seen in
daylight they do give the impression of a yellowish tint,
probably due to contrast with the blue sky. Against a dark
night sky most observers will find them simply white with
perhaps a tinge of greyish or bluish in the fainter star.

A second companion, of magnitude 9½, was discovered
by S.W.Burnham with an aperture of 18½ inches in 1879, and
is difficult to observe even in large telescopes. Hartung
thinks it was possibly visible with 30 cm in 1960, but has
since closed down somewhat, to about 0.5"; the PA increased
from 88° in 1880 to about 132° in 1959. Evidently the two
stars are in slow orbital revolution with a period probably
approaching 1000 years. During an occultation of β Scorpii
by the Moon on July 8, 1976, the faint star was found to be
considerably brighter than expected, probably about magnitude 6.5. Either the star is variable, or the estimate of
magnitude 9½ was a great underestimate resulting from the
difficulty of observation. This close pair, incidentally,
is called "A-B"; the third star at 13.7" is designated "C".

In addition to these two visible companions, the
bright primary is a spectroscopic binary with a period of
6.828145 days. From the known orbit, the masses of the two
components appear to be about 21 and 13 solar masses, and

the orbit of the brighter component is about 8 million miles from the gravitational center of the system, with an eccentricity of 0.27.

A widely observed and rare event was the occultation of the Beta Scorpii system by Jupiter on May 13, 1971. On this occasion the bright component passed behind the disc of the planet very close to its south pole, remaining hidden for about 90 minutes; the 5th magnitude "C" component at 13.7" was occulted nearly centrally for a 2.2 hour interval of "totality". On reappearance, each star was first seen as a faint point of light apparently *inside* Jupiter's limb, the brighter star requiring about 7 minutes to regain full brilliance. Both stars brightened irregularly with "flares", apparently the result of the stratified structure of Jupiter's atmosphere. During this event, an even rarer phenomenon was observed; the occultation of Beta Scorpii C by Jupiter's satellite Io. The star remained hidden for 4^m 11^s as observed in Jamaica, and slightly over 5 minutes as seen from the Observatory of the University of Florida. During this event, strong evidence was found for the close duplicity of Beta Scorpii C; according to astronomers at McDonald Observatory the star is probably a close pair of about 0.10" separation in PA 308° with a brightness difference of about 2 magnitudes. The Beta Scorpii system then becomes a quintuple, though only the bright A-C pair will be seen in most amateur telescopes.

DELTA   Name- DSCHUBBA, thought to be a confused version of the Arabic *Al Jabhah*, the "Forehead" or Front of the Scorpion. Another Arabic term was *Iklil al Akrab*, the "Crown of the Scorpion", which included also several of the nearby stars; this is evidently the source of Riccioli's *Aakrab genubi*. According to R.H.Allen the Babylonian title *Gis-gan-gu-sur* was applied to the group formed by Delta, Beta, and Pi; the translation *The Tree of the Garden of Light* suggests some possible connection with the legend of Eden. This same group was the Persian *Nur*, "Bright", and the Coptic *Stephani*, "the Crown".

Delta Scorpii is magnitude 2.34; Spectrum B0 V; Position 15574s2229, just 7° NW from Antares and 3° SSW from Beta. At a computed distance of 590 light years, the star is at approximately the same distance as Beta, and is

one of the prominent members of the Scorpio-Centaurus
Association. At the derived distance, the true luminosity
of the star is found to be about 3300 times that of the
Sun (absolute magnitude -4.0). Delta Scorpii shows an
annual proper motion of 0.032"; the radial velocity is 8.5
miles per second in approach.

A curious tendril of faint nebulosity connects Delta
Scorpii with the group consisting of the stars 1, 2, 3 and
Pi Scorpii, some 3½° to the SSW. This was first detected on
wide-angle Milky Way plates made by E.E.Barnard in the
1890's, but is virtually impossible to observe visually in
any telescope. The star 2 Scorpii itself is a 2.5" double
showing common proper motion, but only slight relative
change, if any. E.J.Hartung thinks that the separation may
be decreasing slowly, and suggests that the system may be
a binary with the orbit nearly edgewise to the line of
sight. He calls the color of the primary "bright yellow"
and finds that the companion "looks whiter". The pair is
another member of the great Scorpio-Centaurus Association.
"Enchanting environs" says C.E.Barns.

EPSILON   Magnitude 2.28; Spectrum K2 III or IV.
Position 16469s3412. Epsilon Scorpii appears
to have no proper name in any of the standard atlases,
although it is a bright star and occupies an important spot
in the constellation, marking the beginning of the tail of
the Scorpion. It lies about 9° SSE from Antares. This is
one of the nearer stars in Scorpius, lying some 65 light
years distant, and is a subgiant with an absolute magnitude
of about +0.7 (luminosity about 45 suns). The star may be
slightly variable in light. The annual proper motion is
0.66" in PA 247°; the radial velocity is about 1½ miles per
second in approach. This is one of the few bright stars in
the constellation which is not a member of the Scorpio-
Centaurus Association.

The cepheid variable star RV Scorpii lies in the same
wide-angle field, about 1½° to the ENE.

ZETA   Position 16511s4217. A wide naked-eye double
star, forming an interesting color-contrast
pair for binoculars or small telescopes. The components,
about 6.8' apart, are oriented almost due east and west,

and lie about 4½° south of the similar but brighter pair
Mu Scorpii. The components of Zeta Scorpii do not form a
true physical pair, as both the parallaxes and the proper
motions differ widely.

The eastern component, called Zeta-2, is visually the
brighter of the two stars; it is magnitude 3.62, spectrum
K5 III, and is an orange subgiant lying about 155 light
years distant, according to a direct trigonometrical parallax. The annual proper motion is 0.27" in PA 208°; the
radial velocity is about 12 miles per second in approach.

The western star, Zeta-1, is a vastly more distant
and more luminous star; it is a blue supergiant of type B1
and luminosity class Ia; the spectral features suggest that
the energy output probably exceeds that of Rigel. Zeta-1
is believed to be a member of the great cluster NGC 6231
which is centered just ½° to the north, and whose estimated
distance is about 5700 light years. If a true member of
the cluster, the star, at apparent magnitude 4.80, must
have an absolute magnitude of close to -8, and a luminosity
exceeding 100,000 suns. (Refer also to NGC 6231, page 1722)

ETA   Magnitude 3.33; Spectrum given in various
catalogues as F0, F1 or F2, luminosity class
III or IV. This is the southernmost of the bright stars in
the tail of Scorpius, lying a little below the midpoint of
a line drawn from Zeta to Theta Scorpii; as in the case of
a number of other bright stars in the constellation, it
seems never to have been honored with a proper name. Eta
Scorpii is one of the nearer stars in the constellation,
at a distance of about 50 light years; it has an actual
luminosity of about 10 suns, and an absolute magnitude of
+2.3. The annual proper motion is 0.29" in PA 176°; the
radial velocity is about 17 miles per second in approach.

A little more than 1° distant, toward the ENE, the
observer will find a nearly equilateral triangle of three
8th magnitude stars, measuring about 7' on a side, and
neatly enclosing two easy little pairs of faint stars; this
little pyramid-shaped asterism is the galactic cluster NGC
6322. Easily found, it is a rather sparse group, lacking
in faint stars. The position of Eta Scorpii is 17086s4311,
the cluster is at 17152s4250.

THETA   Name- SARGAS, from the ancient Babylonian
title; R.H.Allen suggests that the Persian
Vanant, "The Seizer" or "The Smiter" also referred to this
star. Theta Scorpii is magnitude 1.87; Spectrum F0 Ib;
Position 17337s4258. This star and Eta, 4° apart, are the
two southernmost stars in the curving tail of Scorpius.
Strangely, Allen refers to Theta as a red star, although it
is an early F-type supergiant nearly as white as Altair.
From a spectroscopically derived luminosity the distance
appears to be about 650 light years; the spectral features
suggest an absolute magnitude of about -4.6 and a true
luminosity of perhaps 5800 suns. The annual proper motion
is 0.01"; the radial velocity is about 0.7 mile per second
in recession.

R.H.Allen quotes a statement made by Prof. T.J.J.See
that a 14th magnitude "greenish" companion to Theta was
discovered in 1897, separation 6.77" in PA 316.9°. There
appear to be no references to this star in modern double
star catalogues, though Prof.See, according to Allen, spoke
of it as "a system of surpassing interest; one of the most
difficult of known double stars". Possibly, through some
error, this description actually refers to some other star.
A check of the Lick *Index Catalogue* reveals that there is
a faint double star (h4963) almost exactly 1° north of
Theta, and with virtually the same measurements (6.9" in PA
316°; 1897) although the magnitudes are much fainter, at
about 8½ and 10½. Perhaps somewhere a measurement or line
of description has been accidentally transferred to the
wrong star.

Just 1.7° south of Theta Scorpii, the observer will
find the bright compressed globular star cluster NGC 6388,
easily located, but resolved only with fairly large apertures. It is some 4' in size with a total magnitude of 7,
but the individual stars are fainter than 16th. Light loss
through absorption is estimated at about 2 magnitudes in
the region; allowing for this effect, the true distance of
the cluster may be about 35,000 light years.

Another deep-sky object in the vicinity of Theta is
the faint planetary nebula IC 4663, lying about 2.3° SE
from the star. It appears as a pale greyish disc about 15"
in size, floating spectral-like in a field of faint stars.

IOTA   Magnitude 2.99; Spectrum F2 Ia;   Position 17441s4007. This is the easternmost star in the curve of Scorpio's tail, lying just 3½° NE from Theta. Iota Scorpii is another huge supergiant star with a computed absolute magnitude of about −7.1 and a true luminosity of nearly 60,000 times that of the Sun. The distance is thought to be about 3400 light years. The star shows an annual proper motion of less than 0.005"; the radial velocity is about 16 miles per second in approach, but with variations suggesting that the star is a spectroscopic binary.

A companion of the 13th magnitude was recorded by Prof. T.J.J.See in 1897 at a separation of 37.5" in PA 95°; the pair appears in his lists under the designation λ338. The star is very likely not a true physical companion to Iota; if it is actually at the same distance the projected separation is about 38,000 AU, or 0.6 light year.

Iota itself is often labeled Iota-1 on star charts; the fainter field star called Iota-2 lies 15' distant to the east; it is magnitude 4.80 and spectral type A3. The two stars do not appear to be physically associated. Iota-2 shows an annual proper motion of 0.01"; the radial velocity is 11 miles per second in approach. A faint companion to the star was detected at Harvard Observatory in 1900; it is magnitude 11, separation 32.6" in PA 37°. As in the case of Iota-1, it is not known if the faint star is a true companion to the primary; no measurements for either of these stars are recorded in standard catalogues since 1903!

The faint galactic star cluster called H17 will be found about 1.2° to the west; this is a fairly inconspicuous group containing an 8th magnitude star on its NW edge, and about 20 or so fainter members.

KAPPA   Magnitude 2.39 (slightly variable); Spectrum B2 IV; Position 17390s3900, about 1.4° NW from Iota Scorpii. The computed distance of Kappa is about 470 light years, the actual luminosity about 1900 times that of the Sun, and the absolute magnitude about −3.4. The star shows an annual proper motion of 0.03"; the radial velocity is about 6 miles per second in approach.

From studies made in 1974-1976, Kappa Scorpii is now recognized as a Beta Canis Majoris variable, with the very slight magnitude range of about 2.39 to 2.42, and a period of 0.19987 day, or $4^h 47.8^m$. As in many stars of this type the star shows a secondary period, of 0.20544 day, and the two interfering cycles produce a harmonic oscillation or "beat period" of 14.74 days. Stars of this class are all early B-type subgiants which appear to be in the early stages of stellar evolution away from the main sequence. Beta Cephei and Beta Canis Majoris are the two standard examples of the type.

Kappa Scorpii is located 2½° SE of Lambda, near the end of the Scorpion's tail. The faint galactic cluster H17 is 1° distant to the SSW.

LAMBDA   Name- SHAULA, "The Sting". Magnitude 1.62; Spectrum B1 V; Position 17302s3704. This is the 24th brightest star in the sky, ranking immediately after Castor (α Geminorum) on the list of the 25 brightest stars. The name is thought to be derived from the Arabic *Al Shaulah*, "The Sting", though Al Biruni, in the 11th Century, gave the source as the word *Mushalah*, which appears to mean "Raised", evidently referring to the position of the Scorpion's sting. The Coptic *Minamref* and the Babylonian *Sarur* refer to Lambda and Upsilon together. These two bright stars are 35' apart and mark the end of Scorpio's tail.

The computed distance of Lambda Scorpii is about 310 light years, and the actual luminosity about 1700 times that of the Sun (absolute magnitude -3.3). The star shows an annual proper motion of 0.03"; the radial velocity is very near zero, but with cyclic shifts in a period of 5.6 days. Possibly the star is a spectroscopic binary. From recent studies, it has also been identified as a short-period variable of the Beta Canis Majoris type, with a very slight magnitude range of about 1.59 to 1.65, in a period of 0.21370 day, or $5^h 07.7^m$. A secondary period of 0.106852 day has been identified, and a resulting "beat period" of 10.15 days. As of 1977 this appears to be the brightest known star of its class. The spectral type is given by various authorities as B1 or B2, luminosity class IV or V.

The small galactic star cluster called H16 will be found in the field, 0.5° to the NW; the similar cluster NGC 6400 lies 1.3° east and slightly north.

REGION OF LAMBDA SCORPII. This is the star which marks the tip of the Scorpion's tail. The star cluster in the field is H16. Lowell Observatory 13-inch telescope photograph.

MU   Magnitude 3.12; Spectrum B2 V; Position 16485s
3758. The star forms a wide naked-eye pair with
Mu-2, which is magnitude 3.56, spectrum B2 IV. This is the
conspicuous pair which will be noted about 3.8° south from
Epsilon Scorpii. They are 346" apart, oriented almost due
E-W, with the brighter star, Mu-1, to the west. W.W.Gill
has preserved for us a Polynesian legend which sees in the
two stars the *Piri-ere-ua*, "The Inseparable Ones", two
children fleeing their wicked parents, rather resembling
the Hansel and Gretel story. The whole curve of the southern part of Scorpius, from Mu to Lambda, was in other Polynesian tradition the *Fish-hook of Maui*, with which that god
is said to have dredged up various South-Sea islands from
the depths of the sea, particularly the island of Tongareva
which lies to the west of the Marquesas.

The components of Mu Scorpii may be considered an
actual physical pair, since they show the same proper
motion, and both are members of the Scorpio-Centaurus group
which includes many of the bright stars in Scorpius. The
actual separation, however, is very great; the *minimum*
value, on the assumption that the two stars are at exactly
the same distance, is about 55,000 AU, or 0.88 light year!
The computed distance is about 520 light years; the annual
proper motion is 0.03"; and the radial velocity is about
15 miles per second in approach.

Mu-1, the western component, is a double-line spectroscopic binary, one of the first to be identified, discovered by S.I.Bailey in 1896. In 1920, Miss A.C.Maury at
Harvard found that the components form an eclipsing system
of the β Lyrae type; modern observations give the period as
1.44027 days. The eclipses are of little interest visually,
being of rather small amplitude. Primary minimum has a
depth of about 0.3 magnitude; secondary minimum only about
half as much. The chief facts about the components, from
the combined studies of Maury, Rustnick, Elvey, and Gaposchkin, are given in the table:

	Spect.	Diam.	Mass	Lum.	Abs.Mag.
A	B2 V	5.6	14	1000	-2.7
B	B6 V	6.2	9	700	-2.3

REGION OF MU SCORPII. The bright double star is at lower
right. Star cluster NGC 6242 is at lower left and NGC 6281
is at the top. Lowell Observatory photograph.

As with most binaries of this type, there are considerable uncertainties concerning the dimensions of the system and the exact orbital elements. According to Miss Maury the orbit of the system is very nearly circular, with the slight eccentricity of 0.05; the separation of the two stars appears to be about 6 million miles, which places their surfaces nearly in contact.

Sweeping an area about 2.3° to the east, the observer will find the fairly compact and elongated galactic cluster NGC 6281; another similar group, NGC 6242, lies about 1.5° to the SSE. (See photographs, page 1681)

NU   Quadruple star. Position 16091s1921, about 1.5° east and slightly north from the bright star Beta Scorpii. Nu Scorpii is a 4-star system arranged as two close pairs; it is sometimes referred to as a "fine quaternary system" while other lists describe it as a "double-double" resembling Epsilon Lyrae. Miss Agnes Clerke (1905) calls it "perhaps the most beautiful quadruple group in the heavens, from the narrow limits within which the brilliant objects composing it are crowded. As a wide pair it was noticed by C.Mayer in 1776; after seventy years the smaller star was divided by O.M.Mitchel at Cincinnati, and the larger one of fourth magnitude yielded similarly, in 1874, to the insistence of S.W.Burnham...Both pairs share with several neighboring stars a slow drift through space".

With any good small telescope, the observer will first note the wide pair, called "A-C"; these are 41.4" in separation, PA 336°; magnitudes 4.0 and 6.23; Spectra B2 and AO, both subgiants of luminosity class IV. Then with somewhat higher powers, the "C" component may be resolved into a 2.3" pair (C-D) which has possibly widened somewhat and also shown a slight PA increase since discovery. These were first resolved by Prof.Mitchel in 1846, and discovered independently by Prof.Jacobs the following year. E.J.Hartung finds the CD pair resolvable with 7.5 cm, but Olcott thinks something larger than 3-inch aperture is required. The present author finds them obvious and steady with a 6-inch. The bright A-B pair, on the other hand, is not an easy object, though S.W.Burnham detected them originally with an aperture of only 6 inches: "I examined it several times under the most favorable circumstances, but could not

get rid of an apparent elongation of the principal star in
a direction nearly north and south. I requested Professor
C.A.Young to examine it with the splendid 9.4 inch Clark
refractor of the Dartmouth College Observatory. He examined
it several times, and at last when the air was very steady
he was rather inclined to think it double, although he
could not even notch it.." In presenting the medal of the
Royal Astronomical Society to S.W.Burnham in 1894, in recognition of his unrivalled achievements in the discovery,
cataloguing, and measurements of double stars, Captain W.
de W.Abney stated that *"the discovery of this pair* [ν Sco
AB] *is a remarkable feat with a 6-inch, and the more so as
another companion to ν Scorpii had been measured by many
observers before, and the chief component must have been
well scrutinized..."*

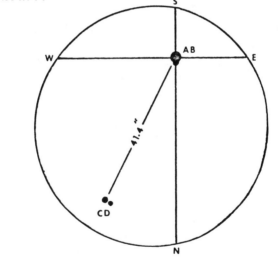

NU Sco

The AB pair has possibly widened somewhat since the
time of Burnham; in 1876 it was claimed to be about 0.7"
but a measurement in 1967 gave 1.2" in PA 2°. According to
the Yale *Catalogue of Bright Stars* (1964) the magnitudes
are AB= 4.6 & 5.6; CD= 7.0 & 7.7. In another list, by A.
Slettebak (1963) the four components are rated 4.4, 6.4,
6.8, and 7.8. Slettebak classes the A-C spectra as B2 V
and B9 Vp, but states that the C component shows silicon
[Si II] lines which are too strong for the assigned type.
All four stars undoubtedly form a vast common motion group
and also appear to be members of the Scorpio-Centaurus

Association; the annual proper motion of about 0.03" very
closely matches that of Beta Scorpii, about 1.5° away.
The radial velocity of the system is about 4 miles per
second in approach, with the CD pair showing a somewhat
higher velocity than the bright primary. V.M.Slipher at
Lowell Observatory found large variations in the radial
velocity of star A, and classed the star as a spectroscopic
binary. No orbit computation, or even a period, seems to
have been obtained for this star. Parallax measurements
for Nu Scorpii have also been rather discordant; the distance is currently thought to be at least 400 light years,
which is the tentative figure obtained from the computed
dynamic parallax of the CD pair. The 41" separation of the
wide pair then corresponds to about 5000 AU, which suggests
the vastness of this system.

Nu Scorpii is located in the midst of a large but
faint nebulosity, IC 4592, first noticed by E.E.Barnard on
wide-angle Milky Way plates made in March 1895. He found
it wing-like, resembling the hazy outline of some great
bird, well defined, and brightest at the western edge; the
total length from east to west being about 2.5°. This
appears to be a portion of the same diffuse cosmic haze
which is seen again a few degrees to the south and east,
gleaming faintly around Rho Ophiuchi and Antares.

XI   Multiple star. Position 16016s1114 in the
            northernmost portion of the constellation,
some 8½° due north from Beta Scorpii. Magnitude 4.17;
Spectra F5 IV. This is one of the most interesting of the
multiple star systems, discovered by Sir William Herschel
in 1782. The close pair, A-B, are magnitudes 4.8 and 5.1,
and form a binary system with direct motion, but with a
very eccentric orbit, in a period of 45.69 years. At periastron in 1951 the apparent separation was 0.2" but at the
widest separation, as in 1976, it increases to about 1.25".
According to orbit computations by P.Baize, the semi-major
axis of the orbit is 0.72", the inclination is about 37°
and the eccentricity has the large value of 0.74. In a
good telescope with fairly high powers this close pair
shines with a bright golden tint; when near maximum separation they form a fair test for a 4-inch aperture. Both
star are subgiants of type F5 IV with computed absolute

magnitudes of 2.9 and 3.1, and computed masses of 1.45 and 1.34 solar masses. The parallax of Xi Scorpii is accurately known from a thorough study at Sproul Observatory, and indicates a distance of about 80 light years. From this the average separation of the AB pair is about 18 AU, more or less comparable to Uranus and the Sun.

The third component of the system is 7.4" distant in PA 52°; it is a dG7 orange dwarf of magnitude 7.2. This star is in slow retrograde motion around the system in a period which probably exceeds 1000 years. A decrease of about 18° has been measured in the PA during the last century. The earliest measurement reported for the A-C pair in the A.D.S.Catalogue was made by F.G.W.Struve in 1825; the PA was then 79°, whereas a measurement made at Lowell in 1968 gave 52°.

The double star Σ1999, located 283" to the south in the field, is now recognized as a physical member of the Xi Scorpii system, showing the same parallax, proper motion and radial velocity. The components are 11.4" apart in PA 100°, magnitudes 7.4 and 8.1, spectra dK0 and about dK3. The projected separation from Xi AB is some 7000 AU.

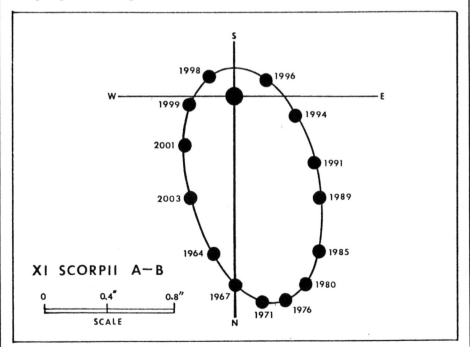

XI SCORPII A—B

Very little change, if any, has taken place in the
Σ1999 system in the last century. The separation has perhaps widened slightly; F.G.W.Struve obtained a measurement
of 10.5" in 1831. E.J.Hartung refers to this as "a deep
yellow pair", while R.H.Allen reports colors of "bright
white, pale yellow, and gray" for the components of Xi
Scorpii itself. The whole vast multiple system shows an
annual proper motion of 0.07"; the radial velocity is 20
miles per second in approach.

PI
Magnitude 2.90; Spectrum B1 V; Position
15558s2558, about 6° west of Antares. The
star is another bright member of the Scorpio-Centaurus
Group, lying at a computed distance of about 570 light
years, and shining with about 1700 times the light of the
Sun. The annual proper motion is 0.034"; the radial velocity is about 2 miles per second in approach. Pi Scorpii
is a double-line spectroscopic binary with a period of
1.571 day. According to a computation by O.Struve, the
orbit of the brighter component has a semimajor axis of
some 3 million kilometers; the eccentricity is 0.05. The
star lies in a region of faint diffuse nebulosity which
also envelops the scattered group of stars lying to the
west, consisting of 1, 2, 3 and 4 Scorpii.

A faint companion star of the 12th magnitude was
recorded by S.W.Burnham in 1878 at 51" in PA 132°. This
pair appears in catalogues as β622, but the published
measurements are not yet sufficient to prove the possibility of common proper motion.

SIGMA
Name- AL NIYAT, of somewhat uncertain meaning
but thought to signify "The Support of the
Heart" or possibly "The Shield of the Heart". Magnitude
2.86 (slightly variable); Spectrum B1 III; Position 16181s
2529, easily found about 1.8° NW from Antares. Sigma is an
unequal but not particularly difficult double star for
moderate telescopes; it seems to have first been measured
by John Herschel in 1822. The 9th magnitude companion is
20" distant in PA 273° and evidently shares the proper
motion of the bright star; the separation and angle have
remained fixed since discovery. T.W.Webb gives no color
estimates for this pair; Olcott calls them white and blue.

DESCRIPTIVE NOTES   (Cont'd)

Hartung finds them "brilliant pale yellow" and white. The small star has a spectral type of about B9. In addition to the visible companion, the bright primary is at least a spectroscopic pair, and quite probably a triple.

F.Henroteau in 1918 first identified the star as a spectroscopic binary with a period of about 33 days; in the 3rd Supplement to the Moscow *Catalogue of Variable Stars* (1976) the period is given as 34.1 days. The computed orbit has the visible star about 9 million miles from the center of gravity of the system; the eccentricity is 0.33. But in addition to the spectroscopic pair, the presence of a more distant companion seems clearly indicated by observations of Sigma Scorpii during occultations by the Moon. This was probably first realized during the occultation of March 12, 1860, observed at the Cape of Good Hope, where the reappearance of the star occurred in two stages, separated by about 0.5 second. On July 21, 1972, an occultation of the star was observed at the South African Astronomical Observatory at Capetown; an analysis by R.E.Nather, J.Churms, and P.A.Wild shows that the star has a companion 2.2 magnitudes fainter than the primary, 0.49" distant in PA 268°. This star is much too distant from Sigma to be the 34-day companion; the derived orbital period is some 300 to 400 years.

In addition to its growing collection of companions, Sigma itself is a pulsating variable of the Beta Canis Majoris type; it has a magnitude range of about 0.1 mag and a period of 0.24684 day, or $5^h 55.4^m$. This is one of those strange stars which shows unpredictable changes in period and unexpected secondary oscillations. In a study made in 1962-1965, A.van Hoof found that the star pulses in two superimposed periods, of 0.2468406 and 0.2396710 day; the interfering oscillations produce a "beat period" of 8.252 days, which is very close to 1/4 of the spectroscopic orbital period. This correlation, if real, suggests that the pulsations may be connected in some way with the tidal effect of the close companion. In the years from 1916 to 1955, the main period of the star increased slightly, from 0.246829 to 0.246846 day; since 1955 it has been slowly decreasing. Secondary oscillations are detected in many of the light curves, reaching amplitudes of up to 0.4 magnitude in the visual, and 0.075 in the ultraviolet.

DESCRIPTIVE NOTES   (Cont'd)

The distance of the Sigma Scorpii system does not appear to be accurately known, as values of from 500 to about 1300 light years have appeared in various catalogues. In a study made at Steward Observatory in 1966, a probable distance of a little more than 900 light years was adopted; this gives the primary star an absolute magnitude of about -4.4 (luminosity = 4500 suns) if no correction is assumed for absorption. The star is another of the bright members of the Scorpio-Centaurus Association. The annual proper motion is 0.03"; the measured radial velocity is very close to zero.

The fine globular star cluster M4 lies about 1° distant, toward the SSE.   (See page 1699)

TAU   Magnitude 2.85; Spectrum B0 V; Position 16328s 2807. Tau Scorpii is the first bright star below Antares, about 2.1° to the SSE; it is another of the bright stars belonging to the Scorpio-Centaurus Association and lies at a computed distance of about 750 light years. The derived absolute magnitude is about -4.0 (luminosity about 3300 suns). The star shows an annual proper motion of about 0.03" and a radial velocity of about 0.5 mile per second in approach.

The surrounding field is rich in faint stars but contains no particular objects of interest. About 1° to the ENE lies the faint double star h4878, an 8" pair with both stars about 8½ mag. The southern component is a very close and equal pair of about 0.1", an excellent test object for large telescopes.

UPSILON   Name- LESUTH or LESATH, from the Arabic *Al Las'ah*, "The Sting". The star forms a bright 35' pair with Lambda Scorpii, the two stars marking the end of Scorpio's tail. Lesuth is magnitude 2.71; Spectrum B3 Ib; Position 17274s3715. The computed distance is about 540 light years, the derived absolute magnitude about -3.4, and the true luminosity about 1900 times that of the Sun. The star shows an annual proper motion of 0.04" and a radial velocity of 11 miles per second in recession with measurable variations. The star may be a spectroscopic binary, but no period has yet been derived. The cluster H16 lies in the field, 0.5° north of the star.

G   Position 17465s3702. Magnitude 3.20; Spectrum K1 III. This is the 3rd magnitude star which lies a little less than 3° due east from Lambda Scorpii in the tail of the Scorpion; on some star maps it marks the tip of the sting. The computed distance is about 100 light years; the actual luminosity must be about 45 times that of the Sun, and the absolute magnitude about +0.7. The star shows an annual proper motion of 0.06"; the radial velocity is 15 miles per second in recession.

Immediately following G Scorpii in the field, about 4.5' to the east, will be found the hazy glow of the faint globular star cluster NGC 6441, difficult to resolve in small telescopes owing to the estimated 3 magnitudes of absorption in this region. In larger apertures, observers may obtain a somewhat clearer view by occulting the bright star with some sort of obscuring bar placed in the focal plane of the eyepiece. With a diameter of about 3', the cluster has a total integrated magnitude (pg) of 8.9, and an integrated spectral type of G2. The computed distance is about 26,000 light years.

T   Nova Scorpii 1860. Position 16141s2252. This is the first known nova to appear in a globular star cluster; it flared up near the center of cluster M80 in the spring of 1860. (Refer to M80, page 1716)

U   Recurrent Nova. Position 16196s1746, about 8½° north of Antares, near the Scorpius-Ophiuchus border, 1.3° NW from Chi Ophiuchi. This is one of the few known examples of a repeating nova, with observed outbursts in 1863, 1906, 1936, and 1979. Owing to its position only a few degrees from the Ecliptic, the region is unobservable for a considerable period in November and December while the Sun is passing through Scorpius. Thus it is possible that one or more maxima of the star have been missed.

The first recorded outburst occurred on May 20, 1863 when the star was discovered by N.Pogson at magnitude 9.1. The nova faded rapidly, however, and had fallen to the 12th magnitude within a week. It seems to have been last seen at magnitude 13.3 on June 10. Pogson's observations are the only ones known for the outburst of 1863, and the light curve is based entirely on his data.

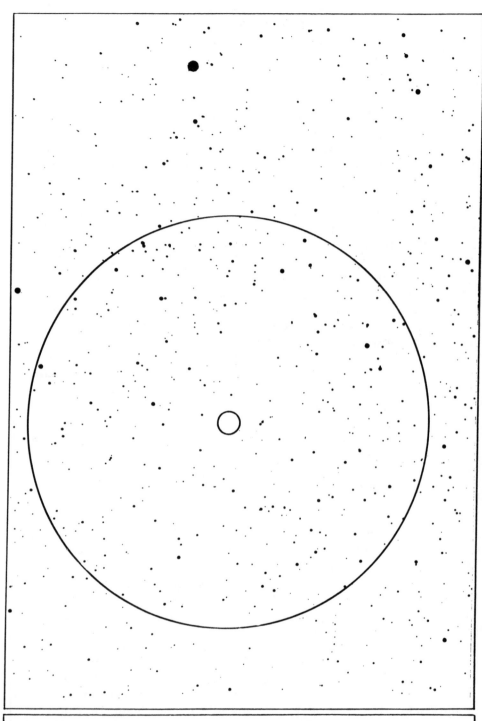

U SCORPII FIELD. The large circle is 1° in diameter with north at the top; stars to about 16th magnitude are shown. From a Lowell Observatory 13-inch telescope plate.

DESCRIPTIVE NOTES   (Cont'd)

On May 12, 1906, U Scorpii blazed up for the second time and reached a magnitude of 8.8 (pg). The outburst went unobserved at the time, but was discovered many years later by H.L.Thomas of Harvard Observatory, through a check of photographic plates. The only other recorded observation of the 1906 outburst was made on June 11, when the star was magnitude 11.7 and still fading rapidly.

The third maximum occurred on June 22, 1936, and is well covered by Harvard plates. The rise of the nova was found to be amazingly rapid, as appears to be characteristic of recurrent novae in general. The increase from magnitude 10.96 to 8.99 was accomplished in an interval of 2^h 56^m and the nova reached a maximum of 8.8 approximately $19\frac{1}{2}$ hours later. The decline was equally rapid; within two days the nova had dimmed to magnitude 10.4, and by July 14 it was down to 15th magnitude. Between outbursts the star remains at about 18th magnitude, and is difficult to observe visually even in large telescopes. Consequently, U Scorpii has not been subjected to the detailed studies with which astronomers have honored T Corona Borealis and RS Ophiuchi; both of these stars may be observed when at minima in relatively small telescopes.

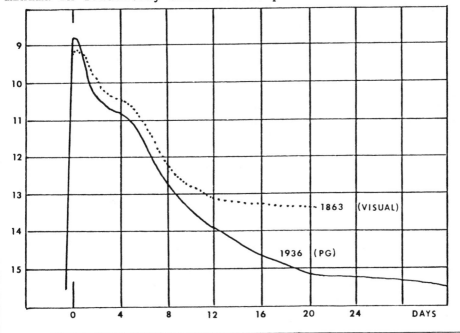

DESCRIPTIVE NOTES   (Cont'd)

The light range of U Scorpii at each explosion was about 9 magnitudes, corresponding to a luminosity increase of about 4600 times. The real brightness of the star is still a matter for investigation, since the distance is not accurately known. Studies of the two brightest recurrent novae have made it virtually certain that these stars attain an absolute magnitude of -7 or -8 at maximum, and thus have luminosities comparable to the normal "classical" novae. If U Scorpii is really a member of this same group, its apparent faintness implies an enormous distance. C.P. Gaposchkin has derived a distance of nearly 45,000 light years, and the resulting absolute magnitudes and luminosities are:

At maximum -6.8 = Luminosity of 50,000 X Sun
At minimum +2.5 = Luminosity of   9 X Sun

These values are consistent with the results obtained for T CorBor and RS Ophiuchi, but lead to another peculiar difficulty. At a distance of 45,000 light years the apparent galactic latitude of 20° corresponds to an actual distance of about 16,000 light years above the plane of the Galaxy. This is an improbably high value, and suggests either that the star is heavily obscured by interstellar haze in the Milky Way, or that the star is a "dwarf nova" and is not comparable in luminosity to the T CorBor stars. In either case the computed distance would be much over-estimated.

The postulated existence of dwarf novae is a most interesting suggestion. The recurrent nova WZ Sagittae is one star, at least, which appears to fall definitely into this category. An investigation by J.L.Greenstein has shown the star to be a white dwarf with an absolute magnitude at minimum of +10 or +11; from the observed range of 9 magnitudes the maximum luminosity is seen to be about 25 times that of the Sun. In range and interval between outbursts WZ Sagittae resembles the recurrent novae, but in actual luminosity it more nearly compares with the explosive dwarf "SS Cygni" or "U Geminorum" stars. Possibly the dwarf novae represent an intermediate stage between these two classes.

It is interesting to note that T Coronae and RS Ophiuchi are both close binary stars, and that all the well studied SS Cygni stars are also close double systems. In each case, the explosive star is a hot dwarf or subdwarf, and, according to one theory, its outbursts are probably "triggered" by the steady accretion of material from the cooler companion. WZ Sagittae was found to be an extremely close binary in 1961; the period is 1.36 hour and the system closely resembles both the SS Cygni stars and the classical nova DQ Herculis. It would therefore be extremely interesting to learn whether U Scorpii is also a close binary. Unfortunately the star is so faint at minimum that it remains virtually unstudied, and no spectra have been obtained during the brief maxima. The observed outbursts have been so short that it is quite likely a number have been missed. This is one star which should be kept under continual survey. Interested observers should become thoroughly familiar with the field (Page 1690) so that any reappearance of the nova will be immediately detected and reported. The position of the star is shown by the small circle at the center of the field; the star itself is beyond the range of the telescope employed, except during an outburst. The brightest star on the chart, near the top, is GC 22018, magnitude 6.7.

(Refer also to T Corona Borealis, RS Ophiuchi, WZ Sagittae, and T Pyxidis. The SS Cygni stars are described under SS Cygni, U Geminorum, SS Aurigae, and Z Camelopardalis. For a brief survey of the "classical" novae, refer to Nova Aquilae 1918)

RR   Variable star. Position 16534s3030, about $6\frac{1}{2}°$ SE of Antares. This is the brightest of the Mira-type stars in Scorpius, and the only one which reaches naked-eye visibility on occasion, rising to magnitude 5.1 or 5.2. It lies approximately 1° WSW from the bright globular star cluster M62, where it may usually be identified without much difficulty from its fine ruddy tint, which resembles that of R Leonis. RR Scorpii was discovered by Mrs. W.Fleming at Harvard in 1894; ten maxima of the star were observed in the interval from July 1889 to August 1903 and the derived period was 281 days. According to the Moscow *General Catalogue of Variable Stars* (1970) the best mean value for the period at present is 279.74 days. The star declines to magnitude 12 or so at minimum, and may then be somewhat difficult to locate among the many faint

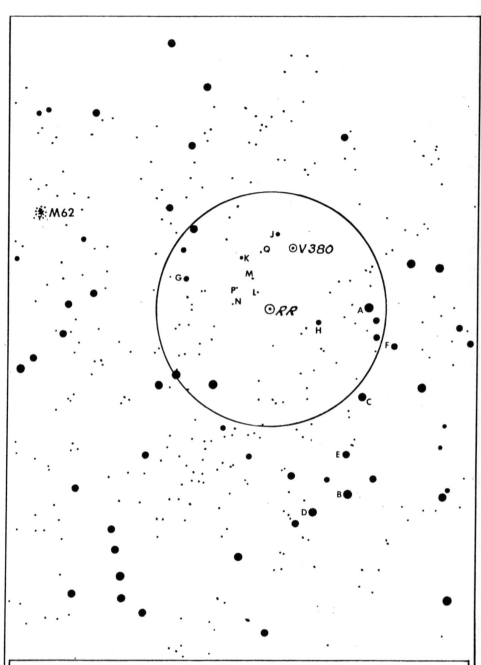

RR SCORPII CHART, adapted from an AAVSO chart. The circle is 1° in diameter with north at the top; stars to about magnitude 12.5 are shown.

COMPARISON MAGNITUDES (AAVSO) A= 6.4; B= 6.8; C= 7.4; D= 7.5; E= 8.1; F= 8.7; G= 9.9; H= 10.2; J= 10.6; K= 10.8; L= 11.7; M= 12.0; N= 12.1; P= 12.2; Q= 12.7.

DESCRIPTIVE NOTES   (Cont'd)

stars in the field. With sufficient aperture, however, the strong red color makes the identification somewhat easier. In period, range, and appearance, the star closely resembles R Leonis, but is not as well known an object owing to its position rather low in the southern sky. The light curve below is based on the observations of Leon Campbell of the AAVSO, and shows that the star has a somewhat more symmetrical cycle than many of the long-period variables; the time required to reach maximum is only slightly less than one-half the mean period. During the rise to peak luminosity the spectral type changes slowly from M8 to M6e, showing the usual bright lines of hydrogen. The spectral features suggest a luminosity class of II or III. From its estimated distance of about 600 light years, RR Scorpii would appear to have an actual peak luminosity of about 250 times that of the Sun; the absolute magnitude may be about -1 or -1.5.   The star shows an annual proper motion of 0.02"; the radial velocity is about 28 miles per second in approach.

To identify RR Scorpii when the star is faint, the chart on page 1696 will be found of value. The circle is 30' in diameter with north at the top, and the scale is close to 20" per mm. Comparison magnitudes, courtesy of the AAVSO, are given with decimal points omitted; thus "121" indicates a magnitude of 12.1. The star V380, about 16' to the north and slightly west, is a semi-regular red variable which ranges from about 9.5 to 10.5 in a cycle of about 187 days; it is a red giant of type M0.

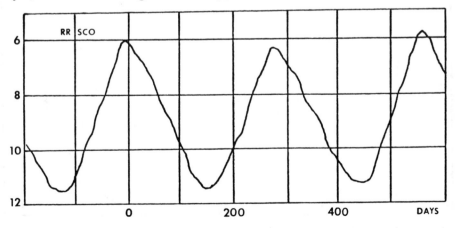

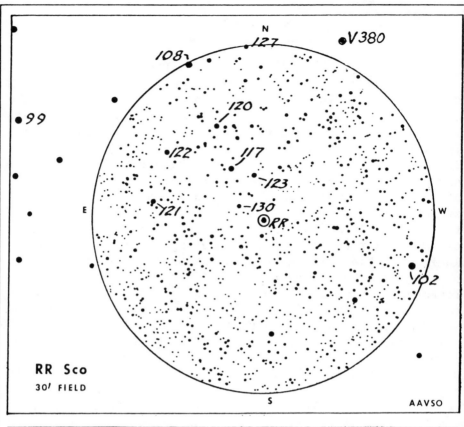

N
V380

108
127

99

120

122
117

123

E
121
130
RR

W

102

RR Sco
30' FIELD

S

AAVSO

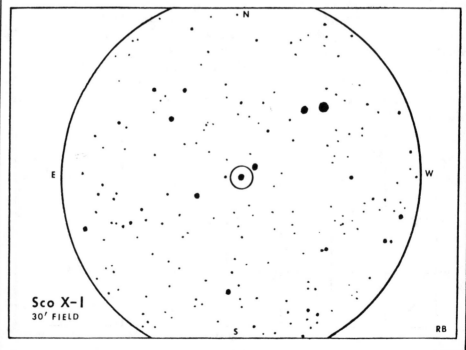

N

E

W

Sco X-1
30' FIELD

S

RB

1696

SCORPIUS X-I   (V818) Position 16171s1531. The strongest discrete X-ray source in the sky,
and the first to be identified with an optically visible
stellar object. It is located in the northern portion of
the constellation, about 5½° NE from Beta Scorpii and some
2.4° north of the recurrent nova U Scorpii. Sco X-1 was
first detected in June 1962, during the flight of an Aerobee rocket launched by a research group headed by R.Giacconi
of American Science and Engineering, Inc., and B.Rossi of
M.I.T. From the results obtained with two separate counters
it was shown that the radiation was in the X-ray region of
the spectrum, in the range of 2 - 8 angstroms. The following year a study made at the Naval Research Laboratory
determined the position to within 1°, placing it in northern Scorpius.

With this information as a guide, astronomers were
able to identify the precise position from observations
made from a NASA rocket launched at White Sands, New Mexico
on March 8, 1966. By this time it was known that the total
X-ray emission of the object was about equivalent to the
visible light of a 5th magnitude star; on any reasonable
theory of X-ray emission the object was expected to appear
7 or 8 magnitudes fainter in the visible range, giving an
expected visual magnitude of about 13. With the position
accurately known, astronomers soon found exactly the object
expected at the computed spot; a hot, bluish, variable star
of the 13th magnitude, resembling a former nova. The identification was probably first made by Prof.M.Oda at Tokyo
Observatory in June 1966, and independently by C.B.Stephen-

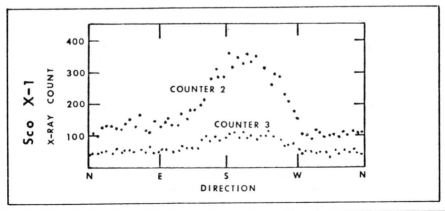

son and H.M.Johnson at Kitt Peak. The object showed bright hydrogen and helium emission lines, irregular light variations of as much as a magnitude in 24 hours, and rapid "flickerings" on a time-scale of minutes. A check of the Harvard plate collection has shown that Sco X-1 has been varying in this way since at least 1896, but the over-all mean brightness has not changed significantly in over 70 years. According to the Moscow *General Catalogue of Variable Stars* (1970) the extreme range is about 3 magnitudes.

Spectra obtained by Dr.A.Sandage with the 200-inch reflector show a bright continuum with hydrogen and helium emission lines and high-excitation features of nitrogen and carbon (plus very conspicuous mercury lines at $\lambda 4047$ and $\lambda 4358$ from the lights of Los Angeles!) The only dark (absorption) lines are those of interstellar ionized calcium; from the strength of these [Ca II] lines the distance of Sco X-1 is thought to be about 500 parsecs (1600 light years); the derived visual luminosity is then about equal to that of the Sun. In color and luminosity Sco X-1 resembles an old nova, but the intense X-ray emission represents about 1000 times more energy than the visible light of the object. Various theoretical models have been proposed to explain this feature, most of them requiring the supposition that the star, like typical former novae, is actually a close binary.

G.R.Burbidge and K.H.Prendergast suggested in 1968 that the object is a very thick shell of extremely hot gas growing through accretion, on the surface of a very dense white dwarf. I.S.Shklovsky (1967) finds evidence that "the source is a neutron star forming a comparatively massive component of a close binary system. A stream of gas flowing out of the second component is permanently incident on the neutron star..... It is suggested that the optical object accompanying the X-ray source might be a cool dwarf with half of its surface heated by a strong flux of hard X-rays from the source..." On this model the actual hot disc or shell emitting the X-ray energy is probably comparable in size to the Earth; if the derived distance is reasonably correct its actual diameter must be 1500 to about 15,000 miles. A similar picture of Sco X-1 has been proposed by F.Pacini and E.E.Salpeter of the Center for Radiophysics and Space Research at Cornell, and K.Davidson of the

Astronomy Department at Princeton (1971). In their "Cocoon
Pulsar Model" a rapidly rotating neutron star is embedded
in an extremely hot accretion cloud or disc having a diameter of about 10,000 km.; its computed temperature is close
to 50 million degrees K. The source of the cloud is evidently a close companion star.

Evidence for the binary nature of Sco X-1 has been
obtained in two studies made in 1975. E.W.Gottlieb, E.L.
Wright, and W.Liller have determined magnitudes from 1068
plates in the Harvard collection, and find a definite periodic variation with an amplitude of 0.22 magnitude and a
periodicity of 0.787313 day; the plates cover the period
from 1889 to 1974. From a study of spectrograms made with
the Kitt Peak 2.1 meter reflector, A.P.Cowley and D.Cramp-
ton find a periodic variation in the radial velocity, with
a range of 120 km/second, in a period of 0.787 day, very
closely matching the magnitude variation. Each component
seems to have a mass of less than 2 suns, and the computed
semi-major axis of the spectroscopic orbit is a little less
than 400,000 miles. The secondary is thought to be a star
which has evolved somewhat off the main sequence; the X-ray
energy originates in the gas stream which is heated to
tremendous temperatures as it falls through the extremely
strong gravitational field of the degenerate star or neutron star. Sco X-1 is also a radio source, of only moderate
intensity, but apparently multiple in nature; two secondary
sources are detected beside the main body, at distances of
1.2' and 1.4'.

Scorpius X-1 also has a variable star designation,
V818; its number in the standard catalogue of X-ray objects
is 3U1617-15. (Refer also to Cygnus X-1, page 793)

M4   (NGC 6121) Position 16206s2624. Fine globular
star cluster, one of the largest objects of
its type, and also one of the nearest. It is probably the
easiest of all the bright globulars to locate; merely point
the telescope to Antares, and then move 1.3° directly west,
and there you are. M4 is located quite easily in binoculars
and may even be detected without optical aid under good sky
conditions. John H.Mallas and Walter S.Houston have both
found it visible to the naked eye from dark sky sites; to
observers in the Southern Hemisphere it nearly equals M13

STAR CLUSTER M4 in SCORPIUS. One of the closest globular
clusters to the Solar System. Lowell Observatory photograph
made with the 13-inch telescope.

in Hercules, with a total visual magnitude of about 6½.
M4 was probably first observed by P.L. de Cheseaux in 1746;
he described it as "close to Antares....white, round and
smaller than the preceding ones. I do not think it has been
found before.." Messier, in May, 1764, described it as a
"cluster of very small stars; with an inferior telescope it
appears more like a nebula; this cluster is situated near
Antares and on the same parallel. Observed by M.de la
Caille and reported in his catalogue....Diam 2½' " Bode
and Lacaille both thought it resembled the hazy nucleus of
a comet. To Admiral Smyth it was "a compressed mass of
small stars with outliers and a few small stellar companions in the field. It is elongated N-S and has the aspect
of a large pale granulated nebula, running up to a blaze in
the centre."
    M4 is a rather loose cluster, showing no great cen-
tral condensation, and begins to show resolution into stars
in a good 4-inch refractor. The brightest detail and the
first to appear is a curious central "bar" or chain of 11th

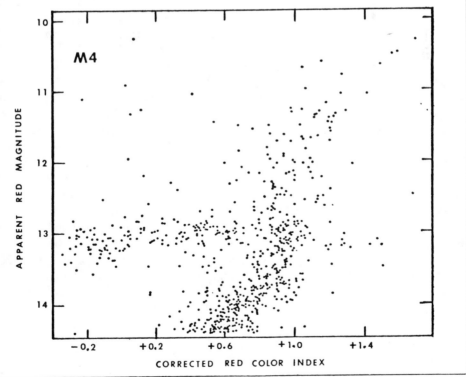

magnitude stars, some 2.6' in length, oriented toward PA
12°; this was noted by Herschel in 1783 and described as a
"ridge of 8 or 10 pretty bright stars running from the
middle to N.f." With larger apertures many other chains of
stars are discerned around the edges of the cluster, flowing outward in the form of curved loops and streams. The
central bar makes the cluster appear quite oblate in small
instruments, but long exposure photographs show that the
outer distribution of stars is very nearly spherical.

The first detailed study of the stars of M4 was made
by J.L.Greenstein in 1939; magnitudes and colors for 660
stars were determined, including all stars brighter than
15.6 (pg) within 13.5' of the cluster center. The H-R diagram on page 1701 is based on this survey, which showed that
M4 is reddened by about 0.8 magnitude by dark nebulosity in
the region. Greenstein called M4 a "dwarfish globular compared to other clusters in the Galaxy", as the population
is relatively poor in giant stars. Only 210 stars brighter
than absolute magnitude +0.5 (pg) are found in M4, whereas
the Hercules Cluster M13 contains at least 1000 such giants.
In a star count reaching to magnitude 19.3, slightly over
10,000 star images have been detected in M4; the number of
fainter stars will probably never be known.

According to H.B.S.Hogg's *Bibliography of Individual
Globular Clusters* (First Supplement, 1963) the total apparent diameter of M4 is 22.8', the total integrated magnitude
(pg) is 7.41, and the average magnitude of the 25 brightest
stars is 13.1. The cluster is receding at a velocity of 39
miles per second. At the time of the Greenstein survey, 35
variable stars were known in M4; the total known in 1973
was 43, all but three of which appear to be classic short
period RR Lyrae type stars. Two semi-regular red variables
have been identified with periods of about 60 and 40 days
and spectral types of G or K. Another object (Variable #6)
has been identified by A.de Sitter as an eclipsing binary
of the short-period W Ursae Majoris class, with a range of
13.6 to 14.3, in a period of 0.64103 day. All the RR Lyrae
stars lie in the very conspicuous "horizontal branch" on
the H-R diagram , at apparent red magnitude 13. From studies of these stars, the apparent modulus of the cluster is
about 12.4 magnitudes, making this definitely one of the
nearest globulars to the Solar System.

Greenstein, in 1939, derived a distance of about 1900 parsecs (6200 light years) but the accuracy of the figure is uncertain owing to the uneven distribution of heavy obscuring clouds in the vicinity. In a study made in 1975 with the 1 meter telescope at Cerro Las Campanas, G.Alcaino obtained a probable distance of 1.75 kiloparsecs (5700 light years) which is in fair agreement with the earlier value; from these results it would appear that M4 may be closer to the Solar System than NGC 6397 in Ara, usually considered the nearest of the globulars. On the other hand, a recent study by C.R.Sturch (1977) gives a somewhat greater distance, of about 2.2 kiloparsecs; he concludes that "within the accuracy of the determinations, M4 and NGC 6397 are at equal distances from the Sun".

The present author had the opportunity, some years ago, to view M4 through the fine 40-inch Ritchey-Chretien reflector at the U.S. Naval Observatory station in Flagstaff. It happened that a few days before, he had obtained a small but fascinating device called a *spinthariscope* in which the effect of radioactivity is made visible to the eye; ever since he has mentally associated M4 with alpha particles! As many users of this book may know, the spinthariscope consists of a short metal tube fitted with a viewing lens at one end, and a fluorescent zinc-sulphide screen at the other. A tiny speck of radium is held about ¼ inch over the screen on the tip of a thin metal wire. As the radium atoms disintegrate, the alpha particles are expelled at a velocity of about 12,000 miles per second, producing an endless rain of star-like flashes on the screen. The observer, after a period of dark-adaption, looks into the lens to see a view resembling M4 "brought to life" with hundreds of microscopic "stars" blazing up and vanishing every second. The experiment is a very beautiful one, and an excellent entertainment for visitors to your observatory on cloudy nights. It is also, incidentally, about as close as one can come to actually seeing an individual atom. The alpha particle, with an atomic weight of 4, contains two protons and two neutrons, and is identical to the nucleus of a helium atom. (Refer also to M13 in Hercules, Omega Centauri, M3 in Canes Venatici, M5 in Serpens, and NGC 6397 in Ara. A series of diagrams illustrating globular cluster age-dating will be found on pages 990- 992)

A FIELD OF STAR CLUSTERS IN SCORPIUS. The bright group at upper right is M6; the cluster involved in the star cloud at the left is M7. Lowell Observatory photograph.

1704

M6   (NGC 6405) Position 17368s3211, some 5° north
of the tail of Scorpius, and about 3½° northwest from the similar cluster M7. These are two of the
largest and brightest of the galactic star clusters, and
among the most suitable objects of the type for the small
telescope. The discovery of M6 is usually credited to P.L.
de Cheseaux in 1746, though the cluster is a definite naked
eye object, and appears to be mentioned, along with M7, in
the catalogue of Ptolemy. They seem to be the *Girus ille
nebulosus* of the 1551 edition of the *Almagest*, and also
appear in Ulug Beg's star catalogue as the *Stella nebulosa
quae sequitur aculeum Scorpionis*, "The Cloudy Ones Which
Follow the Sting". De Cheseaux, however, was probably the
first to identify M6 as "a very fine star cluster". In 1752
it was noticed by Lacaille who found it "an unusual cluster
of small stars disposed in three parallel bands forming a
lozenge 20' - 25' in diameter..." Messier, in 1764, found
it "a cluster of small stars between the bow of Sagittarius
and the tail of Scorpius. To the naked eye it resembles a

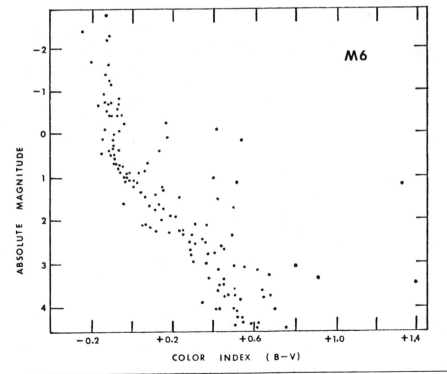

nebula without a star, but even a small telescope reveals
it as a cluster of small stars. Diam 15'." Flammarion saw
here "stars of 7 - 10 mag. very dispersed and arranged in a
remarkable pattern...Three starry avenues leading to a
large square". The present author regards this as cne of
the most attractive clusters in the heavens for veiy small
instruments, a completely charming group whose arrangemen⁞
suggests the outline of a butterfly with open wings. The
main portion of M6 just fills a 25' field, and the cluster
is at its best in a good 6 or 8-inch glass with wide-angle
oculars. Visually, the brightest member is the golden K-
type giant which highlights the NE wing-tip; this is the
semi-regular variable BM Scorpii which has a range of about
2 magnitudes and a long cycle of about 850 days. It appears
as #1 in the list below and the chart opposite.

In a study by A.Wallenquist (1959) some 80 cluster
members were identified in a field 54' in diameter, giving
a computed central density of about 25 stars per cubic parsec, and about 0.6 star per cubic parsec for the over-all
density of the entire cluster. The "Butterfly" figure is
roughly 9 light years in width, but the total diameter of

CLUSTER M6 - MAGNITUDES AND SPECTRA								
1.	6.17	K0-K3	18.	9.06	B8	35.	9.86	B8
2.	6.76	B8	19.	9.06	B9	36.	9.88	G8:
3.	7.18	B5	20.	9.08	B8	37.	9.88	-
4.	7.26	B4	21.	9.19	A7	38.	9.94	B8
5.	7.27	B8	22.	9.20	B8	39.	10.00	B8
6.	7.88	B9	23.	9.34	B8	40.	10.01	A2
7.	8.08	B7	24.	9.41	B8	41.	10.16	-
8.	8.26	B5	25.	9.41	F6	42.	10.17	-
9.	8.34	B8	26.	9.48	B9	43.	10.17	-
10.	8.53	-	27.	9.51	B6	44.	10.19	-
11.	8.62	B9	28.	9.66	B9:	45.	10.23	B8
12.	8.76	B8	29.	9.67	F6	46.	10.23	B9
13.	8.76	B6	30.	9.80	B6	47.	10.28	B9
14.	8.77	B9	31.	9.83	-	48.	10.37	A2
15.	8.78	B6	32.	9.83	B8	49.	10.40	A0
16.	8.91	B7	33.	9.84	B8	50.	10.42	-
17.	8.96	B7	34.	9.85	-	51.	10.51	A1

the entire group is close to 20 light years. In a study made at Harvard's Boyden Observatory in South Africa in 1958, photoelectric magnitudes and colors were obtained for 132 stars in the cluster by K.Rohlfs, K.W.Schrick, and J. Stock, using the Boyden 60-inch reflector. Their results for the cluster members brighter than magnitude 10.5 are presented in the table on page 1706; star numbers are keyed to the chart below. The great majority of the members are B-type main sequence stars, with the striking exception of star #1 which has evolved to the yellow giant stage. The computed age of the cluster is about 100 million years, older than the Pleiades, but less than half the age of the nearby cluster M7. Published distances for M6 show rather large discrepancies, a common circumstance in objects in heavily obscured regions. Rohlfs and his associates found a value of about 630 parsecs (about 2000 light years) but newer studies of the absorption in the region have reduced the figure to something in the range of 400 - 450 parsecs.

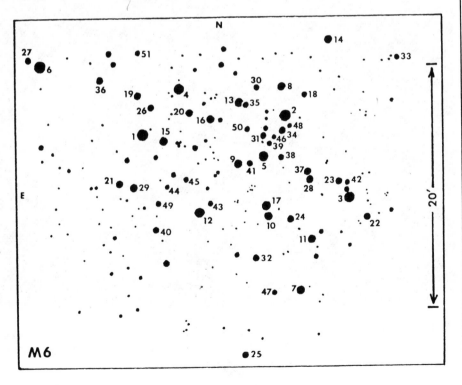

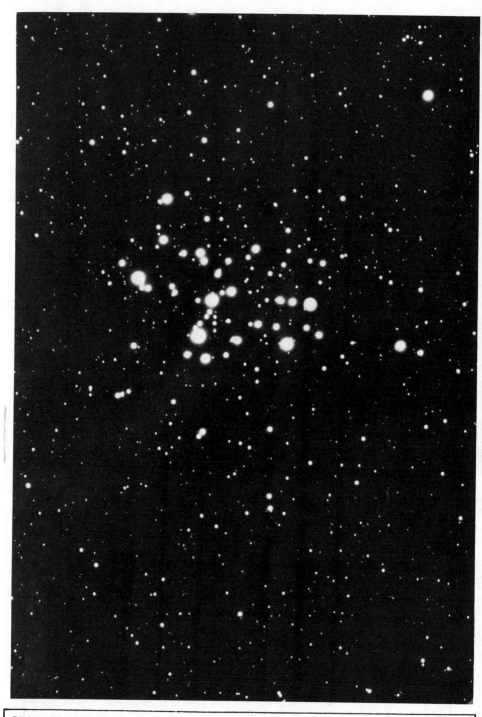

STAR CLUSTER M6 in SCORPIUS, the "Butterfly Cluster". This print is oriented with south at the top. Lowell Observatory photograph made with the 13-inch telescope.

DESCRIPTIVE NOTES   (Cont'd)

A.Becvar, in the 1960 edition of the Skalnate Pleso *Atlas Coeli Catalogue*, gives the distance as 570 parsecs or about 1860 light years; H.B.S.Hogg in her *Catalogue of Galactic Clusters* has a value of 0.4 kpsc (1300 light years); while K.G.Jones in his book on the Messier Objects (1968) gives 405 parsecs as the best modern estimate. There is a general agreement that M6 is perhaps 50% more remote than M7.

Several fainter clusters will be found in the close vicinity of M6. Just 1.2° to the WSW lies the scattered group NGC 6383; the brightest star is the unequal double h4962, magnitudes about 5.5 and 10.5, separation 5.5". Some 50' to the east of M6 will be found another sprinkling of star-dust in the form of cluster NGC 6416. A more difficult object is the group called NGC 6404, just 1° south of M6; it is scarcely more than an uncertain concentration of 13th magnitude stars in a Milky Way field. All these clusters lie within 1.5° of the Galactic Equator. The fine cluster M7 is easily located, about 3½° to the southeast.

M7   (NGC 6475) Position 17507s3448. Fine galactic star cluster lying about 4° NE from Lambda Scorpii in the Scorpion's tail, and about 3½° SE from M6. M7 is a large and brilliant group, easily detected with the naked eye, and one of the few clusters which can be thoroughly appreciated in a good pair of field glasses. It is mentioned in the catalogue of Ptolemy, and in the 16th Century Latin translation of the *Almagest* appears as the *Girus ille nebulosus*, the reference probably including both M7 and M6. The Arabian name *Tali al Shaulah* is the equivalent of the Latin translation of Ulug Beg's title: *Stella nebulosa quae sequitur aceleum Scorpionis*, "That which follows the Sting". Hevelius includes M7 in a list published in 1690, and it appears again in W.Derham's short catalogue of "nebulous stars" in 1730. Lacaille observed it at the Cape of Good Hope in 1751, and found it to be "a group of 15 or 20 stars, very close together in a square figure". Messier, in May 1764, described M7 as "a cluster considerably larger than the preceding [M6]. It appears to the naked eye as a nebulosity; it is situated a short distance from the preceding, between the bow of Sagittarius and the tail of Scorpius. Diameter 30'." The cluster is seen projected on a background of numerous faint and distant Milky

Way stars, while the bright stars of the group are close
to naked-eye visibility. On Lowell Observatory 13-inch
telescope plates the bright central portion of the cluster
just fills a 30' field; the total apparent diameter is possibly about 50'. According to a study by A.Wallenquist in
1959, M7 contains 80 stars brighter than 10th magnitude in
a field 1.2° in diameter. The group as a whole resembles
Praesepe (M44) in Cancer, though somewhat smaller, and
would certainly be as well known if it were farther north.
Incidentally, this is the southernmost object in the catalogue of Messier, and is not too well placed for observers
in Great Britain or the northern U.S.

M7 has been the subject of several detailed studies,
and the distance appear to be well known, at about 800
light years. Some individual results are reported here:

   R.J.Trumpler (1930)---------------- 240 parsecs
   A.Wallenquist (1931)-------------- 212 "
   D.Koelbloed (1959)---------------- 233 "
   A.A.Hoag & N.L.Appelquist (1965)-- 250 "
   M.S.Snowden (1975)---------------- 251 "

In the most recent study reported here, that of M.S.Snow-
den, a precise modulus of 7.06 magnitudes was obtained,

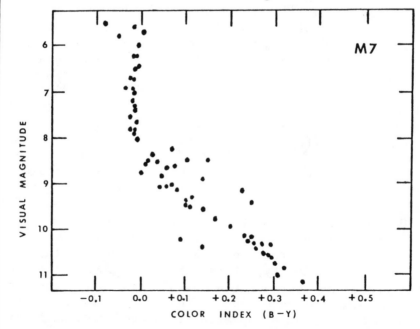

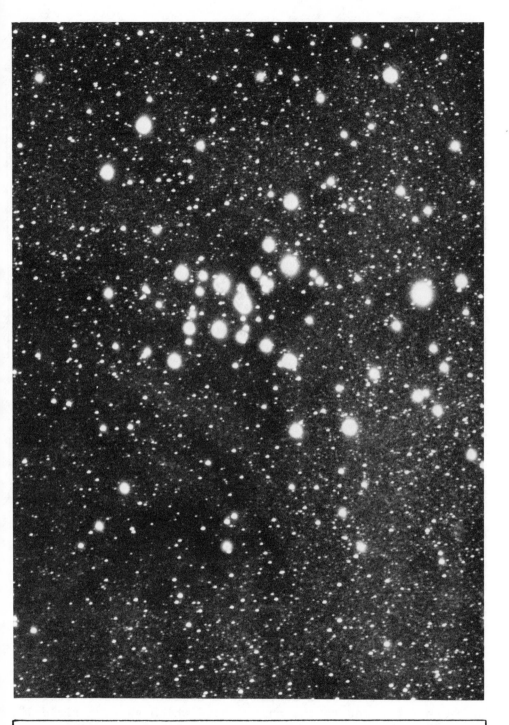

STAR CLUSTER M7 in SCORPIUS. This bright group is easily observed in binoculars or the small telescope. Lowell Observatory photograph made with the 13-inch telescope.

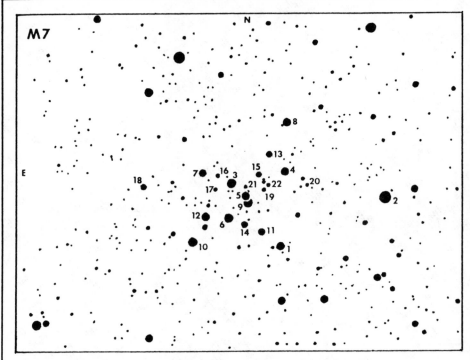

which converts to a distance of 818 light years. The color-magnitude diagram for M7, shown on page 1710, is based on the Snowden study, and shows that the members brighter than apparent magnitude 7.5 are somewhat evolved off the main sequence. The derived age of the cluster is about 260 million years, considerably older than the Pleiades, but much younger than the globular clusters.

In a study made at the U.S.Naval Observatory in 1961, Dr.A.A.Hoag and his associates measured accurate magnitudes and colors for the M7 members brighter than magnitude 11½; the results for the 22 brightest stars are given in the table opposite, with spectral types from various sources added to the list. From a comparison of different catalogues, it appears that the spectral types, in many cases, may be uncertain by 1 or 2 tenths of a spectral class. Star numbers are keyed to the chart above which shows a field 1° wide (E-W) centered on the cluster; north is at the top.

Dr.H.A.Abt at Kitt Peak (1970) has identified eight spectroscopic binaries in this cluster, including the stars numbered 3, 7, 9, 13, 15, and 16. According to his study

CLUSTER M7 - MAGNITUDES AND SPECTRA								
1.	5.60	gG8	9.	7.16	B9	17.	8.80	A2
2.	5.89	B6	10.	7.24	B9	18.	8.78	A1
3.	5.96	B9+B9	11.	7.37	B9	19.	8.93	A8
4.	6.17	B8	12.	7.48	B9	20.	8.99	A4
5.	6.38	B9	13.	7.61	B9	21.	9.01	A3
6.	6.45	Ap	14.	7.77	A1	22.	9.06	A3
7.	6.88	B9	15.	8.21	B9			
8.	6.97	A0	16.	8.49	A1			

the early type stars of the cluster show unusually slow
rotational velocities; whereas the late type members seem
to be rotating faster than normal. For the six numbered
binaries, he derives the following orbital periods:

#3= 2.7754 days	#13= 9.499 days
#7= 6.226 days	#15= 5.4505 days
#9= 3.051 days	#16= 6.052 days

In addition to these stars, the cluster also contains at
least three visual double stars. The brightest of these is
λ342 on the SW edge of the central mass (#1 on the chart),
discovered by T.J.J.See in 1897 and a rather difficult object for moderate telescopes. The components are both about
magnitude 6½, separation 0.4", with a slowly diminishing PA
from 286° in 1897 to 242° in 1959. Star #3 is B1871, a very
close pair of 0.1"; the PA has increased from 87° in 1929
to about 111° in 1957. Equally difficult is Star #4, found
to be a 0.5" pair by S.W.Burnham in 1889, and catalogued by
him as β1123. The separation at discovery was about 0.5" in
PA 213°, but the components have since closed down to about
0.1" in PA 341° (1959). E.J.Hartung (1968) finds that λ342
may be seen elongated, but not truly resolved, with an
aperture of 30 cm. The other two doubles are observable in
large telescopes only. All these stars show the same radial
velocity as the cluster, about 5 miles per second in approach.

A faint globular star cluster, NGC 6453, lies in the
field of M7, just 20' NW from star #2; it appears as an 11^m
fuzzy spot, about 1' in diameter, and was first noticed by
John Herschel in June 1837. Some 45' SE of M7 lies the dim
galactic cluster called H18 with about 80 faint stars.

STAR CLUSTER M62 in SCORPIUS. A very dense cluster in a
rich Milky Way field. Lowell Observatory photograph made
with the 13-inch telescope.

M62   (NGC 6266) Position 16581s3003. Globular star
              cluster, located squarely on the Ophiuchus-
Scorpius border, about 7° SE from Antares, or about 40% of
the way along a line drawn from Antares to cluster M7. In
some observing guides M62 is catalogued under Ophiuchus.
The cluster is one of Messier's discoveries, found in June
1771, and re-observed in 1779 when it was described as "a
very fine nebula; it resembles a little comet..It is bright
in the centre and is surrounded by a faint glow..." The
identification of M62 as a star cluster was first made by
William Herschel; he thought it a miniature of the cluster
M3 in Canes Venatici. Admiral Smyth saw here "A fine large
resolvable nebula; an aggregated mass of small stars running up to a blaze in the centre". The thickest massing of
stars, according to J.E.Gore, is a "perfect blaze, but not
quite in the centre".
        H.B.S.Hogg, in the First Supplement to her *Bibliog-
raphy of Individual Globular Clusters* (1963) gives the
total apparent diameter as 6.3' and the total integrated
magnitude (pg) as 8.16; the integrated spectral type is F8
and the average magnitude of the 25 brightest stars is 15.9.
The cluster is seen against, and is probably embedded in, a
rich Milky Way star field, so that the area, for many degrees around the group, is sprinkled with multitudes of tiny
star-sparks. This is one of the globulars which appears to
be actually immersed in the starry hub of the Galaxy. Light
loss in the area, from absorption, is estimated to be about
2.4 magnitudes.
        M62 is one of the most unsymmetrical clusters; the
non-spherical outline was probably first noticed by John
Herschel in 1847, and remarked upon by S.I.Bailey in 1915.
Shapley called it "the most irregular globular cluster" and
from star counts determined that the major axis is oriented
toward PA 75°. The lack of symmetry was found to be "marked
not only in the distribution of stars but especially in the
distribution of variables, 19 being found north of the centre and 7 to the south..." On Lowell Observatory 13-inch
telescope plates, the powdering of faint stars seem noticeably richer on the north edge of the cluster, as compared
with the south edge; possibly there is greater obscuration
along the south rim. K.G.Jones finds that the central condensation is actually "well to the SE of the centre", and

regards M62 as one of the most "comet-like" of all the
Messier objects; he thinks that the color is "slightly
bluish". E.J.Hartung calls M62 a "charming object" but sees
only beginning resolution with an aperture of 30 cm.

Eighty-nine variables have been catalogued in this
cluster up to 1973; virtually all of which are short-period
RR Lyrae type stars. The derived distance is about 26,000
light years, uncertain because of variable obscuration in
this rich region near the Galactic Hub. The cluster shows
a radial velocity of 46 miles per second in approach.

The bright red variable star RR Scorpii lies just 1°
from M62, toward the WSW. (Refer to page 1693)

M80 (NGC 6093) Position 16141s2252. A small but
very bright globular star cluster located a
little more than 4° NW from Antares, about midway between
Antares and Beta Scorpii. The discovery of M80 has been
credited to both Messier and Mechain, both of whom recorded the cluster in January 1781; Messier's observation seems
to have preceded that of his friendly rival by about three

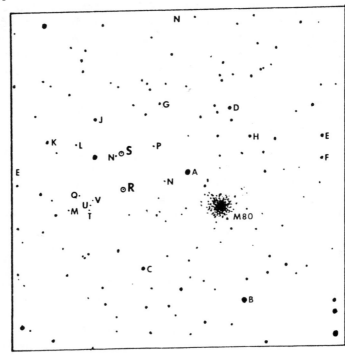

M80

30' FIELD

STAR CLUSTER M80 in SCORPIUS. This very dense cluster is here shown with the Lowell 13-inch telescope (top) and with the 42-inch reflector (below)

weeks. Messier described it as "a nebula without star in the Scorpion, between the stars g.[Rho Ophiuchi] and Delta; compared with g. to determine its position. The nebula is round, the centre brilliant and it resembles the nucleus of a little comet, surrounded with nebulosity.." Sir William Herschel in 1785 referred to it as "the richest and most condensed mass of stars which the firmament can offer to the contemplation of the astronomer". Between the cluster and Rho Ophiuchi begins the large heavily obscured region stretching for about 4° to the east, which Herschel called a "Hole in the Heavens". The cluster, being on the western edge of this starless gap, suggested to Herschel the whimsical suspicion that "the stars of which it is composed were collected from that place and had left the vacancy.." John Herschel found M80 a fine object, brightening to a central blaze, all resolved into stars 14..15 mag. Some hint of resolution is evident in a good modern 6 or 8-inch telescope, but the true splendor of M80 is reserved for the fortunate few who have access to great telescopes.

    H.B.S.Hogg gives the total angular size as 5.1' and the total integrated magnitude (pg) as 8.39; the integrated spectral type is about F4. The average magnitude of the 25 brightest members is 14.9. From current studies the distance of this cluster is believed to be about 11 kiloparsecs

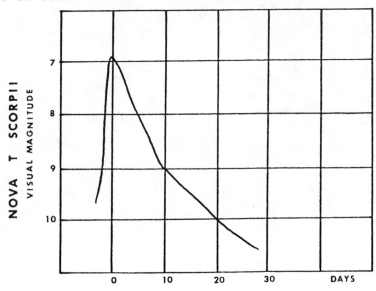

or roughly 36,000 light years; the actual diameter must
then be about 50 light years. M80 shows only a moderate
radial velocity of 11 miles per second in recession. Light
absorption in the region is estimated to be about 0.9 magnitude; if completely unobscured the total visual magnitude
of the cluster would be about 6.8 which corresponds to -8.4
absolute, or about 190,000 times the light of the Sun. At
the same distance, the Sun would appear as a star of magnitude 20!

Only eight variable stars are known in M80, a count
which includes the two long-period red variables R and S
Scorpii, thought to be members of the cluster although they
are 9' and 10' distant, to the E and ENE. These have periods of 223 and 177 days, respectively, and were first detected by J.Chacornac in 1853 and 1854. In each case the
magnitude range is from 9.8 or so to about 15.5. An identification field for these stars appears on page 1716. Comparison magnitudes, courtesy of the AAVSO, are given here:
A= 8.3; B= 8.7; C= 10.2; D= 11.1; E= 11.6; F= 12.0;
G= 12.4; H= 12.6; J= 12.8; K= 13.2; L= 13.4; M= 13.7; N=
14.0; P= 14.2; Q= 14.4; T= 14.5; U= 14.6; V= 15.0.

THE NOVA OF 1860. The famous star which is now designated
T Scorpii was for many years the only nova on record in a
globular star cluster. At least one other case is now known
however, a faint nova which appeared in the globular M14 in
Ophiuchus in 1938, but remained undetected until 1964. Two
such cases favor the probability that these stars were true
cluster members, and not merely foreground stars.

The nova of 1860 was first seen by A.Auwers at Berlin
on May 21, at about the 7th magnitude. Dr.Auwers had observed the cluster three nights previously when, according to
Miss Agnes Clerke, "it presented its usual appearance of a
somewhat hazy ball of light, brightening gradually inward,
and resolvable with difficulty into separate stellar points
together constituting a closely-packed and most likely ex-
cessively remote globular cluster.... he saw that these
minnows [now] had a triton in their midst. A seventh magnitude star shone close to the center of the stellar group".
The nova was independently found by N.Pogson on May 28,
when the magnitude was estimated to be about 7.6; he found
the new star to be nearly central in the cluster, slightly

less than 5" from the geometrical center, The magnitude
estimates are undoubtedly rather uncertain owing to the
bright background of the cluster; the following figures
were published by Dr.Auwers:

May 21= 7.0	May 22= 7.0	May 24= 7.5
May 25= 7.8	June 3= 9.2	June 8= 9.7
June 12= 10.0	June 16= 10.5	

Pogson's account states that by June 10 "the stellar
appearance had nearly vanished, but the cluster yet shone
with unusual brilliancy and a marked central condensation".
The fading was considerably more rapid than usual for novae
in general; the star decreased by 2 magnitudes in the first
11 days. No observations after mid-June are recorded. In
the *Astronomical Journal* in 1902, J.Baxendell stated that
a re-brightening of the nova to the 9th magnitude had been
observed by Pogson in early 1864; but it still seems uncertain whether this report, never published by Pogson,
actually refers to the nova T Scorpii. J.Schmidt at Athens
kept a close watch on M80 for more than 10 years after the
outburst and never saw a reappearance of the star. If a
true member of the cluster, the nova had an actual peak
luminosity of about -8.5, close to 200,000 times the light
of the Sun, and about equal to the combined light of all
the rest of the cluster stars put together!
    The question of the probability of collisions between
stars in a dense globular cluster has been examined by J.G.
Hills and C.A.Day (1976) at the University of Michigan. In
a globular containing a million stars, some 335 collisions
have probably occurred during the cluster's 12-billion year
lifetime. In the case of M80 itself, one of the densest
globulars known, the probability rises to about 2700 star
collisions since the cluster was formed. Owing to the relatively low velocity of collision, it is uncertain whether
a nova outburst might result from such an event; the analysis suggests that the colliding stars "will tend to coal-
esce, with only a small fraction of their combined mass
escaping". In M80 it seems likely that more than 1.4% of
the cluster stars have experienced such an event in the
group's long history, and that the normal pattern of stellar evolution of such stars has been drastically altered.
(Refer also to M13 in Hercules, NGC 5139 in Centaurus, M3
in Canes Venatici, M5 in Serpens, M22 in Sagittarius, etc.)

DARK NEBULOSITY IN THE SCORPIUS MILKY WAY. The field shown here is about 2° SE of cluster M62. Lowell Observatory photograph made with the 13-inch camera.

DESCRIPTIVE NOTES   (Cont'd)

NGC 6231   Position 16507s4143. A brilliant galactic star cluster, a striking and impressive object for both the visual observer and the theoretical astronomer. It is located just half a degree north of the bright star Zeta Scorpii, and would require only a more favorable position in the northern sky to make it one of the most famous objects in the heavens. It was probably first noticed by Lacaille during his studies of the southern sky in 1755, but may be seen without optical aid under good conditions. In the small telescope it resembles a miniature edition of the Pleiades, with a central knot of 7 or 8 bright stars. The size of this central mass is about 6' but the outlying fainter members bring the total size to something like 15'. E.J.Hartung (1968) speaks of it as a "glorious cluster" with many bright stars, pairs and triplets, "which sparkle in patterns of lines and small groups". To the present author, this cluster produces the impression of a handful of glittering diamonds displayed on black velvet. There is very little color in this cluster; all the brighter stars appear brilliantly white.

NGC 6231 is remarkable for its population of high-luminosity 0 and B-type supergiants, including two Wolf-Rayet stars and several eruptive stars of the P Cygni type. Magnitudes and spectra for the bright members are given in the table below, according to studies made in 1953 by W.W. Morgan and G.Gonzales. Star numbers are keyed to the identification chart on page 1725. These stars are all giant and supergiant objects of intense brilliance; the absolute

NGC 6231-	MAGNITUDES	AND SPECTRA			
1.	5.33	08f	11.	7.08	B0
2.	5.54	B0	12.	7.21	09
3.	5.92	07p	13.	7.36	B0
4.	6.37	09	14.	7.49	08
5.	6.38	08	15.	7.70	09
6.	6.42	B1	16.	7.86	09
7.	6.53	WN7	17.	8.33	B0
8.	6.55	09	18.	8.38	B2
9.	6.56	06	19.	8.56	B0
10.	6.70	WC6	20.	8.77	09

NGC 6231 in SCORPIUS. One of the finest of the galactic
star clusters. Zeta Scorpii is the bright star at the lower
edge of the print. Lowell Observatory photograph.

magnitudes range from about −7.0 for star #1 to about −3.5
for star #20. The brightest of these objects is about equal
to Rigel in luminosity, emitting as much light as 60,000
suns. The two Wolf-Rayet members are #7 and #10; these are
strange and rare objects showing enormously wide emission
bands attributed to the violent expansion of a gaseous
shell, or to large-scale atmospheric turbulence. Stars #1
and #3 are P Cygni type objects, showing spectra which have
sometimes caused them to be called "permanent novae". The
cluster is rich in double stars; Numbers 2, 4, 5, 8, 15, 17
and 20 are spectroscopic binaries.

From recent studies (1966) the distance of NGC 6231
is believed to be about 5700 light years, which tells us
that the central concentration of the cluster is about 8
light years in diameter. It is interesting to reflect that
if this group was as near to us as the Pleiades, the two
clusters would appear nearly identical in size, but the
members of NGC 6231 would outshine the Pleiades stars by a
factor of about 50 times; the brightest members would then
shine with about the brilliance of Sirius!

NGC 6231 lies in a remarkable area of the sky. It was
noticed many years ago that the cluster itself appears to
be merely the nucleus of a large field or association of
bright O and B stars. A degree north and slightly east of
the cluster lies the richest portion of this outlying association, often marked on star atlases as a separate cluster
under the designation H12. Modern studies have shown that
the connection is definitely real, and that this great field
of high-luminosity stars, often called the "I Scorpii Association", marks the course of one of the spiral arms of our
Galaxy. This spiral arm, clearly outlined by an extended
stream of giant stars, lies some 5000 to 6000 light years
distant, nearer to the Galactic Center than the arm which
contains our Sun.

A photographic study of this region has been made
with a small wide-angle Schmidt camera at Mt.Stromlo in
Australia. In addition to the faint nebulosity IC 4628,
which lies about 1.5° north of NGC 6231, a much larger loop
of nebulosity was detected which encircles the whole vast
I Scorpii Association in a great irregular ring some 4°
wide. This huge formation, measuring over 300 light years
in diameter, is one of the great "H II" regions which we

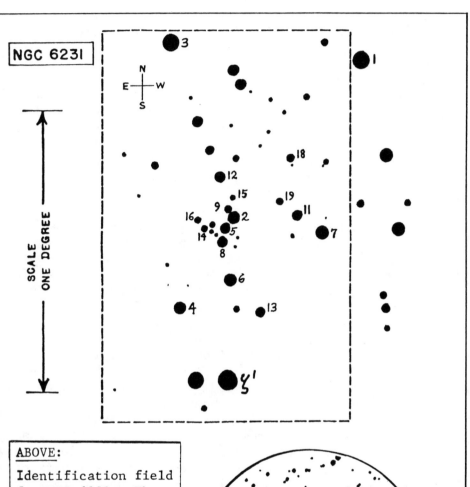

NGC 6231

SCALE ONE DEGREE

ABOVE:

Identification field for NGC 6231. The dashed rectangle is approximately 80' high N–S.

RIGHT:

A field 10' in diameter, showing the cluster center. Star numbers are keyed to the list on page 1722.

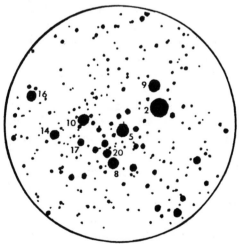

often see in the external galaxies, outlining the course
of the spiral arms like so many "pearls on a string". In
the northern sky the great Double Cluster in Perseus forms
the nucleus of a similar association.

In the small telescope, the stars of the group H12
form a sort of "tail" to NGC 6231, extending north and east
in a curve which ends in the region of brightest nebulosity
about 1.5° NNE from the main cluster. Another 1° to the
north the observer will find the compact little cluster
NGC 6242, an attractive group for a good 6-inch telescope,
and illustrated on page 1681. Traveling in the opposite
direction, to a spot about half a degree south of NGC 6231,
will bring the observer to the wide naked-eye star pair
called Zeta 1 and 2 Scorpii, about 6.8' apart, with a very
noticeable color contrast. Zeta 1, the western component,
is a B-type star (mag 4.80) while Zeta 2 is an orange giant
of type K5 (mag 3.62). These appear at the bottom of the
photograph on page 1723.

Zeta 1 is of considerable interest, as it appears to
be at the same distance as the cluster, and shows the same
radial velocity of about 17½ miles per second in approach.
Recent spectroscopic studies confirm the probable membership of this star in the NGC 6231 complex, which gives the
star an absolute magnitude of about -8, one of the most
luminous stars known in the Galaxy. The spectral class is
B1 Ia with emission features. In a study of the star made
at La Plata, Argentina, in 1972, M.and C.Jaschek found very
conspicuous and unexpected "variations in the strength of
lines belonging to different elements.....the observations
suggest a very turbulent state of the outer layers of this
star which is probably connected with mass loss from the
surface..."

NGC 6231 contains several visual double stars of
interest. Star #3, north of the main group, is I 576, a 5"
pair of magnitudes 6 and 12; the primary has a P Cygni type
spectrum of type O7e. Star #2 is B1833, a 0.4" pair; some
56.6" distant is star #9, a close triple listed by T.J.See
under the designation λ294. Star #5, in the cluster center,
is another wider triple, called λ297. Star #10 is another
difficult pair with a Wolf-Rayet primary; the spectral
types are WC6 and about O8. Data for these stars appear in
the list on page 1647.

FIELD OF NGC 6231. The bright cluster appears at lower right; the large group near center is H12. The compact cluster at top is NGC 6242. Lowell Observatory.

1727

NGC 6334 in SCORPIUS. This curious field of nebulosity lies about 2½° NW from Lambda Scorpii.

Mt. Wilson Observatory

1728

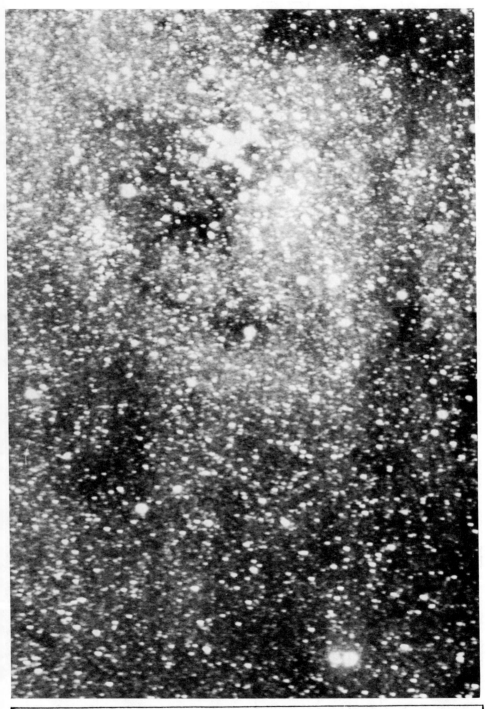

SECTION OF THE SCORPIUS MILKY WAY. This region lies south of star cluster M7 and NE of Lambda Scorpii. Lowell Observatory photograph made with the 5-inch camera.

SCULPTOR

LIST OF DOUBLE AND MULTIPLE STARS

NAME	DIST	PA	YR	MAGS	NOTES	RA & DEC
Hwe 92	7.2	271	51	7½- 10	relfix, cpm, spect K0	23103s3235
Hwe 63	6.4	267	34	7½- 11	relfix, spect B9	23214s2733
λ489	19.9	145	00	7- 11½	spect F0; galaxy IC 5332 in field	23311s3632
λ492	0.5	1	26	7 - 8½	Binary, 69 yrs; PA inc, spect G0	23330s2746
I 693	0.9	54	56	7½- 9	PA inc, spect F5	23343s3705
Hwe 93	5.3	252	37	6½- 10	relfix, cpm; spect K0	23344s3209
h5417	8.9	321	52	6½- 9	relfix, cpm; spect dF5	23419s2631
B613	0.2	133	56	8½- 8½	PA dec, spect F0	23423s3837
δ	3.9	243	56	4½-11½	(h3216) (β1013)	23463s2824
	74.3	297	29	- 9½	PA inc, spect A0; ABC all cpm	
h5423	13.5	307	44	6½-11½	relfix, spect A0	23472s2536
B620	2.5	9	27	7½-14½	spect G5	23475s2941
λ496	16.8	192	27	8- 12½	spect F0	23488s3242
h5429	28.8	226	55	7½-10½	slight dist inc, spect K5	23511s2941
B624	6.4	192	31	8½- 13	spect A0	23513s3741
Δ253	6.8	269	50	6½- 7	(Lal 192) relfix; spect A2	23518s2719
Arg 46	11.1	171	51	8 - 8½	(Lal 193) relfix; cpm, spect F0	23569s2648
ς	3.0	330	27	5 - 14	(B631) spect B4	23598s3000
κ'	1.3	265	54	6 - 6	(β391) PA dec, spect dF2+dF2	00068s2816
Rst 2236	1.7	99	45	7 - 13	spect K0; in field of θ Sculptoris	00082s3508
I 701	0.5	58	59	8½- 8½	spect K0	00177s3610
h3442	21.9	196	48	7 - 11	Dist & PA dec, spect K0	00301s2538
h3377	20.1	60	54	7½- 10	optical, dist & PA inc, spect K0, M5	00311s2622
h3375	5.3	168	54	6½- 8½	cpm, slight PA inc; spect G0	00313s3516
I 705	0.1		44	8 - 8	PA uncertain, Spect G0	00348s3733

LIST OF DOUBLE AND MULTIPLE STARS   (Cont'd)

NAME	DIST	PA	YR	MAGS	NOTES	RA & DEC
h1991	46.5	93	53	6½- 8½	spect K0	00364s2522
λ¹	0.7	3	54	6½- 7	(Hd 182) PA inc, spect A0	00403s3844
I 261	0.3	53	54	8 - 9	no certain change, spect A5	00455s2936
Stn 60	8.7	219	31	6½- 11	cpm, spect K5	01021s3348
h3436	9.8	126	18	7 - 9½	relfix, spect K0	01247s3029
β1230	2.7	226	59	6 - 11½	spect gK4	01280s2628
δ31	0.2			8 - 8	(β1000) AB binary,	01327s3010
	1.4	109	36	- 10	4.6 yrs; PA inc, spect K4. AC also binary, about 115 yrs; PA inc.	
τ	1.1	122	59	6 - 7	(h3447) PA inc, dist dec, spect dF4	01338s3010
I 448	20.3	350	30	6½-12½	spect A0	01364s2517
h3452	20.5	276	30	7 - 8½	relfix, spect K0,K0	01375s3744
ε	4.7	34	59	5½- 9½	(h3461) PA dec, spect F1	01433s2518

LIST OF VARIABLE STARS

NAME	MagVar	PER	NOTES	RA & DEC
R	6.1--8.8	363	Semi-reg; spect N3p	01247s3248
S	6.2--13.5	366	LPV. Spect M3e--M8e	00129s3219
T	8.5--13.5	201	LPV. Spect M3	00267s3811
U	8.3--14..	334	LPV. Spect M5e	01092s3023
V	8.7--14..	296	LPV. Spect M4e--M6e	00061s3930
X	9.8--14..	260	LPV.	00471s3511
Y	7.5--9.0	300:	Semi-reg; Spect M4	23064s3024
RS	9.8--11..		Uncertain; possibly not variable	01248s3310
RT	9.6--10.5	.51157	Ecl.bin; lyrid; spect A5 + F3	00340s2557
RU	9.4--10.9	.49333	Cl.Var; spect A0	00002s2513
SW	8.5--10..	144	Semi-reg; spect M1e--M4e	00037s3306
VV	8.4--8.6	2.4796	Ecl.bin; lyrid, spect A5p	01138s3425

SCULPTOR

LIST OF STAR CLUSTERS, NEBULAE, AND GALAXIES

NGC	OTH	TYPE	SUMMARY DESCRIPTION	RA & DEC
7507	2[2]	⊖	E0; 12.0; 1.0' x 1.0' pB,cS,R,psvmbM	23095s2849
----	I.5332	⊖	Sc; 11.9; 4.0' x 4.0' eF,vL,bM; with delicate pattern of spiral arms	23317s3622
7713		⊖	SB; 11.8; 4.0' x 1.5' pB,L,E,vglbM	23338s3813
7755		⊖	Sc; 12.5; 4.0' x 3.0' B,cL,lE,psbM; asymmetric spiral pattern	23455s3048
7793		⊖	Sd; 9.7; 6.0' x 4.0' ! L,B,vSN; large oval galaxy with ill-defined loose spiral pattern	23553s3251
24	461[3]	⊖	Sb/Sc; 12.2; 4.5' x 0.9' vF,cL,mE, nearly edge-on	00074s2515
55	△507	⊖	I or SBp; 7.8; 25' x 4.0' ! vB,vL,vmE,BN. Member of Sculptor Galaxy group (*)	00125s3950
134		⊖	Sb/Sc; 11.4; 5.0' x 1.0' vB,L,vmE,psbM; nearly edge-on spiral; in field of Eta Sculptoris	00279s3332
148		⊖	E4/S0; 12.9; 1.2' x 0.5' vB,S,lE,smbM; spindle-shaped	00318s3204
150		⊖	SB; 12.2; 2.0' x 1.0' pF,pS,E; thick pattern of spiral arms	00318s2805
253	1[5]	⊖	Sc; 7.0; 22' x 6' !! vvB,vvL,vmE,gbM. Fine, highly tilted spiral (*)	00451s2534
254		⊖	E4/S0; 12.8; 1.3' x 0.5' vB,pS,lE,smbM; 8^m star 5' nf; vF outer arms or ring	00452s3142
288	20[6]	⊕	Mag 7.2; Diam 10'; Class X; B,L,lE; stars mags 12..... loose-structured globular	00502s2652
289		⊖	Sb/Sc; 12.1; 2.0' x 1.5' vB,pL,pmE; gbM	00504s3129

LIST OF STAR CLUSTERS, NEBULAE AND GALAXIES (Cont'd)

NGC	OTH	TYPE	SUMMARY DESCRIPTION	RA & DEC
300		⊖	Sc/Sd; 11.3; 21' x 14' pB,vL,vmE,vgpsbM; large S-shaped, loose structured spiral (*)	00526s3758
---	---	⊖	vvL,eeF;1E; Diam 75'; Dwarf elliptical, member of Local Galaxy Group. The "Sculptor System" (*)	00570s3400
439		⊖	E3; 13.0; 0.7' x 0.5' pB,S,1E; bM	01115s3200
491		⊖	S0/SBb; 13.0; 1.0' x 0.8' B,S,v1E, bM	01191s3419
613	281[1]	⊖	SBc; 11.1; 3.0' x 2.0' vB,vL,vmE, sbM; fine many-armed spiral	01320s2940

DESCRIPTIVE NOTES

NGC 55   Position 00125s3950, near the Sculptor-Phoenix border, about 3.6° NNW from Alpha Phoenicis. This is one of the outstanding galaxies of the southern heavens, but rather low in the sky for observers in the United States. NGC 55 is a large, nearly edge-on system which has been classed as either a loose-structured barred spiral or a flattened irregular galaxy resembling the Large Magellanic Cloud. In small instruments it appears as a long spindle-shaped streak some 20' to 25' in extent, tilted toward PA 110°, the western portion being brighter, but with no definite central nucleus. The fainter eastern part of the system is sometimes classed under a separate number, IC 1537. On long-exposure photographs the total length is close to 30' and densitometer studies increase this to 50'; the total visual magnitude is about 7.8.

A good 10-inch reflector begins to show noticeable mottling across the body of this galaxy, while large telescopes reveal great numbers of individual stars, emission nebulosities, and dark dust clouds; supergiant stars begin

to appear in large numbers at about magnitude 18.4. Near
the center of the main mass, a little to the east of the
nuclear region, a very prominent and sharp-edged dust
cloud may be seen; smaller masses of this type are scattered across the star clouds of the galaxy. NGC 55 does not,
however, show any definite equatorial dust lane which is
often so prominent a feature in edge-on galaxies; this
seems to support the classification as an irregular system
rather than a spiral. G.de Vaucouleurs (1959) classes it
as an asymmetric barred spiral, though the edge-on presen-
tation makes it difficult to detect any spiral structure,
even if present. The Large Magellanic Cloud seems to show
a faint incipient spiral pattern, so it is quite possible
that the two classes of galaxies gradually merge into each
other. NGC 55 is a source of weak radio emission, first
detected by B.Y.Mills in 1959 with the 450-meter cross-type
array at Sydney, Australia.

NGC 55 is undoubtedly one of the nearest galaxies
beyond the Local Group, though the exact distance is still
not known with great accuracy. It is a member of a group
of galaxies called the Sculptor Group, or sometimes the
"South Galactic Pole Group", including NGC 253, 300, 7793,
NGC 247 in Cetus, and possibly NGC 45. This group and the
M81 group in Ursa Major are the two nearest clusters of
galaxies to our Milky Way. In a list of nearby galaxies
published in 1977, S.van den Bergh gives a distance of 7.5
million light years for NGC 55; J.Sersic, in the *Atlas
Galaxias Australes* (1968) adopts a figure of 7.8 million,
while studies by B.M.Lewis (1972) gave a larger result of
about 9.8 million for the group. According to M.L.Humason,
the red shift, from a measurement of the bright emission
patch near the central mass, is about 110 miles per second.
G.de Vaucouleurs obtained a smaller figure, of about 50
miles per second for the nuclear mass itself. From radial
velocity measurements at various points, he finds a total
computed mass of about 46 billion suns for the system, and
a total luminosity of at least 6 billion suns (absolute
magnitude about -19). If a distance of 8 million light
years is accepted, the true diameter of the 30' photographic image is close to 70,000 light years, comparable to
other large galaxies of this type. (Refer also to NGC 253,
page 1736)

NGC 55 in SCULPTOR, one of the nearest galaxies beyond the
Local Group. The photographs were made at Mt.Stromlo (top)
and at Radcliffe Observatory (below).

NGC 253   Position 00451s2534. Large, much-elongated
spiral galaxy, one of the most prominent in
the sky for the small telescope, and probably the most
easily observed spiral with the exception of the Great
Andromeda Galaxy M31. NGC 253 is located near the Cetus-
Sculptor border, about 7½° south from Beta Ceti, and was
discovered by Caroline Herschel in 1783, during one of her
systematic searches for comets. John Herschel, using his
18-inch metallic mirror reflector at the Cape of Good Hope
some 50 years later, described it as "very bright and large
(24' in length); a superb object.... Its light is somewhat
streaky, but I see no stars in it except 4 large and one
very small one, and these seem not to belong to it, there
being many near..." Modern observers will find it readily
visible in binoculars, though, owing to the southern declination, it should be observed when near the meridian. The
main body is about 22' long, oriented toward PA 53°; with
dark skies it may be possible to trace the luminosity out
to about 30' diameter. The central mass is quite bright and
has a mottled appearance in 8 or 10-inch telescopes, but
without a sharp star-like nucleus. This galaxy is usually
classed as an Sc type, though it may possibly be an SBc;
it is a little difficult to be certain about a possible
barred structure owing to the orientation only 12° from the
edge-on position.

Allen Sandage, in the *Hubble Atlas of Galaxies* (1961)
refersto NGC 253 as "the prototype example of a special
subgroup of Sc systems.....photographic images of galaxies
of the group are dominated by the dust pattern. Dust lanes
and patches of great complexity are scattered throughout
the surface. Spiral arms are often difficult to trace....
The arms are defined as much by the dust as by the spiral
pattern."

NGC 253 is the brightest member of the "Sculptor
Group of Galaxies" or the "South Galactic Pole Group", a
small cluster of galaxies including NGC 55, 300, 7793, NGC
247 in Cetus, and possibly NGC 45; the whole aggregation
covers an area about 20° in apparent size. Most of the members are large loose-structured spirals of low surface
brightness with the exception of NGC 55, usually classed
as a much-elongated irregular system. The group is not well
placed for observers in the United States, but is of great

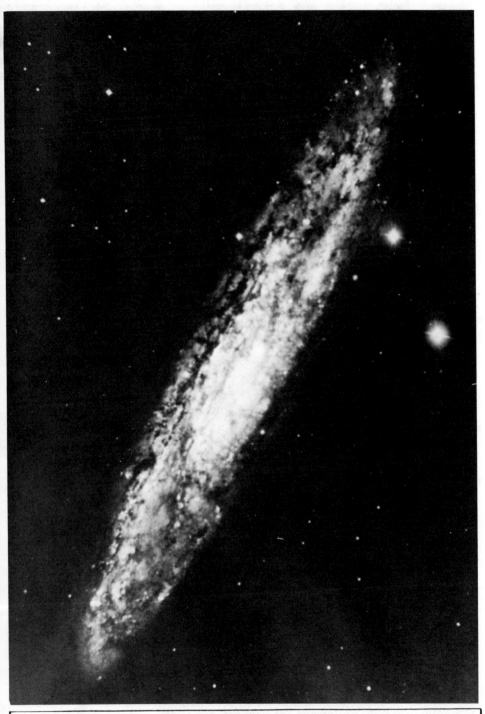

GALAXY NGC 253 in SCULPTOR. One of the largest galaxies of the southern sky. Mt.Wilson Observatory photograph with the 100-inch reflector.

MEMBERS OF THE SCULPTOR GROUP OF GALAXIES. Top: The large spiral NGC 253. Below: The faint and irregular spiral NGC 300. Mt. Wilson Observatory photographs

DESCRIPTIVE NOTES   (Cont'd)

interest, as it may be the nearest aggregation of galaxies
beyond the Local Group; only the M81 group in Ursa Major
appears to be almost equally near. The exact distances,
however, are still being investigated. G.de Vaucouleurs, in
a study made in 1959, concluded that the Sculptor Group
was slightly nearer than the M81 group, and found evidence
that the members form an expanding association with a total
computed mass of at least 150 billion suns.   J.L.Sersic, in
the *Atlas Galaxias Australes* (1968) reports a distance of
about 7.8 million light years for the Sculptor galaxies,
compared to about 7.4 million for M81.   S.van den Bergh, in
a list published in 1977, also places the M81 group nearer
with a distance of about 6.5 million, compared to about 7.5
million for the Sculptor galaxies. On the other hand, B.M.
Lewis and B.J.Robinson (1972) obtained larger distances for
both groups (about 9.7 and 10.5 million light years) with
the Sculptor aggregation somewhat the closer of the two.
There can be no doubt, however, that NGC 253 is one of the
very nearest galaxies beyond the Local Group. This is clear
not only from the large apparent size, but from the lack of
a red-shift in its spectral lines; the corrected radial
velocity is about 42 miles per second in approach.
    D.S.Evans at the Cape Observatory has measured the
rotation of the galaxy, which he finds to reach a maximum
value of about 360 miles per second some $5\frac{1}{2}'$ out from the
central mass. E.M. and G.R.Burbidge, using the 120-inch
reflector at Lick Observatory in 1970, found evidence for
an outflow of gaseous material from the center at about
70 miles per second relative to the main body of the system. The central mass contains a group of strong emission
regions, but no certain stellar-appearing nucleus.   The
rotation measurements make it clear that the NW edge of
the galaxy is the near side, and that the NE end is moving
toward us as the system rotates.   NGC 253 is a fairly
strong radio source, first identified by B.Y.Mills at
Sydney, Australia, and confirmed at other stations. One
supernova is recorded for this galaxy, in 1940; evidently
it was discovered long after maximum, as the observed magnitude was only 14. According to the Moscow *General Cata-*
logue of Variable Stars (1971) the position was 50" west of
the nucleus and 17" south.   (Refer also to NGC 55, page
1733)

SCULPTOR SYSTEM   Position 00570s3400, about 8½° SSE from NGC 253, and 4° south from Alpha Sculptoris. The Sculptor System is an extreme type of dwarf elliptical galaxy, a veritable phantom system which most nearly resembles a giant, extremely rarified, and very dim globular star cluster. Roughly 50 times the size of a large globular, it still remains virtually invisible in any tele- scope owing to its low star density; it covers a field well over 1° in apparent diameter, but the brightest stars are scarcely above 18th magnitude. This galaxy and a similar object called the "Fornax System" were the first galaxies of this type discovered; both were found at Harvard Obser- vatory in 1938, and for a time were regarded as unique. In more recent years, similar systems have been found in Leo, Draco, and Ursa Minor, all members of the Local Group of Galaxies. There is no reason to doubt that such dwarf ' galaxies are actually quite common in the Universe, but are impossible to detect at distances beyond the Local Group. When the Sculptor System was first noticed on a photograph, it appeared as a mere indefinite smudge resembling a minor plate defect; more powerful instruments and longer expos- ures finally revealed the individual stars. The best photo- graphs today show us the sort of object which might result if we could blow up Omega Centauri to about 50 times its present size and then remove 99% of its stars. The star distribution is very smooth, showing only a slight concen- tration toward the center; the brightest members are 17.8 magnitude.

From the presence of a few cepheid variables in the System, the distance is fairly well known, at about 270,000 light years, roughly 35% more distant than the Magellanic Clouds. The apparent size of 75' then corresponds to about 5500 light years, and the total luminosity is close to 3 million suns. S. van den Bergh (1977) gives the total (pg) magnitude of the System as about 10.5; the derived absolute magnitude (pg) is −9.2, one of the intrinsically faintest galaxies known. The very similar Fornax System appears to be about 3 magnitudes brighter in actual luminosity, but is also considerably farther away; the Leo I System, near Regulus, is still more distant, probably about 750,000 light years away. (Refer to Fornax, page 903, and Leo, page 1061)

SCUTUM

LIST OF DOUBLE AND MULTIPLE STARS

NAME	DIST	PA	YR	MAGS	NOTES	RA & DEC
Σ2306	10.2	220	59	7 - 8	(D18) Dist dec, Spect F5.	18194s1507
Σ2306b	1.2	70	50	8½- 9	PA slow inc, dist inc.	
Σ2313	6.1	198	51	7 -8½	relfix, cpm, spect G0	18220s0638
Σ2325	12.4	257	25	6 - 9	relfix, cpm; spect B2	18287s1050
Σ2337	16.3	297	15	8 - 9	relfix, spect both A0	18321s1445
β135	2.3	186	47	6½- 11	relfix, spect B9	18352s1403
	20.0	65	16	-14½		
δ	15.2	46	43	4½- 12	(Rst 4594) (HV 36)	18395s0906
	52.5	130	12	- 10	Spect F3; primary variable (*)	
ε	13.6	97	34	5- 14½	(J104) Spect G8	18408s0819
	37.6	195	34	- 13½		
	15.4	312	34	- 14½		
Σ2373	4.4	337	53	7 - 8	cpm, relfix, spect F2	18431s1033
A1887	3.3	256	26	6 -13½	relfix, spect F2	18440s1011
Σ2391	38.2	333	23	6½- 9½	relfix, spect A2	18460s0605
Σ2391b	12.6	107	07	-14		
S	14.4	238	36	7½- 12	(β969) Primary is N-type variable	18476s0758
	44.1	102	13	-12½		

SCUTUM

LIST OF VARIABLE STARS

NAME	MagVar	PER	NOTES	RA & DEC
δ	4.7---4.8	.19377	Short period pulsating variable; typical "Delta Scuti" star; spect F3 (*)	18395s0906
R	4.9--8.2	140	RV Tauri type; Spect G0-K0 pec (*)	18448s0546
S	7.3--9...	148	Semi-reg; Spect N3; also visual double star	18476s0758
T	8.9--10..	122	Semi-reg; spect N3	18527s0815
U	9.6--10.5	.9550	Ecl.bin; lyrid; spect F0	18516s1240
W	9.4--10.5	10.270	Ecl.bin; Spect B3 + B0	18217s1341
X	9.6--10.7	4.1981	Cepheid; Spect F5--G4	18285s1309
Y	9.2--10.5	10.342	Cepheid; Spect F7--G4	18353s0825
Z	9.4--10.6	12.901	Cepheid; Spect F8--G5	18403s0552
RS	9.8--10.7	.6642	Ecl.bin; Spect F5	18464s1018
RU	9.0--10.0	19.700	Cepheid; Spect F5--G5	18393s0410
RW	8.7--10..	117	Semi-reg; Spect M5	18538s1036
RX	9.0--11..	Irr	Spect N3	18343s0739
RY	9.9--10.6	11.125	Ecl.bin; lyrid; Spect B0	18227s1243
RZ	7.9--9.1	15.190	Ecl.bin; Spect B3	18238s0914
SS	8.1--8.9	3.671	Cepheid; Spect F6--G2	18410s0747
VW	9.0--15..	234	LPV. Spect M4--M6	18296s0958
CW	9.8--10.2	1.7862	Ecl.bin; lyrid; Spect B9	18530s0602
ER	9.1--9.4	1.3610	Ecl.bin; Spect A0	18400s0745
EU	8.4--17..	---	Nova 1949	18536s0416
EW	8.6--9...	Irr	Spect K0	18352s0651
FV	7.0--21..	---	Nova 1960	18317s1258
V368	6.9--17..	---	Nova 1970	18430s0836
V373	6.1--18..	---	Nova 1975	18527s0747

LIST OF STAR CLUSTERS, NEBULAE, AND GALAXIES

NGC	OTH	TYPE	SUMMARY DESCRIPTION	RA & DEC
6649			Mag 9; Diam 7'; Class F; S,1C; about 35 stars mags 10 ...14; Double star Σ2325 about 0.6° to SW.	18307s1026
6664	12[8]		Mag 9; Diam 18'; Class D; L,pRi,vlC; about 25 stars mags 10... In field of Alpha Scuti, 0.5° to east.	18340s0816
6682			Milky Way condensation on NW edge of Scutum Star Cloud; probably not a true cluster.	18390s0449
6683			Rich Milky Way field 15' diam in Scutum Cloud; probably not a true cluster.	18395s0617
6694	M26		Mag 9.5; Diam 9'; Class F; cL,pRi,pC; about 25 stars mags 11...14 (*)	18425s0927
6704			B,S,mC; Diam 2'; about 25 stars mags 12...14	18481s0516
6705	M11		! Mag 6; Diam 12'; Class G; vB,L,iR,vRi; One star 9^m + 200 stars mags 11...14. Fine object (*)	18484s0620
6712	47[1]	$\oplus$	Mag 9; Diam 3'; Class IX; pB,vL,vglbM; rrr; stars mags 15....	18503s0847
----	I.1298	◎	pL,eF; Mag 14; Diam 80" x 60" with 17^m central star; 0.4° ESE from NGC 6712	18519s0851

DELTA   Variable. Position 18395s0906. Mean visual
magnitude 4.74; Spectrum F3 III- IV. Delta
Scuti is a short-period pulsating variable of a rather rare
type, the standard star of its class. It is located near
the center of the constellation, about 2° ESE from Alpha
Scuti and roughly 2/3 of the way along a line drawn from
Alpha to the star cluster M26. As a visual variable, the
star is not of great interest, as the light range is only
about 0.15 magnitude, in a period of 0.193770 day, or 4.65
hours. The variability was first detected by E.A.Fath at
Lick Observatory in 1935; more recent studies have shown
that the star, in addition to its primary cycle, is also
oscillating in at least two superimposed periods. In a
study by W.S.Fitch (1960) a second overtone of 0.095129 day
and a resonance term of 0.186872 day were identified; the
resulting "beat period" is 5.24774 days, during which the
shape of the light curve changes from cycle to cycle. Delta
Scuti shows a light curve resembling those of the RR Lyrae

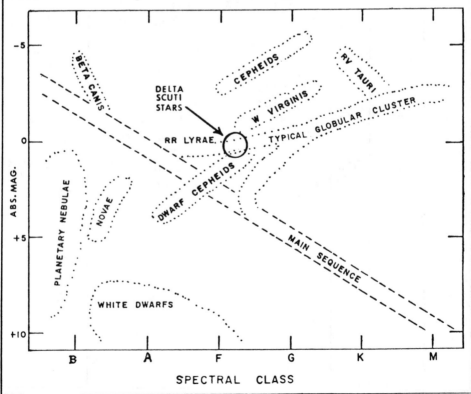

stars, with a rapid rise and a slower fall, but with a much smaller amplitude. Delta Scuti differs also from the RR Lyrae stars and the cepheids in a showing a different type of relationship between the light curve and the radial velocity curve. Greatest brightness occurs about 0.07 period before maximum expansion velocity. In two other stars of the type, Rho Puppis and Delta Delphini, peak brightness occurs at about the midpoint of the ascending portion of the radial velocity curve; in another star, DQ Cephei, it lies about halfway down the descending branch. Apparently there is no definite rule that these stars follow, whereas in typical cepheids the maximum brightness coincides closely with maximum approach velocity.

Stars of the type all appear to be giants of spectral type A or early F, ranging from A3 to about F6; absolute magnitudes range from about 0.0 to +2, though a few stars are known which seem to reach about -2. Delta Scuti itself appears to have an actual luminosity of about +0.3 if the derived distance of 250 light years is reasonably correct; O.J.Eggen in 1960 obtained a somewhat fainter value of +1.1 which corresponds to an actual luminosity of about 30 suns. Delta Scuti stars are roughly comparable to the RR Lyrae stars in actual luminosity, but show a greater spread in individual luminosities. Also, unlike the RR Lyrae stars and the cepheids, Delta Scuti stars remain at very nearly the same spectral type throughout the light cycle. When plotted on the H-R Diagram (see preceding page) they overlap somewhat the realms of both the RR Lyrae stars and the fainter dwarf cepheids; in some cases it is difficult to decide, from period and light range alone, to which class a particular star should be assigned. Dwarf cepheids, however, usually show a larger amplitude of variation.

M.S.Frolov, in his contribution to the book *Pulsating Stars* (1975) has listed 38 Delta Scuti stars known up to that year; the brightest examples being Beta Cassiopeiae, Rho Puppis, Epsilon Cephei, Upsilon Ursae Majoris, Delta Delphini, Delta Scuti, 14 Aurigae, Rho Phoenicis, 1 Monocerotis, and 4 Canum Venaticorum. All the periods of these stars are under 0.2 day; Delta Scuti itself has the longest period known, while 38 Arietis has the shortest, going through its cycle in 50.4 minutes. These very short-period stars, however, show only slight changes in light, in some

cases as small as 0.02 magnitude. Frolov suggests that the
Delta Scuti stars may comprise one of the most numerous
classes of variable, but that as yet only the brightest
examples have been detected; "This is a natural consequence
of the difficulty in discovering variables with low amplitudes and short periods, which is practically achievable
only by photoelectric photometry".

Although the Delta Scuti stars resemble the RR Lyrae
stars in general behavior, current studies suggest that the
two classes of objects have very different histories, and
are not directly related. Delta Scuti stars are definitely
objects of Population I and seem to be objects which have
begun to evolve away from the Main Sequence. Although a few
have spectra of luminosity class V (Main Sequence), the
majority are classed as giants or subgiants of classes III
and IV; a few (as Rho Puppis and 1 Monocerotis) are giants
of class II. Delta Scuti stars are occasionally found in
relatively young star clusters, and seem to be identified
with the flat disc or "plane component" of the Galaxy, not
with the galactic halo. Delta Scuti itself, according to
O.J.Eggen, appears to share the space motion of the "Taurus
Stream" associated with the Hyades cluster, while Delta
Delphini is seemingly a member of the Sirius stream. From
present theories of stellar evolution, the expected mass of
a typical Delta Scuti star is in the range of 1.5 to 2.0
suns; the prototype star itself has a computed mass of 1.9
suns and a diameter of about 3.0 suns. The radial velocity
(variable) has a mean value of 29 miles per second in
approach; the star shows an annual proper motion of 0.01".

Multiple periodicity in a Delta Scuti star, according
to one theory, may indicate that the star is a close pair;
the secondary periodicity resulting from the tidal effect
of a close companion. Beta Cassiopeiae, Delta Delphini, and
14 Aurigae are recognized spectroscopic binaries; Delta
Scuti itself has been listed as a spectroscopic binary but
no period has yet been derived.

An optical companion of the 10th magnitude lies 52.5"
distant in PA 130° and a closer star of the 12th magnitude
will be found 15" distant in PA 46°. The brighter companion
has a spectral type of about gK0. The galactic cluster M26
is easily located near Delta Scuti, about 0.8° distant to
the ESE. (See page 1756)

R   Variable. Position 18448s0546. A peculiar
semi-regular pulsating star, usually considered to be of the RV Tauri class, discovered by the English
observer E.Pigott in 1795. It is located just 1° south of
Beta Scuti, near the north edge of the rich Scutum Star
Cloud, and about 1° NW of the fine star cluster M11. This
star at discovery was one of about a dozen variables then
known; excluding novae, the full list then read: Omicron
Ceti, Algol, Chi Cygni, R Hydrae, R Leonis, Beta Lyrae,
Delta Cephei, Eta Aquilae, Alpha Herculis, and R Corona
Borealis. Two of these, R Scuti and R Coronae, were dis-
covered by Pigott, in the same year.
The discoverer presented a paper on the star to the
Royal Society in 1797, and in an additional report in 1805
suggested that the variations might be attributed to the

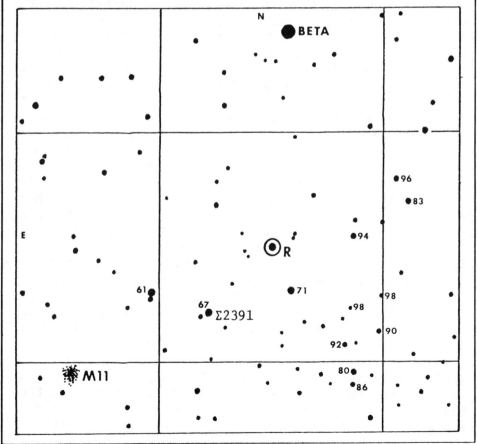

DESCRIPTIVE NOTES (Cont'd)

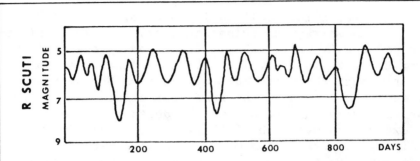

slow rotation of a globe mottled with various light and
dark areas. He found an interval of 62 days to best represent the probable rotation period, and reported that the
star remained at maximum, without any perceptible change,
for 9½ days, falling to minimum in approximately 33 days.
He found, however, that the "times of the periodical return
of brightness are, in general, irregular". L. Campbell and
L. Jacchia in *The Story of Variable Stars* (1914) refer to
R Scuti as "a good friend of many variable star observers"
and "a good example of a very fanciful light curve". The
first comprehensive studies of the star were made by H.D.
Curtis at Lick Observatory in 1903, and continued by D.B.
McLaughlin. A typical section of the light curve, shown
above, is based on observations by Campbell.

The star has a primary period given by various observers as 146 days, 140 or 144 days, but the amplitude is not
at all constant, the star rising at times to magnitude 4.8
or so, and declining usually to about 6.0. Every fourth or
fifth minimum is exceptionally low, the magnitude dropping
to 8.0 or fainter. The star is evidently oscillating in at
least two superimposed periods, as does RV Tauri itself.
It differs from RV Tauri, however, in showing no very long-
period wave-like cycle (about 1300 days) which in the case
of RV Tauri alternately raises and lowers the entire light
curve. McLaughlin found that R Scuti, except for the deep
minima, resembled a cepheid, but that the measured radial
velocity varied greatly in an irregular manner. He classed
the spectrum as about G2, on the average, but found that
typical red-giant features (titanium-oxide bands) appeared
whenever the magnitude dropped below 6.0. At minimum the
bands are as strong as in the spectrum of any typical M3
red giant. Bright hydrogen lines appear in the spectrum

THE MILKY WAY IN SCUTUM. The view is centered on the Scutum Star Cloud, Barnard's "Gem of the Milky Way". Photographed with a 50 mm Takumar lens by David Healy.

DESCRIPTIVE NOTES (Cont'd)

during the rapid rise to maximum; these gradually turn to absorption lines shortly after peak brightness. This very individualistic star seems to share some of the character-istics of both the cepheids and the long-period red vari-ables. From studies of spectral features, it appears that various layers of the star are expanding (or contracting) at different rates, and that the star has unusually great actual dimensions, at least 100 times the diameter of the Sun. Stars of this type are among the largest and most massive variables known, with absolute magnitudes in the range of -4.5 to -5.0, and inferred masses of 20 or 25 times that of the Sun. At peak luminosity R Scuti shines with the light of about 8000 suns; the distance is probably in the range of 2500 to 3000 light years, depending on the size of the correction made for light loss through space absorption. This star has the longest period and probably the highest luminosity of any star of the class. In the Moscow *General Catalogue of Variable Stars* (1970) it is classed as type G0e Ia (maximum light) and K0p Ib when at minimum. J.S.Glasby in his book *Variable Stars* (1969) re-ports a range of G8 to M3, however, illustrating the diffi-culty of accurate classification of stars with complex and anomalous spectral features.

R Scuti shows an annual proper motion of 0.06" and a radial velocity (variable) of about 26½ miles per second in recession.

Observers of the star should not fail to examine the very fine cluster M11 which lies about 1° distant, toward the SE. The brightest star between the variable and the cluster is the double Σ2391, an easy object for any small telescope with a separation of 38", magnitudes 6½ and 9½. The identification chart on page 1747 shows stars to about 10th magnitude; grid squares are 1° high with north at the top. Comparison magnitudes, courtesy of the AAVSO, are shown with decimal points omitted.

M11 (NGC 6705) Position 18484s0620. Exceptionally fine galactic star cluster, lying on the north edge of the prominent Scutum Star Cloud, and one of the outstanding objects of its type for telescopes of moderate aperture. It was discovered by Gottfried Kirch of the Ber-lin Observatory in 1681 as a "small, obscure spot with a

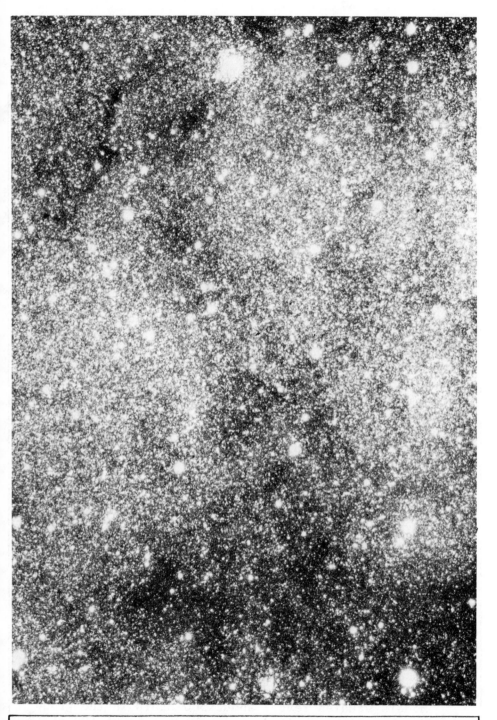

STAR FIELD IN THE SCUTUM STAR CLOUD. The rich galactic cluster M11 appears near the top of the print. Lowell Observatory photograph made with the 13-inch telescope.

star shining through and rendering it more luminous". In
1715 the object was included by Halley in his short list
of "nebulous stars", but the Rev. Wm.Derham of England
seems to have been the first to resolve it into a cluster,
in 1732: "It is not a nebula but a cluster of stars somewhat like that which is in the Milky Way". Le Gentil, in
July 1749 saw here "a prodigious cluster of very small
stars, forming a large, white cloud; six of the principal
stars form a large letter V somewhat similar to the Hyades
but with the opening towards the south..." Messier, in
May 1764 listed M11 as "A cluster of a great number of
small stars which can be seen in a good telescope. In a 3-
foot [F.L.] instrument it looks like a comet. The cluster
is mingled with a faint light; 8 mag.star in cluster..."
William Herschel thought M11 just visible to the naked eye
while in his telescope it consisted of "11th magnitude
stars divided into 5 or 6 groups noted independently in a
5½ inch. An 8 mag.star is a little within the apex." The
seeming division of the cluster was also noted by D'Arrest
who saw M11 as "A magnificent pile of innumerable stars...
Irregular and as if divided into several agglomerations".
To Admiral Smyth the main group resembled "a flight of wild
ducks".

Modern observers will find this one of the richest
and most compact of the galactic clusters. In binoculars
or a low-power telescope it at first resembles a globular,
but with increasing magnification the stars begin to draw
apart, finally revealing M11 as a rich swarm of glittering
star points, somewhat triangular in shape with one brighter
star near the center, but no real central nucleus. Barnard
thought the extreme diameter to be about 35' and the "wild
duck" group about 1/3 of this; Walter S.Houston confirms
this estimate with a 10-inch reflector and describes the
cluster as "a carpet of sparkling suns to the very center
with outlyers swarming on all sides. A good 10-inch shows
hundreds of glittering star points all over the field of
view".

Modern studies reveal about 500 stars in the group
brighter than 14th magnitude; probably at least 400 of
these are true cluster members, while the number of probable members down to magnitude 16½ appears to be about 870
according to a study by J.B.McNamara and W.L.Sanders in

STAR CLUSTER M11 in SCUTUM. One of the richest of the
galactic clusters.   Lowell Observatory photograph made with
the 13-inch telescope.

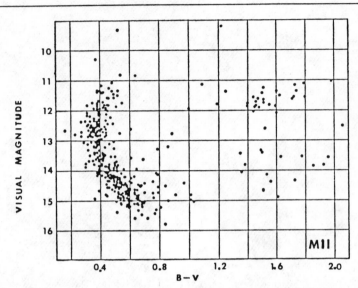

1977. The same investigators found a probable total mass
of about 2900 suns for M11 including an estimated 800 suns
for the contribution from unseen companions. The derived
distance of the cluster is about 5500 light years which
agrees quite well with the figure of 5400 published by H.L.
Johnson and A.R.Sandage in 1956. As the Sun would appear
as magnitude 15.9 at that distance, it is evident that all
the bright stars of M11 are very luminous giants; members
which appear as 11th magnitude stars have a true luminosity
of about 100 suns. The total light of the cluster is close
to 10,000 times that of the Sun. From the plotted color-
magnitude diagram, the majority of the members are found
to be main sequence A and F stars, but at least two dozen
yellow and red giants also exist in the group. With a computed age of about 500 million years, M11 is intermediate
in age between the Pleiades and Praesepe (M44) in Cancer.
        The true density of this cloud of remote suns has
been the subject of several studies. The 10' central mass
is roughly 15 light years in diameter; from this the central star density has been estimated at about 83 stars per
cubic parsec, giving an average separation of less than a
light year. R.J.Trumpler (1932) calculated that an observer at the center of M11 would see several hundred 1st magnitude stars in his sky, and possibly 40 or so with an
apparent brightness ranging from 3 to 50 times the light

STAR CLUSTER M11 in SCUTUM. A shorter exposure than that shown on page 1753; this photograph was made with the 42-inch reflector at Lowell Observatory.

of Sirius! The star density is, in fact, not greatly inferior to some of the less condensed globulars. There are
a few other galactic clusters known with a similar population density (NGC 2158 in Gemini and NGC 6791 in Lyra) but
M11 is probably the closest example known. The rich star
field of the Scutum Milky Way adds, of course, to the
apparent high star density of the cluster, but M11 is considerably nearer to us than the Scutum Cloud and is not
directly involved in it.

The bright and unusual variable star R Scuti will be
found about 1° to the NW; another easy object in the field
is the double star Σ2391 which lies between the cluster and
the variable. (Refer to chart on page 1747)

M26 (NGC 6694) Position 18425s0927. Galactic
star cluster in a rich Milky Way field in
Scutum, located about 0.8° ESE from Delta Scuti and about
3.5° SSW from M11. Discovery of M26 is usually credited to
Le Gentil at some uncertain date prior to 1750; it was
found by Messier in June 1764, who reported it as "not distinguished with a 3½-foot [F.L.] telescope.....it needed a
better instrument...Contains no nebulosity". In a note to
his copy of the catalogue of 1784 he added "I have seen it
very well with a Gregorian telescope magnifying 104 X."
Admiral Smyth described M26 as "A coarse but bright cluster
of stars in a fine condensed part of the Milky Way. The
principal members of this group lie nearly in a vertical
position with the equator and the place is that of a small
pair in the south of the field". Messier gave the apparent
size as 2' but the cluster shows a total diameter of about
9'on modern photographs; the total magnitude is about 9½.
The cluster is possibly slightly nearer than M11; the best
current distance estimate is about 4900 light years.

In the small telescope M26 appears as a small but
tight group with the brightest star (11ᵐ) on the SW corner
and two short curving "prongs" of stars pointing upward
from the north edge, enclosing a nearly circular dark gap
about 2' wide. About 25 stars may be counted in the group
with a 6 or 8-inch telescope; about 70 additional fainter
stars are considered to be true cluster members, and the
true diameter is in the range of 12 - 16 light years. (See
photograph, page 1757)

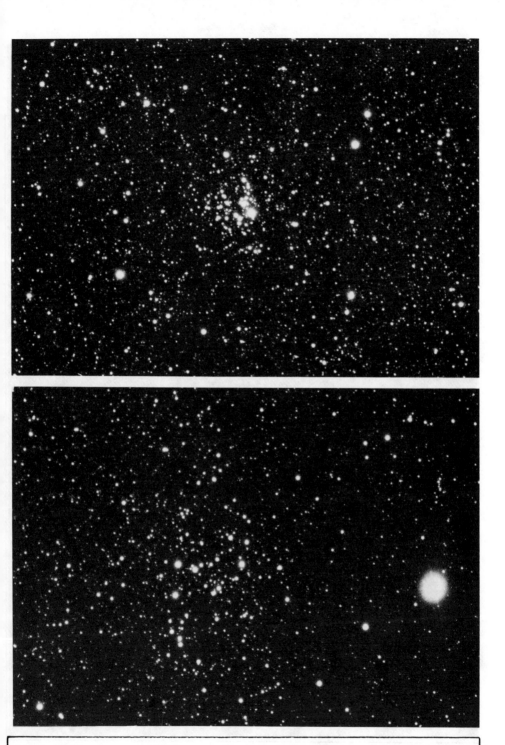

STAR CLUSTERS IN SCUTUM.   Top: The compact cluster M26.
Below: The bright star Alpha Scuti and the scattered group
NGC 6664. Lowell Observatory 13-inch camera photographs.

GLOBULAR STAR CLUSTER NGC 6712 in SCUTUM. This cluster lies
in a rich region of the Milky Way. Lowell Observatory
photograph with the 13-inch telescope.

SERPENS

LIST OF DOUBLE AND MULTIPLE STARS

NAME	DIST	PA	YR	MAGS	NOTES	RA & DEC
A1116	0.6	41	60	8 - 8	PA inc, spect A5	15092n1019
Σ1919	24.0	10	58	6½- 7½	cpm, relfix; spect G5, dG6	15105n1928
A691	0.1	231	58	7½- 7½	Spect K0	15114s0109
β943	2.3	92	40	6½- 12	relfix, cpm, spect K0	15158n0107
Σ1931	13.3	170	52	6½- 7½	Slight PA dec?	15163n1027
	162	93	08	- 12	cpm; spect F5, G0	
5	11.2	36	58	5 - 10	(φ Serp) (Σ1930)	15167n0157
	127	40	24	- 9	AB cpm, slow dist inc, PA dec; spect F8, K4. Globular Cluster M5 in field	
6	3.1	20	58	5½- 9½	(β32) cpm; slight dist inc, PA inc, spect K3	15185n0054
Σ1940	0.6	324	67	8 - 8½	Dist dec, spect F8	15239n1821
A2074	0.3	276	67	8 - 8½	spect F8	15251n1748
Σ1944	0.8	313	66	8 - 8½	PA dec, spect F8	15252n0616
δ	3.9	178	62	4 - 5	(13 Serp) (Σ1954)	15324n1042
	65.2	14	59	- 14	PA dec, dist inc; Spect both dF0; A= δ Scuti variable	
δ c	4.4	339	11	14-14½		
Σ1957	0.6	149	67	8 - 9½	PA & dist dec, Spect F8	15335n1305
OΣ300	15.3	262	58	6½- 10	relfix, spect G8	15378n1213
A2076	0.6	176	60	8 - 8	PA & dist inc, spect A2	15382n1850
ι	0.1	78	62	5 - 5	(Hu 580) Binary, 10½ yrs; PA inc; spect A1	15393n1950
	143	352	60	- 13½		
	151	110	60	- 12½		
A2176	0.2	16	61	8 - 8	Binary, 56 yrs; PA inc, spect A0	15395n0037
β619	0.5	358	60	7 - 7½	relfix, spect G8	15408n1350
ψ	4.2	61	57	6- 12	(A2230) PA dec,	15415n0240
	208	208	18	- 9	Dist slight inc,	
	171	281	18	-10½	spect dG5	
β	30.8	265	40	3½- 9	(Σ1970) cpm, spect	15439n1535
	201	210	60	- 10½	A2, dK3	

LIST OF DOUBLE AND MULTIPLE STARS   (Cont'd)

NAME	DIST	PA	YR	MAGS	NOTES	RA & DEC
A2079	3.6	59	58	6– 12½	cpm, relfix, spect F2	15513n1613
A2080	0.2	69	62	8½– 8½	Binary, 90 yrs; PA inc, spect F2	15519n1708
Σ1985	5.8	347	59	7 – 8	PA inc, Dist slow inc, spect G0	15533s0203
Σ1988	2.1	259	59	7½– 8	cpm, dist dec, PA slow dec, spect F2	15544n1237
Σ1987	10.4	322	54	7– 8½	relfix, spect A0	15548n0333
Σ1993	24.5	40	58	8 – 8	dist dec, spect A0	15576n1731
	242	179	18	– 10		
OΣ303	1.2	162	67	7½– 8	PA & dist inc, spect F5	15586n1325
Σ2000	2.7	229	54	8 – 9	relfix, spect F2	16007n1408
Σ2003	14.4	171	10	7½– 11	relfix, spect K2	16013n1134
Σ2007	36.6	323	58	6½– 8	Probably optical,	16037n1327
	168	138	21	– 10	Slight PA dec, dist inc, spect K0	
Σ2031	21.1	230	22	7½– 9½	cpm, relfix, spect	16138s0132
	93.4	21	22	– 11½	F8	
Σ2033	10.7	174	32	8½– 8½	cpm, relfix, spect A3	16156s0209
ν	46.3	28	59	4½– 8½	(Sh 247) Slight PA dec, spect A1	17180s1248
Σ2204	14.5	24	53	7 – 7	relfix, both spect A0	17435s1318
Hu 189	1.5	241	47	7½– 8½	PA inc, spect F5	17502s1338
h4995	28.8	155	20	6½– 11½	relfix, spect K0	17513s1120
Hn 139	3.7	152	45	7– 10½	relfix, spect F2	17521s1138
h2814	20.8	157	04	6½– 9	AB cpm, spect A0;	17534s1549
	33.7	349	04	– 11½	AC optical	
A2595	2.6	62	25	7 – 14	spect F8	18033s0807
A36	65.6	42	15	7½– 11	spect G0	18097s0718
A36b	1.3	196	25	11– 11		
β131	2.8	278	53	7½– 9½	relfix, spect F5;	18107s1537
	8.4	289	44	– 11½	AC PA & dist inc	
Σ2296	3.4	7	37	6½– 10½	relfix, spect K0	18131s0322
Σ2303	2.1	236	59	6½– 9	cpm; PA inc, dist dec, spect F5	18174s0802

LIST OF DOUBLE AND MULTIPLE STARS   (Cont'd)

NAME	DIST	PA	YR	MAGS	NOTES	RA & DEC
AC 11	0.6	355	61	7 - 7	Binary, about 240 yrs; nearly edge-on, PA dec, spect dF5	18224s0136
β1203	0.3	134	60	7½- 7½	PA inc, spect A3	18235n0045
59	3.9	318	58	5½- 7½	(Σ2316) cpm, slight dist dec? Primary variable; Spect G0 + A6	18246n0010
Σ2321	6.8	190	67	8 - 9½	relfix, spect A0	18274n0109
Σ2324	2.3	146	67	8½- 9	relfix, spect B8	18284n0121
Σ2342	10.3	329	24	6½-12½	(β643) Dist inc,	18331n0454
	30.5	5	30	- 8½	PA dec, spect A0; AC PA dec, dist inc	
A88	0.2	354	60	7 - 7	Binary, 12.2 yrs;	18358s0314
	16.3	120	60	- 14	PA dec, spect dF8; AC optical	
OΣ360	1.5	285	51	6½- 10	Dist slight inc? PA slow dec, spect K0	18362n0449
Σ2369	0.6	78	66	8 - 8½	PA & dist dec; spect G0	18414n0234
Σ2375	0.1	136	60	6½- 6½	(φ332) AB PA inc,	18430n0527
	2.4	116	66	- 6½	Spect A0	
Σ2375c	0.1	135	60	7½- 7½		
A2192	0.2	121	67	7½- 7½	Binary, 72 yrs; PA dec, spect A2	18533n0323
θ	22.2	103	55	4½- 5	(Σ2417) relfix cpm, Fine pair; Spect both A5 (*)	18538n0408

SERPENS

LIST OF VARIABLE STARS

NAME	MagVar	PER	NOTES	RA & DEC
δ	4.2--4.25	.134	δ Scuti type; Spect F0; Also visual double	15324n1042
o	4.2--4.26	.053	(56 Serpentis) δ Scuti type; Spect A2	17386s1251
τ⁴	6.1--7..	Irr	Spect M5	15341n1516
χ	5.4--5.44	1.5958	(20 Serpentis) α Canes Ven type; Spect A1p	15394n1300
59	4.9--5.9	Irr?	(d Serpentis) Class uncertain; Spect G0+A6	18246n0010
R	5.7--14..	357	LPV. Spect M6e--M8e (*)	15484n1517
S	7.7--14.1	368	LPV. Spect M5e--M6e	15193n1430
T	9.1--15..	340	LPV. Spect M7e	18264n0616
U	7.8--13.5	238	LPV. Spect M4e--M6e	16049n1004
V	9.6--10.7	3.4535	Ecl.Bin; Lyrid; Spect B8	18140s1532
W	8.5--10.0	14.159	Ecl.Bin; Spect gF5ep (*)	18070s1534
X	8.9--18.3	---	Nova 1903	16167s0223
Y	8.5--10..	433	Semi-reg; Spect M5e	15114s0142
Z	8.1--9.0	87	Semi-reg; Spect M5	15135n0221
RT	10----16	---	Probably a slow nova; (1910) (*)	17370s1155
VY	9.3--10.5	.71409	Cl.Var; Spect F2--F6	15285n0151
AM	9.9--12..	104	Semi-reg; Spect M2	15271s0014
BC	8.7--13..	245	LPV. Spect M3e--M5e	15584n0219
BQ	9.3--9.9	4.3204	Cepheid; Spect F3	18338n0421
CD	8.6--10..	80:	Irr or Semi-reg; Spect M4; in field of χ Serp.	15414n1248
CT	9.0--16..	---	Nova 1948	15433n1432
CV	9.9--10.2	29.705	Ecl.Bin; Wolf-Rayet type Spect WC7 + B0	18163s1139
DR	8.4--11..	Irr	Spect N	18449n0524
DX	9.0--11..	360	Semi-reg; Spect M	16060s0124
DZ	8.0--17..	---	Nova 1960	17576s1034
EG	8.7--9.5	4.9736	Ecl.Bin; Spect A0	18234s0143
FH	4.5--16..	---	Nova 1970	18283n0234

SERPENS

LIST OF STAR CLUSTERS, NEBULAE AND GALAXIES

NGC	OTH	TYPE	SUMMARY DESCRIPTION	RA & DEC
5904	M5	⊕	!! Mag 6.2; Diam 13'; vB,vL, eRi,eCM; Class V; stars mags 11.... Superb cluster (*)	15160n0216
5921	148[1]	⊖	SBb; 12.0; 4.0' x 3.5' cB,cL,1E,vsbM	15195n0515
5936	130[2]	⊖	SBb; 12.9; 1.0' x 0.8' F,L,R,gbM; tight spiral	15276n1309
5962	96[2]	⊖	Sc; 11.9; 2.2' x 1.3' pF,pL,cE,gbM; faint outer ring, bright oval center	15342n1646
5970	76[2]	⊖	Sc; 12.4; 2.7' x 1.2' pF,pL,E	15361n1220
5984	656[2]	⊖	SBb; 13.0; 2.6' x 0.5' pB,S,E,bM; nearly edge-on	15406n1422
6027		⊖	Curious compact group of 5 S0 & E galaxies; 1.7' x 1.3' Mags 14 & 15 (*)	15570n2055
6070	553[3]	⊖	Sc; 12.5; 3.1' x 1.7' F,pL,pmE,vgbM	16074n0050
6118	402[2]	⊖	Sb/Sc; 12.3; 4.2' x 1.2' vF,cL,cE; multiple arm spiral	16193s0211
6535		⊕	Mag 11, Diam 1.5'; Class XI; pF,vS,R, stars eF	18013s0018
6539		⊕	Mag 12, Diam 2.5'; Class X; vS,F, in heavily obscured area	18021s0735
----	H19	⸪	S,pB; Diam 5'; about 40 stars mags 12... Class G; 30' NW from M16	18145s1318
6611	M16	▢	!! Mag 6½, Diam 25'; L,B, scattered cluster, about 60 stars mags 8.... Class C; Surrounded by extensive nebulosity (*)	18160s1348
----	I.4756	⸪	vvL, Diam 70'; scattered group of 80 stars mags 7.... Class D (*)	18366n0526

DESCRIPTIVE NOTES

ALPHA   Name- UNUK AL HAY or UNUKALHAI, from the
Arabic *Unk al Hayyah*, the "Neck of the Snake".
The star has also been called *Cor Serpentis*. The serpent
in this case is the symbol of the divine healer *Aesculapius*
who became the god of medicine, and who appears in the sky
as Ophiuchus, the Serpent-bearer. Serpens is the only star
group completely divided into two portions: *Serpens Caput*
to the west of Ophiuchus, and *Serpens Cauda* to the east.

Alpha Serpentis is Magnitude 2.65; Spectrum K2 III;
Position 15418n0635. From direct parallaxes the distance is
about 70 light years, giving the actual luminosity as 35
times that of the Sun (absolute magnitude +1.0). The star
shows an annual proper motion of 0.14"; the radial velocity
is a little less than 2 miles per second in recession.

An optical companion of the 12th magnitude was first
measured by John Herschel in 1836; the pair appears as
h1277 in his catalogue. Both the separation and the PA have
decreased somewhat since discovery, the change resulting
from the proper motion of the bright star. The Lick *Index
Catalogue of Visual Double Stars* gives the separation as
58.2" in PA 350°. According to R.H.Allen the colors are
pale yellow and blue.

BETA   Magnitude 3.67; Spectrum A2 IV; Position
15439n1535. Beta Serpentis marks the Head of
the Serpent, but appears to have no proper name on standard
star charts; R.H.Allen states that the Chinese designated
it *Chow* (or Chou), the name of one of the most important of
their ancient dynasties. The parallactic distance is about
95 light years, which leads to an actual luminosity of 23
suns. The annual proper motion is 0.08"; the radial velocity is less than 1 mile per second in approach.

The star should first be examined with lowest power,
advises K.McKready, "The field near it, especially a little
way to the northeast, is a fine spectacle even for the
opera-glass or field glass". Beta itself forms a common
proper motion pair with a 9th magnitude companion 30.8"
distant; the small star is a dK3 yellow dwarf with about
1/6 the solar luminosity. This pair has remained relatively fixed since the first measurements of F.G.W.Struve in
1832, but at a projected separation of about 900 AU no sign
of orbital revolution is to be expected.

Curiously, both McKready and Allen give the colors of the two stars as "both blue"; T.W.Olcott, however, calls them "blue and yellow".

Some 27' from Beta Serpentis, to the west and slightly south, lies another faint double star (Roe 75) which shares the proper motion of Beta and must be a physical member of the system. Located at 15420n1522, it consists of a 6.4" pair, magnitudes 9 and 9½, PA 326°, spectral types near dG0. According to the Yale *Catalogue of Bright Stars* this pair is 1642" distant from Beta, equivalent to a projected separation of about 48,000 AU. This whole multiple system shows very nearly the same space motion as Sirius, and is regarded as a member of the "Sirius stream" for which members have been recruited from virtually every part of the sky.

Slightly over 1° from Beta Serpentis, toward the ESE, lies the fine long period red variable star R Serpentis, one of the most suitable variables of the class for observation with small instruments. (Refer to page 1766)

ETA   Magnitude 3.25; Spectrum K0 III or IV;
Position 18187s0255. This is the brightest star lying between the Scutum Star Cloud and the Head of Ophiuchus. The star has no English proper name, though R.H. Allen states that a Chinese name, translated "The Heavenly Eastern Sea", was applied to both Eta and Zeta Serpentis. Eta is a fairly close neighbor to the Solar System, at about 60 light years distance; its actual luminosity is some 6 times that of the Sun. The star shows a large annual proper motion of 0.89" in PA 218°; the radial velocity is about 5½ miles per second in recession.

THETA   (63 Serpentis) Name- ALYA, probably from the
Arabic name for "serpent". Magnitude 4.05; Spectrum A5 V; Position 18538n0408. The star lies in the NE portion of the constellation, near the Serpens- Aquila border, and marks the end of the Serpent's tail. Theta is one of the most attractive double stars in the sky for a very small telescope, and is generally resolvable in good binoculars, or a 1-inch aperture telescope. The components are separated by 22" and have remained fixed since the early measurements of Struve in 1830. C.E.Barns calls this

an "imperial pair in regal setting"; Webb refers to it as
a "noble pair, in a very fine field; [it] lies to the naked
eye in a dark space between two streams of the Galaxy...."
Olcott calls the stars both yellow, while E.J.Hartung lists
them as "an attractive pale yellow pair". The present auth-
or, however, sees them as a clear sparkling white. In the
Yale *Catalogue of Bright Stars* the individual magnitudes
are given as 4.59 and 4.99; spectra A5 V and A5n; another
study by A.Slettebak (1963) has them as A5 V and A7 V. The
stars share a common proper motion of 0.05" annually, with
a projected separation of about 900 AU. From a direct Yale
Observatory parallax, the distance is about 130 light years
and the actual luminosities about 19 and 13 times the Sun.
The system is approaching at about 28 miles per second.

Admiral Smyth suspected the variability of one or
both components on more than one occasion; B.A.Gould in the
1880's found a total range of about 0.5 magnitude. T.W.Webb
states that a difference of 1.3 magnitudes was noted between
the stars at Harvard on one occasion. No very certain
changes have been detected in recent years.

R   Variable. Position 15484n1517, about 1.2° from
          Beta Serpentis, toward the ESE. The brightest
of the long-period red variable stars in Serpens, and the
only one which reaches naked-eye visibility on occasion,
with a maximum recorded magnitude of about 5.7. The star

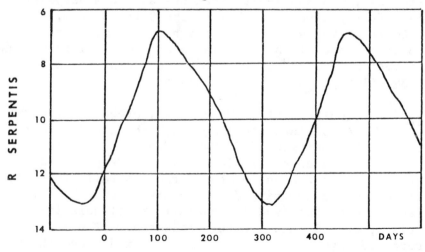

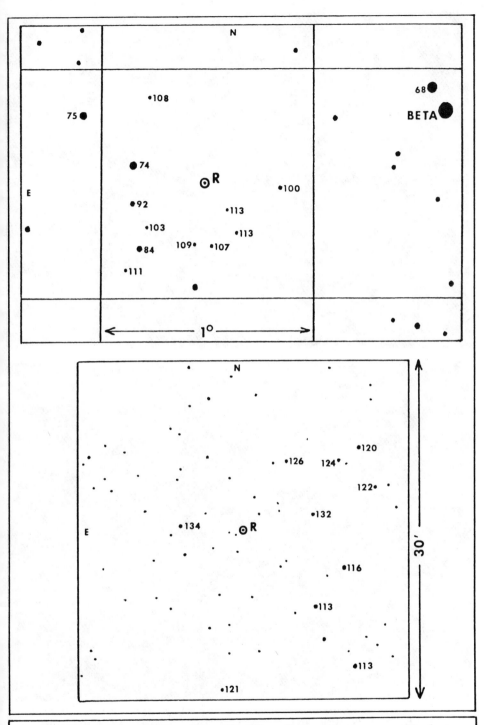

R SERPENTIS Identification fields. Comparison magnitudes are given (with decimal points omitted) according to observations of the AAVSO.

VARIABLE STAR R SERPENTIS. A typical member of the long-period red giant type of variable. Lowell Observatory photographs made with the 13-inch telescope.

DESCRIPTIVE NOTES   (Cont'd)

was first identified as a variable by the German astronomer K.L.Harding in 1826. Harding was also the discoverer of R Aquarii in 1811, and the discoverer of the third known asteroid, Juno, found with a 2-inch refractor in September 1804. R Serpentis, however, had been observed prior to Harding's time, though not recognized as a variable; it appears on a French chart made by D'Agelet in March 1783, thus preserving for us the approximate date of a very early maximum.   Between 1826 and 1868, Argelander thought to find some evidence of a steady decrease in period, of about 1/4 day from cycle to cycle. The present period of 356.75 days, however, does not differ significantly from the figure of 357.2 days published in 1907 in the Harvard *Second Catalogue of Variable Stars*. The light curve shown on page 1766 is based on studies by L.Campbell of 46 maxima of the star, observed between 1904 and 1949. The mean maximum during this period was magnitude 6.9, and the mean minimum about 13.4.   A magnitude brighter than 6 or fainter than 14 is rare for this star; the extreme range appears to be about 5.7 to 14.2.

R Serpentis is a fairly typical member of the Mira class of variables, the spectrum varying from M6e to about M8e as the light declines to a minimum. Stars of the type are all low-density red giants with diameters averaging several hundred times that of the Sun, and with true peak luminosities of about 250 suns. R Serpentis is estimated to be some 600 light years distant; it shows an annual proper motion of 0.05" and a radial velocity of about 14½ miles per second in recession. (For a brief account of the long period red variables, refer to Omicron Ceti, page 631)

W   Variable. Position 18070s1534, near the Serpens Sagittarius border, about 3° SW of the great nebulous cluster M16.   W Serpentis was discovered by Miss A.J.Cannon at Harvard in 1907, and catalogued as an eclipsing binary system with a period of about 14.1 days. According to recent studies the range (photographic) is magnitude 8.9 to 10.3, and the best value for the period is presently 14.15782 days. The very odd light curve of this star makes it virtually unique among eclipsing binaries, and has even raised some doubts as to whether the star is truly an eclipsing system at all! From the first detailed

DESCRIPTIVE NOTES   (Cont'd)

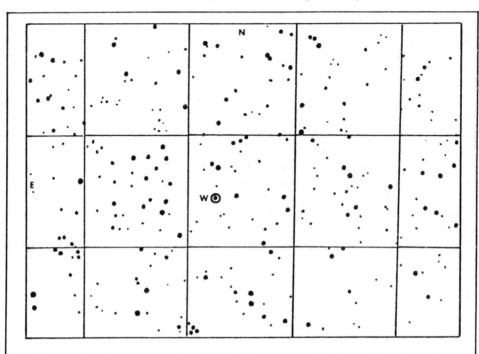

studies of the star, in 1927, W Serpentis has displayed changes and irregularities which can only be described as weird. Not only are there *two* secondary minima between the primary eclipses (giving effectively *three* maxima) but the light of the whole system shows irregular changes and the depth of the minima varies unpredictably. Were it not for the well-defined primary minima, it is doubtful that this star would be considered an eclipsing system at all.   In addition to the eclipsing components (if this interpretation is correct) at least one of the stars must be intrinsically variable.

      In an early study of spectra covering a 9-year interval, Dr.A.H.Joy at Mt.Wilson (1927) found spectral features resembling a cepheid variable, the spectrum varying from G1 to about G4, and showing the sharp-line "c" characteristic, one of the identifying features of the supergiants. "The change in the spectral type....is noticeable mostly near the time of minimum light. At this time the whole spectrum undergoes an interesting transformation. The dark absorption-enhanced lines lose their c-characteristic and seem to approach the condition of emission. The bright

edges of the hydrogen lines increase in strength and several lines of ionized iron, together with a number of unidentified lines, show emission of considerable strength....
The lines of ionized iron which show emission at minimum
change in about a day to absorption lines...." Prof.Joy
found a mean radial velocity of about 14½ miles per second
in approach, reached at minimum light, as is always the
case with eclipsing variables.

In 1937 S.Gaposchkin at Harvard derived an accurate
light curve from 900 measurements of the star on Harvard
patrol plates covering a 40-year span; the result seemed to
confirm the star as an eclipsing binary, but the light
curve is one of the oddest known for any eclipsing system.
In summarizing his studies of this enigmatic star, Dr.
Gaposchkin stated that "W Serpentis displays three types of
variations: it is an eclipsing star with a range of about
0.9 mag., and a mean period of 14.15326 days; it has short-
period intrinsic variation with an amplitude of about 0.2
mag., and a mean period equal to, or half of, the eclipsing
period; and it has a long-period fluctuation, with a range
of about 0.6 mag., and a possible period of about 270 days".
Gaposchkin adopted a probable absolute magnitude of about
-3.7 for the visible star; from the light curve the unseen
secondary would appear to be another giant of somewhat
later spectral type, probably about gK2. The bright spectral lines evidently originate in a huge and extensive
chromosphere of the primary. But in addition to the eclipsing pair, it seems necessary to postulate that at least

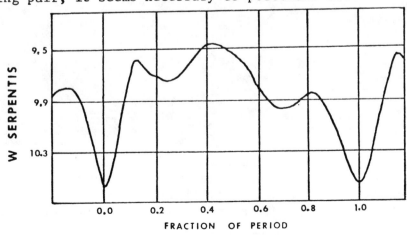

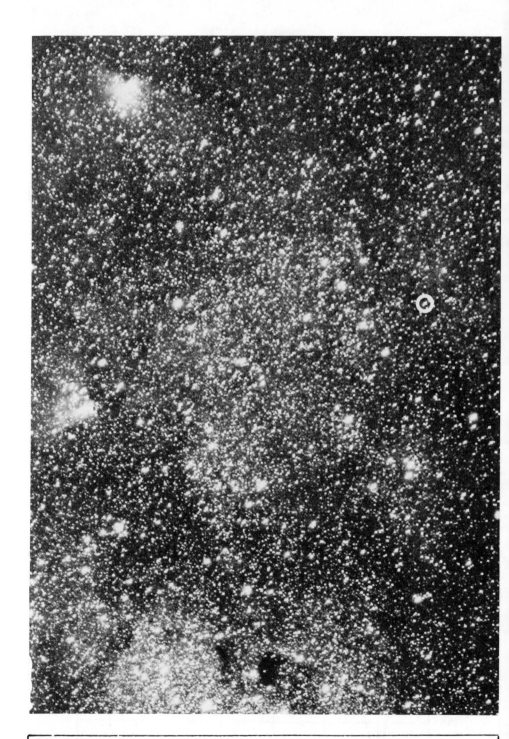

W SERPENTIS FIELD. The variable (indicated by the circle) is located in a rich Milky Way field. The Swan Nebula M17 is at the left edge; M16 at the top. Lowell Observatory

one of the components is intrinsically variable. Some of
the irregular variations of the star may also be due to
interaction with the surrounding gas and dust clouds, as
there appears to be considerable dark nebulosity in the
region at about 3000 light years distance, close to the
accepted distance of the star.

In a study made in the summer of 1956, C.R.Lynds
confirmed the "many peculiarities for which the system is
noted" but in addition found that "the ultraviolet light
curve is not at all like that of an eclipsing binary.....
The B-V color undergoes a strikingly regular variation
which appears to repeat itself from cycle to cycle in contrast to the light variation..." This study showed also
that virtually all the visible light of the system comes
from the primary star which has the color of a reddened
giant of about type F0. M.Hack has classed it as F5, while
J.Sahade and O.Struve (1957) rate it closer to type A. They
find evidence that the W Serpentis primary is probably an
"undermassive late A-type star which ejects matter at high
velocity. The gaseous mass forms an expanding envelope
around this [shell] star, and the whole system is surrounded by a very extended atmosphere, not completely transparent. There is also a gaseous stream from the secondary star
toward the following hemisphere of the primary...." This
gaseous shell probably plays some part in the eclipses of
the system, and is partially responsible for the peculiar
shape of the light curve.

From current studies the two components appear to be
not greatly different in size; each may be about 6 or 8
times the diameter of the Sun; the computed separation is
roughly 14 million miles, with an orbital eccentricity of
about 0.37. The orbital period is presently increasing at
the rate of 20 seconds per year, which suggests that this
star, like Beta Lyrae, is a system which is evolving with
abnormal rapidity.

The identification chart on page 1770 shows stars to
about magnitude 9½; north is at the top and the grid square
pattern indicates intervals of one degree.

The triple star β131 lies just 1° distant to the east
and the "Swan Nebula" M17 in Sagittarius is just over 2°
distant to the ESE. The great nebulous cluster M16 is 3°
to the NE. (Refer to page 1783)

RT   Position 17370s1155. Peculiar variable star,
               usually classed as a slow nova, although some
investigators have questioned whether it can be called a
genuine nova at all. The star lies about 1° NNW from 56
Serpentis, and about 6° SSW from the famous recurrent nova
RS Ophiuchi. RT Serpentis was discovered by Max Wolf in
Europe and independently by E.E.Barnard at Yerkes. It was
first noted as an object of magnitude 13.9 on a plate made
in July 1909. Earlier plates going to 16th magnitude do
not show it. The star brightened slowly, reaching 11th magnitude in 1910 and 9th magnitude in August 1913. It then
remained at a nearly constant maximum for more than ten
years. The fading was equally unhurried; the brightness was
about 10th in 1930, 11th in 1935, and 14th in 1940. By 1963
the star had declined once again to its original 16th magnitude state.
      According to C.P.Gaposchkin the spectroscopic fea-
tures showed the "essential nova character" although it may
be significant that this did not become really evident until rather late in the history of the star. The characteristic bright lines were finally detected after the "nova"
had fallen to nearly 14th magnitude. But perhaps the oddest
feature of RT Serpentis was the complete lack of any high
expansion velocity; the radial velocities were at all times
positive, giving a speed in recession of about 60 miles per
second. A typical nova, of course, shows enormously blue-
shifted spectral lines as matter is blasted off the star at
velocities of several hundred miles per second. RT Serpentis is the only nova on record which showed no sign of this
effect. Are we dealing here with a true nova, or some other
more gentle type of stellar outburst, perhaps one of the
sort which presumably create the planetary nebulae? The
star currently has a spectral type of near A8, but the true
luminosity is unknown, since the distance is uncertain. If
the present magnitude of 16 represents a normal post-nova
luminosity of about 1 sun, then a distance of about 5700
light years is derived, making no correction for light loss
through absorption in space. The rise of only 7 magnitudes
is unusually small for a nova, however, adding to the unusual puzzle of RT Serpentis. (For an account of novae in
general refer to V603 Aquilae, page 213)

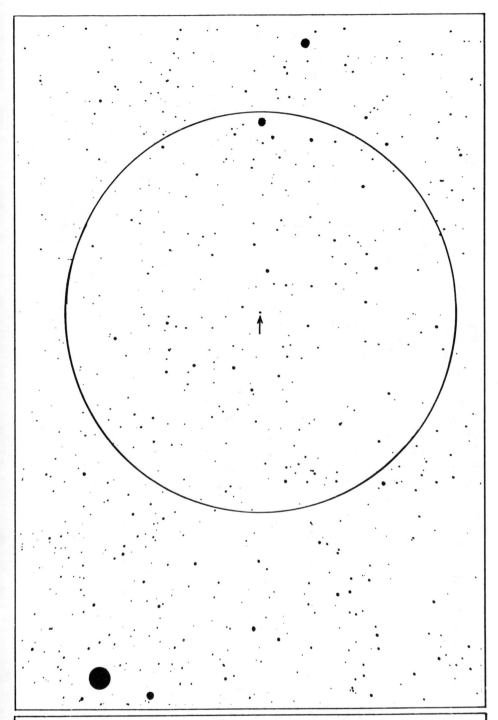

FIELD OF RT SERPENTIS. The field is 1° in diameter with
north at the top; limiting magnitude about 16. Bright star
at bottom is 56 Serpentis, mag 4.4.

DESCRIPTIVE NOTES   (Cont'd)

M5   Globular star cluster, also known as NGC 5904.
Position 15160n0216, a little less than 8° SW
from Alpha Serpentis, and about 20' NNW of the double star
5 Serpentis. M5 is one of the great show objects of the
summer sky, ranking with M13 in Hercules and M3 in Canes
Venatici as one of the three finest globulars in the north
half of the sky. In total integrated light it appears to
hold 5th place in the entire heavens, excelled only by
Omega Centauri, 47 Tucanae, M22 in Sagittarius, and the
great M13 itself.

Discovery of M5 is credited to the German astronomer
Gottfried Kirch, who became director of the Royal Observatory in Berlin in 1705. From an account given by his wife,
Marie Margarethe, it is known that the discovery occurred
on the night of May 5, 1702, while Kirch was looking for a
comet. Kirch was also the discoverer of Chi Cygni, the fine
star cluster M11 in Scutum, and the sun-grazing comet of
1680. Charles Messier rediscovered M5 in May 1764, record-
ing it as "a fine nebula which I am sure contains no star.
Round; seen well in a good sky in a telescope of 1-foot...
Diam 3'." As with nearly all the globulars, resolution was
first definitely achieved by Sir William Herschel; in 1791
he found that the central core of M5 was a thick mass of
about 200 stars "although the middle was so compressed that
it is impossible to distinguish the components".   In the

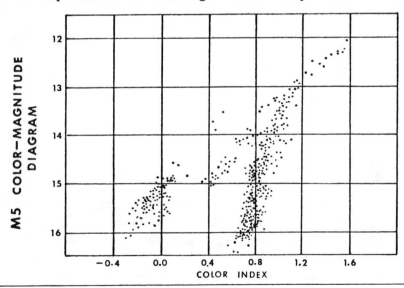

GLOBULAR STAR CLUSTER M5 in SERPENS. One of the finest clusters in the heavens. The photograph was made with the 42-inch reflector at Lowell Observatory.

great reflector of John Herschel the cluster gave the
impression of a cosmic snowball; the richer central part
seemed to be "projected on a loose, irregular ground of
stars". Admiral Smyth spoke of it as a "superb object....
a noble mass, refreshing to the senses after searching for
faint objects, with outliers in all directions and a bright
central blaze which even exceeds M3 in concentration".
Lord Rosse, in 1875, thought the central core about 1' in
diameter, with "stars 12-15 mag., many going out from the
centre in curved lines.." T.W.Webb called M5 "A beautiful
assemblage of minute stars, 11-15 mag. Greatly compressed
in the centre".

Mary Proctor, writing in 1924 in *Evenings With the
Stars*, gave an exquisite description of her impressions of
M5 as seen through the world's largest refractor, the great
40-inch lens at Yerkes Observatory: *"Myriads of glistening
points shimmering over a soft background of starry mist,
illumined as though by moonlight, formed a striking contrast to the darkness of the night sky. For a few blissful
moments, during which the watcher gazed on this scene, it
suggested a veritable glimpse of the heavens beyond...."*

Many users of this book will remember Isaac Asimov's
classic science-fiction story *Nightfall*, dealing with the
fate of a civilization on a planet in a six-star system;
the inhabitants have always known perpetual daylight, and
have no psychological defenses against total darkness which
comes to their world only once every 2,049 years.........
At the climax of the tale, when the last of their six suns
goes into total eclipse, they find that their world is in
the center of a giant star cluster, with thirty-thousand
mighty suns blazing through the darkness in soul-shattering
splendor....

The present author first read Asimov's apocalyptic
tale just two nights after having seen M5 through the fine
40-inch Ritchey-Chretien reflector of the U.S. Naval Observatory station at Flagstaff, Arizona. In that first stunning view it seemed as if the fireflies of a thousand summer nights had been gathered here, frozen forever in time
and suspended among the stars. Asimov very likely had no
definite star cluster in mind as the scene of *Nightfall*,
but to the author of the *Celestial Handbook* it will always
be M5.

GLOBULAR CLUSTER M5, as it appears on a plate made with the 200-inch telescope at Palomar Observatory.

Asimov, incidentally, admits that the original inspiration for the story came from the famous quotation from Emerson in his essay *Nature*:

"If the stars should appear one night in a thousand years, how would men believe and adore, and preserve for many generations the remembrance of the City of God which had been shown!"

With a pair of good binoculars, the observer may obtain a glimpse of M5 as a small fuzzy-looking star-like spot of the 7th magnitude; in a 3-inch refractor it is a bright round nebula about 5' in size. Resolution begins to be apparent in telescopes of 4-inch or larger, and the angular size increases also as fainter outlying stars come into view. Walter S. Houston thinks it attains about 27' in his 10-inch reflector, under exceptionally clear skies; catalogues report various measurements ranging from 10' up to nearly 30' depending on the telescope and the observing techniques employed. H.B.S. Hogg gives the total apparent diameter as 19.9' and the total integrated magnitude as 7.04, photographic; the integrated spectral type is near F5, and the average magnitude of the 25 brightest members is 13.97. The cluster shows a moderate radial velocity of about 30 miles per second in recession. There is a general agreement that the distance of the group is about 26,000 or 27 000 light years; a little more remote than M13. The two clusters appear to be very nearly identical in size and luminosity; the total population of each is probably in excess of half a million stars, allowing for the large numbers of intrinsically faint stars which cannot be seen at all a such a distance. Our Sun, if placed in M5, would appear as a star of magnitude 19.4; this tells us immediately that the brightest cluster stars must be huge giants of absolute magnitude about −3 or −3.5; the true luminosity of such a star ranges up to 2000 times that of our Sun. The entire cluster radiates with the light of about a quarter of a million suns.

M5 is not a completely circular cluster; it shows a slight ellipticity of perhaps 10%, the longer axis tilted toward PA 50°. It also shows a somewhat irregular distribution of the brighter stars, when compared with a typical "smooth" globular such as Omega Centauri. Radiating star

DEEP-SKY FIELDS IN SERPENS. Top: The core of globular star cluster M5 as it appears in a short exposure. Below: Star cluster IC 4756.   Lowell Observatory photographs.

NEBULOUS CLUSTER M16, photographed in blue light (top) and red (below) with the 13-inch telescope at Lowell Observatory. North is at the left, to match page 1786.

streams are prominent around the edges of the cluster,
sometimes giving the impression of a vague spiral pattern;
this appearance is common in many bright globulars, and is
often mentioned in the descriptions of Lord Rosse and his
associates at Birr Castle.

M5 contains an unusually large number of variable
stars, 97 having been identified up to 1975. The great
majority of these stars are "cluster variables" of the RR
Lyrae type with periods of near 0.5 day; these all have
median magnitudes of close to 14.9. Two brighter stars
(11th - 12th mags) appear to be cepheids; their periods
are 25.74 and 26.42 days. One unusual variable seems to be
a faint eruptive star of the SS Cygni class, reaching 17th
magnitude at maximum.

Harlow Shapley, in 1917, made use of the variables
of M5 to check a theory that the velocity of light might
possibly vary slightly with wavelength. By measuring the
maxima of the stars in both blue and yellow light, it was
shown that the greatest possible difference in the velocity could not exceed one part in 20,000,000,000. It is
now generally agreed that there is no difference detectable, and that all radiant energy travels at the same
velocity in space.

M5 is regarded as one of the most ancient clusters
known, and like virtually all the globulars, is thought to
have originated in the very early period of the history of
the Galaxy, probably over 10 billion years ago. According
to H.Arp in 1962, M5 has a computed age of about 13 billion
years. (Refer also to M13, page 978, and the brief outline
of cluster age-dating beginning on page 990).

The bright double star 5 Serpentis will be noticed
in the same low-power field with M5; it is about 20' distant, to the SSE. For descriptive data, refer to page 1759.

M16   (NGC 6611) Position 18160s1348, near the
                Serpens-Scutum-Sagittarius intersection, just
3° north of the Swan Nebula M17 in Sagittarius. M16 is a
large scattered star cluster immersed in a vast diffuse
nebula, a most wonderful object whose full glory is only
revealed on long-exposure photographs. The present author
has introduced the name *The Star-Queen Nebula* for this
exceptional wonder of deep-space; the name "Eagle Nebula"

has also appeared in modern observing guides, but seems
perhaps a little too prosaic for a vista of such cosmic
splendor, aside from the fact that the eagle is already
honored by two first magnitude stars, Vega and Altair.
     The Swiss astronomer P.L.de Cheseaux was probably
the first observer of M16; he recorded it in 1746 as "a
cluster of stars between the constellations of Serpens,
Sagittarius and Antinous". Messier, in June 1764, saw it
as "A cluster of small stars enmeshed in a faint light....
In an inferior telescope it appears like a nebula..." This
early reference to a "faint light" surrounding the stars
is of great interest, as no other early observer recorded
any impression of nebulosity in the cluster. The keen-eyed
Admiral Smyth saw here "A scattered but fine, large stellar
cluster. As the stars are disposed of numerous pairs among
the more evanescent points of more minute components, it
forms a pretty object in a telescope of moderate capacity".
John Herschel thought the stellar membership to be about
100 stars, and T.W.Webb referred to it as "a grand cluster"
but neither observer seems to have detected the diffuse
nebulosity which is quite evident today in a good 6-inch or
8-inch telescope. John C.Duncan rated it as "no more difficult than the nebula in the Pleiades" and found it easily
visible with the 12-inch refractor at the Whitin Observatory. E.E.Barnard, in 1895, was probably the first to record the nebulosity photographically; it also shows on a
plate made by Isaac Roberts with his 20-inch reflector in
1897.   Under good conditions the nebulous haze may be
traced out to 20' or 25'; on the best photographs the full
extent is about 35' x 30'.

CLUSTER M16 – MAGNITUDES AND SPECTRA								
1.	8.24	05e	9.	9.69	B0+B5	17.	10.81	B2
2.	8.27	K0	10.	9.86	09	18.	11.06	08?
3.	8.30	B0+08	11.	9.88	B1:	19.	11.22	B1
4.	8.77	07	12.	9.94	B0	20.	11.24	B1
5.	8.95	08	13.	10.05	B1	21.	11.30	08+05
6.	9.43	09	14.	10.05	06	22.	11.41	09
7.	9.48	B1	15.	10.36	09	23.	11.60	B0:
8.	9.60	08	16.	10.74	B1	24.	11.62	K2:

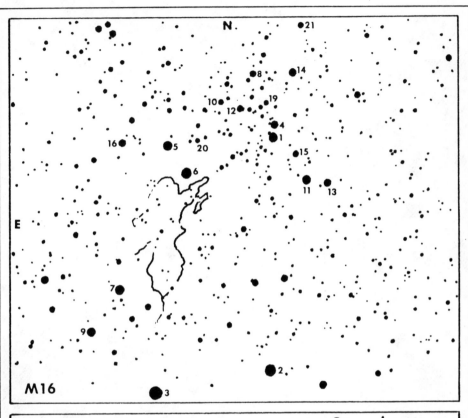

N.

E

M16

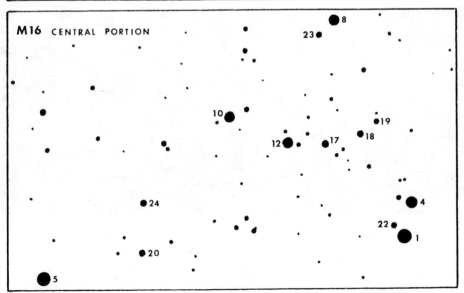

M16 CENTRAL PORTION

NEBULOUS STAR CLUSTER M16 in SERPENS. One of the most spectacular examples of a great nebula. The photograph was made with the 200-inch reflector at Palomar Observatory

DESCRIPTIVE NOTES (Cont'd)

On photographs obtained with great modern telescopes M16 is one of the most spectacular of the diffuse nebulae, and shows an astonishing amount of fascinating detail. Thrusting boldly into the heart of the cloud rises a huge pinnacle like a cosmic mountain, the celestial throne of the *Star Queen* herself, wonderfully outlined in silhouette against the glowing fire-mist, where, as modern star pilgrims have learned, countless new stars are to be born. In the vast reaches of the Universe, modern telescopes reveal many vistas of unearthly beauty and wonder, but none, perhaps, which so perfectly evokes the very essence of celestial vastness and splendor, indefinable strangeness and mystery, the instinctive recognition of a vast cosmic drama being enacted, of a supreme masterwork of art being shown. H.P.Lovecraft, in the opening passages of his *Dream-Quest of Unknown Kadath*, must surely have been seeing some such vision when he wrote:

"It was a fever of the gods, a fanfare of supernal trumpets and a clash of immortal cymbals. Mystery hung about it as clouds about a fabulous unvisited mountain....
...... Vaguely it called up glimpses of a far forgotten first youth, when wonder and pleasure lay in all the mystery of days, and dawn and dusk alike strode forth prophetic to the eager sound of lutes and song, unclosing fiery gates toward further and surprising marvels...."

Here, in one of the great masterworks of the heavens, we can find something of that world-transcending glory which the young art critic Albert Aurier, in 1890, saw in the art of Vincent van Gogh:

"...skies like molten metals and dissolving crystals, with, sometimes, the torrid, incandescent sun-disks; beneath the incessant and formidable stream of every kind of light; in an atmosphere which is heavy, flaming, scorching, as if thrown off from fantastic furnaces which are vaporizing gold and diamond and strange gems - a disturbing, troubling, strange nature, both truly real and half supernatural......"

And Tennyson, the Victorian "Poet of Science", seems to be experiencing a similar vision as he looks into the star-deeps:

THE STAR QUEEN. This strange formation is the central
feature of the great nebulous cluster M16. The print was
enlarged from the photograph on page 1786.

"Yon myriad-worlded way;
The vast sun-cluster's gathered blaze;
World-isles in lonely skies,
Whole heavens within themselves, amaze
Our brief humanities...."

Poets have always made the best prophets, it has been said, and in Tennyson's unerring vision of "whole heavens within themselves" we now find the simplest and most direct statement of astronomical fact. Contemplate for a moment the unearthly scene on the facing page. The great figure of the Star Queen and her Throne rises up to a height of about 6 light years; roughly 36 trillion miles, or about 5000 times the diameter of our Solar System. Surrounding this vast figure, glowing nebulosity extends out in all directions, over an area some 70 light years across. And in this great "world-isle" of star-mist, we are witnessing the birth of new suns, for this is one of the regions of space where star formation is presently in progress.

In the wider view shown on page 1786, we can examine some of the other remarkable features of M16. The major dark projection is one of those sharply-outlined dust clouds which have been called "elephant-trunk" formations by Fred Hoyle; he tells us that in this nebula we are seeing a cloud of hot gas expanding into cooler gas, and that the hot gas is expanding outward like an exploding bomb. Several other major projections are seen at various points in the nebula; all are more or less cone-shaped and oriented in such a way that the dark apex points generally in toward the center of the cloud. On the NE edge of the nebula (lower left on the print) we find the very curious formation shown on page 1791 on a larger scale; this strange feature resembles a towering pillar of black smoke, some 40 trillion miles high, perhaps a cosmic version of Aladdin's genii rising from the magic lamp. A third major projection is the dark triangular wedge which may be seen at the upper left edge on page 1786; this is fairly distinct visually in apertures of 8 or 10 inches, but only from a truly dark-sky site. Many smaller dark nebulae of very odd and intriguing shapes are scattered across the face of the bright cloud, the interplay of bright and dark features producing many a strange effect.

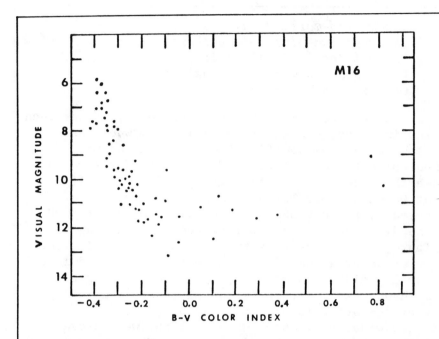

The involved star cluster lies chiefly in the northwest portion of the nebula, and contains a large number of
O and B-type giants, stars of great luminosity and high
temperature. Magnitudes and spectra for the 25 brightest
members are given in the table on page 1784; star numbers
are keyed to the charts on page 1785, and the scale is
given by the separation of stars 5 and 13, which is almost
exactly 7.5'. A color-magnitude diagram for the cluster
appears above, excluding star #2 which is known to be a
foreground object.
        Owing to strong and very variable obscuration and
reddening throughout the region, reaching four magnitudes
in some areas, an accurate distance for M16 has been very
difficult to obtain, and estimates ranging from about 4200
up to 11,000 light years have appeared in various catalogs.
M.F.Walker in 1960 derived a distance of about 3300 parsecs; C.R.O'Dell in 1965 found 2000 parsecs; and Y.Terzian
in a radio-emission study in 1965 adopted 2500 parsecs.
This last figure, of about 8000 light years, is believed to
be the best modern estimate; it places M16 in the great
Sagittarius spiral arm of the Galaxy, which agrees well
with the evidence from radio studies.

THE BLACK PILLAR. This curious formation is located in the outer regions of the M16 nebulosity on the NE side.

Palomar Observatory

Radio emission from the Star Queen was first detected with the 300-foot "dish" at the National Radio Astronomy Observatory of Green Bank, West Virginia.   In a later study by Y.Terzian (1965) it was found that the total mass of M16 out to a diameter of 28' is approximately 12,500 solar masses. At the accepted distance of the nebula 28' corresponds to an actual diameter of about 70 light years; the bright central portion of M16 is about 25 light years wide, or roughly 20,000 times the diameter of the Solar System.

True luminosities for the cluster stars have now been computed. In the case of Star #1, which appears magnitude 8.24, the obscuration has been found to be 2.49 magnitudes; if totally unobscured the star would therefore appear about 5.75; at a distance of 8000 light years the true absolute magnitude must be -6.25 and the luminosity about equal to 26,000 suns. For star #3 the computed absolute magnitude is about -5.7. These are tremendously brilliant stars which must also be relatively young stars, since such giants can maintain their vast outpourings of energy for only a relatively short time.

Extensive photographic and photoelectric studies of M16 have been made by W.F.Walker (1961) and by A.A.Hoag and H.L.Johnson and their associates (1961) at the U.S. Naval Observatory. All the studies show that the cluster has a color-magnitude diagram characteristic of extremely young clusters; the brighter stars are main sequence objects but the fainter members lie above the normal main sequence positions, by up to several magnitudes in some cases. The effect begins to show at about spectral type B5 and virtually all stars later than type B8 still appear to be in the pre-main sequence stage. This interpretation is supported by several other observations: the presence of highly luminous blue giants which are known to be necessarily short-lived stars, the presence of small dark nebulae or "globules" which are thought to be newly-forming "protostars", and the presence of many faint variables of the T Tauri type, which are thought to be still in the process of gravitational contraction. An age of about 800,000 years has been derived as an average for the cluster, though, as star formation is still underway, some of the members may have begun to shine as recently as 50,000 years ago. (See also NGC 2264 in Monoceros and NGC 2362 in Canis Major)

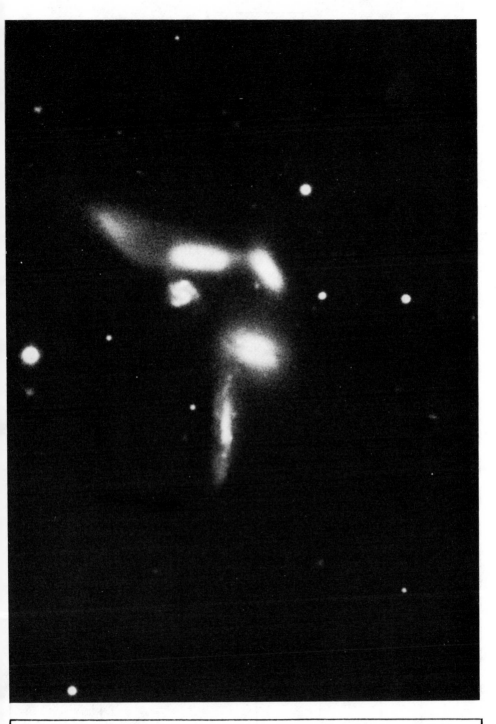

NGC 6027 in SERPENS. An unusual group of galaxies which appear to be linked by luminous bridges. Palomar Observatory photograph made with the 200-inch telescope.

LIST OF DOUBLE AND MULTIPLE STARS

NAME	DIST	PA	YR	MAGS	NOTES	RA & DEC
Σ1377	3.9	136	44	8 - 11	PA slow dec, Spec F8	09408n0251
γ	0.2	113	67	5½- 6	(8 Sext) (h4256)	09500s0752
	35.8	325	40	- 12	Binary, 75 yrs; PA dec, spect A2	
A1767	1.8	19	33	7½- 11	PA inc, spect G5	09552s0142
β25	1.5	150	67	8½- 9	PA dec, spect G5	10192s0931
β1322	10.2	312	54	7 - 13	PA dec, dist inc, spect K0	10216n0237
A2570	0.3	311	66	8 - 8	PA dec, spect A0	10234n0311
β1280	116	192	00	7 - 9	Spect A2	10236n0411
β1280b	0.9	19	24	9- 11½		
OΣ218	0.5	104	67	8 - 9½	PA inc, spect K0	10249n0349
Σ1441	2.7	167	58	6½- 10	cpm, relfix, spect	10285s0723
	64.7	312	25	- 12	K5, F6; AC optical	
β1073	3.5	49	40	7 - 11	cpm, relfix, spect K0	10300s0549
J84	1.0	61	58	8½ -8½	Spect F8	10336n0057
A556	1.3	102	60	7 - 10	PA inc, cpm; spect G0	10345s0835
Σ1456	13.6	46	54	8- 9½	relfix, spect F5	10357n0130
Σ1457	1.8	331	67	7½- 8½	PA & dist inc; spect F2	10361n0600
A2768	0.5	207	57	8- 9½	PA dec, spect F5	10400n0351
35	6.4	236	58	6½- 7½	(Σ1466) Slight PA dec? Spect K3, K0	10407n0501
Σ1470	1.2	189	67	8- 8½	spect A0	10437s0530
40	2.2	10	67	7 - 7½	(Σ1476) PA inc, spect A2	10467s0346
41	27.3	306	34	6- 11½	(h838) relfix, spect A2	10478s0838

SEXTANS

LIST OF VARIABLE STARS

NAME	MagVar	PER	NOTES	RA & DEC
S	8.3--13..	261	LPV. Spect M3e--M5e	10324s0005
T	9.8--10.3	.32470	Cl.Var; Spect A8--F4	09509n0218
W	9.0--10..	40:	Semi-reg; spect N	09484s0148
Y	9.3--9.9	.41982	Ecl.Bin; W U.Maj type; Spect F8	10002n0120
Z	9.1---9.7	57:	Semi-reg; spect M4	10084n0248
RR	8.5--10..	Irr	Spect M4	09523n0526
RT	8.0-- 8.5	96:	Semi-reg; spect M6	10099s1004
RX	6.6--6.62	.0799	δ Scuti type; spect A3	10236n0411

LIST OF STAR CLUSTERS, NEBULAE AND GALAXIES

NGC	OTH	TYPE	SUMMARY DESCRIPTION	RA & DEC
2967	275[2]	⊖	Sc; 12.4; 2.0' x 1.8' pF,pL,R,vglbM; face-on spiral	09395n0034
2974	61[1]	⊖	E4/S0; 12.0; 1.0' x 0.6' B,cS,R; 10^m star at SW tip	09400s0329
2990	624[2]	⊖	Sc/Sd; 13.0; 1.0' x 0.5' F,pS,1E	09436n0557
3044	254[3]	⊖	Sc/SBc; 12.6; 5.0' x 0.6' vF,vL,vmE; edge-on; flat ray	09510n0149
3055	4[6]	⊖	Sc; 12.5; 1.4' x 0.8' F,pL,vlE, vgbM	09527n0431
3115	163[1]	⊖	E7/S0; 10.0; 4.0' x 1.0' vB,L,vmE,mbMN "Spindle Neb" Edge-on galaxy (*)	10028s0728
3156	255[3]	⊖	E5/S0; 13.1; 1.0' x 0.5' F,cS,R,psbM	10101n0322
3166	3[1]	⊖	Sα; 11.5; 4.0' x 1.5' B,pS,R,psmbM; 9' pair with NGC 3169	10112n0340
3169	4[1]	⊖	Sb; 11.4; 3.9' x 1.7' B,pL,mE, mbM; pair with 3166	10117n0343
3423	6[4]	⊖	Sc; 11.7; 3.5' x 3.0' F,vL,R,vgbM; fine face-on spiral	10487n0607

NGC 3115 in SEXTANS. The bright spindle-shaped galaxy as
photographed with the 13-inch Lowell astrograph (top) and
the 200-inch reflector at Palomar (below).

NGC 3115   Position 10028s0728. A bright lens-shaped galaxy, apparently seen nearly edge-on, but somewhat difficult to locate from its position in a blank portion of the sky to the south of Leo. With equatorially mounted telescopes it is most easily located by simply moving the telescope a little less than 20° to the south from Regulus in Leo. Another method is to first locate, with binoculars, the little pair 17+18 Sextantis, just 11° due east from Alphard (Alpha Hydrae); the galaxy is about 1.5° NW from this pair.

NGC 3115 is noted for its unusually high surface brightness which permits the use of fairly high powers; it appears to the eye in amateur telescopes much as it does on the photographic plate. The total magnitude is about 9.8 and the extreme length about 4'. Popularly called the "Spindle Nebula", this galaxy has long been classed as a much-flattened elliptical system of type E7. However, the best of modern photographs obtained with the 200-inch telescope appear to show a dual structure; there is a bright oval nuclear hub, and a thin flat equatorial plane about 15" thick. From the presence of these two sub-systems, the classification has been changed to an S0, or possibly a transition type E7-S0. There is no evidence for any true spiral pattern, nor for any equatorial dust lane which is so often seen in edge-on galaxies.

The system does not appear to be a member of the great Virgo Galaxy Cluster which is centered some 40° away toward the NE. It has a much smaller red-shift, of about 260 miles per second, which suggest a distance of close to 25 million light years and an actual diameter of 30,000 light years. The derived absolute magnitude for the same distance is -19.6, and the actual luminosity about 5.8 billion times that of the Sun. R.Minkowski, at Palomar, has made an independent distance estimate of 21 million light years from observations of the globular clusters in the system, agreeing well with the distance derived from the observed red shift. In another study made at Palomar in 1975, T.B.Williams derived a rotation curve for NGC 3115 which indicates a total mass of approximately 24 billion suns out to the furthest point measured, 11,000 light years from the nucleus. No supernovae have yet been recorded in this galaxy.

TAURUS

LIST OF DOUBLE AND MULTIPLE STARS

NAME	DIST	PA	YR	MAGS	NOTES	RA & DEC
β1178	1.0	349	54	6½- 12	relfix, spect G0	03210n0442
Σ383	5.5	121	40	8 - 8½	cpm, relfix, spect G5	03214n1723
β879	24.7	71	25	6½- 11	cpm; spect G5	03259n1113
A983	0.2	20	66	8½- 9	PA inc, spect G0	03279n2927
Σ406	9.4	124	40	7 - 9	relfix, spect F0	03282n0459
Σ401	11.0	270	49	6½- 7	relfix, cpm, spect A0	03283n2724
Σ403	2.6	175	41	8½- 8½	PA slightly dec, spect F8	03284n1937
A1931	0.8	55	66	8½- 9	PA dec, spect F0	03291n0739
AG 68	17.6	249	25	7 - 10	spect A0	03294n1123
7	0.6	11	66	6½- 6½	(Σ412) Binary, PA dec, about 600 yrs, spect A3. AC PA dec slowly	03315n2418
Σ414	7.4	185	49	8 - 8	relfix, cpm; spect A0	03316n1938
A1933	1.1	140	65	8½- 10	spect A0	03328n0615
β1040	3.6	337	28	7½-11½	spect A0	03331n2949
Σ422	6.2	260	57	6 - 8½	cpm, PA inc, spect dG9, dK6	03342n0026
Hu 813	3.4	290	20	7- 14½	spect A5	03374n2115
Σ427	6.9	209	54	6½- 7½	cpm, relfix, spect A0	03376n2837
Σ430	26.1	55	35	6 - 9	relfix, spect G5	03378n0458
	37.0	301	35	- 10		
Σ435	13.0	2	34	7½- 8½	cpm, relfix, F5	03401n2531
β1041	127	43	20	6 - 6	(0Σ38) wide pair; Spect F0, G0	03416n2745
β1041b	15.2	331	23	6- 13½	BC PA dec, dist inc	
29	64.7	65	27	5½-11½	(h2004) spect B3	03430n0554
β536	0.6	190	61	8½- 9½	binary, about 1000 yrs; PA dec, spect	03433n2402
	39.1	306	55	- 8	dA6; AC dist inc; In Pleiades	
β536c	18.1	8	15	- 12		
β537	0.9	175	54	8½- 10½	PA dec, spect A2; In Pleiades	03441n2440
Σ449	6.8	330	11	8½- 11	relfix, spect A0; In Pleiades	03444n2430

LIST OF DOUBLE AND MULTIPLE STARS (Cont'd)

NAME	DIST	PA	YR	MAGS	NOTES	RA & DEC
Σ450	6.1	265	52	7 - 9	cpm, relfix, B9	03444n2346
30	9.2	59	33	4½- 9½	(Σ452) relfix, cpm	03455n1059
					spect B3	
A831	0.5	5	65	8½- 9½	PA inc, spect F0	03461n1133
27	0.4	39	29	5 - 8	(Σ453) Spect B8;	03462n2354
					In Pleiades	
Ho 324	0.8	333	68	8 - 8½	PA dec, Spect F8	03462n1449
OΣ64	3.1	239	43	7 - 10	relfix, spect B9;	03470n2342
	10.1	234	37	- 9	In Pleiades	
OΣ65	0.6	203	61	6½- 6½	Binary, 62 yrs; PA	03473n2526
					inc, orbit nearly	
					edge-on; spect A3	
Σ457	1.1	95	44	8½- 8½	PA dec, spect F5	03474n2232
	18.1	340	11	-12½		
31	0.4	207	60	6½- 6½	(Kui 15) PA dec,	03493n0623
					spect B9	
Σ479	7.2	128	54	7 - 8	relfix, cpm,	03580n2304
	58.3	241	36	- 9	spect B9	
OΣ70	12.0	226	14	6- 11½	relfix, spect B8	03590n0952
Σ481	2.4	106	58	7½- 11	relfix, spect G5;	03592n2759
	15.5	326	58	- 9½	AC dist dec	
Σ491	2.6	102	42	8½- 9	PA dec, spect F5	04031n1050
Σ493	1.4	91	66	8½- 9	PA slow dec,	04041n0533
					spect F8	
Σ495	3.8	221	42	6 - 8½	relfix, cpm;	04048n1502
					spect dF2	
OΣ72	4.4	326	38	6 - 9	relfix, spect K0	04051n1712
β1232	0.3	354	66	8½- 9½	spect G0	04058n2904
Σ494	5.3	187	55	7½- 7½	neat relfix pair,	04059n2258
					cpm, spect A3	
Σ510	11.3	302	25	6½- 9½	relfix, spect G5	04096n0036
β1278	8.4	303	38	6½-13½	Dist inc, spect A3,	04108n0846
	54.3	255	59	-12½	AC PA inc	
46	0.1	166	62	6 - 6½	Binary, 7.2 yrs;	04108n0735
					PA inc, spect F3	
Σ515	3.5	41	41	8½- 8½	relfix, spect F5	04108n0245
47	1.1	351	58	5½- 8	(β547) PA dec,	04112n0908
	30.0	226	59	- 12	spect gG5; AB cpm,	
					AC optical	

LIST OF DOUBLE AND MULTIPLE STARS   (Cont'd)

NAME	DIST	PA	YR	MAGS	NOTES	RA & DEC
HVI 98	65.5	315	37	$6\frac{1}{2}$- 7	cpm, spect G0, G5,	04128n0604
	214	47	07	- 10	M2	
OΣ78	2.4	247	42	$7\frac{1}{2}$- $9\frac{1}{2}$	relfix, spect G0	04129n2955
Σ517	3.4	12	34	$7\frac{1}{2}$- 9	relfix, spect A0	04134n0020
Ho 328	0.2	342	66	$7\frac{1}{2}$- 8	binary, 60 yrs; PA	04141n1933
					dec, spect dF5	
Σ520	0.3	154	60	$8\frac{1}{2}$- $8\frac{1}{2}$	PA inc, spect F5	04153n2241
Σ523	10.5	163	14	$7\frac{1}{2}$- $9\frac{1}{2}$	relfix, spect A0	04168n2337
	108	48	15	-		
55	0.8	54	61	7 - 8	(OΣ79) Binary, 91	04170n1624
					yrs; PA dec, spect	
					dF7	
φ	52.1	250	25	5 - $8\frac{1}{2}$	(Sh 40) Optical,	04173n2714
					dist dec, PA slow	
					inc, spect K1	
β87	2.0	171	46	6 - 9	cpm, relfix, nice	04194n2042
					colors, spect gM0,	
					A0	
χ	19.5	25	31	$5\frac{1}{2}$- $7\frac{1}{2}$	(Σ528) cpm, spect	04195n2531
					B9	
OΣ82	1.3	13	61	$7\frac{1}{2}$- $8\frac{1}{2}$	binary, about 240	04199n1456
					yrs; PA dec, spect	
					dF9, dG1	
Σ535	1.1	300	66	$6\frac{1}{2}$- 8	PA & dist dec,	04205n1116
					spect A2	
62	29.0	290	32	6 - 8	(Σ534) cpm, relfix,	04210n2411
62b	110	336	11	8 - 12	Spect B3, A0	
66	0.3	33	61	6 - 6	(Hu 304) Binary,	04211n0921
					52 yrs; PA inc,	
					spect A3	
ϰ	339	173	00	5 - 6	(65+67 Tauri)	04224n2208
					wide cpm pair,	
					spect A7, A5. Σ541	
					in field	
Σ541	5.6	329	56	$9\frac{1}{2}$- 10	relfix	04224n2208
68	1.4	333	58	$4\frac{1}{2}$- $7\frac{1}{2}$	(Kui 17) PA inc,	04226n1748
	77.1	233	25	- 9	All cpm, spect A2	
β1185	0.3	38	55	8 - $8\frac{1}{2}$	Binary, 29 yrs; PA	04229n1845
					dec, spect dG4	
Σ546	6.8	184	54	8 - $9\frac{1}{2}$	PA dec, spect G0	04241n1901

LIST OF DOUBLE AND MULTIPLE STARS (Cont'd)

NAME	DIST	PA	YR	MAGS	NOTES	RA & DEC
Σ545	18.8	57	24	7½- 9	relfix, spect A0	04242n1806
β1186	0.5	166	52	6 - 8½	PA dec, spect B7	04247n1106
A2033	0.7	251	19	7½- 10½	spect A0; in field	04253n0958
					with R Tauri	
Σ548	14.6	36	32	6 - 8	relfix, spect dF4	04257n3015
	121	194	08	- 10½		
Σ548c	16.6	52	08	10½-13½		
θ	337	346	17	3½- 4	(78 + 77 Tauri)	04258n1546
					wide cpm pair in	
					Hyades, spect A7,	
					K0; Hu 1080 in field	
Hu 1080	0.4	84	65	7 - 8	binary, 40.5 yrs;	04261n1603
					orbit nearly edge-	
					on, spect dF7	
80	1.6	18	67	5½- 8	(Σ554) binary, 170	04273n1532
					yrs; PA dec, spect	
					A6; in Hyades	
OΣ84	9.5	254	32	7½- 8	relfix, cpm;	04284n0641
					spect G5	
Σ559	3.1	276	66	7 - 7	relfix, spect B9	04306n1755
h5461	25.5	102	33	5½-10½	Spect B9 or Ap	04315n2852
	50.0	133	24	-11½		
Σ562	2.0	274	42	7 - 10½	relfix, spect F2	04318n2235
L 4	0.3	192	02	7 - 7½	Uncertain; spect F8	04328n1947
88	69.7	299	20	4½- 8½	(Sh 45) relfix,	04329n1004
					cpm, spect A3	
α	31.4	112	34	1 - 13½	ALDEBARAN (β550)	04330n1625
	121	34	23	- 11	AB cpm, AC dist inc	
α c	1.7	274	23	11-13½	Spect K5, dM2 (*)	
Σ569	8.2	133	25	8 - 8½	relfix, spect A3	04335n0907
OΣ86	0.6	25	65	7½- 7½	PA dec, spect A2	04336n1940
Σ567	2.0	335	66	8½- 9	PA & dist slow inc,	04338n1924
					spect G0	
Σ572	3.9	194	59	6½- 6½	PA dec, dist inc,	04354n2651
					spect dF2, dF3	
τ	0.1		09	4 -8	(94 Tauri) (OΣΣ54)	04392n2252
	62.9	212	26	- 9	(Ho 642) AB uncert.	
					Spect B3	
96	29.3	57	25	6 - 11	(h3261) (β551)	04469n1549
96b	6.3	205	24	11- 13	Spect gK3	

LIST OF DOUBLE AND MULTIPLE STARS   (Cont'd)

NAME	DIST	PA	YR	MAGS	NOTES	RA & DEC
A1843	0.4	280	55	7½- 10	PA dec, spect A0	04505n2517
β1237	4.5	59	41	7½- 10	relfix, spect K0	04507n2328
99	6.3	9	38	6 - 12	(β1045) cpm, spect	04548n2352
	103	353	38	- 13	gG8	
Σ623	20.7	204	34	7 - 8½	relfix, spect B9	04568n2715
ΣI 12	0.3	294	54	7 - 9½	(A1844) (S461) AB	04586n2636
	78.7	159	21	- 8½	binary, 25 yrs; PA inc, all cpm; spect F5	
OΣ95	0.9	307	65	6½- 7	cpm, dist inc, PA dec, spect A2	05026n1944
104	0.1	40	61	6 - 6	(A3010) binary, 5.3 yrs; PA inc, spect G4	05044n1835
103	14.4	150	25	5½- 12	(HV 114) relfix;	05051n2412
	35.3	197	24	- 9	spect B2	
Σ645	11.9	27	52	6 - 8½	(β1047) relfix, spect A3	05066n2758
Σ645b	0.3	72	60	8½- 9	BC slight PA dec?	
Σ665	1.5	255	66	8½- 9	relfix, spect F0	05128n1940
Σ670	2.5	164	65	7½- 8	PA dec, cpm, spect B3	05138n1823
Σ674	9.7	150	40	6½- 9½	cpm, relfix, spect F5	05144n2005
Σ680	9.0	204	38	6½- 10	relfix, spect K0	05163n2005
Σ686	9.3	223	37	8 - 8	spect F2	05179n2359
Σ694	1.1	6	43	8 - 8	relfix, spect A2;	05210n2455
	10.2	347	22	- 15	AC PA inc	
115	10.3	306	14	5½-10½	(OΣ107) Probably	05242n1755
	9.6	341	08	- 12	all cpm; spect B5; BC= 5.6" pair	
118	5.1	204	57	6 - 6½	(Σ716) cpm, PA slow	05262n2507
	141	99	12	- 12	inc, spect B9	
OΣΣ64	10.6	129	17	7½- 13	(β891) AB cpm;	05269n1824
	54.2	21	32	- 8	spect A0; C= A2433	
A2433	1.1	248	31	8 -13	spect F	
Σ730	9.8	141	22	6½- 7	relfix, spect A2	05293n1701
Σ742	3.6	268	53	7 - 7½	PA inc, spect F8; ½° west from Crab Nebula M1	05334n2158

LIST OF DOUBLE AND MULTIPLE STARS   (Cont'd)

NAME	DIST	PA	YR	MAGS	NOTES	RA & DEC
Σ740	21.7	119	18	8 - 9	relfix, spect B5	05334n2109
Σ749	1.0	333	66	7 - 7	PA inc, dist inc, spect B8	05340n2654
Σ755	5.7	318	22	8½- 9	relfix, spect F8	05361n2316
	146	34	07	- 8		
126	0.3	238	63	6 - 6	(β1007) binary, 78 yrs; PA inc, spect B3	05384n1631
Σ770	0.9	333	68	8½- 10	relfix, spect A0	05386n1912
Σ776	2.4	100	55	8 - 9	relfix, spect A0	05400n2520
Σ777	4.8	85	22	8½- 9	relfix, spect A0	05403n2211
OΣ115	0.6	119	58	7½- 8½	perhaps slight PA dec, spect G0	05417n1503
	92.8	256	11	- 11		
Σ785	14.2	348	43	7 - 8	(OΣ116) relfix, spect B9	05428n2554
	17.8	66	08	- 12		
	201	10	11	- 10		
Σ785d	6.1	256	11	10½-12½		
Σ787	0.8	66	67	8 - 8½	PA dec, spect F2	05430n2118
	12.7	40	11	- 13		
β91	1.6	85	43	7½- 10	relfix, primary spect composite, F + A	05446n2056
OΣ118	0.4	314	53	6 - 7½	(OΣΣ67) All cpm, relfix, spect B7; C spect= A0	05454n2051
	75.5	161	33	- 7		
Ku 23	0.9	106	53	7 - 9	no certain change, spect B9	05480n1426
Σ805	12.2	49	32	8 - 9	relfix, spect A2	05486n2827
136	15.0	232	58	6 - 12	(β1054) Spect A0	05502n2736

TAURUS

LIST OF VARIABLE STARS

NAME	MagVar	PER	NOTES	RA & DEC
λ	3.4--4.1	3.9530	(35 Tauri) Ecl.Bin; Spect B3 + A4 (*)	03579n1221
R	8.1--14.7	324	LPV. Spect M5e--M7e	04256n1003
S	9.4--16..	373	LPV. Spect M7e; in field with R Tauri	04265n0950
T	9.4--13..	Irr	Erratic; In diffuse Nebula NGC 1555 = "Hind's Variable Nebula"; Spect dG5e (*)	04191n1925
V	8.5--14.1	170	LPV. Spect M0e--M4e	04491n1727
W	9.0--13..	261	Semi-reg; Spect M5; In Hyades Cluster, near θ Tauri	04251n1556
X	7.4--	---	Variability uncertain; Spect F5	03505n0737
Y	7.1---9.5	241	Semi-reg; Spect N2; 2° ESE from ζ Tauri	05427n2040
Z	9.2--13..	494	LPV. Spect M7e; On Taurus-Orion border	05496n1547
RR	9.9--13..	Irr	RW Aurigae type; Spect A2e (*)	05364n2621
RU	9.9--14.8	568	LPV. Spect M4e; In field of Z Tauri	05497n1558
RV	9.5--13..	79	Peculiar, multiple period; typical "RV Tauri" star; Spect G2e--K3 (*)	04440n2606
RW	7.6--12.0	2.7688	Ecl.Bin; Spect B8+K0; In field of 41 Tauri (*)	04008n2800
RX	9.1--14.6	335	LPV. Spect M7e	04355n0814
RY	8.6--10.6	Irr	RW Aurigae type; Spect dF8e--dG2e	04189n2820
ST	8.0--8.6	4.0342	Cepheid; W Virg type; Spect F5--G5	05422n1334
SU	9.5--16..	Irr	R Cor.Bor type; Spect G0e p	05461n1903
SV	9.5--11.0	2.1669	Ecl.Bin; Spect B9+A0	05490n2806
SW	8.6--9.7	1.5836	Cepheid; W Virg type; Spect F4--F8	04219n0401
SY	9.6--10.4	Irr	Spect K3	03458n2322

TAURUS

LIST OF VARIABLE STARS   (Cont'd)

NAME	MagVar	PER	NOTES	RA & DEC
SZ	6.8---7.4	3.1488	Cepheid; Spect F5--F9; 2° north from Aldebaran	04343n1827
TT	8.0--10..	166	Semi-reg; Spect N3	04484n2827
UX	9.8--12.5	Irr	RW Aurigae type; Spect dG5e; 1° south from ε Tauri in N.Hyades	04271n1807
VX	9.8--15..	299	LPV. Spect Me; In Hyades	04226n1627
WW	8.3--12..	116	Semi-reg; Spect G2e--K2	03586n3007
XX	6.0--16.5	---	Nova 1927	05165n1640
AB	9.1--11..	142	Semi-reg; Spect M6	05379n2805
BL	8.5--11..	Irr	Irr or Semi-reg; M5	03501n2004
BP	9.6--12..	Irr	T Tauri type; Spect dK5e	04161n2859
BU	5.0--5.5	Irr	(Pleione) (28 Tauri) In Pleiades; Spect B8	03462n2359
CD	6.7--7.2	3.4352	Ecl.Bin; Spect F2+F2	05146n2005
CE	4.7---5.1	165	(119 Tauri) Semi-reg; Spect M2	05293n1834
CH	9.4--10.7	97	Semi-reg; Spect M6	03423n0946
CM	-4.5---16	---	Supernova July 1054 AD; = Pulsar PSR 0531+21; "Crab Nebula" NGC 1952 (M1) is remnant of this nova (*)	05315n2159
CQ	8.2--10..	Irr	RW Aurigae type; Spect F2; Field of 121 Tauri	05329n2443
DV	8.4--9.5	Irr	Spect M6; in field with CE Tauri	05282n1832
DY	9.4--10..	Irr	Spect M4	05391n1831
ET	9.1--10.0	5.9969	Ecl.Bin; Spect B8; in field of nebula S147	05346n2715
EU	7.6--8.0	2.1025	Cepheid; Spect G5	05427n1838
GS	5.15-5.22	11.94	α Canes Ven type; Spect A pec	04035n2730
HU	6.0--6.8	2.0563	Ecl.Bin; Spect B9	04353n2035
IM	5.37-5.44	.132	δ Scuti type; Spect dF3	04078n2621

LIST OF STAR CLUSTERS, NEBULAE AND GALAXIES

NGC	OTH	TYPE	SUMMARY DESCRIPTION	RA & DEC
1435		☐	vL,vF Neby surrounding 23 Tauri (Merope) in Pleiades	03433n2342
----	M45	⦂⦂	PLEIADES Cluster; Diam 100'; vvB,vL, Class C; brilliant naked-eye star cluster (*)	03439n2358
1514	69^4	◎	L,F, Mag 11, Diam 120" with 10^m central star (*)	04061n3038
----	I.359	☐	S,F, comet-like wisp length 1' with 12^m star at south tip	04171n2806
1555		☐	vF,S; Diam 30"; "Hind's Variable Nebula" associated with variable star T Tauri (*)	04190n1925
----	L.2087	☐	S,F neby Diam 2' in large dark obscuring cloud 28' wide	04369n2538
1647	8^8	⦂⦂	vL,pB; Mag 6½, Diam 40'; 25 stars mags 8...13 in widely scattered group; Class C (*)	04432n1859
1746 1758	21^7	⦂⦂	L,nC,Irr group Mag 6; Diam 45'; about 50 stars mags 8... Class E; E portion is 1758	05006n2344
1807		⦂⦂	pL,pRi; Mag 7½, Diam 10'; about 15 stars mags 8.....9 Class E	05078n1628
1817	4^7	⦂⦂	L,Ri,1C; Mag 8, Diam 15'; 50 stars mags 10...14; Class D; Pair with NGC 1807	05092n1638
1952	M1	☐	!! Mag 9; Diam 5' x 3'; with outer filaments; The "Crab Nebula"; Supernova remnant (1054 AD) (*)	05315n2159
----	S147	☐	eL,eF; Diam 2° x 3°; Huge oval cloud with filaments & nebulous shreds; probably supernova remnant (*)	05360n2800

ALPHA   Name- ALDEBARAN, from the Arabic *Al Dabaran*, "The Follower", or *Na'ir al Dabaran*, "The Bright One of the Follower", supposedly from the fact that the star follows the Pleiades across the sky. Aldebaran is the 13th brightest star in the sky. Magnitude 0.86 with possible slight variability; Spectrum K5 III; Position 04330n1625, on the left or eastern tip of the V-shaped Hyades Star Cluster. Opposition date (midnight culmination) is December 3.

Aldebaran has been from remote times considered the Eye of the Bull in the constellation, and has been honored with a variety of names and titles. The ancient Roman name *Palilicium* commemorates the *Palilia* or Feast of Pales, the special deity of Roman shepherds; this celebration also marked the traditional date of the founding of Rome, on April 21, 753 BC. Ptolemy refers to it under a name which has been translated "The Torch-bearer"; in Babylonian lands it was *I-ku-u* or "The Leading Star of Stars"; an even more ancient Akkadian title, *Gis-da*, has been translated "The Furrow of Heaven". The Persian name *Paha* and the Sogdian title *Baharu* both seem to mean "The Follower". Aben Ezra, in the 12th Century, identified Aldebaran with the Biblical *Kimah*, and the familiar reference to "Arcturus and his sons" in the *Book of Job* is thought by some modern scholars to actually refer to Aldebaran.

On the charts of Riccioli the name usually appears as *Aldebara*, sometimes as *Aldebaram;* Chaucer has *Aldeberan*, while Edmund Spenser, addicted to deliberate archaisms, refers to the star as *Aldeboran* in his curious epic poem *The Faerie Queen*. Schickard has the unusual form *Addebiris*, while Bayer's *Uranometria* of 1603 titles the star *Subruffa* in recognition of its "rose-red" tint. According to R.H. Allen, native Arabian titles included *Al Fanik*, "The Stallion Camel", *Al Fatik*, "the Fat Camel" and *Tali al Najm* or *Hadi al Najm*, referring to its position as the "Leader" or "Driving One" of the Pleiades. The Latin *Stella Dominatrix* expresses the same concept. An old English name, *Oculus Tauri*, refers of course to the Bull's Eye.

Aldebaran was one of the four "Royal Stars" of old Persia, the other three being Antares, Regulus, and Fomalhaut. In various cultures it was connected with the spirits or gods of the rain, and with the fertility of the Earth.

"..........go forth at night,
And talk with Aldebaran, where he flames
In the cold forehead of the wintry sky..."

SIGOURNEY

1808

The Hindu title *Rohini*, the "Red One" or "The Red Deer"
was applied both to Aldebaran and the entire constellation.
It was near the banks of the Rohini River in Nepal, incidentally, where the famed *Lumbini Gardens* once existed, the
traditional birthplace of the Buddha, at the "time of the
May Full Moon", probably 563 BC. Modern scholars, however,
do not seem to know whether the Rohini River was named in
honor of Aldebaran or the constellation Taurus, or if the
name in this case simply means "The Red Stream". According
to the English missionary Joseph Edkins, Aldebaran was
known to the Hindus as *Sataves*, which is said to mean "The
Leader of the Western Stars".

Aldebaran appears from the Earth as a member of the
Hyades Star Cluster, but the "membership" is accidental;
it is roughly twice as close to us as the cluster, and has
a different motion in space. The cluster, however, makes it
impossible to mistake Aldebaran for any other 1st magnitude
star. *"The locality is still more distinctly marked,"* says
Martha E. Martin, *"if one chances to know the shimmering
light of the Pleiades, which are in the same constellation
.....and constitute the one group of stars that is usually
familiar to the youth who celebrate Hallowe'en. Aldebaran
shoots its ruddy face above the horizon just an hour after
the hazy little dipper of the Pleiades has appeared, and
the star is then to the east of, and, hence, almost directly under, the Pleiades........ In his section of the sky
Aldebaran reigns throughout all the lovely autumn evenings,
with beautiful Capella in her own realm to the north of
him and Fomalhaut far to the south...."*
Aldebaran is a fair-sized giant star, not an actual
supergiant like Betelgeuse, but a moderate K5 giant some
40 times the solar diameter, with about 125 times the solar
luminosity. It lies some 68 light years distant, shows an
annual proper motion of 0.21" in PA 160°, and has a radial
velocity of about 33 miles per second in recession. The
color of the star is usually described as "rosy" or "pale
reddish-orange"; the poet W.R.Thayer evidently saw a deeper
tint in the great star:

> *"....I saw on a minaret's tip
> Aldebaran like a ruby aflame, then leisurely slip
> Into the black horizon's bowl..."*

DESCRIPTIVE NOTES   (Cont'd)

Aldebaran, like many reddish or orange giants, may be slightly variable; the maximum recorded range appears to be about magnitude 0.78 to 0.93. For the owners of a visual spectroscope, Aldebaran shows a finely detailed absorption spectrum with many dark lines and bands. E.J. Hartung finds some of these visible with an aperture of 7.5 cm. A star of this class has a fairly low surface temperature, of about 3400°K, and a relatively low density of perhaps 0.00005 the solar density.

A red dwarf companion of the 13th magnitude shares the proper motion of Aldebaran at a distance of 31.4" in PA 112°; this is a rather difficult object owing to its feeble light and field glare from the dazzling primary. Its spectral type has been measured as dM2, and the computed absolute magnitude is near +12 or about 0.001 of the solar luminosity. The projected separation of the two stars is about 650 AU.

At a distance of 121" in PA 34° the observer may find another faint star, of the 11th magnitude, which, however, is not a true companion to Aldebaran. This is the object called β1031, resolved into a close pair of 1.7" by S.W. Burnham in 1888, and known to be a physical pair with a slow decrease in both PA and distance. Both stars are faint dwarfs, with spectral classes of dK5 and dM2, and the gap between this pair and Aldebaran itself is gradually widening from the individual motions of the two objects.

Aldebaran is one of the few stars of the 1st magnitude which may be occulted by the Moon. Such events occur in "seasons" in which the star may be occulted repeatedly, month after month, as in 1978 when twelve such eclipses of the star were scheduled to occur. Owing to the lack of a lunar atmosphere, the disappearance of the star on such occasions is startlingly abrupt, one of the most nearly instantaneous phenomena which the eye can observe. Predictions for these events are carried in the standard *Nautical Almanac* for each year, as well as in monthly astronomical magazines. Occultations which occur at the moon's dark edge are particularly intriguing, as the star may be observed in comfort, without lunar glare, right up to the very last fraction of a second. One of the "classic" occultations of Aldebaran was that seen at Athens in March of 509 AD; in studying the records of this event more than a thousand

THE BULL'S EYE. Aldebaran as it appears on a photograph in red light, made with the 13-inch camera at the Lowell Observatory.

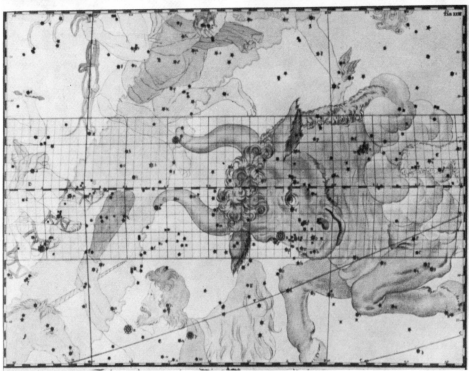

TAURUS APPEARS ON THE Egyptian zodiac at Denderah (upper
left) and on the astronomical ceiling of the tomb of Seti I
at Thebes (upper right). Below: Taurus as depicted on the
1750 edition of John Bevis' star atlas.

DESCRIPTIVE NOTES   (Cont'd)

years later, Edmund Halley concluded that the phenomenon
could not have occurred unless the position of Aldebaran
had been several minutes of arc further north at the time.
From a comparison of his positions with those reported in
the ancient records, Halley found that Sirius, Arcturus,
and Aldebaran had measurably changed their positions since
ancient times. Halley, in 1718, announced the discovery of
what we now call "proper motion". Modern measurements
show that the motion of Aldebaran, in 2000 years, amounts
to about 7', or roughly a quarter of the apparent size of
the Moon's disc.

Mankind, for at least the last several thousand
years, has seen the V-shaped Hyades group as the Head of
the Bull, with ruddy Aldebaran as the Bull's Eye. Taurus
is one of the very earliest constellations to be recognized
and was probably named as early as 4000 BC when it marked
the Vernal Equinox, and its meeting with the Sun heralded
the beginning of the agricultural year. In later Greek
myth Taurus was identified with the snow-white Bull which
carried off Europa, and was later revealed (surprise!) as
Zeus in disguise. This story is the subject of the well
known painting by Veronese, one of the treasures of the
Ducal Palace in Venice. In another legend, Taurus is the
famed "Cretan Bull" (Fig.1) or the "Bull From the Sea" who
was eventually conquered by Hercules, as portrayed on this
silver coin from Selinus in Sicily, minted about 460 BC.
The Bull was a very popular subject on ancient Greek coins;
we sense his restless strength and fierce power on a silver
coin of Thurium dating from about 425 BC (Fig.2); he is
seen again, alert and wary, on a silver stater minted at
Gortyna in Crete, a century later (Fig.3). It is in the
Minoan culture of ancient Crete, of course, that we find
the most completely "Bull-oriented" civilization known to

DESCRIPTIVE NOTES   (Cont'd)

us. In the great capital at Knossus were held the strange
"bull-leaping" games depicted in the wall-paintings of the
king's palace; here the huge carved "horns of consecration"
might be seen, stark against the sky atop the palace roof;
here in the mysterious *Labyrinth* was kept the terrifying
Bull-monster, the *Minotaur*, reputed to be a child of the
king himself, and eventually slain by the Greek warrior-
hero Theseus. The Labyrinth appears on many coins of the
city of Knossus (Fig.4) minted, however, some 900 years
after the legendary days of King Minos.

Sir Arthur Evans, excavator of the Minoan palaces,
possibly found one basic explanation of the Minoan bull-
cult when he experienced a sudden earthquake at his villa
near Knossus in 1926:

".......*Perhaps I had hardly realized the full awe-
someness of the experience, though my confidence in the
full strength of the building proved justified, since it
did not suffer more than slight cracks. But it creaked and
groaned, and rocked from side to side, as if the whole must
collapse.....A dull sound rose from the ground like the
muffled roar of an angry bull.....*"

"*It is something to have heard with one's own ears
the bellowing of the bull beneath the earth who, according
to primitive belief, tosses it upon his horns......*"

In other civilizations, much older than that of
Crete, Taurus was associated with royalty and the concept
of divine power. The Sumerian high god *Enki* is referred to
in inscriptions as "The Bull" or "The Divine Bull"; in the
lands of the Canaanites a supreme deity called *El* is shown
in carvings as a figure wearing a crown or helmet adorned
with bull horns, and is hailed as "The Great Bull, Creator
of all Things". In the oldest of the Egyptian inscriptions,
the *Pyramid Texts* of the 5th Dynasty, we find the following
lines, addressed to the king in his role as divine being:

"*Behold, thou art the Enduring Bull of the wild bulls
of the gods who are on earth.....of the gods who are in the
sky.... Endure, O Enduring Bull, that you may be infinite
in strength as the ruler of all, at the head of the spirits
forever.....*"

In an 18th Dynasty hymn to Amon-Re, dating from about
1500 BC, the god is addressed as:

"Lord of the Thrones of the Two Lands;
Ruling over Karnak,
Bull of his Mother, Presiding over his fields;
Far-reaching of stride,
Eldest of Heaven, First-born of Earth,
Lord of all that is, Enduring in all things,
The Goodly Bull of the Assembly of the Gods...."

On the touchingly beautiful and exquisitely carved *Chair of Millions of Years* in the tomb of the boy-Pharaoh Tutankhamon, the central figure of the God of Eternity is flanked by two carvings of the Bull; the accompanying inscription gives the young king's name and hails him as *"the Strong Bull, Beautiful of Birth":*

In the ancient cult of Mithras, once a serious rival to Christianity, the central rite was the mystic sacrifice of the Bull, by which the god redeemed the world.   There seems to have been considerable astronomical symbolism in this mysterious faith, as in the ancient Mithraic grotto still preserved beneath the church of St.Clemente in Rome, we find a ceiling inscribed with zodiacal figures, and an altar flanked by two figures representing the ascending and the descending Sun. Vastly more ancient, though, are the enigmatic bull-paintings found on the walls of the caves of Altamira and Lascaux in Spain and France, the wonderfully "modern" work of the Paleolithic cultures which existed in southern Europe some 15,000 to 20,000 years ago. Here the Bull is the subject of some of the earliest religious art produced by Man. As the paintings adorn the walls of some of the darkest and most inaccessible caverns, it seems unlikely that they were intended to be admired by tourists and art-lovers of the 20th millenium BC.   Anthropologists

DESCRIPTIVE NOTES   (Cont'd)

believe that their function was religious or magical; that they were used in rites intended to insure the success of the hunt, an early example of (as J.R.Conrad phrased it) "art for meat's sake". The animal portrayed in the Lascaux caves is generally claimed to be the ancient *auroch* or wild bull of paleolithic Europe, probably the ancestor of the fighting bulls of modern Spain.

By late Neolithic times the association of the Bull with the star of Taurus had become a permanent tradition; he is mentioned on Babylonian cylinder seals as the "Bull of Light" or the "Heavenly Bull"; R.H.Allen tells us that the Assyrians named their second month of the year in his honor, as "The Directing Bull". The first letter of both the Phoenician and the Hebrew alphabet appears to be a symbol of the head of a Bull or Ox, and has descended to us as the Greek *Alpha*. The symbol is said to have been the tribal sign of Ephraim, from the sentence in the 33rd chapter of *Deuteronomy*, *"his horns are the horns of the wild ox.."* As a symbol of divine strength and virility, the Bull became the central figure of the *Apis* cult of Egypt; whole series of tombs were prepared for these sacred bulls at the mysterious *Serapaeum* near Sakkara. In Chaldaean legend, the "Star-of-Taurus" was associated with the name of the mythical King *Alaparos,* who was said to have ruled in the ages before the great Flood. In later Christian tradition, however, Taurus became associated with the very popular Saint Andrew, or occasionally with the patriarch Joseph.

Some 25 miles northwest of Dublin stands the revered hilltop of *Tara,* site of the stronghold of the high kings of ancient Ireland. Here was held the *Tarbfeis* or Bull-

Feast of legend, during which the name of the destined
next king was revealed by a mysterious rite called the
Imbras Forosnai, "The Knowledge Which Illuminates". In the
tiny churchyard at Tara may still be seen the curious
Adamnan's Stone, a carved pillar some five feet high upon
which is engraved the enigmatic figure of a horned god...
He is thought to be *Cernnunos*, an important deity of the
ancient Celts. In Medieval Scotland, Taurus was the traditional *Candlemas Bull*, who journeyed across the sky on
New Year's Eve.

R.H.Allen tells us that the constellation was the
Al Thaur of Arabia; it was *Taura* in Syria, *Tora* in Persia,
and *Shor* in Judaea. In some Latin writings it is referred
to as *Princeps Armenti*, "The Leader of the Herd", or as
Bubulcus, the "Driver of the Oxen". Ovid's titles *Europae*
and *Agenoreus* refer to the maiden Europa and her father;
Martial calls the constellation *Tyrius*, evidently from the
tradition that Europa was a princess of Tyre in Phoenicia.
The Chinese, however, saw here a part of their large star
pattern *The White Tiger*; it also appears as a *Great Bridge*
on some oriental star maps. In later times, after Jesuit
missionaries had introduced the European constellations,
it became the *Golden Ox*.

On traditional star-charts, only the head and horns
of Taurus are usually shown, with the V-shaped Hyades outlining his face, Aldebaran his fiery eye, and the tips of
his horns extending to Beta and Zeta Tauri, some 15° to
the northeast. The rest of the Bull's body is presumably
hidden beneath the waters of the sea as Taurus swims away
with Europa. To the south and east, the great figure of
the celestial warrior *Orion* , with shield held high on one
arm, stands ready as he faces the thundering charge of the
Bull, a mysterious sky-drama which is re-enacted each year
in the stars.

THE HYADES   Prominent star cluster in Taurus, seen
as a distinct V-shape group to the naked
eye, marking the outline of the Bull's Head. It extends
from Alpha (Aldebaran) to Gamma Tauri, and then from Gamma
to Epsilon, each side of the "V" being some 4¼° in length.
A larger though less conspicuous group than the Pleiades,
the Hyades Cluster forms an attractive region for small

THE HYADES. A portrait of one of the nearest star clusters. The small cluster at upper left is NGC 1647. Lowell Observatory photograph made with the 5-inch Cogshall camera.

telescopes; as a cluster it appears at its best in a good pair of binoculars. Aldebaran, which adorns the eastern tip of the "V" figure, is not a true member of the Hyades, but lies roughly midway between us and the cluster. The bright stars of the cluster are listed, with magnitudes and spectra, in the table on page 1820.

In the lore of the ancients, the Hyades were associated with wet and stormy weather; the name itself is said by some to be derived from an archaic Greek word meaning "to rain". Pliny speaks of them as

"......*a star violent and troublesome;*
bringing forth storms and tempests
raging both on land and sea.... "

Homer refers to the "rainy Hyades" in very nearly the same words, and Tennyson, in modern times, faithfully followed the ancient tradition when he wrote in *Ulysses:*

"Thro' scudding drifts
the rainy Hyades vext the dim sea.... "

It is rather strange to find that the same tradition was a part of the lore of ancient China. In an anonymous poem in the pre-Confucian *Shih Ching* or "Book of Songs" we find the following passage, dating from at least the 6th Century BC:

"The mountains and streams never end;
The journey goes on and on......
....
The Moon is caught in the Hyades;
There will be great rains.
The soldiers who are sent to the east
Think only of this.... "

In Greek myth the Hyades were the daughters of Atlas and Aethra, half-sisters of the Pleiades; they were entrusted by Zeus with the care of the infant Bacchus, and afterwards rewarded by a place in the heavens. Hesiod has given us their names - Eudora, Koronis, Phaeo, Kleea, and Phaesula; at least two names appear to have been lost as the traditional number was seven. Pherecydes, in another version, gives the names as Diona, Ambrosia, Thyrene, Aesula, Polyxo, Koronis, and Eudora. Other lists preserve

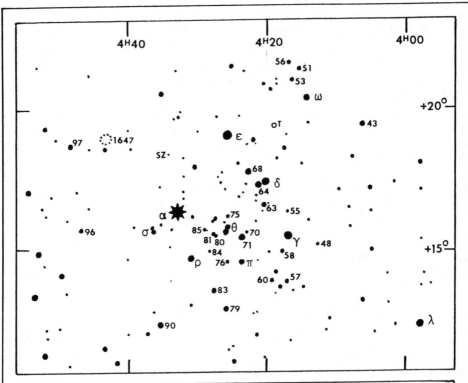

Star		Mag.	Spectrum	Star	Mag.	Spectrum
θ^2 =	78	3.34	A7 III	58	5.17	sgA8
ε =	74	3.53	K0 III	67	5.19	A5
γ =	54	3.68	G9 III	81	5.40	A7m
δ =	61	3.77	K0 III	57	5.49	A9
θ^1 =	77	3.87	K0 III	80	5.50	A6
κ =	65	4.11	A7 V	63	5.56	A1:
υ =	69	4.20	F0 III	51	5.57	dA8
	90	4.22	A5 V	60	5.64	A3:
	68	4.24	A2 IV	45	5.65	dF4
	71	4.43	F0 V	85	5.95	A9
ι =	102	4.52	A7 V	48	6.35	dF2
σ^2 =	92	4.59	A5 V	101	6.70	F5
ρ =	86	4.66	F0 V	55	6.87	dF7
	64	4.72	A7 V			
	79	4.93	A6			

LIST OF BRIGHT HYADES STARS

DESCRIPTIVE NOTES (Cont'd)

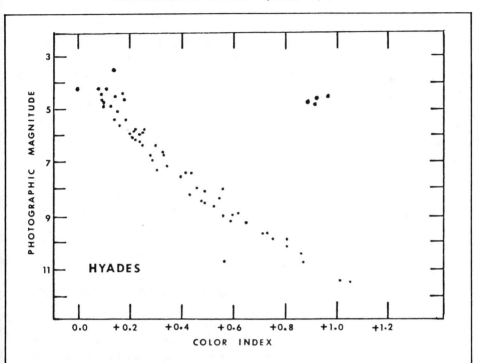

PHOTOGRAPHIC MAGNITUDE

HYADES

COLOR INDEX

slightly different versions, but no list is known in which the names are identified with individual stars. The name Thyrene, Thyene, or Thyone was used by Ovid for the whole group; Pliny designates them as *Parilicium*, evidently from one of the ancient names for Aldebaran. Another name, in use in Roman times, was *Sidus Hyantis*, preserving for us the legend of the great grief of the sisters at the death of their brother Hyas, killed in Libya by a wild boar. In Arabic lands the group was *Al Mijdah*, the "Triangular Spoon", or *Al Kilas*, the "Little She-Camels"; the classic tradition of the "rainy Hyades" appears again in another Arabic name, *Al Kallas*, "The Boiling Sea". Bayer's strange title *Succidae* is derived from a name popular among the country-folk of Roman times, *Suculae*, "The Little Pigs"; Isidorus Hispalensis, however, in the 7th Century, thought the source to be the Latin *sucus*, "moisture". Admiral Smyth speculated that Aldebaran and the Hyades might have suggested a sow with her litter to the rural people of the ancient world. The meaning of the Anglo-Saxon name *Raed-gastran*, according to R.H.Allen, is not definitely known.

In the dark tales of his *Cthulhu Mythos*, H.P. Lovecraft has made his own contribution to the mythology of
the Hyades; he tells us that the Lake of Hali *"which is on
a dark star near Aldebaran in the Hyades"* is the present
lair of the *Old Ones*, those mind-freezing, shambling
horrors from beyond the barriers of space and time, who
once held dominion over the Earth, and seek ever to re-
conquer it..

To modern astronomers the Hyades Cluster is one of
the most important star groups in the sky, as it seems to
be the nearest of the galactic clusters with the single
exception of the Ursa Major moving group. According to the
best summary of present evidence, the center of the Hyades
is at a distance of 40 parsecs or about 130 light years;
the bright V-shaped concentration measures about 3.5° in
width, centered at about 4h 25m, +16½°. This central group
is some 8 light years in diameter, but forms only the nucleus of a much larger group called the "Taurus Moving Cluster", a loose aggregation containing many fainter members
scattered over a large portion of the constellation. The
entire group is drifting slowly through space toward a
point a few degrees east of Betelgeuse in Orion, a fact
first demonstrated by Lewis Boss in 1908; the true space
motion of the cluster is about 26 miles per second. The
Taurus Moving Cluster was closest to the Solar System some
800,000 years ago; in the course of the next 50 million
years it will have receded so far as to appear as a dim
telescopic cluster about 20' in diameter, east of the present position of Betelgeuse.

The diagram on page 1824 illustrates the motion of
the Hyades and demonstrates the apparent convergence of the
group as it recedes from us. This convergence is, of course
the result of perspective, but it makes it possible to
search for new Hyades members by proper motion studies.
Stars at the cluster center show an annual proper motion
of 0.11" but the motions of other members may be larger or
smaller, depending upon the position in the cluster. The
area of the sky covered by the known bright members is over
24° in diameter; the over-all size of the cluster must be
at least 80 light years. Recent studies give the convergent
point as RA 6h 08m; Dec +9.1°.

THE FACE OF THE BULL. The entire V-pattern of the Hyades
appears in this print. Aldebaran is at upper left, ε Tauri
at upper right, and γ Tauri at bottom center.

LOWELL   OBSERVATORY

1823

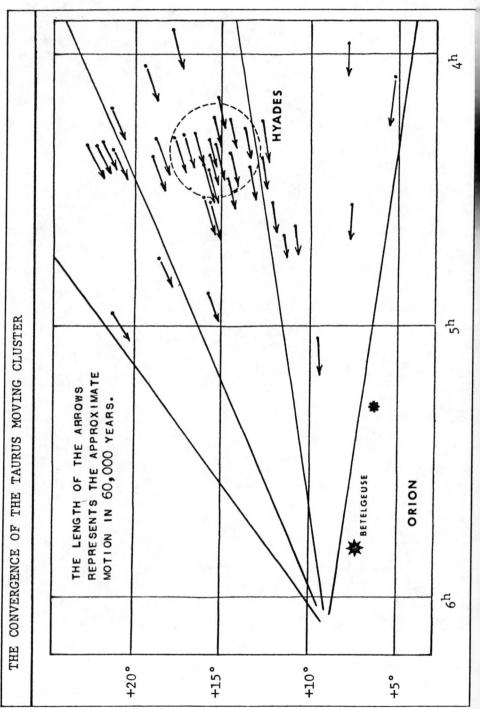

THE CONVERGENCE OF THE TAURUS MOVING CLUSTER

THE LENGTH OF THE ARROWS
REPRESENTS THE APPROXIMATE
MOTION IN 60,000 YEARS.

HYADES

BETELGEUSE

ORION

1824

DESCRIPTIVE NOTES (Cont'd)

According to a survey by H.G.van Bueren (1952) there are 132 members of the cluster brighter than apparent magnitude 9. The number of fainter members is unknown, but is probably at least several hundred. In 1962 a preliminary survey was made at Lowell Observatory of a region covering 160 square degrees, centered on the southwest section of the Hyades; photographic plates made some 30 years apart were compared on the Lowell projection-type blink comparator. In this survey, 259 probable Hyades members were detected, all fainter than 9th magnitude, the majority ranging from 12th to 16th magnitude. As a standard of comparison it should be remembered that our Sun, as a Hyades star, would appear of magnitude 7.7. A member which appears as a 16th magnitude star is thus a very faint dwarf, some 2100 times fainter than the Sun. Stars of such low luminosity cannot be detected in the majority of clusters because the distances are too great; the nearby Hyades offers a rare opportunity to study the faint end of a cluster "family portrait".

The brightest Hyades star is Theta-2 Tauri, spectrum A7 III, magnitude 3.34, about 50 times the luminosity of the Sun. It forms a wide double with Theta-1, a K0 giant of magnitude 3.87; the separation is 337". Another wide pair for binoculars or the small telescope is Sigma Tauri, just 1° SE from Aldebaran; the components differ by about 0.45 magnitude and are 429" apart. In this case, however, only the brighter and more northerly star, Sigma-2, is a true Hyades member. There are four yellow giants in the Hyades: Epsilon, Gamma, Delta, and Theta-1; all are of luminosity class III, with spectra of K0 or G9. All the rest of the Hyades population are main sequence stars of types A,F,G,K and M; we find no B-type giants in this star group. At least a dozen white dwarfs are recognized members of the cluster; the brightest of these are 14th magnitude objects. A number of others will doubtless be found as the fainter cluster stars are more thoroughly studied.

Visual and spectroscopic doubles are common in the Hyades. Among the visual pairs are 55 Tauri (period about 91 years); 80 Tauri (period about 170 years); OΣ82 (about 240 years); and β552, across the border in Orion, with a period of about 101 years. Among spectroscopic doubles are 57, 63, 80 and 90 Tauri. Periods are known for 63 Tauri

JUPITER IN TAURUS. A composite photograph made by the
author with a 1.7-inch Xenar lens, April 7, 1977. Star
cluster M45 (The Pleiades) appears at upper right.

(8.4178 days) and the bright star Theta-1 (140.751 days).
The primary component of 0Σ82 is a spectroscopic double
with a 4 day period.

A number of bright stars at various positions in the
sky appear to show approximately the same space motion as
the Hyades, and are often regarded as members of a very
extensive "Taurus Stream" possibly associated with the
cluster. Among the supposed members are Cor Caroli (Alpha
in Canes Venatici) and the 1st magnitude star Capella.

The Hyades Cluster is considered to be a moderately
old star group, as it contains no visible nebulosity or
high temperature giants; four of the members have evolved
to the K-giant stage. From present knowledge of stellar
evolution, the cluster age is estimated to be about 400
million years. The color-magnitude diagram of the cluster
(page 1821) is virtually identical to that of the Praesepe
Cluster (M44) in Cancer, which also shows very nearly the
same space motion and velocity. The two clusters seem so
similar in type, size, motion, and age, that a common
origin has been proposed. The present separation, however,
is something like 450 light years. (A brief outline of
cluster age-dating appears on pages 990 - 993)

BETA   Name- EL NATH, from the Arabic *Al Natih*, "The
            Butting One". This is the Babylonian *Shur-*
narkabti-sha-iltanu, the "Star in the Bull Towards the
North". In Hindu legend it is associated with *Agni*, the
god of fire, whose name we take in vain whenever we *ignite*
anything. Beta Tauri is magnitude 1.65; Spectrum B7 III;
Position 05231n2834. The star is located on the Taurus-
Auriga border, and has been claimed by both constellations;
in modern catalogues it is officially awarded to Taurus,
and marks the tip of the Bull's northern horn.

The computed distance of the star is about 300 light
years, and the actual luminosity about 1700 times that of
the Sun (absolute magnitude -3.2). The star shows an annual
proper motion of 0.18" and a radial velocity of 5 miles per
second in recession.

Beta Tauri lies about 3° from the central line of the
Milky Way, and is the closest bright star to the position
of the Galactic Anti-center, the point in the sky exactly
opposite to the nucleus of our Galaxy in Sagittarius.

ZETA   Magnitude 3.00; Spectrum B2 III or IVp;
Position 05347n2107. Zeta Tauri lies about 8°
SSE of Beta, and marks the tip of the Southern Horn of the
Bull. The star seems to have no English proper name; R.H.
Allen, on the authority of the English astronomer John
Reeves, states that it was the Chinese *Tien Kwan*, The Gate
of Heaven.

The star is too remote to yield a measurable trigonometric parallax. From various spectral features the true
luminosity is estimated to be about 4400 times that of the
Sun (absolute magnitude -4.2); the computed distance on
this basis is about 940 light years. The star shows an
annual proper motion of 0.023" and a radial velocity of
about 15 miles per second in recession; periodic variations
in the measurement indicate that the star is a close binary
with a period of 132.91 days. The semi-major axis of the
orbit of the visible star is about ten million miles, with
an eccentricity of 0.16. Evidently, the unseen star is an
object of much smaller mass than the giant primary.

Zeta Tauri is one of the best known "shell stars",
with spectroscopic features that reveal the presence of an
extensive and turbulently expanding atmosphere. Since the
early years of the 20th Century, the strength of the shell
spectrum has varied considerably; it was very strong in
1914 and in 1950, weakening in 1951 and 1952, then once
again strengthening until 1959. Although Zeta Tauri is
often considered a fairly "placid" example of a shell star,
it has shown considerable activity in recent years, with
notable changes sometimes occurring within a few days. At
such times the star's outer layers may show turbulent motions, with huge amounts of material flowing outward or
inward with velocities of up to 100 km per second. The
process suggests something resembling solar prominences on
a vastly greater scale, sufficent to keep a permanent ring
or shell of gas around the star. As in other stars of the
type, there often seem to be several different layers in
action at the same time, rising or falling at different
speeds. Chemical stratification is detected in many of the
stars of the type; the spectral lines of different elements
evidently originate at different levels above the surface.
While some stars of the class are optically variable, Zeta
Tauri shows no certain changes in light.

FIELD OF ZETA TAURI. The diffuse object at the upper right
is the famous "Crab Nebula" M1. Lowell Observatory photograph made with the 13-inch telescope.

Very little is known about the operating mechanism of the shell stars, but it is thought that a series a shockwaves may be generated in a region of instability beneath the visible surface, keeping the outer layers of the star in an eruptive state. Rapid rotation may also generate turbulence, since the various layers of the extended shell undoubtedly rotate at different rates. Continuous shell formation may be one of the factors in the evolution of a star to the giant or supergiant state. (Refer also to 48 Librae, 17 Leporis, Gamma Cassiopeiae, and P Cygni)

The celebrated "Crab Nebula" M1 will be found near Zeta Tauri, just 67' to the NW. (Refer to page 1842)

LAMBDA   (35 Tauri) Position 03579n1221, about 6° SW from Gamma Tauri, which is the star at the sharp point of the "V" pattern of the Hyades. Lambda Tauri is one of the brightest of the eclipsing variable stars, and among the first to be identified, found in 1848 by J. Baxendell in England. The variations are caused by the partial eclipse of the B-type primary by a faint companion, and may be easily followed without optical aid. Excellent comparison stars are Gamma Tauri (Mag 3.68, Spect G9) and Xi Tauri (Mag 3.74, Spect B8), located 6° and 9° distant on opposite sides of the variable.

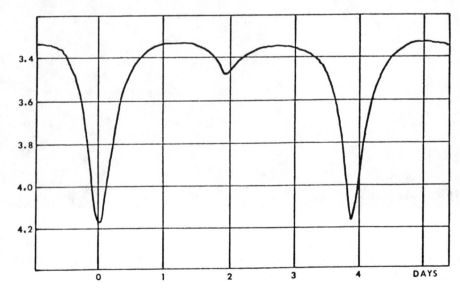

Lambda Tauri was established as a spectroscopic binary by A.Belopolsky in 1897, and the first photoelectric measurements were made by J.Stebbins at Mt.Wilson in 1916. This star has a light curve of the Beta Lyrae type with a well-marked secondary minimum and continuous variation outside of eclipse; the orbital period is 3.9530153 days. As in many stars of this type, the exact dimensions and orbital elements are uncertain; in an analysis by D.B.McLaughlin the following results were obtained:

	Spect.	Diam.	Mass.	Lum.	Abs.Mag.
A	B3 V	6.0	6.1	525	-2.0
B	A4	4.0	1.6	70	+0.2

The computed separation of the two stars is about 8.5 million miles, and the calculated orbit has an eccentricity of 0.055. The eclipses of this system are partial, with about 40% of the diameter of the primary being occulted at mid-eclipse. From a study of the radial velocity measurements, it appears that a third body is present, with an orbital period of about 30 days; no spectral features attributed to this star have yet been detected.

Lambda Tauri has a computed distance of about 400 light years; the annual proper motion is 0.01" and the mean radial velocity is 9 miles per second in recession. (See also Beta Lyrae, Beta Persei, U Cephei, and U Sagittae)

T Position 04191n1925, about 1.8° west and a little north from Epsilon Tauri, the northernmost bright star in the "V" figure of the Hyades. (See chart on page 1820) T Tauri is a remarkable irregular variable, associated with the peculiar "variable nebula" NGC 1555. Star and nebula both were discovered with a 7-inch refractor by J.R.Hind of London in October 1852. Hind described the nebula as very faint and not exceeding 30" in diameter, and announced that none of the catalogues listed any object in that position. He likewise observed that a 10th magnitude star just north of the nebula was not shown on current charts, and correctly surmised that the star was probably a variable.

Hind's nebula was seen by several observers during the next few years, but by 1861 it had faded noticeably

and by 1868 could not be detected in any telescope. The
nebula was not seen again until 1890 when it was rediscovered by E.E.Barnard and S.W.Burnham with the 36-inch telescope at Lick Observatory. Five years later it had vanished
again, but was detected photographically in 1899, and has
been followed regularly since that time. Photographic
studies have not only confirmed the variations in the light
of the nebula, but have also revealed definite changes in
its apparent size and form. The early observers described
the nebula as lying about 40" SW of T Tauri; at the present
time it appears directly west of the star, and has also
brightened significantly in the last 50 years. On modern
photographs the nebulosity appears as if composed of three
or four overlapping curved filaments which form a rough 60°
arc curving around the star. The appearance is suggestive
of a portion of a luminous shell surrounding the star. A
much smaller nebulosity about 5" in diameter closely envelops the star itself; this was seen as a fairly bright disc
of elliptical shape in 1890, but is now represented by a
few faint patches which can be detected only with the largest telescopes.

T Tauri itself is an unpredictable and enigmatic
star. It varies erratically from magnitude 9 or so to about
13, sometimes passing through the entire range in a few
weeks, at other times remaining nearly constant for months.
The spectrum varies from type G4 to about G8, but not
necessarily in correlation with the light changes; another
peculiarity is the presence of numerous bright lines,

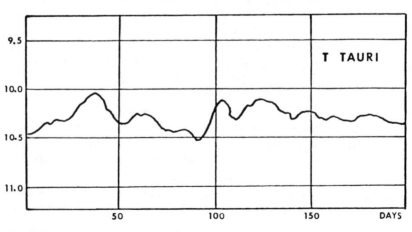

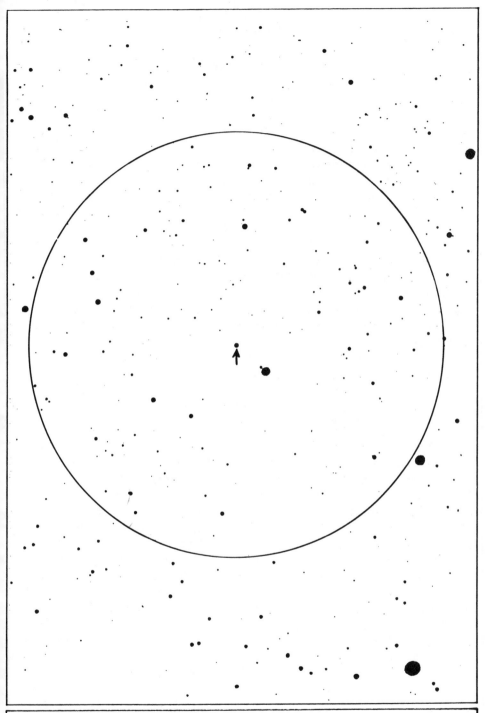

T TAURI Identification field, from a Lowell Observatory 13-inch telescope plate. The circle is 1° in diameter with north at the top; limiting magnitude about 15.

DESCRIPTIVE NOTES (Cont'd)

resembling the spectrum of the chromosphere of the Sun.
The distance has been determined to be about 450 light
years, and the resulting absolute magnitude at intermediate
brightness is found to be about +5. T Tauri is thus a dwarf
star of the main sequence, about equal to our Sun in average luminosity. At maximum it is some 4 or 5 times brighter
but shows a somewhat later spectral type than the luminosity would suggest. To put it another way, the star is distinctly over-luminous for its spectral type.

T Tauri is usually considered the typical star of a
class of peculiar dwarfish variables which appear rather
similar to the erratic RW Aurigae stars, and are sometimes
considered to be a sub-class of that group. A.H.Joy defined
the distinctive features of the T Tauri stars as follows:

(1) Irregular and erratic light variations of about
3 magnitudes.

(2) Spectral types of F and G with emission lines,
resembling the solar chromosphere.

(3) Low luminosities, comparable to that of the Sun,
but somewhat over-luminous for the spectral type.

(4) Association with bright or dark nebulosity.

The last characteristic is an important one, since
it suggests an intrinsic connection between these stars and
diffuse nebulosity. In recent years, large numbers of T
Tauri stars have been detected in the great dark obscuring
clouds of the Milky Way, where they often seem to lie on
the edges of obscuring patches or on the borders between
bright and dark nebulosity. Many stars of the type are
found in the region of the Great Nebula in Orion; they
occur also in such great nebulous complexes as NGC 2264 in
Monoceros, M8 in Sagittarius, and M16 in Serpens. There is
now a general agreement that these stars are newly born
citizens of the galaxy, only recently formed from the huge
nebulous clouds where we find them. The erratic variations
of the T Tauri stars are seen as symptoms of their relative
youth; they have not yet reached a stable main sequence
state, and may in fact still be growing through accretion.
One of the odd features of the T Tauri spectra is the high
over-abundance of lithium, some 70 to 100 times that found
in the Sun. No satisfactory explanation is known, but the
possible theories would seem to fall into two basic groups:
Either the stars were formed from lithium-rich material

DEEP-SKY OBJECTS IN TAURUS. Top: The star T Tauri and the nebula NGC 1555, photographed at Mt.Wilson. Below: Star Cluster NGC 1647 as it appears on a plate made with the 13-inch camera at Lowell Observatory.

in the beginning, or some lithium-generating nuclear reactions are active in stars of this type. Presumably, much
of this material will be transformed into other elements
by the time the star has reached the main sequence, about
10 million years from now. In addition to the unusual lithium content, T Tauri also shows emission lines of hydrogen,
neutral oxygen, and singly ionized silicon and calcium, as
well as the more usual absorption lines whose broad and
diffuse appearance indicates high rotational velocity. It
is in stars of this type, according to one theory, that we
may be seeing an early stage in the formation of a planetary system. (Refer also to M16 in Serpens, M42 in Orion,
and NGC 2264 in Monoceros. For brief accounts of two other
variable nebulae, refer to NGC 2261 in Monoceros, and NGC
6729 in Corona Australis)

RR   Position 05364n2621, about 3° SE of Beta Tauri
                and some 30' north of 125 Tauri. This is an
irregular variable star of uncertain class, discovered by
L.Ceraski at the Moscow Observatory in 1900. Observations
by E.Hartwig suggested a period of about 95 days, while a
period of over 200 days was indicated by later studies at
Harvard; during the last 50 years the variations have been
more or less erratic. At times the fluctuations are slow
and of small amplitude; at other times there may be changes
of more than a magnitude in only a few days. The light
variations resemble those of T Tauri, but the spectral type
of A2e is much earlier; G.H.Herbig (1960) finds some evidence for shell activity in the spectrum of the star, and

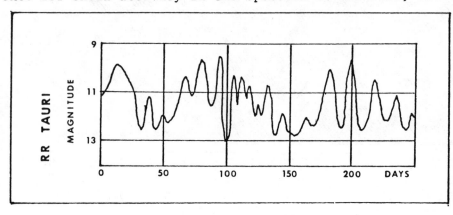

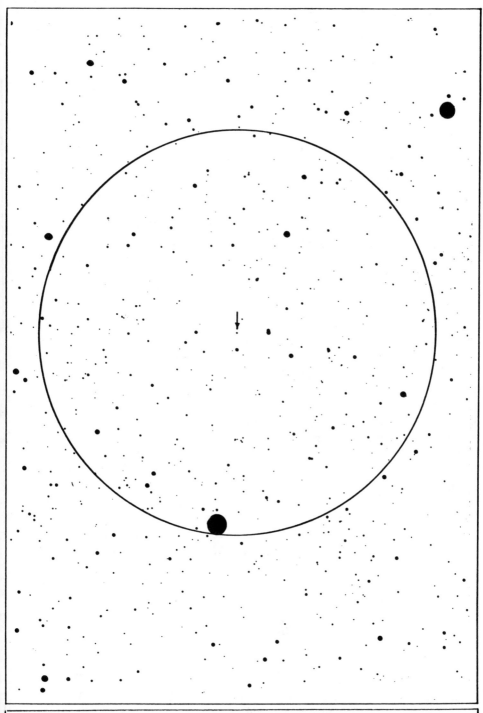

RR TAURI Identification field, from a Lowell Observatory
13-inch telescope plate. The circle is 1° in diameter with
north at the top; limiting magnitude about 15. The bright
star is 125 Tauri.

classes the spectrum a little earlier, about B8 or B9e.
The classical T Tauri stars are usually type G, though a
few are known of late class F.

RR Tauri is located on the southern edge of a great
ring-shaped filamentary nebulosity known as S147, invisible
on most ordinary photographs but appearing clearly on red-
sensitive plates. This great cloud, believed to be a very
old supernova remnant, measures over 2° in diameter, and
is estimated to be some 3000 light years distant. If it is
actually involved in this nebula, RR Tauri must have a peak
absolute magnitude of about 0.0, the luminosity ranging
from 4 suns up to about 85. As yet, however, there is no
definite proof that the star is actually at the same distance as the nebula and physically associated with it. The
apparent association might be merely a chance line-up,
though the apparent brightness of the star seems to place
it at about the same distance as the nebula. However, even
if the association should be found to be accidental, there
is no doubt that the star should be classed as a nebular
variable. It lies precisely upon a small knot of nebulosity
from which a small loop extends northward about 30"; this
knot may or may not be connected with S147, but it is definitely connected with the star. Observations have not yet
shown whether the nebula varies in light, as does that
connected with T Tauri.

The Moscow *General Catalogue of Variable Stars* (1970)
classes RR Tauri as an irregular nebular variable of the
"Orion" type, resembling T Orionis; other references list
it as an RW Aurigae type star, a class to which T Orionis
itself is often assigned! The nebular variables display so
many individual traits and peculiarities, that any attempt
at a strict classification often seems pointless. (Refer
also to T Tauri, RW Aurigae, T Orionis (page 1331), NGC
2261 in Monoceros, and NGC 6729 in Corona Australis.)

RV   Position 04440n2606. A variable star of a rare
            type, discovered by L.Ceraski at the Moscow
Observatory in 1905, and located in a region of fairly
heavy obscuration about 8° WSW from Beta Tauri. The star
shows a semi-regular variation with a period of 79 days,
during which there are two maxima of nearly equal height,
and two minima of unequal depth. These fluctuations are

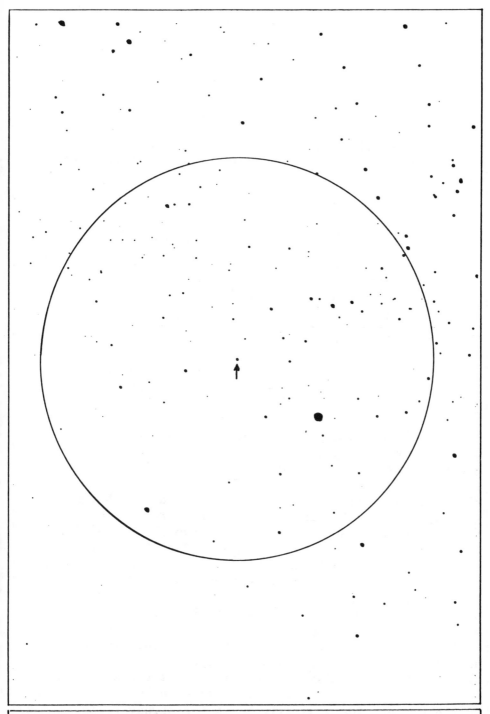

RV TAURI Identification field, from a Lowell Observatory
13-inch telescope plate. The circle is 1° in diameter with
north at the top; limiting magnitude about 14.

superimposed on a longer cycle of about 1300 days, which causes the entire light curve to rise and fall in a long wave of about 3½ years. The 79-day changes are most conspicuous when the star is near the maximum of the long cycle; when near minimum the variations are irregular and of small amplitude.

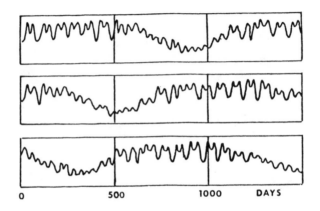

RV Tauri is a yellowish giant of spectral type G2e Ia which alters to type K3 or so as it fades; when near the minimum brightness the features of a red giant appear in the spectrum, and the star is classed by some observers as M2 Ia at that time. Stars of this class are high luminosity giants, and seem to combine the characteristics of the long period variables and the cepheids. From their position on the H-R Diagram, the absolute magnitudes at maximum appear to reach -4.5 or -5.0, and the computed masses lie in the range of 20 or 25 suns. Although only a few dozen stars of the type are known, the existence of at least two sub-types seems evident. One class, of which RV Tauri is itself a member, shows the superimposed long cycle of about 3 years; in the other class, typified by R Scuti, this feature is not present. All the stars of the type show rather peculiar spectra in which emission lines appear as the star rises to maximum, bordering the absorption lines on both sides, that on the violet side generally being the stronger. As in many pulsating variables, there is clear evidence that various layers of the star's atmosphere are oscillating at different rates. Among fairly bright stars of the type are SS

Geminorum, AC Herculis, U Monocerotis, TX Ophiuchi, TT
Ophiuchi, R Sagittae, AR Sagittarii, V Vulpeculae, and R
Scuti. Only a few stars are known with light curves of the
RV Tauri type (showing the superimposed long cycle) and the
brightest examples are SX Centauri, SU Geminorum, IW Cari-
nae, AI Scorpii, and RV Tauri itself. (Refer also to R
Scuti, page 1747)

RW   Position 04008n2800, about 5° NE of the bright
               Pleiades star cluster, and a little less than
1° NW from 41 Tauri. RW Tauri is an eclipsing binary system
of unusual type, popularly called "the star with a luminous
ring". It was discovered by Mrs. W.Fleming at Harvard in
1905. The primary star, of spectral type B8, is totally
eclipsed by a K-type subgiant companion, at intervals of
2.768846 days; each eclipse lasts 9 hours, with the total
phase lasting 84 minutes. The depth of primary minimum is
one of the greatest known, 3.50 magnitudes visually and
4.49 photographically. In a study of the system in 1959,
G.Grant obtained a distance of about 1370 light years, and
derived the following properties for the components.

	Spect.	Diam.	Mass.	Lum.	Abs.Mag.	Dens.
A	B8 V	2.25	2.55	130	−0.5	0.22
B	K0 IV	3.00	0.55	3½	+3.4	0.02

The computed separation of the two stars is about 5
million miles. The peculiar feature of this system is the
"Saturn-like" ring of glowing hydrogen which rotates around
the B-type primary, and is eclipsed with it. The ring, of
course, cannot be seen visually in any telescope, but is a
conspicuous spectroscopic feature. In a study of the star
in 1942, A.H.Joy found that the bright spectral lines of
the ring appear just after the total phase of the eclipse
begins; at this time the lines show a radial velocity of
about 210 miles per second in recession. Some 27 minutes
later the lines vanish as the ring is hidden completely by
the eclipsing K-star. Then 27 minutes before the end of
totality, the spectral lines of the ring reappear, this
time showing a radial velocity of 210 miles per second in
approach. From the Doppler shift of the lines, Joy finds
that the ring is rotating in a period of about 14 hours,

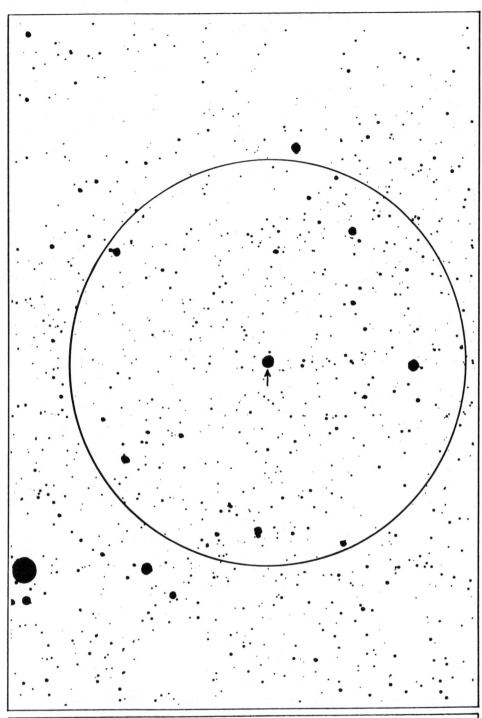

RW TAURI Identification field, from a Lowell Observatory
13-inch telescope plate. The circle is 1° in diameter with
north at the top. Bright star at lower left is 41 Tauri.

and that its size is about 4 times the diameter of the Sun. Were it not for the presence of the eclipsing companion, the ring could not have been detected because of the over-powering glare of the bright B-type star. The situation is comparable to observing the corona of the Sun, visible in its full glory only during a total solar eclipse. The light of the ring appears to be about 0.001 of the luminosity of the B-star. Gaseous rings of this sort evidently originate in material which is being lost by one of the stars, very probably from the strong tidal effects between the components of close pairs. We find evidence for gaseous streams between the two stars of such pairs as Beta Lyrae, Algol, U Cephei, and U Sagittae.

RW Tauri is one of the eclipsing systems in which slight but definite changes in period have been detected. The exchange of mass between the two stars, through the mechanism of the ring, may be partially responsible. It is also probable that at least one unseen star exists in the system. A faint visual companion was in fact discovered by Prof.Joy; it is magnitude 12½ and the separation is about 1". This star is too distant (projected separation = 400 AU) to be the hypothetical perturbing body which still remains unseen and unproved. If real, it is possibly a white dwarf star, and too faint and close to the variable to be detected.

M1   (NGC 1952) (TAURUS X-1) (TAURUS A) (3C 144) (PSR0531+21) Position 05315n2159. The famous "Crab Nebula" in Taurus, the most conspicuous known supernova remnant, and one of the most interesting objects in the heavens. It is located 67' NW from Zeta Tauri, the star which marks the southern tip of the Bull's horn.(Refer to photograph on page 1829) The nebula was discovered by the English physician and amateur astronomer John Bevis, in 1731, but was independently found by Charles Messier some 27 years later. In September 1758 he detected what he called a "nebulosity above the southern horn of Taurus.... It contains no star; it is a whitish light, elongated like the flame of a taper; discovered while observing the comet of 1758". It was this discovery which prompted Messier to compile his famous catalogue of nebulae and clusters in order that other observers might not confuse them with comets:

"*What caused me to undertake the catalog was the nebula I discovered above the southern horn of Taurus on Sept. 12, 1758, while observing the comet of that year... This nebula had such a resemblance to a comet, in its form and brightness, that I endeavored to find others, so that astronomers would not confuse these same nebulae with comets just beginning to shine. I observed further with the proper refractors for the search of comets, and this is the purpose I had in forming the catalog...*"

J.E.Bode, in 1774, observed Messier's nebula as "a small nebulous patch without stars", but John Herschel's impression was that M1 was probably a barely resolvable cluster. Admiral Smyth saw it as a "large nebula, pearly-white, of oval form, with a major axis N.p. & S.f. and the brightest portion toward the south." It was Lord Rosse, in 1844, who detected the extending filaments, and referred to them as resembling the legs of a crab; he seems to have regarded the filaments, however, as something in the nature of star chains: ".....It is no longer an oval resolvable nebula; we see resolvable filaments singularly disposed, springing principally from its southern extremity, and not, as usual in clusters, irregularly in all directions. Probably greater power would bring out other filaments, and it would then assume the ordinary form of a cluster..."

Lassell, observing with his large reflector at Malta in 1852, confirmed the filamentary structure: "With 160X it is a very bright nebula, with two or three stars in it, but with 565X it becomes a much more remarkable object.... Long filaments run out from all sides and there seem to be a number of very minute and faint stars scattered over it; the outlying claws are only just circumscribed by the edge of the field of 6' diameter...."

In Lick Observatory *Publications XIII*, dealing with observations of the nebulae, M1 is described as a "very complex and interesting object.....nearly 6' x 4' in PA about 125°....It is not a typical planetary in form and it is doubtful whether it is properly to be included as a member of the class....Two stars of mag. 16 are close together near the centre but it is not certain that either of them is the central star...."

For the modern telescope, M1 is a fairly easy object, detectable in 3 and 4-inch apertures, appearing irregularly

THE CRAB NEBULA. This expanding gas cloud is the brightest
known supernova remnant in the sky. Palomar Observatory
photograph, made with the 200-inch reflector.

oval in a 6-inch glass, and showing some hint of detail in a 10-inch or larger instrument. The integrated magnitude is about 9, and the visual size about 5' x 3'. A rather large telescope is required to make the filamentary structure plainly evident, and the finest details are revealed only on the photographic plate.

EXPANSION OF THE CRAB NEBULA. Messier's "Number 1" has now the best documented history of any nebula in the heavens. Motion inside the cloud was first detected by C.O.Lampland at Lowell Observatory in 1921, through a study of a series of plates made with the 42-inch reflector. Various nebulous details were seen to vary in intensity, and it soon became evident that the cloud itself was slowly growing in size. Photographs obtained at Mt.Wilson by J.C.Duncan revealed that the expansion rate varied somewhat for different parts of the nebula, but averaged about 0.2" per year. At the derived distance of the nebula, this corresponds to a true velocity of over 600 miles per second, or some 50 million miles per day. This is one of the highest velocities known in the Galaxy, and seems to indicate that the nebula – which is now nearly 6 light years in diameter – had its beginning in a colossal explosion centuries ago, almost certainly a supernova outburst. In 1942, Dr.W.Baade computed a probable age of about 760 years for the cloud; newer estimates allow for the fact that the expansion rate is accelerating somewhat, and the best present figure is about 900 years. The chronicles of medieval China contain a very intriguing account of a "guest star" which appeared near Zeta Tauri in July 1054 AD. According to a translation by J.J.Duyvendak (1942) the text, in the *Sung Shih*, or Annals of the Sung Dynasty, reads as follows:

"....*In the 1st year of the period Chih-ho, the 5th moon, the day chi-ch'ou, a guest star appeared approximately several inches south-east of Tien-Kuan....After more than a year it gradually became invisible..*"

The precise date, according to this account, was July 4, 1054 AD, and the position "several inches" south-east of Zeta Tauri. The use of the term which has been translated "inches" may suggest that the writer was referring to the position on a celestial globe or armillary sphere, rather than the actual sky. A later reference to

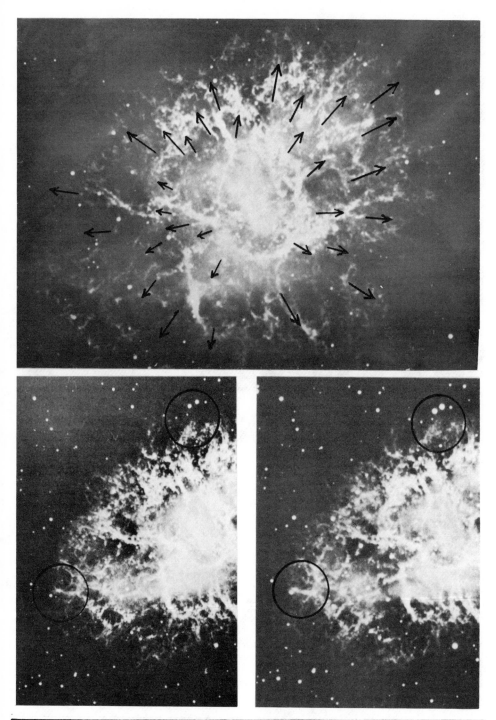

EXPANSION OF THE CRAB NEBULA. Top: The arrows indicate the outward motion of the filaments in about 250 years. Below: Two photographs made 34 years apart, illustrating the outward motion of the nebulous filaments.

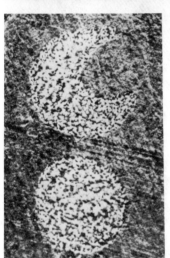 

THE TAURUS SUPERNOVA. Above: The author's imaginary recon-
struction of the view on the morning of July 5, 1054 AD.
Below: American Indian pictographs which may depict this
event. (Courtesy of William C. Miller)

the nova appears in the *Sung hui-yao*, compiled by Chang
Te-hsiang, and giving us a bit of further information about
the appearance of the object:

"....*During the third month in the first year of the
Chia-yu reign period* [March 19- April 17, 1056] *the Direct-
or of the Astronomical Bureau said, 'The Guest Star has
become invisible, which is an omen of the departure of the
guest'. Originally, during the fifth month of the first
year of the Chih-ho reign period [it] appeared in the morning in the east guarding T'ien-Kuan. It was visible in the
day like Venus, with pointed rays in all four directions.
The color was reddish-white...It was seen altogether for
twenty-three days."* [as a daylight object]

There are no known European records of the nova, and
Prof. Fred Hoyle has suggested that religious prejudices of
the time may have forced medieval historians to ignore the
event. Since the nova of 1006 AD was mentioned in the con-
temporary account of a Swiss monk, however, it might be
more charitable to surmise that possibly records were made
but have not survived. Some additional verification may
come from several American Indian pictographs found in
ruins in northern Arizona, one in a cave at White Mesa and
the other on a wall of Navajo Canyon. Both pictographs
show a crescent moon with a large "star" nearby and below.
Computations reveal that the crescent moon was located just
2° north of the present position of the Crab Nebula on the
morning of July 5, 1054. If the Chinese chronicles are
correct, this was the morning after the nova appeared, and
it must have formed a most striking spectacle with the
Moon. The identification cannot be regarded as certain, of
course, but the ruins in question are known to have been
occupied during the 11th to 13th centuries. The fact that
the White Mesa pictograph shows the moon turned the wrong
way is not particularly disturbing; even a modern artist,
when portraying the moon in a painting, is capable of making quite surprising errors of position and orientation.
Cartoonists are possibly the worst offenders, but even a
respected artist seems to think it quite permissible to tip
the lunar crescent any way he pleases. There is no reason
to suppose that an Indian artist, in the 11th Century,
would be overly concerned with scientifically accurate
representations of the strange star and the Moon.

A much more serious objection is that the "star and crescent" symbol is not really very rare in history; aside from being the time-honored symbol of the Islamic faith, it appears on Sumerian cylinder seals, Roman coins, crusader's shields, national flags, and on every other imaginable type of object from medieval coats-of-arms to modern Hummel collector's plates. The conjuction of the crescent moon and Venus is a fairly common occurrence of course, and always attracts much attention. According to students of mystical symbolism, the "star-and-crescent" motif represents the *Star of Isis between the Horns of Taurus*. On the shield or banner of King Richard I, it is said to have represented the Star of Bethlehem. On Roman coins it quite often seems to have no certain meaning, and evidently does not refer to any actual astronomical event as the star is depicted impossibly placed *inside* the lunar crescent, the same arrangement which Coleridge's *Ancient Mariner* speaks of: ".....*the horned moon, with one bright star within the nether tip...*"

The coin shown in Fig.1 is a denarius of Emperor Hadrian, minted about 130 AD. In an earlier Roman coin, dated to 41 BC, the crescent appears with *five* stars; this issue is said to honor Diana, the moon-goddess, whose worship was introduced into Rome by the Sabines. Neither of these coins appear to refer to any specific astronomical event. On the other hand, the "star" which appears on a denarius of the Emperor Augustus (Fig.3) is known to represent the bright comet of 44 BC, which was visible for a week following the assassination of Julius Caesar. In still another example, the classic "owl tetradrachm" of Athens, the crescent moon appears as a symbol of the Battle of Marathon in 490 BC, when the vastly outnumbered Athenians drove back the invading Persian hordes "under a waning moon" to save Athens. (Fig.4) In view of the universal popularity of these

astronomical symbols, there seems to be no real reason to
connect any particular "star and crescent" rock carving
with a specific celestial event, unless strong supporting
evidence is available. Since the North American Pueblo cultures left no written records, the identification of the
Arizona carvings with the nova of 1054 will probably always
remain an unproven - though plausible - speculation. Even
the identification of the Crab Nebula with the 1054 nova
cannot be said to be absolutely proven; there is a definite
discrepancy in position, since the Chinese records describe
the "guest star" as being *southeast* of Zeta Tauri, whereas
the nebula is actually about 1° northwest of the star. The
relative alignment of the two objects has changed somewhat
in 900 years, of course, but by less than 1'. Since the
Crab Nebula is definitely recognized as a supernova remnant,
and its age *must* be close to 900 years, most students of
the problem have preferred to regard the identification as
definite, dismissing the discrepancy of position as the
result of a copyist's error or a mis-translation.

    In the past thirty years, M1 has been one of the most
intensively studied objects in the heavens. In a review of
12 different distance estimates obtained through various
methods, Virginia Trimble (1974) finds that a figure of
1930 parsecs or about 6300 light years is the best modern
value. This tells us immediately that the supernova at
maximum had an absolute magnitude of at least -16½; possibly closer to -17 or -17½ if the effect of absorption is taken
into account. For a few weeks the star was blazing with the
light of about 400 million suns. Only on two occasions in
the 9 centuries since, has man witnessed a comparable cataclysm in our Galaxy. The supernova of 1572 (Tycho's Star in
Cassiopeia) was comparable in apparent brilliance, while
Kepler's Star of 1604 in Ophiuchus appeared a magnitude or
so fainter. An earlier supernova, in Lupus in 1006 AD, may
have been the most dazzling of the four. Unfortunately, all
of these appeared before the invention of the telescope;
modern astronomers now await the next comparable display,
which, it would seem, is already long over-due.
    The Crab Nebula has a peculiar dual structure which
is strikingly evident on the accompanying photographs made
with the 100-inch reflector at Mt.Wilson Observatory. The

BLUE λ3100-λ5000

RED λ6300-λ6750

THE CRAB NEBULA, photographed in blue light (top) and in red light (below). Note the prominence of the filaments on the red plate. Mt.Wilson 100-inch telescope photographs.

inner part of the nebula consists of an intricate network
of fine filaments, conveying an impression of chaotic turbulence; this structure is best seen on red-sensitive
plates. Surrounding and enveloping it is a larger, more
transparent, virtually formless cloud, best seen on blue-
sensitive plates. This nebulosity is computed to have about
1/1000 the density of the filaments, which in turn are less
than a trillionth the density of ordinary air. It is in the
filaments that we find the emission features of the nebula,
bright lines of neutral and ionized helium, hydrogen, and
a few other atoms; the spectrum resembles that of a bright
planetary nebula. The amorphous nebulosity, however, shows
a continuous spectrum in which the light is strongly polarized, a discovery made in 1954. In describing a series of
photographs made through polaroid filters with the 200-inch
telescope, Dr.W.Baade pointed out that the appearance of
the structural details changed noticeably as the polaroid
filter is rotated in PA: "Take any of these structures....
and follow its intensity changes...as the analyser is turned in position angle....One notices at once that the structural features disappear when the electric vector is parallel to them. They obviously reflect the run of the underlying magnetic field..." In this way it was shown (photographs on pages 1854-1855) that the structural features
outlined the magnetic lines of force in the Crab.

At the time that astronomers began to be aware of
the very unusual nature of the Crab Nebula, Dr.Baade had
studied the pair of faint stars near the apparent center,
and had identified the SW component as the probable nova-
remnant; it appeared to be a hot bluish dwarf of the 16th
magnitude. The identification of this star, however, seemed to raise new problems; it is about 1000 times fainter
than the whole nebula, and it is therefore difficult to
see how it can supply the exciting radiation for the entire
cloud unless it is at a fantastically high temperature.
The observed radiation would require a surface temperature
in excess of 500,000°K, with a total energy output of some
30,000 times that of the Sun. At such an extremely high
temperature the star would radiate almost entirely in the
ultraviolet portion of the spectrum, and this would explain
its visual faintness as well as its strong exciting effect
on the gases of the nebula. This relatively simple picture

E V
0°

E V
90°

THE CRAB NEBULA photographed in polarized light. In each print the arrow indicates the orientation of the electric vector.

Hale Observatories

1854

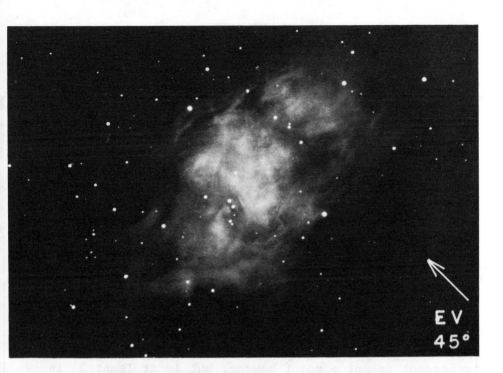

EV
45°

EV
135°

POLARIZED LIGHT PHOTOGRAPHS of the Crab Nebula demonstrate
the existence of strong magnetic fields in the cloud.

Hale Observatories

1855

DESCRIPTIVE NOTES   (Cont'd)

soon became obsolete, however, as new facts emerged. First, in 1948, the nebula was identified as a strong source of radio energy. As measured at Jodrell Bank in England, the radiation has about 10 times the intensity of the Cassiopeia source which is presumably connected with the great supernova of 1572. The Crab Nebula is among the four brightest "stars" in the radio sky, and the radio emission, like the visible light, shows polarization. Then, in April 1963, X-ray energy was first detected from the region during the launch of an Aerobee rocket equipped with an X-ray detector designed at the Naval Research Laboratory. The X-ray energy might be coming from the whole nebula or from only the tiny central star; to settle the question M1 was observed during several occultations by the Moon, beginning with that of July 5, 1964. The observations, repeated during a series of occultations in 1974 and 1975, demonstrated that the X-ray generator is not a point source, but is at least 2' in size. Possibly about 5% of the X-ray energy may originate in the star itself; the rest is produced in some way in the surrounding nebula.

These discoveries have led to a general acceptance of a theory first proposed in 1953 by I.S.Shklovsky, and independently by J.H.Oort and T.Walraven, attributing the radiation of the Crab Nebula to the motion of high-speed electrons being accelerated or decelerated in the nebula's magnetic field. This process is rather comparable to the activity inside a cyclotron, and is known as "synchrotron radiation". It is also thought likely that objects such as the Crab Nebula create much of the high-energy cosmic radiation that reaches the Earth. In X-ray energy alone, the nebula generates about 100 times the energy emitted in the form of visible light.

The total mass of the Crab Nebula is not well known, and some very discordant figures have been published. From a study by R.Minkowski in 1968, it appears that the mass of the bright filaments is almost certainly less than 1 solar mass; Shklovsky and Oort have shown that the observed radiation might be generated by a mass as small as 0.1 the Sun if the conditions are just right. The mass of the amorphous nebulosity is even more uncertain, and different studies have resulted in figures ranging from about 1 solar mass up to 10 or more. Presumably, the original pre-supernova

YELLOW λ5200-λ6600

INFRARED λ7200-λ8400

THE CRAB PULSAR. Arrow indicates the fantastically dense neutron star identified as the stellar remnant of the great supernova of 1054 AD.

Mt.Wilson Observatory

star had a mass of at least several suns, so it appears unreasonable that the total mass of the Crab should be as small as 1 sun. The supernova, in this case, is now thought to have been one of Type II, a star which is younger and more massive than Type I. (Both Tycho's Star of 1572 and Kepler's Star of 1604 have been identified as supernovae of Type I.)

THE CRAB PULSAR. At the time that the Crab Nebula was first recognized as the debris cloud of a supernova, several theoretical studies had suggested that the final product of such a stellar explosion might be a fantastically compressed "core" with a density a few hundred million times that of an orthodox white dwarf star. The Russian physicist L. Landau in 1932 was among the first to postulate the actual existence of such "neutron stars", while F.Zwicky, G.Gamow, W.Baade and others, in studies going back to 1934, found strong theoretical reasons for believing that such objects should be the end-product of the collapse of a very massive star. Up until the late 1960's, the possible identification of the central star of the Crab (now called "CM Tauri") as a genuine neutron star remained uncertain. Then, in 1968, the first example of a new type of object, the "pulsar", was announced, and the neutron star was promoted from theoretical possibility to definite fact.

The first known example was detected during the program of radio studies at the Cambridge University Observatory in 1967, by J.Bell and A.Hewish. Located near the star 2 Vulpeculae at 19196n2147, the mysterious object was found to be emitting perfectly regular pulses of radio energy, at intervals of 1.337301 seconds. Although no visible object could be located at the radio position, the identification of "CP1919" as a neutron star seemed reasonable, as not even the smallest white dwarf star could have a rotation period, or pulsation period, as short as 1 second. With one such object identified, radio astronomers went on to find dozens more, the number totalling over 100 by 1976. Among the most important of the new pulsars was "NP0532", that famous little star at the center of the Crab Nebula. The radio pulses, first detected in 1968, have the astonishingly short period of 0.033089 second; the frequency is thus close to 30 pulses per second!

As the number of known pulsars continued to grow, a standard numbering system was adopted in which the designation gives the approximate coordinates in the sky.   CP1919 in Vulpecula is thus usually referred to as PSR1919+21; the Crab Pulsar has the official designation PSR0531+21.

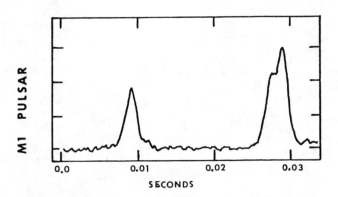

Since the Crab Pulsar, at the time of its discovery, was the only example known which coincided with an optically visible object, it seemed particularly important to determine whether the neutron star was optically variable as well. Although the observations were hampered somewhat by the light of the nebulosity, astronomers at the University of Arizona at Tucson, on January 15, 1969, detected the first optical pulsations from the object, matching the same ultra-short cycle previously found by the radio astronomers at Cambridge. The observations were made with the 36-inch reflector at Tucson by W.J.Cocke, M.J.Disney, and D.J. Taylor; the findings were confirmed by similar studies at Kitt Peak National Observatory and at McDonald Observatory in Texas.   In addition to the optical and radio pulses, the object shows the same "pulsar" effect in its X-ray output.

  With specially designed computer equipment, it has been possible to analyse the form of individual pulses in this strange object. The main pulse, shown at the right on the above graph, is preceded by a brief hump or "precursor" while a lower "interpulse" occurs between the chief peaks. Unusually strong pulses, up to 1000 times more intense than the average, occur at intervals of about 5 minutes, or about one in every 10,000 pulses. The type of pulse is also associated with the degree of polarization. From studies

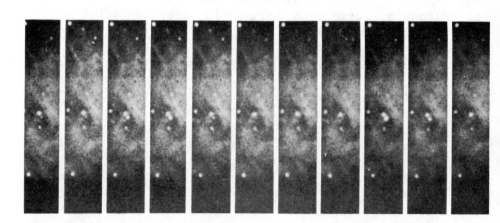

THE CRAB PULSAR. Arrow indicates the 16th magnitude neutron star. Below: A series illustrating the rapid pulsations, from observations obtained at Kitt Peak National Observatory.

made with the giant radio "dish" at Arecibo, Puerto Rico,
it is known that a main pulse is 18% linearly polarized,
while the "precursor" pulse is 100% polarized; the interpulse appears to show only slight polarization, if any. In
this bizarre object, it seems that we are observing a
remarkably tiny neutron star, probably not more than a few
miles in diameter, and with an extremely strong magnetic
field, rotating at the astonishing rate of 30 times per
second. Such an object apparently emits both radio energy
and visible light in the form of a narrow beam at the
magnetic poles; the pulses are detected when the beam, like
that of a rotating searchlight, sweeps across the observer's position. Occasional "glitches" or abrupt changes in
period are attributed to a shrinkage or shifting of portions of the neutron star's "crust"; it can be calculated
that a shrinkage of less than a thousandth of an inch produces a measurable change in the rotation period. At the
present time, the rotation of this pulsar is slowing down
by a slight but measurable amount; the nature of the so-
called "braking mechanism" is not definitely known. As one
research worker has said, ".....at the moment we are in the
exciting, but frustrating position of having data which
appears to be considerably better than our understanding
of the underlying physics..."

A "Crab Nebula Symposium" was held in Flagstaff,
Arizona in June 1969, and a detailed summary and review of
the discussions appeared in the *Publications of the Astronomical Society of the Pacific* (Vol.82) in May 1970. The
Crab Nebula is also the subject of an entire book *The Crab
Nebula* (Symposium No.46) published on behalf of the International Astronomical Union by D.Reidel Publishing Company
in 1971. Another book, *Pulsars*, by F.Graham Smith, was published by Cambridge University Press in 1977, and is probably the best and most informative book on the subject for
the serious student; it was reviewed in *Sky & Telescope*
magazine in February 1978.

(Refer also to "Tycho's Star" of 1572 in Cassiopeia,
"Kepler's Star of 1604 in Ophiuchus, and the supernova of
1006 AD in Lupus. A brief account of the supernova of 1885
in the Andromeda Galaxy will be found on page 144; the
supernova of 1972 in NGC 5253 is described and pictured on
pages 570-571. See also "Neutron Stars" on page 411.)

".....The Pleiads, rising thro' the mellow shade,
Glitter like a swarm of fireflies
Tangled in a silver braid...."
TENNYSON

M45 The Pleiades Star Cluster. Position 03439n2358 or about 12° NW of the Hyades Group. This is the group often called the "Seven Sisters", undoubtedly the most famous galactic star cluster in the heavens, known and regarded with reverence since remote antiquity. The name is said to be derived from the Greek πλειν, "to sail", from the tradition that the heliacal rising of the Pleiades was the sign of the opening of the navigational season in the Mediterranean world; an alternate suggestion is that the name is simply a form of the Greek πλειος, "full" or "many" which is also the meaning of the Arabic *Al Thurayya*, and probably the Biblical *Kimah*. In various Greek and Roman writings they are referred to as *The Starry Seven*, the *Net of Stars*, *The Seven Virgins, the Seven Atlantic Sisters*, *the Daughters of Pleione*, or the *Children of Atlas*. In Greek myth they were half-sisters of the Hyades, and were saved by Zeus from the pursuit by the giant Orion, so the legend goes, by being transformed into a group of celestial doves. Aratus, in a poem written in the 3rd Century BC, gives us their names:

> "...*These the seven names they bear:*
> *Alcyone and Merope, Celaeno,*
> *Taygeta, and Sterope, Electra,*
> *And queenly Maia, small alike and faint,*
> *But by the will of Zeus illustrious all*
> *At morn and evening, since he makes them mark*
> *Summer and winter, harvesting and seed-time...*"

Manilius, in the days of Augustus, speaks of them as the *Narrow Cloudy Trail of Female Stars*, and in another passage as *Glomerabile Sidus*, the "Rounded Asterism" or "Circular Group of Stars". Valerius Flaccus has a similar title: *Globus Pleiadum*, while Seneca refers to them as the *Pleiadum greges*. Hesiod tells us that the Pleiades marked the agricultural seasons in his time:

> "*When Atlas-born, the Pleiad stars arise*
> *Before the Sun above the dawning skies,*
> *'Tis time to reap; and when they sink below*
> *The morn-illumined west, 'tis time to sow..*"

The connection of the Pleiades with agricultural activities is immortalized also in such titles as "*Virgins of Spring"*, the *Stars of Abundance*, the *"Stars of the Season of Blossoms"*, etc., all these names referring to the fact that the Sun reaches conjunction with the Pleiades in mid-May (now about May 20), the time of blossoming flowers. In Buddhist scriptures we read that the birth of the holy child was prophecied to occur *"When the Flower Star Shines in the East"*, very possibly a reference to the Pleiades. In actual fact his birth (in a garden!) in mid-May, about 563 BC, would coincide closely with the yearly Sun-Pleiades conjunction.

The Pleiades are mentioned several times in Hebrew scriptures, particularly in the book of Amos, where (Ch.V Verse 8) we find the reference to

Him that maketh the Seven Stars and Orion...

And in the more famous lines from the 38th Chapter of the *Book of Job*, which refers also to Orion and the mysterious *Mazzaroth*:

31. *Canst thou bind the sweet influence of the Pleiades, or loose the bands of Orion?*
32. *Canst thou bring forth Mazzaroth in his season? Or canst thou guide Arcturus with his sons?*

Milton evidently had these lines in mind when he wrote of the Creation:

*".........the gray Dawn
And the Pleiades before him danc'd,
Shedding sweet influence...."*

In the works of Middle-Eastern poets, the Pleiades often occur in glittering phrases which compare them to celestial jewelry. Hafiz of Persia, in the 14th Century, wrote to a poet-friend:

"To thy poems Heaven affixes the Pearl Rosette of the Pleiades as a seal of immortality...."

In the 13th Century *Gulistan* or "Rose-Garden" of the Persian poet Sadi we read:

"The ground was as if strewn with colored enamel, and necklaces of Pleiades seemed to hang upon the branches of the trees...."

And in his lines in the *Mu allakat*, composed in the

THE PLEIADES. The most famous of the galactic star clusters and a favorite target of amateur telescopes. Photographed with a 5-inch Cogshall camera at Lowell Observatory.

7th Century, the Arabian poet Amr al Kais evokes for us
*"The hour when the Pleiades appeared in the firmament like
the folds of a silken sash variously decked with gems..."*
Edgar Allan Poe, in his poem *Israfel*, inspired by a verse
in the *Koran*, pictures the Pleiades as seven maidens of a
starry choir who have paused in Heaven to hear the voice of
the angel Israfel, "whose heart-strings are a lute", and
who sings with the sweetest voice of all God's creatures.
The most famous reference to the Pleiades in English literature occurs in the opening passages of Tennyson's prophetic *Locksley Hall:*

> *"Many a night from yonder ivied casement,*
> > *ere I went to rest,*
> *Did I look on great Orion, sloping slowly*
> > *to the west.*
> *Many a night I saw the Pleiads,*
> > *rising thro' the mellow shade,*
> *Glitter like a swarm of fireflies*
> > *tangled in a silver braid."*

According to R.H.Allen, one mention of the Pleiades in
Chinese annals appears to refer to an observation made in
2357 BC; in the ancient Hindu Lunar Zodiac the group was
apparently the central feature of the 1st *nakshatra* called
Krittika, the *General of the Celestial Armies*. The cluster
appears on some Hindu charts as the *Flame of Agni*, and was
apparently an object of veneration during the star-festival
of *Dibali*, the "Feast of Lamps", celebrated in the "Pleiad-
Month" of *Kartik* (October-November) each year. W.T.Olcott
in his *Field Book of the Skies* calls attention to the interesting fact that the Pleiades are associated with such
festivals as Halloween, All Saint's Day, and other memorial
services to the dead traditionally held in many lands and
ages in October and November. The date of their midnight
culmination was observed with solemn ceremony in countries
as widely separated in space and time as Pre-Columbian
Mexico and ancient Persia. Olcott suggests that " these
universal memorial services commemorate a great cataclysm
that occurred in ancient times, causing a great loss of
life...." The legendary sinking of the mythical *Atlantis*,
we might speculate, is possibly the catastrophe in question
although, for geological reasons, it is no longer possible

to accept Plato's story literally. Modern students of ancient lore have thus placed Atlantis- since they feel the need to place it *somewhere*- at a wide variety of sites ranging from the Azores Islands to Spitzbergen and from Gibraltar to Patagonia. The currently fashionable theory is that the original inspiration for the legend was the cataclysmic eruption of the Santorin volcano on the island of Thera in the Aegean Sea, some 70 miles north of Crete. This catastrophe is dated to about 1450 BC, and was evidently one of the most violent volcanic explosions ever recorded on the earth, exceeding even the famous eruption of Krakatoa in 1883. According to some theorists, the Santorin eruption may have been one of the chief causes of the downfall of the Minoan civilization; others have attempted to connect it with the stories of the Ten Plagues of Egypt and the events of the Biblical Exodus. In both cases it remains uncertain as to whether the chronology can be made to fit. However, if the old traditions are "true" and the Pleiades were indeed near midnight culmination, we have the consolation, at least, of knowing the approximate time of year of the great cataclysm (mid-November) although there is no hint of the exact *year!*

In American Indian legend the Pleiades are connected with the *Mateo Tepe* or Devil's Tower, that curious and wonderfully impressive rock formation which rises like a colossal petrified tree-stump to a height of 1300 feet above the plains of northeastern Wyoming. According to the lore of the Kiowa, the Tower was raised up by the Great Spirit to protect seven Indian maidens who were pursued by giant bears; the maidens were afterwards placed in the sky as the Pleiades cluster, and the marks of the bears' claws may be seen in the vertical striations on the sides of the Tower unto this day. The Cheyenne had a similar legend.

The Pleiades, in some classical legends, shared the watery reputation of their half-sisters, the Hyades, and were associated with the affairs of seamen. A popular name in Germany some 200 years ago, according to Christian Ideler, was *Schiffahrts Gestirn*, the "Sailor's Stars"; King James I refers to them in his *Tables of some Obscure Wordis* as the *Seamens Starres*. The Russian name, *Baba*, translated "The Old Wife", possibly refers to the legendary Old Witch *Baba Yaga*, whose flight through the sky in a fiery mortar is so vividly interpreted by Moussorgsky in the next-to-last scene of his orchestral suite *Pictures at an Exhibition*. The Anglo-Saxon name *Sifunsterri* is the equivalent of the German *Siebengestirn*; both simply mean the "Seven Stars". Milton calls them the "Seven Atlantic Sisters" and Chaucer has them as the *"Atlantes doughtres sevene"*. R.H. Allen tells us that various European peoples, as the Finns and the Lithuanians, saw them as a "seive" or a net; the curious Swedish title *Suttjenes Rauko* has been translated "Fur in Frost". The name *Killukturset*, in use in Greenland at the time of the arrival of the first Norse missionaries, seems to mean "The Dog Pack" or possibly "The Dogs Attacking the Bear"; the Welsh had a similar tradition with their *Y twr tewdws*, the "Close Pack". J.R.R.Tolkien, in his epic fantasy *The Hobbit*, tells us that the Pleiades were known in the ancient days of Middle-earth as *Remmirath*, or "The Netted Stars".

In ancient Aztec and Mayan tradition, the midnight culmination of the Pleiades was an event of great and ominous significance, particularly at intervals of 52 years when the two Pre-Columbian time-counts (the sacred and the secular calendars) once again came into coincidence. The world, it was believed, would come to an end on one of these dates; it had already been destroyed and re-created on four such occasions in the past. At the sacred site of *Teotihuacan*, some 28 miles northeast of Mexico City, we find that the west face of the great Pyramid of the Sun is oriented to the setting of the Pleiades, and that all the west-running streets of the city point to the same spot on the horizon. These structures, however, are far more ancient than either the Aztec or the Toltec culture; they date to about the 1st Century and, unless the orientation is dismissed as a coincidence, tell us that even at that time the Pleiades cluster was regarded with special reverence. R.H.Allen states that a number of ancient Greek temples were oriented to the rising (or setting) of the Pleiades. In Egypt too we find them honored as a symbol of the goddess *Neith*, and reverenced as *The Stars of Athyr* or Hathor, one of the many forms of Isis.

On the unique *Phaistos Disk*, found in the ruins of a Minoan Age palace on Crete in 1908, we find a puzzling series of strange symbols, stamped into the original clay

DESCRIPTIVE NOTES   (Cont'd)

with carefully made punches before the disc was ceramically
fired. Among the many symbols are a fair number which very
possibly have an astronomical significance; eagles, twisted
serpents, and a frequently repeated pattern of *seven dots
enclosed in a circle*. Of the many possible interpretations
of the disk, the most convincing appears to be the idea
that it represents a sort of early almanac or sky-calendar
to be used by pilgrims visiting sacred sites, or by farmers
who could determine the correct times for sowing and plow-
ing by the stars. The "seven-dot" symbol, according to this
theory, can hardly represent anything other than the cluster of the Pleiades.

In addition to the reverence paid to the Celestial
Sisters in many lands, it is interesting to find them
associated with more mundane matters among rural peoples
everywhere, usually as a *Hen and her Chicks*, as in the
Massa Gallinae of medieval times. Farmers in Germany saw
in them the *Gluck Henne;* the Danes had them as *Aften Hoehne*
or the "Eve Hen", while a Russian peasant name *Nasedha* is
translated "The Sitting Hen". In an early translation of
the Bible, dating from 1535, a marginal note to the *Book of
Job* thoughtfully identifies the "Seven Stars" of the text
as *"The Clock Henne with her Chickens"*. In other traditions
they are a flock of pigeons or doves, a group of goats, or
as in one Arabian text, a Herd of Camels. The identification with doves is probably derived from the Greek myth of
the Seven Doves who carried ambrosia to the infant Zeus;
in another legend we are told of the Seven Sisters who were
placed in the heavens that they might forget their grief
over the fate of their father Atlas, condemmed to support
the sky upon his shoulders. On a silver coin of Thurium in
Lucania, dating from about 430 BC, Taurus is depicted as an
imposing bull with a dove perched on his shoulders, evidently an allusion to the Pleiades. Miss Agnes Clerke, in
her classic work *The System of the Stars* (1907) excellently
summarizes the reverence with which the Pleiades have been
honored in all ages:

*This is the immemorial group of the Pleiades, famous
in legend, and instructive, above all others, to exact in-
quirers - the meeting-place in the skies of mythology and
science. The vivid and picturesque aspect of these stars
rivited, from the earliest ages, the attention of mankind;*

THE PLEIADES. This striking photograph of the cluster was made with the 13-inch astrographic telescope at Lowell Observatory.

1870

*a peculiar sacredness attached to them, and their concern
with human destinies was believed to be intimate and direct.
Out of the dim reveries about them of untutored races,
issued their association with the seven beneficient sky-
spirits of the* Vedas *and the* Zendavesta, *and the location
among them of the centre of the universe and the abode of
the Deity, of which the tradition is still preserved by the
Berbers and Dyaks. With November, the "Pleiad-month," many
primitive people began their year; on the day of the midnight culmination of the Pleiades, 17th November, no peti-
tion was presented in vain to the ancient kings of Persia;
and the same event gave the signal at Busiris for the com-
mencement of the feast of Isis, and regulated less immediately the celebration connected with the fifty-two year
cycle of the Mexicans. Savage Australian tribes to this day
dance in honor of the "Seven Stars", because "they are very
good to the black-fellows". The Abipones of Paraguay regard
them with pride as their ancestors. Elsewhere, the origin
of fire and the knowledge of rice-culture are traced to
them. They are the "hoeing-stars" of South Africa, take
their place as a farming-calendar to the Solomon Islanders,
and their last visible rising after sunset is, or has been
celebrated with rejoicings all over the southern hemisphere
as betokening the "waking-up time" to agricultural activity..."*

The English missionary William Ellis has recorded
that the natives of the Tonga and Society Islands called
the Pleiades *Matarii,* the "Little Eyes"; a similar name,
Matariki, was in use among other Polynesian tribes, who
also had a legend that the group was once a single bright
star, broken into six during a battle of the gods. Hence
the name *Tauono,* "The Six", which bears a suspicious resemblance to the Latin "Taurus". The natives of the Solomon
Islands, according to Allen, called the group *Togo Ni Samu,*
a "Company of Maidens". The Abipones revered them as their
great spirit *Groaperikie,* or Grandfather. A more sinister
influence was attributed to the Pleiades in the Middle Ages
as their midnight culmination became the traditional date
of the fearsome *Witch's Sabbath* or *Black Sabbat,* the night
of unholy revelry held amid high crags in the Caucasus or
on the Brocken in Germany; the date evidently preserves the
memory of the ancient Druids' rites of November 1st, still

observed in the modern world as *All Hallows' Eve* or Halloween. The traditional date has also been kept, although the
actual midnight culmination of the Pleiades now occurs on
November 21.

   Alcyone, the brightest star of the Pleiades, has been
honored with a variety of titles. The Arabians called it
Al Wasat, the "Central One", and *Al Nair,* the "Bright One";
other names have been translated *The Light of the Pleiades*
and *The Leading One of the Pleiades*. The Hindu name was
Amba, the Mother, but J.F.Hewitt relates that an earlier
name was *Arundhati,* identified as the wife of the *Chief of
the Seven Sages,* worshipped particularly by newly-married
couples. A.Lampman paid tribute to this chief star of the
cluster in his poem *Alcyone:*

> "......*the great and burning star,*
> *Immeasurably old, immeasurably far,*
> *Surging forth its silver flame*
> *Through eternity.....Alcyone!*   "

   One of the oldest traditions concerning the cluster
is the persistent myth of a *Lost Pleiad*. The Greeks identified her as Electra, who is said to have veiled her face in
grief at the burning of Troy; another version casts Merope
in the role, as she reputedly hid her face in shame at having married a mortal, the King of Corinth, while all her
sisters had been wedded to gods. Celaeno is another possible candidate, as she is reported to have been struck by a
thunderbolt. Aratus refers to the tradition of the "Lost
Pleiad" when he writes:

> "*Their number seven, though the myths oft say,*
> *And poets feign, that one has passed away..*"

   This tradition, however, is not confined to the Greek
world; the story of a lost Pleiad appears also in Japanese
lore, and in the legends of Australian aborigines, natives
of the Gold Coast of Africa, and head-hunters of Borneo.
If the legend has its origin in astronomical fact, the star
Pleione is the most likely suspect; it has a peculiar shell
spectrum, and is known to be variable by at least half a
magnitude.

> "*The Sister Stars that once were seven*
> *Mourn for their missing mate in Heaven..*"

THE PLEIADES, showing swirls of nebulosity about the stars.
The photograph was made with the 36-inch Crossley reflector
at Lick Observatory.

The Pleiades Cluster remains a group which commands attention and admiration even in the days of modern astronomy. Otto Struve has stated that the cluster has been the target of more astronomical cameras than any other object in the heavens beyond the Solar System. America's foremost non-professional astronomer, Leslie C.Peltier, tells us in his delightful autobiography *Starlight Nights* that a view of the Pleiades remains his earliest memory of a celestial object. Fittingly, he closes his final paragraph with one last view, through the eyes of memory, of the *seven little stars that sparkle in a long-gone autumn sky"*.

To the average eye the Pleiades Cluster appears as a tight knot of 6 or 7 stars, but some observers have recorded 11 or more under excellent conditions. Miss Agnes Clerke tells us that "Maestlin, the tutor of Kepler, perceived 14, and mapped eleven Pleiades previously to the invention of the telescope; Carrington and Denning counted fourteen; Miss Airy marked the places of twelve with the naked eye". William Dawes was able to distinguish 13 stars without optical aid; Carl von Littrow claimed 16. There are at least 20 stars in the group which might be glimpsed under the finest conditions, having a brightness just below usual naked-eye range; the crowded massing of the stars, however, makes this impossible. In the small telescope the Pleiades becomes one of the most attractive celestial objects, and

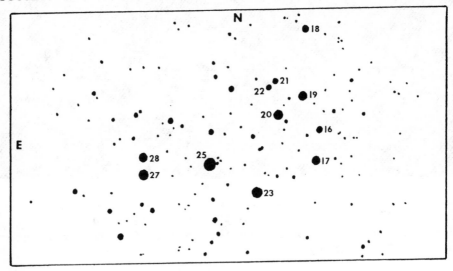

makes its best impression with low-power wide-angle type
instruments; opera glasses, binoculars, or rich-field
reflectors. High powers are absolutely useless on such an
object; obviously the impression of a cluster will be totally lost if only one star can be seen in the field at a
time! The present author finds the most dazzling view of
the Pleiades to be obtained with 20 X 70 binoculars, though
the group is superb even in 7 X 50s. In a dark sky the 8 or
9 bright members glitter like an array of icy blue diamonds
on black velvet; the frosty inpression is increased by the
nebulous haze which swirls about the stars and reflects
their gleaming radiance like pale moonlight on a field of
snow crystals. With larger instruments the number of faint
stars will increase noticeably; Robert Hooke reported a
count of 78 with a 2-inch refractor in 1664; Max Wolf recorded 625 to the 14th magnitude on a photograph made at
the Paris Observatory in 1876; modern photographs reveal
over 2000 stars in the field. Possibly about 250 of these
are known to be true cluster members; further research and
study of the fainter members will probably increase the
total to at least 500.

The nine brightest stars are concentrated in a field
slightly over 1° in diameter, so it is possible to include
the whole group in telescopes up to 10-inch of aperture if
a wide-angle ocular is used. These 9 bright stars form a

THE BRIGHT MEMBERS OF THE PLEIADES			
STAR	NAME	MAGNITUDE	SPECTRUM
η = 25	Alcyone	2.86	B7e III
27	Atlas	3.62	B8 III
17	Electra	3.70	B6e III
20	Maia	3.86	B7 III
23	Merope	4.17	B6 IV
19	Taygeta	4.29	B6 V
28 = BU	Pleione	5.09 v	B8e p
16	Celaeno	5.44	B7 IV
21 + 22	Asterope	5.64; 6.41	B8e V; B9 V
18	--------	5.65	B8 V

pattern resembling a stubby dipper with the short handle at
27 and 28 Tauri (Atlas and Pleione). The cluster is thus
sometimes erroneously called the "Little Dipper", but this
name properly belongs to the constellation Ursa Minor.

The Pleiades group is one of the nearest of the open
or "galactic" star clusters, and appears to be slightly
over 3 times the distance of the nearby Hyades. The best
modern value for the distance still appears to be that
derived in 1958 by H.L.Johnson and R.I.Mitchell in a photometric study of the known members; their result was about
126 parsecs, or 410 light years. The nine brightest stars
are all B-type giants, and are concentrated in a region
about 7 light years in diameter. A few outlying members are
found as distant as 20 light years from the center.

Eta Tauri or *Alcyone*, the central star and brightest
member of the Pleiades, is nearly 1000 times more luminous
than the Sun, and probably about 10 times greater in size.
The apparent magnitude is 2.86, spectral type B7e III; the
absolute magnitude is about -2.6. In contrast, the faintest
known members are less than 1/100 the solar luminosity.
Between these extremes we find the usual Main Sequence containing all intermediate types and spectra, down to faint
red dwarfs of the 16th magnitude. The Pleiades cluster contains no red giant stars, and is evidently a much younger
group than the Hyades. An age of about 20 million years is
suggested by present knowledge of stellar evolution. Only
a few white dwarfs have been detected in the field of the
cluster; the most convincing candidate for true membership
is a 16th magnitude DA star called LB 1497, discovered by
W.J.Luyten; it seems to show approximately the same proper
motion as the cluster.

Much interest attaches to the discovery of a number
of faint red dwarf variables in the Pleiades. These erratic
objects, often called "flash stars", appear to belong to
the same general class as the rapidly flickering faint
variables in the Orion Nebula. In 1963, G.Haro reported on
the presence of seven of these stars in the Pleiades. The
spectra ranged from dK2 down to dM3; the amplitudes of the
sudden "flashes" varied from 0.8 to 3.7 magnitudes, and
the duration of the outbursts varied from several minutes
up to about 3 hours. These stars are all faint objects;
the magnitudes at normal light ranged from 13.3 to 16.9.

NEBULOSITY IN THE PLEIADES. Maia and Taygeta appear near the top of this print; Merope is at lower left. Mt.Wilson Observatory photograph made with the 100-inch telescope.

DESCRIPTIVE NOTES (Cont'd)

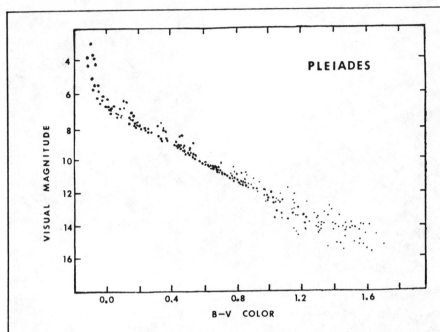

It is believed that such stars are still in the process of
gravitational contraction and have not yet reached a stable
state. Haro suggests that the "flash stars" are objects of
the well known T Tauri type in a somewhat later stage of
evolution. The presence of these stars adds support to
other evidence that the Pleiades cluster is relatively
young, and that new stars may have been formed in the group
fairly recently.

      The Pleiades Cluster is drifting through space in a
SSE direction at an apparent rate of about 5.5" per century. Though this corresponds to an actual space velocity of
about 25 miles per second, the cluster will require over
30,000 years to drift a distance equal to the apparent
width of the Moon. The group was somewhat closer to us in
the remote past; the measured radial velocity of the bright
members is about 4½ miles per second in recession, but the
individual velocities are not in precise agreement owing
perhaps to motions within the cluster, and also to observational difficulties. The bright Pleiades stars all show
rapid rotation which widens and blurs the lines of the
spectrum, making exact radial velocity measurements rather
difficult. The star Pleione (28 Tauri) is an exceptionally

remarkable example of this effect, rotating about 100 times
as fast as does our Sun. The star is definitely variable,
with a range of about 0.5 magnitude, and may be the "Lost
Pleiad" of legend. As a variable it bears the designation
BU Tauri, and is also known as 28 Tauri.

The spectroscope reveals curious turbulence in the
atmosphere of Pleione; at times it seems to eject expanding
gaseous shells which eventually disperse into space. Bright
hydrogen lines were first noticed in the spectrum in 1888;
these had vanished by 1903, but reappeared again in 1937-
1938 when a period of major activity began. During the next
10 years the shell spectrum weakened, vanishing in 1951.
A decrease of about 0.5 magnitude occurred between 1937 and
1939, probably from the absorption of light by the shell.
In recent years only slight variability has been detected,
but the mean brightness has remained about 0.2 magnitude
below that recorded from 1897 to 1936. A new outburst of
Pleione occurred in late 1972.

According to O.Struve, the gaseous shell which grew
in 1938 was about 1.5 times the size of the star, or about
7.5 times the diameter of the Sun. Shell activity in this
star may be chiefly the result of the unusually great rotational velocity; if so the "shell" may be actually more
like a thick equatorial ring.

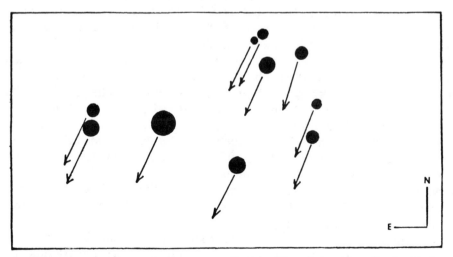

THE PROPER MOTION OF THE PLEIADES. The length of the
arrows shows the motion in approximately 20,000 years.

Studies of the proper motion of the Pleiades led in 1846 to one of the most curious misinterpretations in the history of astronomy. J.H.von Maedler at the Observatory of Dorpat, finding that the members of the Pleiades showed no motion relative to one another, rashly concluded that the cluster, and Alcyone in particular, represented the fixed center of the entire stellar system. The notion that Alcyone was the "central sun" of the Universe gained some popularity, but became entirely obsolete within a few decades as actual knowledge concerning the structure of the Galaxy became available.

THE PLEIADES NEBULOSITY. A remarkable fact about the cluster is that the entire star-swarm is enveloped in a faint diffuse nebulosity of vast extent (Tennyson's "silver braid") which appears to shine by reflected light. This cosmic cloud is elusive visually, but shows much peculiar detail on long-exposure photographs. The spectrum of this nebulosity is identical to the spectra of the involved stars, a fact first discovered by V.M.Slipher at the Lowell Observatory in 1912. The light is apparently star-light, reflected from dust and perhaps larger solid particles.

The brightest portion of this nebulosity envelops the star *Merope*, and extends about 20' to the south; it was first noticed by Prof.W.Tempel with a 4-inch refractor, observing at Venice on October 19, 1859. He described it as resembling a faint stain of fog, like the effect of "a breath on a mirror". Lewis Swift found it detectable with 25X on a 2-inch refractor in 1874; T.W.Backhouse confirmed its reality with a 4½-inch refractor, "though to see it well it was necessary to have Merope out of the field". On the other hand, S.W.Burnham found no sign of it with the great 18-inch Dearborn refractor in Illinois; for some years the Merope Nebula was considered to be variable. No certain changes are evident on photographs dating from 1885 when the nebula was first successfully recorded by the brothers Paul and Prosper Henry of Paris; a plate made by Isaac Roberts in the following year also shows the nebula much as it appears today.

In January 1894 a 4-hour exposure of the Pleiades was obtained by H.C.Wilson with an 8-inch lens at the Goodsell Observatory of Carleton College in Minnesota. "As a result a very fine picture was obtained of the nebula involving

THE MEROPE NEBULA. Fine details are seen on this photograph of NGC 1435, made with the 42-inch reflector at
Lowell Observatory.

nearly the whole group of bright stars and exhibiting marvelous details of structure resembling those of the great
nebula of Orion......The structure of the nebula about
Merope is very curious. There are many approximately parallel and slightly curved lines of light passing the star at
an angle of about 30° to the meridian. The same structure
is apparent, though less marked, in the diffuse part of the
nebula which extends to the south and east of Merope. From
the southern edge of the image of Electra a bright streak
of nebulosity about 20" in width proceeds on a straight line
toward Alcyone. It extends about one-third of the way to
Alcyone, tapering to a point. About 1' south of this a narrower streak extends half the distance of the former, and
5' south another streak, almost as bright as the first and
nearly parallel, extends from a point 2' west of Electra to
a point 5' or 6' northwest of Merope, taking in on the way
the two brightest stars between Electra and Merope. Through
the middle of the group and especially around Alcyone the
background is filled with mottled patches of nebula........
......The region about Maia is especially interesting. A
very bright hornshaped patch of nebula runs out from the
west edge of the star image, curving immediately northward,
and extends to a distance of 3' north of the star. The
nebula here is full of irregularly parallel streaks similar
to those about Merope......Some of them run to and beyond
the bright stars north of Maia......"

Both the Merope and Maia clouds are very remarkable
for their fine structure, resembling in their wind-torn
appearance the typical streamers of a high-altitude cirrus
cloud on the earth. Other portions of the nebulosity run in
the form of peculiar straight wisps, occasionally seeming
to link the brighter stars. It is not known what forces may
have acted to produce such effects. The diameter of the
Merope cloud is about 2 x 3 light years; the nebulosities
surrounding Maia and Alcyone are somewhat smaller and considerably fainter.

Modern observers will find the Merope Nebula, now
catalogued as NGC 1435, only moderately difficult in good
6-inch or 8-inch instruments; the present author finds it
surprisingly distinct with low-power oculars and a dark sky,
but totally invisible in moonlight. Walter Scott Houston,
observing with an 8-inch reflector in the exceptionally

DESCRIPTIVE NOTES   (Cont'd)

clear mountain air of southern Arizona, reported a superb view of the Pleiades in 1974: *"When I looked into the eyepiece, expecting to see a few faint wisps, the field was laced from edge to edge with bright wreaths of delicately structured nebulosity.."* Miss Agnes Clerke mentions the impression that "the masses of nebula, in numerous instances, seem as if *pulled out of shape* and drawn into festoons by the attractions of neighboring stars. But the strangest exemplification of this filamentous tendency is in a fine thread-like process, 3" or 4" wide, but 35' to 40' long, issuing in an easterly direction from the edge of the nebula about Maia, and stringing together "like beads on a rosary", seven stars met in its advance. Two similar rectilinear nebulae run parallel to the first..." E.E.Barnard, in 1890, detected a tiny but very bright nebulous nucleus, resembling the head of a small comet, very close to Merope, and nearly hidden in the radiance of the star; this may be the central nucleus of the Merope nebula.

DOUBLE AND MULTIPLE STARS are common in the Pleiades. With binoculars the observer will first notice the bright pair 27 and 28 Tauri (Atlas and Pleione) on the east edge of the cluster; they are just 5' apart and often appear fused into a single elongated image if no optical aid is employed. The simplest field-glass, however, will show them separate. The brighter star, 27 Tauri, is a very close and difficult pair first catalogued by F.G.W.Struve in 1827; it has a reported separation of 0.4" and a brightness difference of three magnitudes. From 1843 to 1935 the companion was not seen at all, except for one reported glimpse during an occultation by the Moon in January 1876; modern observations are fairly discordant but the reality of the companion seems to be reasonably certain.

Another wide and easy pair is *Asterope* (21 + 22 Tauri) on the north edge of the cluster, with a separation of 2.8' and referred to on some star maps as "Sterope I and II".

The field of Alcyone (Eta Tauri) is attractive for small instruments with a neat 1' triangle of 9th magnitude stars lying just 3' to the NW; Olcott calls this "a quadruple star and a very beautiful object". From observations made in 1972 during a grazing occultation by the Moon, it appears that Alcyone itself may be a very close pair, with a companion closer than 1" and about a third the brightness

of the primary. Taygeta (19 Tauri) is another bright Pleiad
with a visual companion; the pair was first listed by John
Herschel under the designation h3251. The separation is 69"
in PA 330°; various catalogues list the companion as magnitude 8, 9, or 10; R.H.Allen calls the colors "lucid white
and violet". According to a study by H.A.Abt at Kitt Peak,
the bright primary shows radial velocity variations and may
be a spectroscopic binary with a period of about 1300 days.
Maia (20 Tauri) is sometimes listed as a spectroscopic pair
but the radial velocity variations are rapid and possibly
not periodic, suggesting that the star, like Pleione, shows
atmospheric pulsations or eruptions.

  In addition to the bright pairs, the cluster also
contains several fainter visual doubles. These are indicated on the chart below; for data on each star refer to the
list on page 1798 and 1799:

  A= β536 Position 03433n2402 Triple system
  B= β537   03441n2440 Double
  C= Σ449   03444n2430 Double
  D= Σ450   03444n2346 Double
  E= 0Σ64   03470n2342 Triple

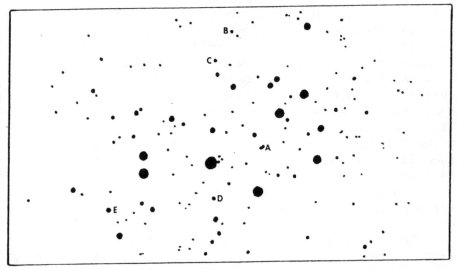

The Pleiades Cluster lies about 4° from the Ecliptic, and
thus is subject to occasional occultations by the Moon. As
R.H.Allen points out, the mean angular diameter of the Moon
(about 31') is several minutes of arc *less* than the space

DEEP-SKY VIEWS IN TAURUS. Top left: Star cluster NGC 1746.
Top right: The faint planetary nebula NGC 1514. Below: The
bright star Alcyone in the Pleiades.   Lowell Observatory.

between Alcyone and Electra, or between Merope and Taygeta, so that the Moon "could be inserted within the quadrangle formed by those four stars with plenty of room to spare; although in looking at the cluster the impression is that our satellite would cover the whole. An occultation of the Pleiades by the moon gives a vivid realization of this fact; and as this is a not infrequent phenomenon, I commend its observation to any unbeliever".

S147   Position about 05360n2800. One of the strangest nebulae in the sky, a huge, nearly circular cloud composed entirely of many slender thread-like filaments forming a great ring about 2° x 3° in size. It is located in the eastern portion of the constellation, just beyond the tips of the horns of Taurus, about 1° NE from 125 Tauri and about 5° NNE of the Crab Nebula. The major portion of the cloud lies in Taurus; the northern edge crosses the border into Auriga. S147 resembles the famous Veil Nebula in Cygnus, but is apparently a much fainter and much older object. It has not been detected visually in any telescope, and appears clearly only on red-sensitive photographs. Its appearance, when compared with the Veil Nebula, is illustrated on page 808.

S147 was discovered in 1952 by G.A.Shajn and V.E.Hase at the Crimean Astrophysical Observatory at Simeis in the U.S.S.R. It was found in the course of a photographic program begun in 1950, using the 25-inch Schmidt camera at Simeis. At about the same time the nebula was discovered independently on red plates made with the 48-inch Schmidt at Palomar Observatory. More recently, radio observations by D.E.Harris of the California Institute of Technology have shown that S147 is a source of radio energy. Measured at a frequency of 960 mc/sec, the nebula has about half the intensity of the Veil Nebula in Cygnus. The Veil itself is now generally recognized as a supernova remnant, and it can scarcely be doubted that the Taurus feature had the same origin. Although the expansion rate of the cloud has not been accurately measured, a comparison with the Veil seems to suggest that S147 is the older of the two objects, and probably dates back well over 50,000 years.

The exact distance of S147 is not accurately known. From a measurement of the apparent widths of the fine

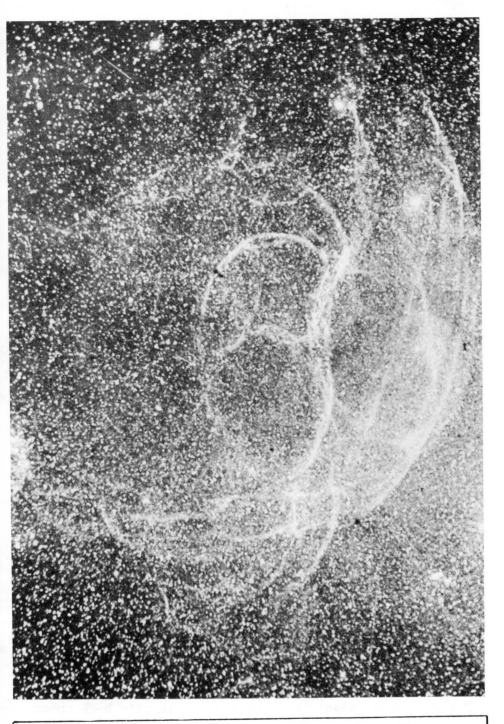

NEBULA S147 in TAURUS. The filamentary structure is well shown in this photograph made at the Crimean Astrophysical Observatory in the U.S.S.R.

filaments, Harris has estimated a distance of about 2600
light years, while the Russian astronomer I.S.Shklovsky
obtained about 3600 light years. If we accept about 3000
light years as a working value, the actual diameter of the
cloud is found to be about 165 light years, somewhat greater than the computed diameter of the Veil, and vastly larger than the size of the famous Crab Nebula, which is known
to be a very recent supernova remnant. There is, however,
at least one object known and identified as a probable
supernova remnant, which exceeds even S147 in size; this is
the great ring-cloud in the Nubecula Major, shown on page
849; its estimated diameter is over 400 light years. From
these examples, it appears that a supernova cloud may remain visible even after it has expanded to fill a volume of
space several hundred light years in diameter.

The filamentary structure, like that of the famous
Veil Nebula (page 801) has excited much speculation. Well-
defined filaments typically have widths of 1" to about 5",
or about 1000 to 5000 AU. There is a growing belief that
electrical and magnetic fields are involved in some way
with their formation. It has also been suggested that a
filament may mark the "front" of an advancing shock wave.
The source of the illumination of S147 remains a mystery
for the present. No object has been identified as a central
illuminating star, nor does any sufficiently bright hot
star exist in the vicinity. Curiously, the source of the
light of the Veil has never been positively identified
either.

An interesting circumstance is the presence of the
erratic nebular variable star RR Tauri on the south edge of
S147. It lies just off one of the bright filaments on a
small nebulous patch of its own. The association with S147
may not be physically real, but the question remains open.
From current studies of the star, it seems that it lies at
about the same distance of the nebula. (Refer to RR Tauri,
page 1836)

(See also NGC 1952 (M1) in Taurus, and NGC 6960 +
6992 in Cygnus. Supernovae are treated in this book under
"Tycho's Star"(Nova 1572) in Cassiopeia, "Kepler's Star"
(Nova 1604) in Ophiuchus, and "Nova 1006" in Lupus. See
pages 570 and 1388 for examples of supernova in other
galaxies)

TELESCOPIUM

LIST OF DOUBLE AND MULTIPLE STARS

NAME	DIST	PA	YR	MAGS	NOTES	RA & DEC
B1879	1.7	266	34	6- 11½	spect K1	18073s4732
Hd 288	1.8	149	43	7- 10½	PA & dist dec, spect F0	18090s5035
I 429	0.1	146	52	8½- 8½	spect A0	18111s5020
h5033	17.3	115	13	7 - 9½	spect K0	18116s4853
	18.2	9	13	- 11		
	27.9	64	13	- 9½		
I 111	1.0	295	44	7 - 9½	PA dec, spect K0	18118s5640
h5034	2.4	97	59	7½- 8½	PA inc, spect A2; Nebula IC 4699 in field	18125s4602
I 1361	2.6	254	35	7½- 12	spect A2	18165s4909
△220	31.1	178	38	8 - 8½	cpm, relfix, spect G0	18180s5535
h5041	2.6	259	43	7½- 9½	relfix, spect F0	18218s5341
h5045	8.3	23	33	7- 10½	spect G0	18272s4803
h5049	19.9	264	13	7 - 11	spect F2	18337s4707
h5053	0.2	154	52	8 - 8	(B398) AB PA dec, spect F5	18393s5549
	32.6	197	13	- 10		
h5055	7.5	76	52	8½- 9	relfix, spect A2	18394s5255
h5059	25.6	238	13	7 - 11	spect B9	18435s4941
	32.0	201	13	- 11		
I 112	1.6	186	48	7½- 9½	(△ 224) PA slow inc; AC spectra F5, A0	18503s4721
	84.0	63	49	- 7½		
I 113	2.6	227	48	6½- 10	relfix, spect K5	18551s4834
h5100	19.4	152	15	7 - 12	spect B5	19146s5614
I 1397	1.5	336	28	6½- 11	spect A0	19157s5140
h5114	69.0	254	13	5½- 7	optical, dist inc, PA dec, spect K2,G0	19238s5426
	70.5	96	00	- 11		
h5114b	8.3	195	32	7 - 11	BC PA inc, dist dec	
I 118	1.2	132	59	7½- 10	relfix, spect G5	19293s4653
Cor 238	3.3	50	53	7½- 8½	relfix, spect A3	19384s5304
h5148	14.2	318	14	7 - 12	spect K0	19466s4531
△227	22.9	149	52	6 - 6½	relfix, nice cpm pair; spect G5, A2, color contrast	19487s5506
φ31	1.8	54	27	8- 12½	spect K0	19512s5156
I 256	1.0	194	42	7½- 9½	PA dec, spect K0	19578s4732

LIST OF DOUBLE AND MULTIPLE STARS (Cont'd)

NAME	DIST	PA	YR	MAGS	NOTES	RA & DEC
I 123	8.2	184	54	7 - 11	PA inc, dist dec, spect K0	20071s4653
λ415	11.1	97	46	7½- 12	spect F5	20174s4542
h5187	17.8	323	14	8- 13½	spect K0	20188s5426
I 664	1.0	280	59	8 - 9	PA dec, spect F0	20254s5115

LIST OF VARIABLE STARS

NAME	MagVar	PER	NOTES	RA & DEC
R	7.8--14..	462	LPV. Spect M5e--M7e	20112s4707
T	9.9--12..	256	LPV. Spect M4e	18229s4941
U	8.8--12..	445	LPV. Spect M7e	19043s4859
V	9.0--10..	125:	Semi-reg; Spect M4?	19144s5032
W	8.3--12..	304	LPV. Spect M5e--M8e	19468s5008
X	9.5--12.5	310	LPV. Spect M5e	20150s5246
Y	8.5--9.5	Irr	Spect M7	20166s5052
Z	9.5--12..	230	LPV. Spect M4e	29360s4542
RR	6.5--16..	---	Nova-like; Spect F5ep (*)	20003s5552
RS	8.5--12..	---	R Cor.Bor Type, Spect R8	18151s4634
RT	9.5--10..	Irr	Spect M3	18402s4718
RU	9.0--14..	271	LPV. Spect M4e	19044s4820
RW	8.0--10..	127	Semi-reg; spect M6e	18397s4550
RX	7.8--9..	350	Semi-reg; Spect M	19033s4603
SV	8.8--12..	226	LPV. Spect M4e--M6e	18525s4932
TY	9.2--14..	361	LPV.	18179s5250
BL	6.9--9.0	778	Ecl.Bin; Spect gF8+M	19027s5130
BM	9.5--15..	415	LPV.	19039s5008
BN	9.0--13..	270	LPV.	19059s4814
BQ	8.7--15..	290	LPV. Spect M5e	20172s5621
BR	8.8--10.2	142	Semi-reg; Spect G5	20202s5302
HO	7.8--8.5	.8918	Ecl.Bin; Spect A2	19484s4659
MT	8.68--9.3	.3164	Cl.Var; Spect A0	18585s4643

TELESCOPIUM

LIST OF STAR CLUSTERS, NEBULAE AND GALAXIES

NGC	OTH	TYPE	SUMMARY DESCRIPTION	RA & DEC
6584	△376	⊕	Mag 8½, Diam 6'; Class VIII; cB,cL,R,gmbM; stars mags 15..	18146s5214
----	I.4699	◎	Mag 12, Diam 5"; S,F, stellar appearance	18148s4601
----	I.4797	⊖	E5; 12.2; 1.3' x 0.6' F,S.bM	18523s5422
----	HA85	⊖	S0; 12.4; 0.7' x 0.5' S,1E,bM	18529s5436
6754		⊖	SBb; 13.1; 1.4' x 0.6' pF,pL,mE,glbM	19075s5044
6758		⊖	E1; 12.7; 1.0' x 0.8' pB,S,R	19098s5624
----	I.4837	⊖	SBc; 12.9; 1.4' x 1.0' F,cS,R,bM	19113s5446
6780		⊖	Sc; 13.2; 0.9' x 0.8' F,L,R,vglbM	19187s5553
----	I.4889	⊖	E5; 12.5; 1.6' x 0.8' S,pF,bM	19413s5429
6851		⊖	E4; 12.8; 1.0' x 0.7' pF,S,vlE,lbM	19599s4825
6854		⊖	E2; 13.2; 0.6' x 0.4' F,S,vlE,glbM	20018s5432
6861	△425	⊖	E5/S0; 12.3; 1.3' x 0.7' B,S,cE,gpmbM	20037s4831
6868		⊖	E2; 12.1; 1.4' x 1.0' vB,S,R,psvmbM	20063s4831
6875		⊖	E6; 12.6; 1.0' x 0.5' F,vS,R,vgmbM, BN	20096s4619
6887		⊖	Sb; 12.8; 3.3' x 1.3' pF,cL,pmE,glbM	20134s5256
6893		⊖	S0; 12.5; 1.5' x 1.0' pF,S,1E,svmbM	20172s4825
6909		⊖	E6; 12.8; 0.8' x 0.6' pB,pL,gbM	20241s4712

DESCRIPTIVE NOTES

RR   Position 20003s5552, in the southern portion of the constellation, about 3° WNW from Alpha Pavonis. RR Telescopii is a peculiar nova or nova-like variable, possibly one of the most extreme cases of the "slow nova" type. It is the only nova on record which was recognized as a periodic variable before the outburst, and according to S.Gaposchkin (1930) had a period of about 387 days with a magnitude range of 12½ to 15. In the following fourteen years the minima became somewhat deeper, falling to about 16½; the total range thus increased to about three magnitudes, the period remaining approximately the same. In 1944 the star rose in a nova-like fashion to about 7th magnitude, where it remained for some 6 years. The maximum brightness was about 6.5 in July 1949. Two years later the magnitude was about 8, and the star was still fading slowly with slight fluctuations in which the presence of the 387-day period could still be detected. This very unorthodox nova required some 1600 days to reach maximum, one of the slowest on record.

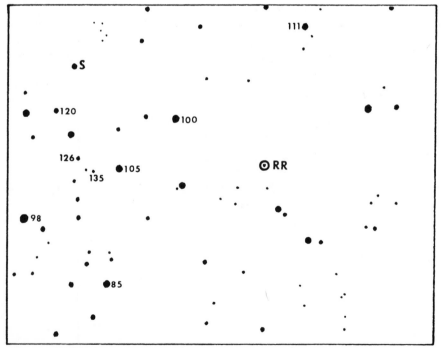

RR TELESCOPII FIELD, based on an AAVSO chart. Scale = 20" per mm. North is at the top.

The nova-like features of the star's spectrum appeared rather late in the history of the outburst, as was also true of the similar star RT Serpentis, described on page 1774. In 1949, when the brightness was near maximum, the spectrum resembled that of an F5 supergiant, but with weak emission features to the red side of the hydrogen lines. Absorption lines of helium (He I) were measured in 1951, and revealed approach velocities of 425 miles per second; in 1952 the same lines gave a measurement of 536 miles per second. The distance of the star is not accurately known. If the actual luminosity at maximum was near -8, a computed distance of about 25,000 light years is derived; if it was closer to -6, the figure is reduced to about 10,000 light years. The true luminosity is uncertain, but, from a study of many different novae, it seems that a typical slow nova is 1 or 2 magnitudes fainter than one of the very fast types. A maximum luminosity of about -6 seems reasonable for RR Telescopii.

The light curve of this enigmatic star resembled that of RT Serpentis, which had an even slower "outburst", remaining at maximum for more than ten years. In the case of RR Telescopii, however, the periodic variations before the outburst suggests a connection with stars like the odd R Aquarii and Z Andromedae. Unfortunately the star was not well observed before its nova-like maximum and it is not even certain that the 387-day period represents the variations of a Mira-type red giant. The star's observed rise of 9 magnitudes corresponds to a light increase of about 4000 times, equal to some of the full-scale novae. Should the star be a close pair like R Aquarii, the true range of the nova component probably exceeded 10 magnitudes. RR Telescopii has also been compared to the puzzling star Eta Carinae, which, however, had a much higher actual luminosity and a much slower spectroscopic development. The two stars seem to be operating on time-scales which differ by a factor of about 10.

A study by A.D.Thackeray at Radcliffe has shown that the rich emission-line spectrum of the star has "increased steadily in ionization level with time....Between about 1951 and 1960 the spectrum also displayed much broader emission bands resembling those in Wolf-Rayet stars..." (Refer also to RT Serpentis)

LIST OF DOUBLE AND MULTIPLE STARS

NAME	DIST	PA	YR	MAGS	NOTES	RA & DEC
Σ158	2.1	267	62	8½- 9	cpm; PA slow inc;	01439n3255
	55.0	256	12	- 12½	spect F8	
	100	70	12	- 11½		
Σ183	0.2	217	66	7½- 8½	AB PA inc, spect F2	01523n2833
	5.6	164	40	- 8½	AC relfix	
A819	0.6	170	58	8 - 9	PA inc, spect F5	01541n3047
	66.5	270	17	- 9½		
Σ197	28.1	234	23	7½- 8½	optical; dist inc; spect F5	01581n3505
ε	3.9	118	35	5 - 11	(Σ201) (3 Tri) relfix, cpm; spect A2	02000n3303
Σ219	11.6	183	35	8 - 9	relfix, spect A0	02073n3308
ι	3.8	72	61	5 - 6½	(6 Tri) (Σ227) cpm; PA slow dec, Spect G5, F6 (*)	02095n3004
Σ232	6.5	246	55	7½- 7½	neat pair, relfix; cpm; spect B8	02118n3010
Σ239	14.0	211	61	7 - 8	cpm, relfix; spect F5	02145n2831
Σ246	9.2	123	55	7½- 8½	cpm, relfix; spect F8	02156n3416
Σ269	1.6	345	61	7½- 10	slight PA inc, spect G0, near 13 Tri.	02252n2938
h653	23.0	42	25	7½- 11	relfix, spect K2	02306n3111
Σ285	1.6	166	66	7 - 7½	PA slow dec, spect K0	02359n3312
Σ286	2.9	254	38	8- 10½	cpm, relfix; spect G5	02368n3341
Σ310	2.6	91	38	7½- 11	cpm, relfix; spect A2	02464n3344

TRIANGULUM

LIST OF VARIABLE STARS

NAME	MagVar	PER	NOTES	RA & EC
R	5.7--12.5	266	LPV. Spect M4e--M8e	02340n3403
S	8.7--12.4	248	LPV. Spect M2e	02243n3231
T	9.9--14..	323	LPV. Spect M8e	01537n3347
W	7.3--8.8	108	Semi-reg; spect M5	02384n3418
X	8.9--11.5	.97153	Ecl.Bin; spect A3+G3	01577n2739
RS	9.6--10.5	1.9089	Ecl.Bin; spect A5	01320n2920

LIST OF STAR CLUSTERS, NEBULAE AND GALAXIES

NGC	OTH	TYPE	SUMMARY DESCRIPTION	RA & DEC
598	M33	⊝	Sc; 6.5; 60' x 40' !! B,eL,R,vglbM; One of the nearest spirals; "Pinwheel Galaxy" (*)	01311n3024
670	611[2]	⊝	E5; 13.0; 0.9' x 0.5' F,S,1E	01445n2738
672	157[1]	⊝	SBc; 11.6; 4.5' x 1.7' F,pL,mE; coarse spiral	01450n2711
750	222[2]	⊝	E1; 13.0; 0.4' x 0.3' cB,pL,R; Contact pair with 751 = E0; 13.8; 0.2'; Separation 0.4'	01546n3258
777	223[2]	⊝	E1/E2; 13.0; 0.9' x 0.7' pB,pL,1E; glbM	01573n3112
890	225[2]	⊝	E4/S0; 12.7; 1.0' x 0.5' B,S,E, bM	02191n3302
925	177[3]	⊝	Sb/Sc; 12.0; 9.4' x 4.0' cF,cL,mE; coarse spiral	02243n3322
949	154[1]	⊝	Sb; 12.8; 1.1' x 0.4' cB,L,E,vgbM; compact center	02276n3656

DESCRIPTIVE NOTES   (Cont'd)

ALPHA Name- CAPUT TRIANGULI, the Latin equivalent of the earlier Arabic *Ras al Muthallath.*
Position 01502n2920; Magnitude 3.49; Spectrum F6 IV. From direct trigonometrical parallaxes the distance of the star is about 65 light years, and the actual luminosity about 13 times that of the Sun (absolute magnitude +2.0). Alpha Trianguli shows an annual proper motion of 0.23 in PA 178°; the radial velocity is 7 miles per second in approach. The star is also a close spectroscopic binary with a period of 1.73652 days and an eccentricity of 0.12.

BETA Magnitude 2.99; Spectrum A5 III; Position 02066n2542. The computed distance of the star is about 140 light years; the actual luminosity about 90 times that of the Sun, and the absolute magnitude about -0.1. The star shows an annual proper motion of 0.16"; the radial velocity is about 6 miles per second in recession. Beta Trianguli is a double-line spectroscopic binary with a period of 31.4009 days and an eccentricity of 0.46; the computed semi-major axis of the orbit of the brighter star is about 6 million miles.

IOTA Position 02095n3004; Magnitude 5.05; Spectra G5 III + F6 V. An attractive double star with a noticeable color contrast, probably first observed by Herschel in 1781, and measured repeatedly since the time of Struve in 1836. The components form a binary in slow retrograde motion, with a gradual decrease in PA of about 7° per century. The separation has remained for many years at 3.8", and the two stars differ in brightness by 1.4 magnitudes. Admiral Smyth thought the color contrast of this pair "exquisite"; they are usually seen as strong yellow and pale blue, though E.J.Hartung refers to Iota Trianguli as "a bright golden yellow pair". C.E.Barns has the colors reversed, apparently, since he calls them "sapphire and gold".
Both components of this fine pair are double-line spectroscopic binaries; the primary pair has a period of 14.732 days and an eccentricity of 0.04; the fainter star has a period of 2.236 days and an eccentricity of 0.01. The actual dimensions of the system are somewhat uncertain but the computed semi-major axis of the orbit of the G5-

star is about 7 million miles, and of the F6 star about
1.8 million miles. Figures obtained in this way are computed on the assumption that the orbit is seen nearly edge-
on; in the case of most spectroscopic binaries the true
inclination remains unknown, so the correction factor also
remains unknown. A system which shows stellar eclipses, of
course, must be oriented edge-on, but in all other cases
the apparent tilt might be anything from nearly edge-on
to nearly face-on.

The computed distance of the Iota Trianguli system
is about 200 light years; the visual pair then has a projected separation of about 250 AU, and the total luminosity must be about 30 suns. The whole system shows an
annual proper motion of 0.08"; the radial velocity is 11
miles per second in approach.

M33   (NGC 598)   Position 01311n3024. This is the
            great "Pinwheel Galaxy", one of the bright
members of the Local Group of Galaxies, and probably the
nearest spiral galaxy after the Andromeda system M31. It
lies on the extreme western portion of the constellation,
about 14° SE from M31, and a little less than 7° SE from
the bright star Beta Andromedae. M33 was one of the
notable discoveries of Messier, found in August 1764, and
described as a "whitish light of almost even brightness.
However, along two-thirds of its diameter it is a little
brighter. Contains no star. Seen with difficulty in a 1-
foot telescope..." Messier thought the apparent size to
be about 15'. J.E.Bode, in 1775, saw it as "a faintly
illuminated nebulous patch of disorderly shape" while Sir
William Herschel recorded a "mottled aspect" in his great
reflector. John Herschel remarked on the rather low surface
brightness of the object, and found it "only fit for low
powers, being actually imperceptible from want of contrast
with my 144 X". The same remark is credited to T.W.Webb.

Admiral Smyth described M33 as "A large, distinct,
but faint and ill-defined, pale white nebula, with a bright
star a little N.p. and five others following at a distance;
between them and the nebula there is an indistinct gleam
of mere nebulous matter..." "Very large, faint, and ill-
defined," says T.W.Webb, "Visible from its great size".
Lord Rosse was probably the first to detect some sign of

THE PINWHEEL GALAXY. M33 in TRIANGULUM is one of the large
members of the Local Group. Lowell Observatory photograph
made with the 13-inch telescope.

the spiral structure; using his great 6-foot reflector at Birr Castle, Parsonstown, Ireland, he found a pattern like an "S" in the center, which he ascribed to the crossing of two similar curves; the arms of the "S" were described as "full of knots" and mottled patches of luminosity. In the *Publications* of Lick Observatory, Volume XIII, the galaxy is described as one of the most beautiful spirals known. "With its faint extensions it covers an area of at least 55 x 40'. It is uncertain whether there is an actual stellar nucleus. A multitude of stellar condensations in the whorls. Best example of resolution into stars".

M33 is a challenging object for the small telescope because of its great size and rather low surface brightness. Among amateur observers it has been the source of the most discordant reports to be found in astronomical literature. While some find it easily visible in field glasses, or even to the naked eye, others report complete inability to locate the galaxy at all, and conclude that its position must be incorrectly charted. In the majority of the cases of failure, the observer is looking for a much smaller and brighter object, rather than a dim glow comparable in apparent size to the Moon. Sky conditions are also critical, and inexperienced observers often use too high a power on small instruments. Faint and extended luminous objects are visible only because of contrast against the sky background; dark skies and wide-angle oculars are essential. For a 2-inch glass, a magnification of about 15X is sufficient; about 30X may be used on a 6-inch glass. To secure these low powers, a wide-field Erfle eyepiece of about 1½ inch focal length will be found ideal, but for the budget-minded beginner a single double-convex lens of 1 inch diameter and 1½ or 2-inch focal length will serve nearly as well. Good binoculars will also be found ideal for locating large but faint objects of the M33 class. The author's favorite is a pair of 20 X 70s, but he finds M33 easily visible in a good field glass. From a truly dark sky site, it may be glimpsed faintly without optical aid; this has been independently confirmed by Walter Scott Houston, Leslie C. Peltier, and other experienced observers.

While the visual diameter is somewhat over half a degree, the best photographs show details out to a full size of about 60' X 35'. The total photographic magnitude

is 7.8, the visual magnitude about 6.0. With a good 6-inch
glass, the observer will notice a distinct nuclear mass,
but without the sharp stellar nucleus displayed by the M31
system in Andromeda. The two galaxies are at fairly similar
distances; M33 is generally thought to be the nearest of
the visible spirals after M31, at a distance of 2.4 million
light years. It is also, evidently, one of the most distant
members of the Local Group, and is both a smaller and more
irregular spiral than the great M31.

With modern telescopes, M33 is possibly the easiest of
all galaxies to resolve; masses of star clouds and even
bright individual stars show clearly on exposures made with
12-inch reflectors. With the world's great telescopes, M33
is revealed as a huge double-armed spiral of star clouds
curving about the bright nuclear mass; the arms are thick
and clumpy, dotted with nebulous regions, and the spiral
structure is loose and irregular when compared with such
symmetrical galaxies as M81 in Ursa Major. Allan Sandage,
in the *Hubble Atlas of Galaxies*, calls attention to the
fact that M33 is the nearest of the Sc spirals, and that
the stellar content can therefore be studied down to stars
of absolute magnitude -1.5. The first detailed study of the
stars of the Pinwheel was made by E.Hubble in 1926; among
the known members we find high luminosity O and B stars,
open and globular star clusters, cepheids and irregular
variables, novae, and fields of bright nebulosity. Spiral
arms are well resolved into masses of stars, most of which,
according to Sandage, are blue supergiants, "but there are
also at least 3000 red supergiants of absolute magnitude =
-5 which are similar to those found in h and χ Persei".
Owing to the large numbers of B-type supergiants, the light
of M33 is "bluer" than that of many galaxies; the integrated spectral type is about A7, and the measured color index
is close to +0.2. In the northern sky, the galaxy most
closely resembling M33 is the nearby NGC 2403 in Camelopardalis, thought to be a member of the M81 group which is
centered in Ursa Major.

The major diameter of 60' corresponds to 42,000 light
years at the accepted distance; microdensitometer tracings
increase this to 90' or about 60,000 light years. From a
study by E.Holmberg, the total mass, out to the visible
radius, is some 8 billion times the solar mass, and the

THE PINWHEEL GALAXY shows much detail on this 12½-inch
reflector photograph, made with a cooled emulsion camera
by Evered Kreimer of Prescott, Arizona.

DESCRIPTIVE NOTES (Cont'd)

total light is about 3 billion times that of the Sun, with
a computed absolute magnitude of about -19. The rotation of
the Pinwheel is clockwise, with the northern arms moving
slowly to the right or west; the rotational period in the
region of the rim has been measured at about 200 million
years.

It was not until the early years of the 20th Century
that telescopes became sufficiently powerful to resolve
individual stars in M33. In 1922 the first variable star in
the system was identified; within four years some 25 stars
of the cepheid class, many irregular variables, and two
novae had been found. Studies of the cepheids made possible
a distance estimate for the system, which was first given
as about 750,000 light years. As in the case of the great
Andromeda Galaxy this estimate was more than doubled by
later studies, but was at least sufficient to demonstrate
that both M31 and M33 were actually external galaxies. M33
is one of the few galaxies that does not show a red shift
of its spectral lines; M.L.Humason gives the measured value
as 117 miles per second in approach, but the correction for
the motion of the Sun in our own rotating galaxy reduces
the figure to a mere 7 miles per second in approach. It is
interesting to note that M33 and M31 are only about 570,000
light years apart, four times closer than either is to our
own Milky Way.

The Pinwheel continues to be the subject of modern
investigations. In a study made at the Asiago Observatory
in 1972, L.Rosino and A.Bianchini discovered five novae in
the galaxy which had appeared in the interval between 1960
and 1972. In a radio investigation of M33, made in 1971 at
Green Bank with the 300-foot paraboloidal antenna, the
system was found to contain at least 1.6 billion solar masses of neutral hydrogen, in addition to the approximately 80
emission nebulae already identified in the spiral arms. Of
these the most prominent is NGC 604, a giant nebulosity and
star association on the NE edge of the galaxy, 10' out from
the center in PA 50°; it is the bright spot at the upper
left edge of the galaxy in the photograph on page 1901. One
of the largest "H II" regions known in any galaxy. This
emission nebulosity measures about 320 parsecs in diameter
(about 1000 light years) and shows a spectrum resembling
that of the Orion Nebula. (See also M31 in Andromeda)

CENTRAL PORTION OF M33. The heart of the Pinwheel Galaxy,
from a plate made with the 60-inch reflector at Mt.Wilson
Observatory.

TRIANGULUM AUSTRALE

LIST OF DOUBLE AND MULTIPLE STARS

NAME	DIST	PA	YR	MAGS	NOTES	RA & DEC
T	41.3	41	00	7 - 10	(Hd 243) Spect A0;	15050s6832
	55.0	311	00	- 11	Primary suspected	
					variable	
I 332	1.1	107	29	6½- 9	Spect B3	15160s6718
I 240	2.4	190	27	7- 10½	cpm, spect B9	15317s6457
B842	0.6	263	60	7½- 8½	no certain change,	15360s6341
					spect A0	
L6477	1.9	149	47	6½- 6½	(Rmk 20) cpm, PA	15436s6517
					dec, spectra both	
					A5	
h4809	1.2	97	43	6½- 8½	(△194) (Slr 11)	15507s6036
	45.0	48	17	- 9	AB slight PA inc,	
	48.1	257	17	- 8½	spect B8	
B854	9.7	284	27	6 - 12	spect B8	15543s6454
R275	3.8	349	37	8½- 9	relfix, spect A0	16146s6427
(	19.6	16	18	5½- 10	(△201) Optical,	16233s6356
					PA & dist dec,	
					spect dF4	
I 336	1.2	198	46	8 - 8½	spect B9	16273s6210
△203	27.4	262	30	8½- 8½	Optical, Pa inc,	16287s6048
					dist dec, spect	
					A0, F2	
Rst 5063	1.4	147	42	7 - 12½	spect A0	16414s6135
Gls 230	7.0	205	20	8½- 8½	relfix, spect G5	16464s6729

TRIANGULUM AUSTRALE

LIST OF VARIABLE STARS

NAME	MagVar	PER	NOTES	RA & DEC
R	6.0--6.8	3.389	Cepheid; spect F6--G4	15153s6619
S	6.1--6.7	6.323	Cepheid; spect F6--G2	15567s6338
T	7.0--		Visual double; variability uncertain; Spect A0	15050s6832
U	7.5--8.3	2.568	Cepheid; spect F4--F8	16029s6247
V	8.3--9.0	Irr	Spect N	16450s6742
W	8.1--11..	249	LPV. Spect Me	16467s6754
X	6.2--7..	Irr	Spect N	15095s6954
Z	8.4--10..	151	LPV. Spect M5e--M6e	16499s6507
RS	8.5--10..	436	LPV.	16176s6120
RT	9.4--10.1	1.9461	Cepheid; spect G2	16299s6302
RU	8.3--12..	327	LPV. Spect M4e	16332s6826
EN	7.7--8.1	36.90	Cepheid; spect G5	14525s6838
EP	9.4--9.9	2.1416	Ecl.Bin; Spect A0	15450s6407
EQ	8.3--8.9	2.7095	Ecl.Bin; spect F5	16052s6602

LIST OF STAR CLUSTERS, NEBULAE AND GALAXIES

NGC	OTH	TYPE	SUMMARY DESCRIPTION	RA & DEC
5979		◎	Mag 13, Diam 8" vS,R, appearance nearly stellar	15434s6102
6025	△304	⁘	Mag 6, Diam 10'; about 30 stars mags 7.... Class D	15594s6022

TRIANGULUM AUSTRALE

DESCRIPTIVE NOTES

ALPHA   Name- ATRIA. Position 16433s6856; Magnitude
1.91; Spectrum variously given as K2, K3, or
K4, luminosity class III. The star is a yellow giant some
80 light years distant, and about 90 times the solar luminosity (absolute magnitude about -0.1). The star shows an
annual proper motion of 0.04"; the radial velocity is 2.5
miles per second in approach.

The modern name *"Abraham's Star"* is based on the
title *The Three Patriarchs* for Alpha, Beta, and Gamma; the
name, according to R.H.Allen, was first suggested by the
Dutch writer P.Caesius in the 17th Century. This Southern
Triangle, with sides of 8°, 8° and 6½°, is a larger and
also more nearly equilateral triangle than its northern
counterpart. Its introduction as a constellation was probably first suggested by Pieter Theodor early in the 16th
Century, though, as Allen says, it first appeared in print
in Bayer's famous *Uranometria* of 1603.

BETA   Position 15507s6317; Magnitude 2.84; Spectrum
F2 IV or V. The star lies some 8° NW from
Alpha, and is apparently about half the distance of the
brighter star, with a computed distance of about 42 light
years. This gives it an actual luminosity of about 10 suns
and an absolute magnitude of +2.3. Beta Tri.Aust shows a
fairly large proper motion of 0.45" annually in PA 205°;
the radial velocity is close to zero.

The cepheid variable S Tri.Aust will be found in the
field, about 35' distant to the ESE; this 6th magnitude
star has a range of about 0.6 magnitude in a cycle of 6.323
days.

GAMMA   Position 15142s6830; Magnitude 2.88; Spectrum
A0 or A1p, luminosity class V. Gamma is probably the most distant of the *Three Patriarchs*, at a computed distance of about 115 light years. It is also the bluest of the three stars, with an A-type spectrum showing
abnormally strong lines of the rare element europium. The
computed luminosity is about 70 times that of the Sun and
the absolute magnitude about +0.2. The star shows an annual
proper motion of 0.07"; the radial velocity is close to
zero. The easy double and suspected variable T Tri.Aust
lies 1° distant to the west.

LIST OF DOUBLE AND MULTIPLE STARS

NAME	DIST	PA	YR	MAGS	NOTES	RA & DEC
B554	4.3	110	30	7 - 12	Spect M	22128s5749
I 20	0.3	243	59	7½- 8	PA dec, spect F5	22146s6304
δ	7.0	282	28	4½- 9	(h5334) cpm relfix, spect B8	22238s6513
I 137	2.0	335	43	7 - 10	spect A2	22241s5815
I 340	0.9	359	49	6 - 9	PA dec, spect K0	22490s6327
Hu 1643	1.8	12	43	7- 11½	PA inc, spect A3	22580s5842
h5373	47.2	96	16	7½- 10	(△244) slight dist dec, spect F0	22591s6434
B590	0.1	355	28	7½- 8	spect G0	23055s6408
△ 245	13.7	291	20	7 - 9	Slight dist dec, spect F2	23056s6000
I 143	3.1	9	43	7- 10½	no certain change, spect K2	23126s5958
△247	46.6	291	20	7 - 8	Optical pair, dist inc, spect G5, A3; color contrast	23151s6116
Hu 1648	1.9	283	34	7- 11½	relfix, spect A3	23235s6322
h5402	36.3	198	17	7 - 9	spect G0	23280s6922
h5403	37.7	45	17	7½- 10	spect F0	23324s6457
h5415	38.2	126	17	7½- 11	spect G5	23405s7106
B1020	0.3	213	49	8 - 8½	spect A3	23429s5833
I 695	8.3	23	14	7 - 13	spect K0	23446s7135
Cor 261	5.6	100	30	8 - 8½	relfix, spect K0	23449s6047
B626	7.5	258	27	7½-13½	spect A3	23535s6309
Gls 289	3.7	267	38	7½- 9	relfix, spect F5	23580s6657
I 43	0.5	320	31	7 - 8½	PA & dist dec, spect A5, G	00082s7330
β1-β2	27.1	169	52	4½- 5	Fine pair; cpm, spect B8, A2	00293s6314
β1	2.4	151	32	4½- 13	spect B8	
β2	0.2	2	26	5 - 6½	(I 260) Binary, 43 yrs; PA dec, spect A2	
β3	0.1	143	54	6 - 6	cpm with β1 and β2 at 10' S.f.; spect A2	00305s6318
I 440	0.1	354	44	7½- 8	PA inc; spect A3; in field of Rho Tucanae	00405s6553

LIST OF DOUBLE AND MULTIPLE STARS   (Cont'd)

NAME	DIST	PA	YR	MAGS	NOTES	RA & DEC
Cor 3	2.4	68	42	6½– 8½	spect F5	00424s6247
Δ 2	20.7	81	52	6½– 7½	cpm, spect F8; in field of Lambda Tucanae	00505s6947
I 48	0.5	339	33	7½– 8½	spect G0	00559s6650
h3416	5.0	128	51	7½– 7½	neat relfix pair; spect F5	01013s6022
Gls 9	3.9	355	28	7 – 10	relfix, spect A0; on edge of Small Magellanic Cloud	01085s7314
I 27	1.0	317	26	8 – 8½	Binary, 81 yrs; PA inc, spect G5; cpm with κ Tuc at 320"	01133s6905
κ	5.4	336	54	5 – 7	(h3423) PA dec, spect F6. Multiple system with I 27.	01140s6909
h3426	2.5	337	28	6½– 9½	cpm, slight PA dec, spect A0	01153s6640
I 263	0.6	233	59	8 – 8½	PA inc, spect F2	01205s6959

LIST OF VARIABLE STARS

NAME	MagVar	PER	NOTES	RA & DEC
θ	6.06--6.15	.052	δ Scuti type? Spect A4m	00313s7133
R	9.5--14.5	286	LPV. Spect M5e	23548s6540
S	8.7--14..	241	LPV. Spect M3e--M5e	00208s6157
T	7.7--13.8	250	LPV. Spect M3e--M6e	22373s6149
U	8.3--14..	259	LPV. Spect M3e--M5e	00557s7516
TZ	9.6--15..	239	LPV. Spect M3e	00512s7009
UU	9.2--12..	335	LPV. Spect M4e	22190s6108
AN	9.9--11.0	5.4609	Ecl.Bin; Spect A5	23276s5842
BQ	5.6--5.8	Irr	Spect M5	00516s6309
BS	7.43--7.57	.065	δ Scuti type, Spect A5	01061s6208

TUCANA

LIST OF STAR CLUSTERS, NEBULAE AND GALAXIES

NGC	OTH	TYPE	SUMMARY DESCRIPTION	RA & DEC
104	△18	⊕	(47 Tucanae) !!! Mag 4½; Diam 25'; Class III; stars mags 11...... vB,vL,eRi, vmCM; Splendid cluster (*)	00219s7221
292	S.M.C.	⊖	Small Magellanic Cloud; Irr Galaxy, member of Local Group; Total Mag 1.5; Diam 3½° (*)	00500s7330
362	△ 62	⊕	Mag 6, Diam 10'; Class III; vB,vL,vmC; stars mags 13...	01006s7107
406		⊖	Sc; 12.9; 3.5' x 1.5' F,vL,E	01058s7009
434		⊖	SB/ S0; 13.0; 1.5' x 1.0' B,S,1E,psbM, BN	01102s5831
7205		⊖	On Indus-Tucana border Refer to Indus	22051s5740
7329		⊖	SBb; 13.0; 2.5' x 1.5' pB,pS,mE	22370s6644

1909

ALPHA Position 22151s6031. Magnitude 2.85; Spectrum K3 III or IV. The star lies near the western border of the constellation, about 20° distant from the Small Magellanic Cloud. Tucana itself was introduced as a constellation by Bayer in 1603, honoring the large -beaked bird of tropical South America. The name sometimes appears as *Toucana*, as on the star charts of E.H.Burritt. According to R.H.Allen it was referred to by both Kepler and Riccioli as *Anser Americanus*. Alpha Tucanae itself has no common proper name.

The star is a yellow K-type giant or subgiant with a moderate luminosity of about 20 suns. From direct parallaxes the distance appears to be about 62 light years. The star shows an annual proper motion of 0.08" and a radial velocity of 25 miles per second in recession.

Alpha Tucanae is a spectroscopic binary whose orbital elements have also been confirmed by astrometric plate measurements. The period is 4197.7 days, or close to 11.5 years, and the orbital eccentricity is 0.38 with periastron in December 1909. According to a study by S.Jones (1928) the semi-major axis of the orbit of the visible star is about 240 million miles, or about 0.05", suggesting that the companion might be within range of visual detection when the pair is near maximum separation.

NGC 104 (47 Tucanae) Globular star cluster. Position 00219s7221, about 2½° to the west of the Small Magellanic Cloud. This is a magnificent globular star cluster, usually regarded as the finest in the heavens with the single exception of the great Omega Centauri. It seems to have been first noticed by the Abbe de Lacaille in 1755, during the course of his observations at the Cape of Good Hope. Unfortunately, this splendid object is situated so far south that it is never observable in the United States; observers in the Southern Hemisphere will find it easily visible to the naked eye as a hazy star-like object of magnitude 4½, only slightly inferior to Omega Centauri itself. Miss Agnes Clerke in *The System of the Stars* (1905) gave a striking description of her impressions of 47 Tucanae:

"The loveliness of the cluster 47 Toucani near the Lesser Magellanic Cloud was, to Herschel's view, set off by a diversity of colour between an interior mass of rose-

GLOBULAR STAR CLUSTER NGC 104 in TUCANA. One of the finest
of the globulars. The photograph was made with the 60-inch
reflector at Harvard's Boyden Station in South Africa.

1911

DESCRIPTIVE NOTES   (Cont'd)

*tinted stars and marginal strata of purely white ones. But
the effect was doubtless subjective; it met with no later
recognition; and to the present writer, in 1888, the sheeny
radiance of this exquisite object appeared of uniform quality from centre to circumference.....Perhaps no other cluster exhibits an equal degree of compression. Within a
sphere of 11' radius are included nigh upon 10,000 stars,
of which 5019 have been actually counted....The blankness
of the surrounding sky renders 47 Toucani all the more obvious to unaided sight; it was, indeed, for several nights
after his arrival in Peru, mistaken by Humboldt for a comet. Only eight variables have been detected in it - a scan-
ty gleaning compared with the rich harvests gathered in ω
Centauri, and in the starry spheres M5 and M3, situated in
Serpens and Canes Venatici respectively..."*

Amateur telescopes of 4-inch aperture or larger will
achieve partial resolution of NGC 104, as the brightest star
images are magnitude 11½. In larger instruments the cluster
becomes a shimmering globe of thousands of star points,
crowding toward a rich central blaze. The diameter is about
25' visually, but on the best photographic plates the full
size is close to 45'. H.B.S.Hogg, in her *Bibliography of
Individual Globular Clusters* (First Supplement 1963) gives
the total integrated magnitude as 4.68, the integrated
spectral type as G3, and the mean magnitude of the 25
brightest stars as 13.44. The cluster, unlike many other
globulars, shows a fairly small radial velocity, of about
15 miles per second in approach.

NGC 104 is definitely among the nearest of the globular clusters, though the precise distance is a little uncertain. Shapley in 1930 derived a distance of about 22,000
light years, but more recent studies suggest a value somewhere between 15,000 and 20,000 light years; evidently the
cluster is at very nearly the same distance as the huge
Omega Centauri. It is also among the intrinsically largest
and most brilliant of the globulars. Accepting a distance
of about 16,000 light years as the probable value, the true
diameter is found to be about 210 light years, and the true
luminosity about 270,000 times that of the Sun. Among the
brightest members are red giant stars with absolute magnitudes of about -2; the total light emitted by the cluster
is close to absolute magnitude -8.8.

DESCRIPTIVE NOTES   (Cont'd)

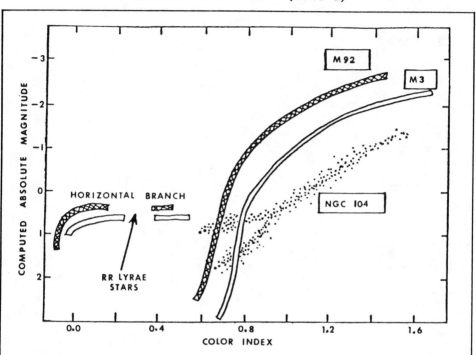

NGC 104 is, in several ways, a rather unusual globular star cluster. The integrated spectrum shows a much higher abundance of the atoms of the metals than is normal in typical globulars. Regarded as an effect of the cluster's history or evolution, this may indicate that NGC 104 is a much younger globular than usual. The lack of bluer stars which usually form the so-called "horizontal branch" near absolute magnitude +0.5 on the color-magnitude diagram is another peculiar feature. The extreme scarcity of RR Lyrae type variables is a third peculiarity which may be connected with the lack of the "horizontal branch" in general. Only two suspected members of this class have been detected in NGC 104, with periods of 0.7365 and 0.3714 day. Several irregular red variables have been identified, and five stars of the Mira class with periods ranging from 155 to 212 days. About a dozen other faint probable variables remain to be studied and classified. (Refer also to Omega Centauri, M13 in Hercules, M22 in Sagittarius, M3 in Canes Venatici, and M5 in Serpens. A brief outline of cluster age-dating will be found on pages 990- 993)

THE SMALL MAGELLANIC CLOUD. One of the companion galaxies
to the Milky Way. The photograph was made with Harvard's
Bruce telescope at Arequipa, Peru.

1914

THE SMALL MAGELLANIC CLOUD (NUBECULA MINOR) (NGC 292) A famous irregular galaxy and nearby member of the Local Group, centered at about 0h 50m, $-73°30'$, and visible to the naked eye as a misty cloud some $3\frac{1}{2}°$ in diameter. It forms a pair with the Large Magellanic Cloud, about 22° distant in the constellation Dorado. The two clouds are the nearest of the visible external galaxies, and are sometimes regarded as satellite galaxies of the Milky Way. The distances are about 190,000 light years for the Large Cloud, and 200,000 light years for the Small Cloud. The actual separation of the two systems is some 80,000 light years.

On the best photographic plates the dimensions of the Small Cloud increase to about 9° x 8°, giving the over-all diameter as about 30,000 light years. The bright central mass measures some 10,000 light years across, and the total integrated magnitude is in the neighborhood of -17 absolute, a total luminosity of at least 600 million suns. Radio observations reveal the existence of a great gaseous envelope of hydrogen which surrounds both clouds; the two systems are further linked by a scattered population of stars and clusters. Both clouds are flattened systems which seem to be inclined considerably from the face-on position; in the case of the Small Cloud the orientation is probably about 30° from the edge-on position.

The flattening of both clouds suggests rotation, which reveals itself, in the Large Cloud, at least, by the faint indication of a spiral pattern. The evidence for such a pattern in the Small Cloud is not very clear; an extension or "arm" has been detected which reaches out toward the Large Cloud, but this feature appears to be a tidal effect or a portion of a "bridge" linking the two systems, rather than an incipient spiral arm. The two Clouds may well be in slow orbital motion about each other, and about our own Galaxy also, though the periods would be a matter of many millions of years. The Small Cloud shows a recession velocity of nearly 100 miles per second; the Large Cloud about 170. Both of these measurements are heavily affected by the rotation of our own Galaxy; applying the proper correction we find that the Large Cloud is nearly stationary with respect to the Milky Way, while the Small Cloud is receding at nearly 50 miles per second.

DESCRIPTIVE NOTES (Cont'd)

The Large Magellanic Cloud is unquestionably the richest astronomical treasure-house of the southern sky, containing such magnificent objects as the Great Looped Nebula or "Tarantula Nebula" NGC 2070, and the most luminous individual star known, S Doradus. By contrast, the Small Cloud remains somewhat neglected, overshadowed by the glory of its larger neighbor. It contains few regions of bright nebulosity, and its stellar population is noticeably less flamboyant. Through the telescope it is resolved into immense numbers of stars, and resembles a Milky Way star cloud. At least a million of its stars are brighter than Sirius, while the fainter members are quite literally beyond counting. As a standard of comparison it should be remembered that our Sun, at the distance of the Small Cloud, would appear as a star of nearly 24th magnitude, and would be just at the limit of detectability with the greatest telescopes in the world.

The number of star clusters in the Small Cloud is not definitely known, since no survey yet made can be regarded as reasonably complete. In a list published in 1956, G.E. Kron catalogued 69 clusters which could be identified on plates made with Harvard's "ADH" Baker-Schmidt telescope in South Africa. Fifteen of these were regarded as globular clusters, the others as galactic clusters and associations of varying degrees of concentration. Among the better known and larger globulars is NGC 419, with a diameter of about 115 light years, and NGC 121 which lies about 2½° distant from the main body of the Cloud.

In the photograph opposite we see several of the brightest clusters and diffuse nebulosities of the Cloud. The bright mass near center is NGC 346, lying about 1° NE of the main mass of the Cloud; it is a great nebulous star cluster measuring about 4' visually, with extensions to 5' X 10'; the true dimensions must be about 580 X 290 light years. This surpasses in size and splendor many of the best known diffuse nebulae in our own galaxy. Above it on the print is the large cluster NGC 371, nearly 300 light years in diameter, and containing myriads of giant stars; the smaller cluster near it is NGC 395. Near the bottom of the photograph is still another group, NGC 330, about 1' across and described by E.J.Hartung as "a very bright knot of stars, irregularly round.....in a field sown with faint

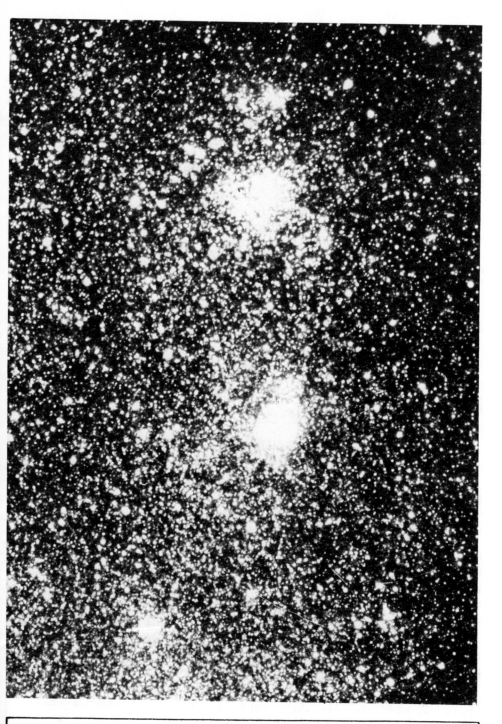

STAR CLUSTERS IN THE SMALL MAGELLANIC CLOUD. NGC 346 is just below center; NGC 371 appears above. The photograph was enlarged from the print on page 1914.

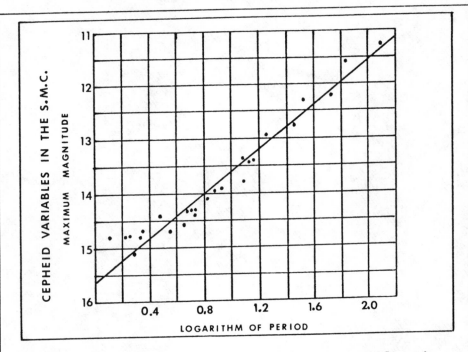

stars; it is well resolved with sufficient magnification and 7.5 cm shows it plainly with some star sparkle.....The stars must be very luminous to be evident in such a remote object".

Another interesting region is the field of the chain of clusters called NGC 456- 465, some 2° east of the main mass of the Cloud, and shown on the photograph opposite. Some nebulosity exists in this stellar association which measures about 20' in length, over 1100 light years in actual extent.

Special attention must be given to the pulsating variable stars of the Small Magellanic Cloud, since their discovery opened a new age in astronomical research. The first such stars were found by Henrietta Leavitt of Harvard in the opening years of the 20th Century, after the Bruce 24-inch refractor had been placed in operation at Arequipa, Peru. A study of the Magellanic Clouds was begun in 1899 and many variable stars were detected and catalogued; close to 1000 were known by 1906. As these stars were studied in detail, a peculiar relationship between their periods and apparent brightness was noted, illustrated by the graph at

STELLAR ASSOCIATION IN THE SMALL MAGELLANIC CLOUD. This is
the group numbered NGC 456- 465, east of the main mass of
the Cloud. Harvard Bruce telescope photograph.

the top of page 1918. The periods ranged from a day or so to over a month, but the apparently brighter stars showed the longer periods. Since the stars of the Small Cloud were virtually all at the same distance, it became evident that the relationship must be a real one. The stars were later identified as cepheid variables, pulsating giants of the Delta Cephei type. The period-luminosity relation of these stars was an important discovery in astronomy since it gave a key to the measurement of vast stellar distances; if the true luminosity of a cepheid can be determined from the observed period, then the star may be used as a distance indicator. The period-luminosity relation was first announced at Harvard in 1912, and a preliminary calibration of the cepheid rule had been made by Harlow Shapley by 1917. Within a few years, such stars had been identified in the Andromeda spiral M31, allowing the distance to be determined, and revealing once and for all that the spirals were actually other galaxies at immense distances. (Refer to Delta Cephei, page 583; M31 in Andromeda, page 129, and the Large Magellanic Cloud in Dorado, page 837)

Two very fine globular clusters will be found in the vicinity of the Small Magellanic Cloud: NGC 104 lies about 2½° to the west, and NGC 362 is about the same distance to the north. These are not truly associated with the Cloud, however; they are foreground objects located in our own Galaxy.

PROMINENT CLUSTERS IN THE SMALL MAGELLANIC CLOUD			
NGC	OTH	DESCRIPTION	RA & DEC
121		Small globular cluster	00244s7148
330	△23	Galactic cluster 1' diam	00545s7244
346	△25	Large nebulous cluster 4' diam= over 500 lt.yrs.	00574s7227
371	△31	Cloud of stars 5' diam	01019s7220
376	△36	Small irregular cluster diam 0.5'	01023s7305
419		Globular cluster 1½' diam	01068s7309
456	△ 7	Large galactic cluster, portion of 20' chain of clusters	01124s7334

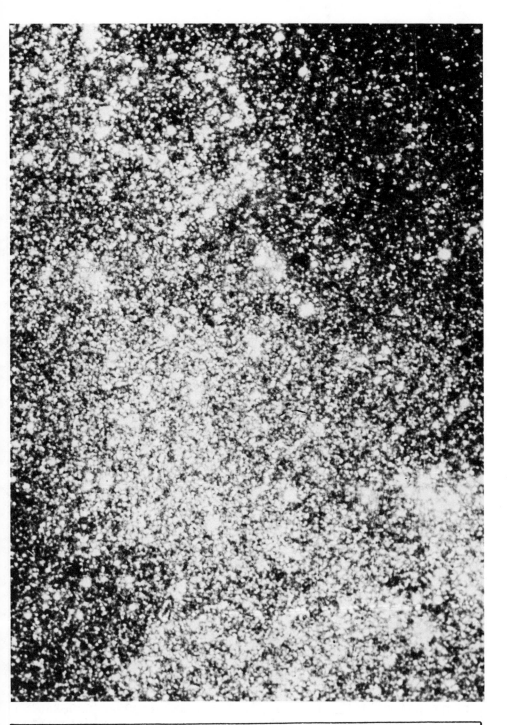

HEART OF THE SMALL MAGELLANIC CLOUD. A dense star field in the main mass of the Cloud; this view resembles a portion of the Milky Way. Enlarged from a Harvard photograph.

URSA MAJOR

LIST OF DOUBLE AMD MULTIPLE STARS

NAME	DIST	PA	YR	MAGS	NOTES	RA & DEC
Σ1192	2.8	258	35	7 - 10½	relfix, spect F0	08116n6032
	48.6	224	21	- 10		
	99.5	195	21	- 12		
Σ1193	43.1	87	25	6 - 9	relfix, spect K5	08152n7234
1	7.1	192	00	3½- 15	(Omicron U.Maj)	08262n6053
	143	152	23	- 11	(β1067) Spect G0	
	173	207	24	- 10½		
Σ1258	9.7	331	55	7 - 7½	relfix, spect F0	08399n4903
Σ1279	1.3	269	58	8½- 8½	relfix, spect G0	08467n3947
Σ1275	2.0	197	58	8 - 8	relfix, spect F0;	08476n5743
	39.8	79	19	-12½	AC slight dist inc	
A1584	0.5	117	62	8 - 8	Binary, 85 yrs; PA	08494n5508
					inc, spect G0	
Σ1280	1.9	77	63	7½- 7½	Binary, about 1200	08510n7100
					yrs; PA inc, dist	
					dec, spect K5, dM1	
β408	2.9	343	46	8- 10½	cpm, relfix,	08549n6337
					spect G5	
ι	4.5	16	58	3 - 10	(h2477) (Hu 628)	08558n4814
ι b	0.2	254	68	10½-11	(OΣ196) All cpm, AB	
					PA inc, dist dec;	
					BC binary, 39 yrs;	
					PA dec, AB spect A7	
					+ dM1 (*)	
χ	0.2	286	62	4½- 4½	(12 U.Maj) (A1585)	09002n4721
					Binary, 58 yrs; PA	
					dec, spect B9 (*)	
σ²	2.4	20	62	5 - 8½	(13 U.Maj) (Σ1306)	09060n6720
	205	148	19	- 9½	AB binary, about	
					700 yrs; PA dec,	
					spect F7 (*)	
Σ1312	4.5	149	51	7½- 8	relfix, cpm, spect	09068n5235
					A5	
τ	57.1	48	06	5- 10½	(14 U.Maj) (HV 73)	09068n6343
					optical, spect Am	
Σ1315	24.8	27	25	7 - 7	relfix, spect A0,A0	09088n6153
Σ1318	2.9	237	58	7½- 8½	PA & dist slow dec,	09103n4712
					spect F8	
Σ1321	18.1	82	62	7½- 7½	Binary, about 700	09114n5255
	28.4	283	07	- 14	yrs; PA inc, dM0	

LIST OF DOUBLE AND MULTIPLE STARS   (Cont'd)

NAME	DIST	PA	YR	MAGS	NOTES	RA & DEC
Σ1331	1.0	153	54	8 - 8	relfix, spect A5;	09169n6134
	13.2	186	54	- 11½	AC dist inc, PA dec	
	20.4	118	54	- 13½		
	139	223	10	- 10		
OΣ199	5.7	132	55	6 - 10	(37 Lyncis) PA inc,	09173n5129
	142	5	24	- 10	spect dF3	
Σ1340	6.2	319	25	6½- 8½	(39 Lyncis) relfix,	09192n4946
	135	83	09	-	spect B9	
OΣ200	1.6	337	59	6½- 8½	relfix, spect G0	09215n5147
21	5.7	311	37	7 - 8	(Σ1346) relfix,	09221n5415
	97.3	189	08	-11½	spect A2	
OΣΣ99	77.3	162	24	5½- 7	(S 598) dist dec,	09254n4550
	83.5	80	23	- 10	Spect G5, F8	
Σ1349	19.2	166	24	7 - 8	relfix, spect A2	09269n6746
23	23.0	271	58	4 - 9	(Σ1351) relfix, AB	09276n6317
	99.6	231	57	- 10½	cpm, spect F0; AC	
					dist inc	
Σ1358	23.5	164	20	7½- 8½	slow PA inc, spect	09277n4454
					M	
θ	4.1	101	58	3 - 13	(β1071) cpm, PA inc	09296n5155
					dist dec, spect F6	
					(*)	
Σ1350	10.4	248	58	7 - 7	relfix, spect F5	09302n6701
	130	214	21	- 9		
Σ1363	10.7	360	04	7½- 11	spect F0	09315n6107
Σ1376	5.1	311	50	8 - 8	relfix, spect F5	09419n4328
	68.9	332	13	- 13½		
υ	11.6	293	34	4- 11½	(OΣ521) (29 U.Maj)	09475n5917
					cpm, relfix, spect	
					F2	
Σ1381	1.1	201	54	8½- 8½	PA & dist dec,	09476n6051
					spect G0	
φ	0.4	48	66	5 - 5½	(30 U.Maj) (OΣ208)	09487n5418
					Binary, 107 yrs; PA	
					inc, spect A3 (*)	
OΣ209	4.7	309	34	7 - 10	relfix, spect G5	09500n5051
OΣ522	14.2	124	15	7½- 11	spect K0	09501n6502
Σ1386	2.1	112	55	9 - 9	spect F5; In field	09510n6908
	144	355	00	- 10	of galaxy M82	
Σ1398	2.2	125	60	8 - 11	PA dec, spect F0	09576n6858

LIST OF DOUBLE AND MULTIPLE STARS   (Cont'd)

NAME	DIST	PA	YR	MAGS	NOTES	RA & DEC
OΣ210	1.0	260	68	7½- 8½	Slight PA dec, spect G5	09594n4636
Σ1402	27.2	102	24	8 - 9½	optical, dist inc, spect K5, G0	10016n5544
Σ1407	5.2	53	40	9 -9½	relfix	10054n6441
Σ1415	16.7	167	58	6½- 7	relfix, cpm, both	10139n7118
	150	14	56	- 10½	spect A3	
Σ1427	9.4	213	28	7 -7½	relfix, cpm, spect F5	10190n4409
Σ1428	3.1	87	54	7½- 8	dist dec, slight PA inc, spect F5	10228n5253
Es1905	47.3	66	21	7½-11½	spect K0	10270n6337
Es1905b	4.0	200	21	-13½		
β1074	2.5	208	26	7 - 11	spect K0	10325n4555
OΣ222	4.4	340	57	6½-10½	relfix, cpm, spect F8	10351n6024
Σ1460	3.5	164	44	8 - 8	relfix, spect F2	10377n4225
Σ1462	8.0	172	34	8 - 9½	relfix, spect A3	10398n5104
	194	56	09	- 10		
Σ1463	7.6	259	17	8½- 9	relfix, spect G5	10399n4657
Σ1465	2.0	10	58	8½- 9	relfix	10403n4453
Σ1469	11.0	322	24	7 - 10	relfix, spect F8	10444n6543
OΣ229	0.7	290	66	6½- 7	PA dec, spect A3	10452n4122
Σ1483	2.8	67	55	9 - 9	relfix, spect G5	10516n4746
Σ1495	34.3	37	24	6 - 8½	relfix, spect K2	10570n5911
ADS8015	13.4	201	34	7 - 11	spect K0	10579n4300
α	0.6	285	44	2 - 5	(β1077) Binary, 44½ yrs; PA dec, spect K0 (*)	11007n6201
51	8.1	251	58	6 - 12	(Ho 377) relfix, cpm, spect A3	11017n3831
Ho 378	0.8	234	60	8 - 8	PA inc, spect F2; in field of 51 U. Maj.	11022n3841
Σ1510	5.0	332	54	7 - 8½	cpm, dist inc, PA dec, spect F5	11051n5306
Σ1512	9.9	52	24	8 - 8½	relfix, spect A2	11062n6246
OΣ231	35.0	264	35	8 - 9	AB cpm, spect K5;	11083n3043
	174	332	33	- 9	AC dist inc; K5+G0	
Ho 254	2.5	163	23	6½-12½	spect K0	11106n3343

URSA MAJOR

LIST OF DOUBLE AND MULTIPLE STARS   (Cont'd)

NAME	DIST	PA	YR	MAGS	NOTES	RA & DEC
A1353	0.2	60	58	7½– 8½	spect F2; in field of nebula M97	11108n5541
Ho 50	3.1	32	58	7 – 10	spect K0	11109n4122
Hu 639	0.1	281	58	7½– 7½	spect A5	11127n4745
Σ1520	12.6	344	58	6½– 8	relfix, cpm pair; spect dF2	11132n5303
ξ	2.9	129	68	4½– 5	(53 U.Maj) (Σ1523) Binary, 60 yrs; PA dec, spect G0 (*)	11155n3149
ν	7.2	147	58	4 – 10	(Σ1524) relfix, spect K3 ,cpm	11158n3322
0Σ233	4.9	335	34	7 – 10	relfix, spect F0	11158n6657
	41.7	303	12	– 12		
Σ1525	2.1	179	39	9 – 9	relfix, spect F5	11167n4745
A1592	4.4	63	28	7 – 14	spect F0	11235n5224
Σ1541	7.7	29	16	8 – 10	relfix, spect F8	11249n4631
Σ1542	3.0	262	39	7– 10½	cpm, relfix, spect F0	11252n4450
57	5.4	359	58	5 – 8	(Σ1543) AB cpm, PA dec, spect A1	11264n3937
	217	9	10	–11½		
L 11	0.9	331	60	7 – 11	PA dec, spect F2	11266n3042
0Σ234	0.3	345	66	7½– 8	Binary, 86 yrs; PA inc, spect dF1	11281n4134
Σ1544	12.3	91	27	7 – 8	relfix, spect A5	11285n5959
0Σ235	0.8	95	68	5½– 7	binary, 72 yrs; PA inc, spect F6	11295n6122
Hu 1134	0.2	94	66	7 – 7	PA dec, spect K0	11295n3631
Σ1553	6.0	168	53	7½– 8	Slight dist inc, PA dec, galaxy NGC 3780 in field	11339n5625
Σ1559	1.9	322	69	7 – 8	relfix, spect A2	11360n6437
Σ1561	9.7	253	24	6½– 8½	AB cpm, PA dec, spect G0; AC dist inc	11361n4523
	121	90	31	– 9½		
Hu 729	1.3	351	55	7– 11½	PA dec, spect A0	11452n5006
Σ1576	5.1	243	40	8 –8½	relfix	11503n3106
65	0.3	59	58	7 – 9	(Σ1579) (A1777) AB PA inc, spect B9	11525n4645
	3.7	40	58	– 8½		
	63.1	114	53	– 6½		
0Σ241	1.6	138	66	6½– 8½	Slight PA inc, F2	11537n3544

LIST OF DOUBLE AND MULTIPLE STARS (Cont'd)

NAME	DIST	PA	YR	MAGS	NOTES	RA & DEC
β918	7.5	231	14	7 - 13	cpm, spect F0	11556n3233
β919	4.5	16	58	6 - 12	relfix, spect K2	11567n3327
OΣ243	0.8	8	67	8 - 8½	cpm, spect F8	11572n5341
Σ1600	7.5	92	41	7 - 8	relfix, cpm; spect F5	12030n5213
Hu 1136	2.0	223	58	6 - 11	relfix, spect K2	12031n6313
Σ1603	22.3	82	55	8 - 8½	relfix, cpm, spect both F8	12056n5545
Σ1608	12.7	221	55	7½- 8	cpm, slow dist inc, slight PA dec, spect K0	12090n5342
Σ1630	2.5	168	38	8½- 9	relfix, spect G0	12165n5639
OΣ249	0.4	280	59	7 - 8	PA dec, spect G5	12214n5426
	13.2	149	25	- 11		
A1601	2.0	67	16	7½- 14	spect K2	12339n5706
Σ1691	18.8	277	37	8 - 9	relfix, spect F0; galaxy NGC 4814 in field	12528n5826
Σ1695	3.7	283	58	6 - 8	PA slight dec, spect A2	12541n5422
	124	142	09	- 10½		
78	1.0	21	67	5- 7½	(β1082) binary, 116 yrs; PA inc, spect F2	12586n5638
ζ	14.4	151	67	2½ - 4	MIZAR (Σ1744) fine pair; PA slow inc, spect A2 (*)	13219n5511
A1361	3.4	37	31	8 - 14	spect F5	13256n5617
S 649	182	111	24	5½- 8½	cpm, spect A0, F8;	13270n6011
S 649b	1.1	128	58	8½- 11	(Σ1752) PA dec	
Σ1770	1.9	121	56	7 - 8	relfix, spect M2	13357n5058
Σ1774	17.6	134	58	6½- 10	cpm, spect F8	13384n5046
β802	3.6	223	50	8 - 11	relfix, spect F0	13466n4836
Σ1795	7.7	3	16	7 - 10	relfix, cpm; spect A2	13571n5321
A1097	0.3	100	24	8½- 8½	binary, about 125 yrs; PA inc, spect F5	14003n5728
	28.0	20	12	- 10½		
Σ1820	2.3	105	67	8 - 8½	PA inc, spect K2	14114n5533
Σ1830	8.1	303	56	8½- 10	PA & dist inc, spect G5	14142n5654
	35.5	81	23	- 13½		

LIST OF DOUBLE AND MULTIPLE STARS   (Cont'd)

NAME	DIST	PA	YR	MAGS	NOTES	RA & DEC
Σ1831	139 6.2 108	63 139 222	56 56 56	– 6½ 6½– 9 – 6½	(Σ1830 A–D) Slight PA dec, spect F0; galaxy NGC 5585 in field	14144n5655

LIST OF VARIABLE STARS

NAME	MagVar	PER	NOTES	RA & DEC
ε	1.76– 1.79	5.0887	α Canes Ven type; Spect A0p (*)	12518n5614
υ	3.77–3.86	.133	(29 U.Maj) δ Scuti type Also visual double; Spect F2	09475n5917
R	6.7--13.4	302	LPV. Spect M3e--M9e (*)	10412n6902
S	7.4--12.4	226	LPV. Spect S1e--S5e	12418n6122
T	6.5--13..	257	LPV. Spect M4e--M6e	12341n5946
U	6.3--	---	Uncertain; possibly not variable; Spect M0	10117n6014
V	9.3--11..	193	Semi-reg; Spect M5-M6	09047n5119
W	7.9---8.6	.33364	Dwarf Ecl.Bin; Typical "W U.Maj" Star; Spect dF8 (*)	09403n5611
X	8.1--14.8	249	LPV. Spect M4e	08373n5019
Y	8.5--9.6	168	Semi-reg; Spect M7	12381n5607
Z	6.8--9.2	196	Semi-reg; Spect M5e	11539n5809
RR	8.6--14.2	231	LPV. Spect M4e	13241n6239
RS	8.4--14..	260	LPV. Spect M4e-- M6e	12367n5846
RT	8.6--9. 6	Irr	Spect N	09149n5137
RU	8.5--14..	252	LPV. Spect M3e--M5e	11390n3845
RV	9.5--10.6	.46806	Cl.Var; Spect F4--F5	13314n5415
RW	9.5--10.5	7.3282	Ecl.Bin; Spect dF9+dG9	11381n5216
RX	9.8--12.2	195	Semi-reg; Spect M5; In field of 17 U.Maj	09101n6728
RY	7.0--8.2	311	Semi-reg; Spect M2--M3e	12181n6135
RZ	8.5--10.3	115	Semi-reg; Spect M5--M6	08064n6522

LIST OF VARIABLE STARS   (Cont'd)

NAME	MagVar	PER	NOTES	RA & DEC
SS	12....	---	Supernova (1909) in NGC 5457 (Galaxy M101)	14003n5442
ST	6.7--7.9	81:	Semi-reg; Spect M4	11251n4528
SU	11.0--14..	Irr	U Geminorum type (*)	08081n6246
SV	9.0--10.3	76	Semi-reg; Spect G1e- K3p	10435n5518
SY	5.3---	---	Uncertain, possibly not variable; Spect A2	09524n5003
TT	8.9--9.5	Irr	Spect M6	09013n6029
TU	9.2--10.2	.55766	Cl.Var; Spect A8--F8	11272n3021
TV	7.1--8.2	50	Semi-reg; Spect M5	11430n3610
TX	7.0--8.8	3.0632	Ecl.Bin; Spect B8 + F2	10424n4550
TZ	8.3--9.0	116	Semi-reg; Spect M4	12071n5840
UX	12.7--13.8	.19667	Ecl.Bin; Possibly former nova (*)	13374n5210
VW	6.9--7.8	125	Semi-reg; Spect M2	10556n7016
VY	6.0--6.6	Irr	Spect N0	10416n6740
WW	9.0--10.4	Irr	Spect M6	11444n5831
WY	8.6--9.4	Irr	Spect M5	10389n5154
WZ	9.5--10.0	Irr	Spect F5	12033n5419
YZ	9.5--11..	Irr	Spect dM5	09522n4414
AC	9.1--11..	6.8548	Ecl.Bin; Spect A2	08516n6510
AW	6.8--7.1	.4387	Ecl.Bin; W U.Maj type; Spect F0 + F2	11274n3015
AZ	8.2--9.0	Irr	Spect M6	11446n4344
CF	8.5--12..	?	Flare star? Spect M4	11504n3805
CG	5.9--6.0	Irr	Spect M4	09181n5655

LIST OF STAR CLUSTERS, NEBULAE AND GALAXIES

NGC	OTH	TYPE	SUMMARY DESCRIPTION	RA & DEC
2639	204[1]	⊖	S0/Sa; 12.5; 1.1' x 0.6' cB,S,E,psmbM	08401n5024
2654		⊖	Sa/Sb; 12.7; 3.0' x 0.7' pF,S,E, edge-on, spindle- shaped	08443n6028
2681	242[1]	⊖	S0/Sa; 11.3; 2.8' x 2.5' vB,vL,vgvsmbM; compact center and distinct spiral arms	08500n5131
2685		⊖	S0/Sb? 12.2; 2.5' x 0.8' pF,1E, unusual system with outer ring (*)	08522n5859
2693	823[2]	⊖	E3p; 13.0; 0.7' x 0.5' pB,pL,1E,psbM; 32' east from NGC 2681	08535n5133
2701	66[4]	⊖	Sc; 12.5; 1.8' x 1.0' pB,1E,bM	08555n5359
2742	249[1]	⊖	Sc; 12.5; 2.5' x 1.1' cB,cL,E, bM	09037n6041
2768	250[1]	⊖	E5; 11.6; 1.6' x 0.8' cB,cL,mE,psbM, LBN	09078n6016
2787	216[1]	⊖	S0/Sa; 12.0; 2.1' x 1.3' B,pL,1E,mbM	09149n6925
2841	205[1]	⊖	Sb; 10.3; 6.2' x 2.0' vB,L,vmE,vsmbM; fine spiral with symmetrical whorls (*)	09186n5112
2880	260[1]	⊖	E3/S0; 12.7; 1.2' x 0.7' B,cS,1E,mbM; in field of 23 U.Maj.	09257n6244
2950	68[4]	⊖	S0/Sa; 12.0; 1.3' x 0.9' B,pS,1E,vgvmbMN	09391n5905
2976	285[1]	⊖	Sc/Sd/Irr? 10.8; 3.4' x 1.3' B,vL,mE; member of M81 group; Spiral pattern uncertain (*)	09432n6808
2998	717[2]	⊖	Sc; 12.8; 3.0' x 1.1' pF,pL,E,bMN	09458n4419
2985	78[1]	⊖	Sa/Sb; 11.3; 4.0' x 3.0' vB,cL,1E,psmbM, with fine spiral whorls	09460n7231
3031	M81	⊖	Sa/Sb; 8.0; 18' x 10' ! eB,L,E,BN fine spiral (*)	09515n6918

NGC	OTH	TYPE	SUMMARY DESCRIPTION	RA & DEC
3034	M82	⊖	Ip; 9.2; 8.0' x 3.0' ! vB,vL,vmE, fine pair with M81 (*)	09519n6956
3043	835[2]	⊖	Sb; 13.2; 1.1' x 0.4' pB,S,vlE,vgbM	09528n5932
3065	333[2]	⊖	SO/Sa; 12.9; 0.5' x 0.5' pF,vS,R,bM; Pair with 3066= SBc galaxy 2.9' SSE	09577n7225
3079	47[5]	⊖	SBb; 11.4; 8.0' x 1.0' vB,L,mE; bright edge-on, long streak tilted N-S	09586n5557
3077	286[1]	⊖	E2/Ip; 11.0; 2.6' x 1.9' cB,cL,mbM; member of M81 group (*)	09594n6858
3184	168[1]	⊖	Sc; 10.5; 5.5' x 5.5' pB,vL,R,gbM; fine face-on spiral in field of Mu U.Maj.	10152n4140
3198	199[1]	⊖	Sc; 11.0; 9.0' x 3.0' pB,vL,mE,vgbM; nearly edge- on spiral, elongated NE-SW	10167n4549
3206	266[1]	⊖	SBc; 13.0; 2.0' x 1.2' pB,cL,E,glbM	10185n5711
----	I.2574	⊖	I/Sd; 13.0; 9.0' x 4.0' vL,eF,Irr;mE; stellar group on N edge; member of M81 group (*)	10250n6843
3259	870[2]	⊖	I/Sc; 12.7; 1.2' x 0.7' F,S,1E,gbM	10292n6518
3310	60[4]	⊖	I/Sc pec; 11.0; 3.0' x 2.0' cB,pL,R,vgvsmbMN	10357n5346
3319	700[3]	⊖	SBc; 11.8; 6.0' x 2.8' cF,L,mE; flattened central region elongated NE-SW; coarse spiral arms	10364n4156
3320	745[2]	⊖	Sc; 12.9; 1.9' x 0.9' F,pS,mE	10367n4740
3353	842[3]	⊖	Sc? 13.0; 0.8' x 0.5' F,cS,R,pgbM	10423n5614
3359	52[5]	⊖	SBc; 11.0; 6.0' x 3.0' pB,L,E,vglbM	10434n6330

1930

LIST OF STAR CLUSTERS, NEBULAE AND GALAXIES   (Cont'd)

NGC	OTH	TYPE	SUMMARY DESCRIPTION	RA & DEC
3348	80[1]	⊖	E1; 12.0; 0.9' x 0.8' B,S,1E,psbM	10435n7307
3415	718[2]	⊖	E4/E5; 13.1; 0.8' x 0.5' pB,pS,vlE	10489n4359
3445	267[1]	⊖	Sc; 12.9; 1.1' x 1.1' cB,pL,R,vglbM; NGC 3458 in field, 14' NE	10516n5715
3448	233[1]	⊖	Sc/I; 12.6; 1.8' x 0.3' B,pL,mE,glbM; edge-on, thick streak with outer extensions; type uncertain, in field of 44 U.Maj.	10517n5434
3458	268[1]	⊖	S? 13.0; 0.8' x 0.6' vB,vS,R; in field of 3445	10530n5722
3478	705[3]	⊖	Sb; 13.2; 2.2' x 0.8' eF,S,1E	10565n4623
3516	336[2]	⊖	SO/Sa; 12.4; 1.0' x 0.8' pB,vS,1E,psmbM	11034n7250
3549	220[1]	⊖	Sb; 12.8; 2.7' x 0.9' cB,cL,cE, nearly edge-on galaxy, flattened oval	11082n5339
3556	46[5]	⊖	Sc; 10.8; 7.8' x 1.4' cB,vL,vmE,pbM; nearly edge- on; Owl Nebula M97 in field 48' to SE (*)	11087n5557
3583	728[2]	⊖	Sb/Sc; 12.2; 2.2' x 1.8' pB,pL,R,vgpmbM	11114n4839
3587	M97	◎	! B,L,R, Mag 11; Diam 150"; 14[m] central star. The "Owl Nebula" (*)	11120n5518
3610		⊖	E4; 11.6; 1.4' x 0.9' vB,pS,1E, vsmbM, SN	11156n5904
3614	729[2]	⊖	Sc; 12.9; 3.5' x 1.6' F,pL,1E, glbM	11156n4602
3613	271[1]	⊖	E5; 11.7; 1.7' x 0.8' vB,cL,mE,smbMN; pair with NGC 3619	11157n5817
3619	244[1]	⊖	SO/Sa; 12.7; 1.0' x 1.0' cB,cL,R,vgmbM; 15' pair with NGC 3613	11165n5802

LIST OF STAR CLUSTERS, NEBULAE AND GALAXIES   (Cont'd)

NGC	OTH	TYPE	SUMMARY DESCRIPTION	RA & DEC
3631	226[1]	⊖	Sc; 11.5; 4.5' x 4.0' pB,L,R,svmbM; bright face-on spiral	11183n5328
3642	245[1]	⊖	Sc; 11.9; 5.5' x 4.2' pB,pL,R,vgbM; outer arms faint and delicate	11196n5921
3665	219[1]	⊖	E2/S0; 12.1; 1.3' x 1.0' cB,cL,lE, pgmbM	11221n3902
3675	194[1]	⊖	Sb; 11.4; 3.5' x 1.3' vB,cL,vmE,vsmbMN; elongated N-S; in field of 56 U.Maj.	11235n4352
3683	246[1]	⊖	SBc; 13.2; 1.3' x 0.5' cB,pL,E	11248n5709
3687		⊖	Sb; 13.0; 1.3' x 1.3' pB,pS,R,lbM	11253n2947
3690	247[1]	⊖	Sp? 12.1; 1.4' x 0.4' pB,pS,vlE,pgbM; appears as two galaxies in contact	11260n5849
3718	221[1]	⊖	SBa; 11.8; 3.0' x 3.0' cB,vL,R,lbM; central dust lane, long spiral arms; NGC 3729 foll 12' to ENE	11299n5321
3726	730[2]	⊖	Sc; 11.3; 5.0' x 3.4' pB,vL,lE,vsmbM; well defined spiral pattern	11307n4719
3729	222[1]	⊖	SBp; 12.5; 1.8' x 1.3' pB,pL,lE; In field with 3718	11310n5324
3738	783[2]	⊖	I; 12.1; 1.2' x 0.9' pB,pL,bM; compact irregular; NGC 3756 in field 15' SE	11331n5448
3756	784[2]	⊖	Sc; 12.3; 3.4' x 1.5' pF,L,mE N-S; 3738 in field	11341n5434
3769	731[2]	⊖	Sb; 12.5; 2.6' x 0.6' pB,S,pmE; nearly edge-on	11351n4811
3780	227[1]	⊖	Sc; 12.6; 2.4' x 2.0' pF,L,vlE,vgbM	11367n5633
3782	732[2]	⊖	SBc/pec; 12.9; 1.2' x 0.6' F,S,E; asymmetric spiral	11369n4644
3813	94[1]	⊖	Sb; 12.6; 1.9' x 0.8' cB,pL,pmE,bM	11387n3649

LIST OF STAR CLUSTERS, NEBULAE AND GALAXIES   (Cont'd)

NGC	OTH	TYPE	SUMMARY DESCRIPTION	RA & DEC
3877	201[1]	⊖	Sb/Sc; 12.0; 4.4' x 0.8' B,L,mE; edge-on spiral; 16' south from Chi U.Maj.	11435n4746
3888	785[2]	⊖	Sc; 13.0; 1.3' x 0.9' pB,S,1E,pgbM; in field with NGC 3898	11450n5615
3893	738[2]	⊖	Sc; 11.0; 3.9' x 2.5' B,pL,1E,mbM	11461n4900
3894	248[1]	⊖	E3/E4; 13.0; 1.2' x 0.8' pB,pL,1E,pgmbM; NGC 3895 2' NE; pF,pL,v1E; Mag 14, Sa	11462n5942
3898	228[1]	⊖	Sa/Sb; 11.9; 2.6' x 0.9' B,pL,mE,svmbM; NGC 3888 in field 15½' SW	11467n5622
3917	824[2]	⊖	Sc/Sd; 12.8; 4.5' x 0.8' F,L,vmE,vgbM; nearly edge-on spiral, flat streak	11483n5206
3938	203[1]	⊖	Sc; 11.2; 4.5' x 4.0' B,vL,R,bM,pBN; fine face-on spiral	11502n4424
3941	173[1]	⊖	E3/S0; 11.3; 1.9' x 1.1' vB,pL,E,smbM	11503n3716
3945	251[1]	⊖	S0/SBa; 11.9; 5.0' x 2.0' B,pL,E,gmbM; outer ring	11506n6057
3949	202[1]	⊖	Sb/Sc; 11.5; 2.2' x 1.1' cB,pL,pmE,vglbM	11511n4808
3953	45[5]	⊖	SBb; 11.1; 6.0' x 2.8' cB,L,E,vsbM; fine multiple- arm spiral	11512n5237
3963	67[4]	⊖	Sb/Sc; 12.7; 2.0' x 1.9' pF,cL,R,vgsbM; in field with Z U.Maj.	11524n5846
3982	62[4]	⊖	Sb; 11.8; 2.0' x 1.8' B,pL,R,gbM	11539n5524
3985	707[3]	⊖	SBd; 12.9; 0.9' x 0.5' vF,cS	11541n4837
3992	61[4]	⊖	SBb; 10.9; 6.4' x 3.5' cB,pL,pmE,sbM,BN, θ Structure barred spiral; in field of Gamma U.Maj., 40' to SE	11550n5339

LIST OF STAR CLUSTERS, NEBULAE AND GALAXIES   (Cont'd)

NGC	OTH	TYPE	SUMMARY DESCRIPTION	RA & DEC
3995		⊖	Sc/p; 12.9; 2.2' x 0.6' F,pL,cE,bM	11552n3235
3998	229[1]	⊖	E2/SO; 11.6; 1.6' x 1.2' cB,pS,R,vgsbM	11553n5544
4013	733[2]	⊖	Sb; 12.7; 4.2' x 0.5' B,cL,vmE,vsmbM; edge-on	11560n4413
----	I.749	⊖	Sc; 13.2; 1.8' x 1.1' pB,L,E,1bM; 3.3' pair with I.750	11560n4301
----	I.750	⊖	Sa/Sb; 13.0; 1.7' x 0.6' pB,L,1E,bM; pair with I.749	11563n4300
4026	223[1]	⊖	E8/SO; 11.9; 3.3' x 0.7' vB,cL,mE,vsvmbM,BN; edge-on lens-shaped galaxy	11569n5114
4036	253[1]	⊖	E6/SO; 11.6; 3.0' x 1.0' vB,vL,E; spindle-shaped; NGC 4041 in field	11589n6210
4041	252[1]	⊖	Sc; 11.7; 2.2' x 1.9' B,cL,R,psvmbMN; in field with NGC 4036; separation 16'	11597n6225
4047	741[2]	⊖	Sa? 12.8; 1.1' x 1.0' pB,pS,R	12002n4855
4051	56[4]	⊖	Sb/Sc; 11.2; 4.2' x 3.0' B,vL,E,gvsmbM; thick spiral arms	12006n4448
4062	174[1]	⊖	Sb/Sc; 12.2; 3.2' x 1.0' pB,vL,mE E-W; vgbM	12015n3210
4085	224[1]	⊖	Sb/Sc; 12.8; 2.2' x 0.5' B,pL,pmE,vsbM; in field with NGC 4088	12028n5038
4088	206[1]	⊖	Sb/Sc; 11.1; 4.7' x 1.5' B,cL,E,1bM; distorted spiral arms, extending mass on NE tip; NGC 4085 in field 11' to south	12030n5049
4096	207[1]	⊖	Sc; 11.5; 5.8' x 1.0' pB,vL,mE, nearly edge-on	12035n4745
4100	717[3]	⊖	Sb/Sc; 11.9; 4.5' x 1.1' pB,vL,vmE,vglbM; nearly edge- on spiral	12036n4951

LIST OF STAR CLUSTERS, NEBULAE AND GALAXIES   (Cont'd)

NGC	OTH	TYPE	SUMMARY DESCRIPTION	RA & DEC
4102	225[1]	⊖	Sb; 12.2; 2.2' x 1.2' B,pS,1E,bM,BN	12038n5259
4111	195[1]	⊖	E7; 11.6; 3.4' x 0.8' vB,pS,vmE; lenticular or edge-on SO; on Canes Venatici border	12045n4321
4144	747[2]	⊖	Sc/Sd; 12.4; 5.3' x 0.7' pF,cL,vmE,vgbM; edge-on	12075n4644
4157	208[1]	⊖	Sb; 12.0; 6.0' x 0.9' pF,cL,vmE; edge-on spiral, narrow streak	12086n5046
4290	805[2]	⊖	SBa/SBb; 12.7; 1.5' x 0.9' pB,pL,1E,gmbM; In field of 70 U.Maj.	12185n5822
----	M40		(Wcn 4) Error; no nebula at this position (*)	12198n5822
4605	254[1]	⊖	Sc/pec; 10.9; 4.0' x 1.2' B,L,vmE,glbM	12378n6153
4814	243[1]	⊖	Sb; 12.5; 2.5' x 2.3' B,pS,vlE,vgbM	12533n5837
5204	63[4]	⊖	I/Sd; 11.9; 3.9' x 2.6' pB,cL,1E,gmbM	13283n5840
5308	255[1]	⊖	SO/E8; 12.6; 2.0' x 0.6' B,pL,mE,psbM, BN; edge-on system, narrow spindle	13454n6114
5322	256[1]	⊖	E2/E3; 11.3; 1.4' x 1.0' vB,pL,iR,psmbM	13476n6026
5376	238[1]	⊖	Sb; 13.0; 1.2' x 0.8' cB,pL,vlE,vgmbM; NGC 5389 in field 15' NNE= F edge-on spiral	13536n5945
5389	240[1]	⊖	Sa? 13.0; 1.7' x 0.7' pB,pL,mE,mbMN; edge-on spiral in field with NGC 5376	13545n6009
5422	230[1]	⊖	SO/E8; 13.0; 2.9' x 0.4' pB,S,pmE,vsmbMN; edge-on	13590n5524
5430	827[2]	⊖	Sb; 12.8; 2.0' x 1.0' pB,S,E,mbM; asymmetric outer arm pattern	13591n5934

LIST OF STAR CLUSTERS, NEBULAE AND GALAXIES   (Cont'd)

NGC	OTH	TYPE	SUMMARY DESCRIPTION	RA & DEC
5448	691[2]	⊖	Sa; 12.3; 3.5' x 1.0' pB,cL,vmE,smbMN; with outer ring	14009n4925
5457	M101	⊖	Sc; 9.0; 22' x 20' pB,vL,R,vsmbM, BSN; splendid large face-on spiral (*)	14014n5435
5473	231[1]	⊖	E2/SO; 12.6; 1.1' x 0.7' pB,S,1E,gbM; in field with M101	14030n5508
5474	214[1]	⊖	I/Sc/pec; 11.5; 4.0' x 3.0' pB,L,bM	14032n5354
5480	692[2]	⊖	Sc/SBc; 12.6; 1.3' x 0.9' F,pS,vgbM; NGC 5481 is 3' to east = SO/E2 galaxy 0.8' diam	14046n5057
5485	232[1]	⊖	SO/Sa; 12.6; 0.9' x 0.8' cB,R,vgbM	14055n5514
5585	235[1]	⊖	Sc/Sd; 11.6; 4.5' x 2.3' pF,L,1E,vgmbM; coarse spiral pattern	14180n5657
5631	236[1]	⊖	SO/Sa; 12.5; 0.8' x 0.8' B,S,R,psbMN	14251n5648

DESCRIPTIVE NOTES

URSA MAJOR GROUP   Among all the constellation patterns of the heavens, there is none so well known as the seven-starred fig-ure of the *Big Dipper* or the *Great Bear*, prominent in the star-lore of all lands, and famous as the traveler's guide to the North Star. To modern observers, the Great Dipper holds an additional interest in the knowledge that the pattern of the "Seven Stars" is not entirely an arrangement of chance. The five central stars of the figure share a common proper motion, and form a true moving group in space. The group is not a rich one, but is of great interest since it is the nearest of all the star clusters.

"Although the group has many titles and mythical associations," says R.H.Allen, "it has almost everywhere been known as a Bear... All classic writers, from Homer to those in the decline of Roman literature, thus mentioned it, - a universality of consent as to its form which, it has fancifully been said, may have arisen from Aristotle's idea that its prototype was the only creature that dared invade the frozen North". Ursa Major, however, was identi-fied as a Bear not only in the Greek and Roman world, but also by the Iroquois and Algonquin tribes of ancient North

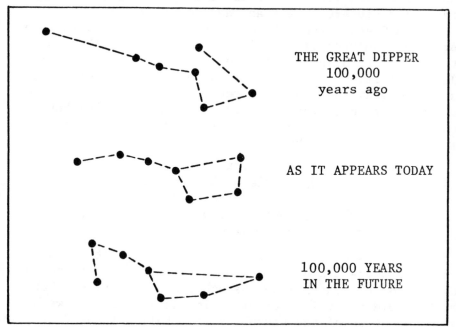

THE GREAT DIPPER
100,000
years ago

AS IT APPEARS TODAY

100,000 YEARS
IN THE FUTURE

1937

DESCRIPTIVE NOTES   (Cont'd)

America. The American historian George Bancroft states that among the Narragansetts and the Illinois, the *North Star* was called the Bear; elsewhere the Bowl of the Dipper was identified as the Bear and the three stars of the handle were three hunters tracking the animal and carrying the pot (Alcor near Mizar) with which to cook it. W.T.Olcott thought that "by no stretch of the imagination can the figure of a bear be traced out of the stars in this region, and it is one of the great mysteries as to how the constellation came to be so named". A fairly tolerable bear may be constructed, however, out of the surrounding stars, as shown in the diagram opposite. The bear's nose is at the star Omicron Ursae Majoris, his ears at Sigma, his front feet at Iota and Kappa, and his back paws at Mu and Lambda. Since the Dipper pattern is slowly changing, it has been suggested that possibly the name was given at some remote time when the pattern resembled a bear more than it does today. The change throughout all human history has been so slight, however, that the explanation does not seem very likely. The inimitable Will Cuppy has suggested the charmingly different Groucho-Marx-type theory that perhaps Ursa Major received its name at some very distant time when bears resembled dippers more than they do today.

In Greek legend the bear was originally Callisto of Arcadia, daughter of King Lycaon; she was transformed into the animal to disguise her from the jealous and wrathful Juno. Callisto's young son Arcas, while out hunting, was about to slay the bear, not recognizing his mother in her strangely altered form; the gods then placed them both in the heavens as the Big and Little Bear. Ovid in his *Metamorphoses*, tells us how Zeus

"........*flung them through the air,*
In whirlwinds to the high heavens,
and fix'd them there,
Where the new constellations nightly rise,
Lustrous in the northern skies."

It was because of the enmity of Juno, it was said, that the Bears are not allowed to go to their rest beneath the rim of the earth like the other constellations, but must circle the Pole eternally. In the same legend we have an explanation of the unusually long tails of the Bears; in

a memoir attributed to Thomas Hood and dating to about 1590
a master and his pupil are debating the question:

Scholar: *I marvell why (seeing she hath the forme of a
    beare) her tayle should be so long..*
Master: *I imagine that Jupiter, fearing to come too nigh
    unto her teeth, layde holde on her tayle, and there
    by drewe her up into the heaven; so that shee of
    herself being very weightie, and the distance from
    the earth to the heavens very great, there was
    great liklihood that her tayle must stretch. Other
    reason know I none..*

James Russell Lowell pictures the titan Prometheus,
chained to his rock on a summit of the Caucasus, watching
the eternal circling of the stars:

*"One after one the stars have risen and set,
Sparkling upon the hoar-frost of my chain..
The Bear that prowled all night about the fold
Of the North Star hath shrunk into his den,
Scared by the blithesome footsteps of the dawn.."*

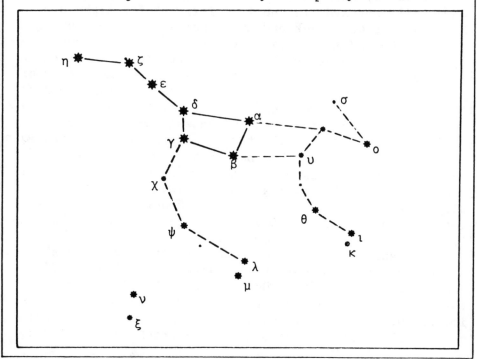

DESCRIPTIVE NOTES (Cont'd)

Milton, burdened by his own unbreakable chains, seems
to find solace in imagining himself as a watcher of the
stars:

"Let my lamp at midnight hour
Be seen in some high lonely tower
Where I may oft outwatch the Bear..."

R.W.Buchanan, in the *Ballad of Judas Iscariot*, seems
to suggest that the Great Bear of the north must be, as
logic would appear to require, a *polar* bear:

".....round and round the frozen Pole
glideth the lean white bear..."

And Bryant again alludes to the fact that the Bear
remains in the company of the eternally circling stars:

"The Bear that sees star setting after star
In the blue brine, descends not to the deep.."

According to R.H.Allen, the Arabian name for Ursa
Major was *Al Dubb al Akbar,* The "Greater Bear", evidently
derived from the Phoenician *Dub,* the Hebrew *Dobh.* Bayer has
the curiously garbled version *Dubhelacbar;* a portion of the
name still survives in the familiar *Dubh* for Alpha Ursae
Majoris. Various Greek and Roman names refer to the legend
of Callisto and Arcas, as in the *Arcadium Sidus* mentioned
by Sophocles, honoring the land of Arcadia, home of the
"people of the Bear". The towns, tribes, and mountains of
Arcadia are commemorated in such titles as *Tegae Virgo,*
Maenalis Ursa, and *Parrhasides Stellae,* the last possibly
having some connection with the Phoenician *Parrasis,* the
"Guiding Star". Ovid calls the Two Bears the *Magna minorque*
ferae; Propertius has them as *Geminae Ursae,* while Horace
and Virgil use the name *Gelidae Arcti.* The Greek *Arctos* is
seen frequently in writings of the Middle Ages; occasionally it appears as *Lycaonia Arctos,* honoring Callisto's
father. Ben Jonson, in 1609, refers to Callisto as *"a star*
Mistress Ursula in the heavens".

In ancient China the *Tseih Sing* or "seven Stars" were
associated with the celestial palace of the *Lord on High,*
sometimes with the "Star God of Longevity" *Shou Lao,* and
the heavenly mountain of *Tien Shan,* paradise of the Taoist
immortals. The star Sirius was the *Heavenly Wolf* who guard-
ed the celestial palace; modern star watchers may find a

strange significance in the choice of Sirius, since it is
now known that the star shares the space motion of the Ursa
Major Group, and is regarded as an outlying member of that
association.

The three stars of the Great Dipper's handle are
referred to in some Chinese writings as the "Jade Scales",
possibly from the connection of the concepts of weighing
and measuring with those of celestial government and divine
justice, as well as the "measuring of the seasons". In an
early poem of the Han Dynasty, written some 2000 years ago,
we find them referred to as a sign of the coming of winter:

> *"The brilliant moon shines splendid in the night;*
> *One may hear the house cricket singing*
> *   on the east wall...*
> *The Jade Scales speak of the beginning of winter,*
> *Scattering a million stars across the sky!*
> *Already the fields are white with hoarfrost;*
> *Now comes the sudden change of season....."*

The *Celestial Palace of the Immortals* appears in a
famous painting attributed to the Sung Dynasty artist Li-
Lung Mien (about 1100 AD), one of the treasures of the
Freer Gallery of Art in Washington. In this mystic and
evanescent work, the temple of the Immortals floats among
drifting clouds in an atmosphere of nirvanic serenity; in
the distance, great craggy peaks loom dimly through the
mist....

It was a similar painting, undoubtedly, which inspired one of the most delightful legends of Old China, the
story of the greatest of all T'ang Dynasty painters, that
magical genius Wu Tao Tzu. Having created a marvelous celestial landscape for the imperial palace, the artist then
vanished forever by *walking into the painting* in the presence of the Emperor and his entire court. Unfortunately, we
have no opportunity to experience the cosmic power which
Chinese mystics saw in the paintings of Wu Tao Tzu, as not
a single one of his works is known to have survived up to
the present. His name, incidentally, signifies *The Master
of the Mystic Way*.

Another painter of legendary skill was that strange
artist Lieh-I, who, it was said, could create vast images

THE COSMIC VISION OF VINCENT VAN GOGH is expressed in his *Starry Night on the Rhone* (top) and the dynamic *Starry Night* (below), painted in 1888 and 1889.

of the Mountain of the Immortals by filling his mouth with
colored water and spitting it out in the form of paintings.
"We have no modern paintings produced by such a method,"
mused that irrepressible astronomer-detective Burt Reglund,
"but many of them *look* like they were".

Our own master of visionary art, Vincent van Gogh,
has given us a number of paintings which literally shimmer
with cosmic power. In his iridescent *Starry Night on the
Rhone*, painted in 1888, the Great Dipper is the central
subject; in his widely reproduced *Starry Night*, completed
the following year, he makes us *feel* the infinite energy
of the rolling spheres, the celestial power that surges
through all things. It was of such transcendent visions
that he wrote:

*"I have a terrible lucidity at moments, these days
when nature is so beautiful... I am not conscious of my-
self any more and the picture comes to me as in a dream.."*

The Seven Stars of the Great Dipper have often been
connected in legend with *Seven Sages, Seven Wise Men,* or
other groups of illustrious persons; the same traditions
surround the cluster of the Pleiades. J.F.Hewitt thought
that the original Sanskrit name signified *The Seven Bears*
though another possible translation is *The Seven Bulls* or
even simply *The Seven Shining Ones*. Prof.W.D.Whitney states
that the term *riksha* may signify both "Bear" and "star"
and also carries the meaning of "to shine". In the great
Finnish national epic, the *Kalevala,* Ursa Major is the
Seitsen Tähtinen; Chaucer has an equivalent title with his
"Sterres seven". In Welsh lore, the constellation is seen
as a symbol of King Arthur; his name, it is claimed, is
derived from *Arth-Uthyr*, "the wonderful Bear". So it was
that the early Britons saw Arthur's Chariot in the great
constellation:

*"Arthur's slow wain his course doth roll
   In utter darkness, round the Pole..."*

In Ireland it was King David's Chariot; in Denmark
and Sweden it was the *Stori Vagn* or Great Wagon, and to the
Teutonic tribes the *Karls Vagn* or Chariot of Thor. Similar-
ly, the Vikings saw it as the Chariot of Wotan or Odin.
A medieval Polish name, *Woz Niebeski,* has been translated
"The Heavenly Wain", which is the exact equivalent of the

German *Himmel Wagen;* the charts of Bayer have it labeled *Horwagen,* while a Saxon name, *Waenes Thisl,* seems to refer to a wagon-pole. To medieval Christians the Wagon was, of course, that heavenly Chariot in which Elijah was taken up to heaven.

In England the constellation has for centuries been called *Charles Wain;* the older forms *Cherlemaynes Wayne* and Charel-Wayne appear in many old writings. Sir John Davies, the Elizabethan Age poet, refers to

> *"Those bright starres*
> *Which English shepherds, Charles his waine do name;*
> *But more this Ile is Charles, his waine,*
> *Since Charles her royall wagoner became..."*

And King James, in his royal survey of the constellations, wrote:

> *"Heir shynes the charlewain,*
> *There the harp gives light,*
> *And heir the Seamans starres,*
> *And there the Twinnis bright...."*

The name, according to the *New English Dictionary,* "appears to arise out of the verbal association of the star name *Arcturus* with *Arturus* or Arthur, and the legendary association of Arthur and Charlemagne; so that what was originally the wain of Arcturus or Bootes.....became at length the wain of Carl or Charlemagne..." In later tradition, however, it became associated with the two British King Charles, particularly the martyred Charles I. In the days of Shakespeare, the constellation was popularly used as a variety of natural time-piece; in Act II of his play *King Henry IV,* one of the porters at the Inn-yard exclaims:

> *"Heigh-Ho! An't be not four by the day,*
> *I'll be hanged; Charles' Wain is over*
> *the new chimney,*
> *And yet our horse not pack'd...."*

Tennyson, in *The Princess,* uses the slow polar circling of Ursa Major as a symbol of the passing of time:

> *" I paced the terrace, till the Bear had wheel'd*
> *Thro' a great arc his seven slow suns...."*

DESCRIPTIVE NOTES   (Cont'd)

Diodorus of Sicily wrote that travelers in the great deserts of Arabia "direct their course by the Bears, in the same manner as is done by us at sea". Edmund Spencer, is his archaically-worded epic *The Faerie Queen*, speaks of the use of Ursa Major as a navigational guide:

> *"By this the northern wagoner had set*
> *His sevenfold team behind the steadfast starre*
> *That was in ocean waves never yet wet,*
> *But firme is fixt, and sendith light from farre*
> *To all that in the wide deep wandering arre..."*

Ursa Major in English tradition is also the *Plough*, derived possibly from the *Triones* or *Teriones*, the Plough Oxen or Threshing Oxen of Roman fable; Cicero mentions them as the *Septentriones*, which later became a term for the north wind, the northern heavens, and polar things in general. Michael Drayton speaks of *"septentrion cold"* while Milton, in *Paradise Regained*, uses the phrase *"cold Septentrion blasts"*. R.H.Allen suggests that the origin of the *Triones* may be the *Hapto-iringas*, or "Seven Bulls" of the Persians. Al Biruni, in the 11th Century, suggested that the word is a form of *tarana*, referring to the "passage" or circling of the Seven Stars. In other traditions, the Seven Stars are a bier followed by mourners, as in the Arabic *Banat Na'ash al Kubra*, the "Daughters of the Great Bier"; Arabian Christians identified it as the Bier of Lazarus. Flammarion thought that the tradition possibly arose from "the slow and solemn motion of the figure around the pole". In the seemingly tomb-obsessed culture of Egypt, however, the constellation was merely a *Bull's Thigh*, and is so mentioned in inscriptions at the Ramesseum at Thebes and in the *Book of the Dead*. The ancient Egyptian name, *Mesχet*, suggests some connection with the malignant deity *Set;* in late Egyptian times the star group became the boat or chariot of Osiris.

In the modern world it is almost universally the Big Dipper, whose position marks the current season in the cycle of the year; Tennyson tells us that

> *"We danced about the May-pole and in the hazel copse*
> *Till Charles's Wain came out above*
> *The tall white chimney tops..."*

MEMBERS OF THE URSA MAJOR MOVING CLUSTER

HD #	STAR	MAG.	SPECT	ABS.MG.	NOTES	RA & DEC
112185	Epsilon U.Maj.	1.79	A0p	-0.3	Spectrum variable	12518n5614
95418	Beta U.Maj.	2.37	A1	+0.3		10588n5639
116656	Zeta U.Maj. (A)	2.40	A2	+0.3	Mizar; fine double	13219n5511
116657	Zeta U.Maj. (B)	3.96	A7	+2.0		" "
103287	Gamma U.Maj.	2.44	A0	+0.2		11512n5358
106591	Delta U.Maj.	3.30	A3	+1.2		12130n5719
116842	80 U.Maj.	4.02	A5	+1.9	Alcor	13232n5515
97696	21 Leo Minoris	4.47	A7	+2.2		10045n3529
124752	GC 19195 U.Min.	8.2	K0	+6.8		14113n6749
113139	78 U.Maj.	4.89	F2	+3.0	Close binary	12586n5638
91480	37 U.Maj.	5.16	F1	+2.9		10320n5720
111456	GC 17404 U.Maj.	5.87	F5	+3.7		12465n6036
129798	Σ1878 Draco	6.17	F2	+2.9	Binary	14408n6129
115043	GC 17919 U.Maj.	6.74	G1	+4.7		13116n5658
109011	HD 109011 U.Maj.	8.1	K2	+6.1		12288n5524
110463	HD 110463 U.Maj.	8.4	K3	+6.3		12395n5601
139006	Alpha Cor.Bor.	2.23	A0+dG6	+0.4	Membership uncertain	15326n2653

DESCRIPTIVE NOTES   (Cont'd)

The Hindu astronomer Varaha Mihira, some 1400 years earlier, had created a similar picture, imagining the Seven Stars as dancing maidens:

"The northern region is adorned with these stars, as a beautiful woman is adorned with a chain of pearls strung together, and a necklace of white lotus flowers, a most beautifully arranged garland....Thus adorned, they are like maidens who dance and revolve about the Pole as the Pole does direct them...."

And B.F.Taylor saw in their approach to the high place of the north, a welcome herald of soft summer nights:

"From that celestial Dipper, or so I thought, the dews were poured out gently upon the summer world..."

THE URSA MAJOR MOVING CLUSTER. The pattern of the Dipper is not entirely the result of the chance alignment of totally unrelated stars. As early as 1869, R.A.Proctor determined that the five central stars of the figure shared a common proper motion, and must form a true moving group in space. This conclusion was verified by W.Huggins in 1872 by a series of radial velocity measurements, showing that the actual space motions were the same. In the years following this discovery of the Ursa Major moving cluster, a number of additional members have been identified. The group is a fairly sparse one, but is of special interest since it is the nearest known star cluster, at slightly more than half the distance of the Hyades group in Taurus.

In a list published in 1949 by N.G.Roman, there were 13 stars recognized as definite members. In a discussion of the cluster in 1958, D.L.Harris listed 17 stars as accepted members. Magnitudes, spectra, and positions for these stars are given in the table on the facing page. The most uncertain entry on the list is Alpha Coronae Borealis, which is more than 30° from the main concentration, but appears to be nearly at the same distance and moving in the same direction. It has been accepted as a member by some investigators, but rejected by others.

The center of the Ursa Major cluster is located at a distance of about 75 light years, and the group occupies a roughly ellipsoidal volume of space measuring some 30 light years in length and about 18 light years in width.   The computed space velocity of the cluster is about 9 miles per second, and the individual proper motions range from 0.07" to about 0.12" annually. Radial velocity measurements differ somewhat, according to the position of the star in the cluster, and range from 4.5 to about 9.5 miles per second in approach. The motion of the group is eastward and south, giving a computed convergent point about 130° away in the eastern portion of Sagittarius at about RA 20^h24^m and $-37°$.

The members of the cluster all appear to be main sequence stars, ranging in type from A0 down to K3; the stellar population resembles those of the Hyades cluster and Praesepe (M44) in Cancer. Epsilon Ursae Majoris is the brightest member of the group; magnitude 1.79, spectrum A0; it is about 85 times more luminous than our Sun. The star is a spectrum variable of the Alpha Canum type with a 5.09 day period, possibly somewhat overluminous for its spectral class. Another notable member of the cluster is Zeta Ursae Majoris, a fine visual double, and accompanied also by a remote companion, Alcor, nearly 12' distant. The double system 78 Ursae Majoris is a well observed binary; recent computations give the period as 116 years. Among the four probable new members listed by Harris is another binary, Σ1878 in Draco. This orbiting pair has shown only a slight change in PA in the last hundred years, and no period can yet be derived.

When the stars of the Ursa Major group were first studied, many years ago, it was noticed that a number of bright stars in various parts of the sky have very nearly the same space motion. The list of such stars eventually grew to include Sirius, Alpha Ophiuchi, Delta Leonis, Beta Aurigae, and Alpha Coronae Borealis, as well as about 100 other fainter stars. It is now generally thought that the majority of these objects, with the possible exception of Alpha Coronae, are not true cluster members, but are part of a general stream of stars which share the cluster motion fairly closely, and may or may not be physically associated with it. This larger "Ursa Major Stream" occupies a region of space at least several hundred light years in extent, enveloping the region of the Sun, so that stream members are scattered all over the sky. In a list published by N.G. Roman in 1949 the number of probable members was given as 135, with about 70 stars regarded as definitely proven.

Although the actual Ursa Major cluster contains only main sequence stars, the larger stream contains a number of red and yellow giants as well. Neither group contains any stars earlier than spectral type A0. The true connection, if any, between the smaller cluster and the extended "stream" may remain speculative for many years to come.

ALPHA   Name- DUBHE, DUBH, or DUBB, from the Arabian *Thahr al Dubb al Akbar*, "The Back of the Great Bear". Magnitude 1.81, Spectrum K0 II or III; Position 11007n6201. This star and Beta form the two "Pointers" at the front of the Dipper's Bowl; a line extended through them will direct the observer to Polaris, the North Star, some 28° north from Alpha. Dubhe, however, is not a true member of the Ursa Major cluster; its motion, and that of Eta Ursae Majoris, are nearly in the opposite direction, so that the "Dipper figure" will become quite distorted in the course of time.

For the small telescope or field glass, Dubhe has a 7th magnitude common motion companion at a distance of 6.3'

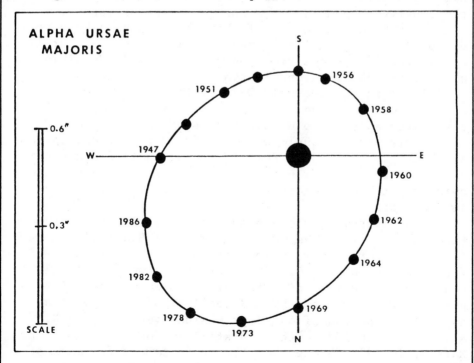

ALPHA URSAE MAJORIS

in PA 204°. This is GC 15179. Although the spectral class
is about F8 (some catalogues list it as F5) the star often
looks bluish in small instruments, a contrast effect resulting from the bright golden color of the primary. This
is a remarkable example of a wide common motion pair; the
true separation must be at least 12,000 AU. R.M.Petrie, in
1959, announced the small star to be a spectroscopic binary
with a period of 6.035 days. (Photograph on page 1961)

Alpha itself is a close visual binary, but a very
difficult object even in large telescopes. The companion
was discovered in 1889 by S.W.Burnham with the 36-inch
refractor at Lick Observatory. The separation was then 0.9"
in PA 326°, and the rapid decrease in PA during the next 10
years seemed to indicate a fairly short period. From 1900
to 1933 the companion was not seen with certainty by any
observer, but was recovered by R.G.Aitken in 1933 and has
been fairly well observed since. According to an orbit computed by P.Couteau (1959) the period is 44.66 years with a
semi-major axis of 0.77" and an inclination of 127°; the
two stars were at periastron in 1965. From studies in 1977
the elusive companion has a spectral type of late A.

Magnitude estimates of the companion have been very
discordant, ranging from Burnham's visual estimate of 11
to G.Kuiper's photometric measurement of 4.9. Possibly the
star is variable, which might account for its apparent invisibility in the years 1900-1933. From an astrometric
study of Greenwich and Allegheny plates, the total mass of
the system appears to be about 3 suns, and the individual
masses are nearly equal, with the fainter star having the
slightly greater mass. Here is one of those odd cases in
which the usual mass-luminosity law is seriously violated;
evidently the companion is some strange variety of underluminous star.

The system has been suspected of variability by more
than one observer. The German astronomer H.J.Klein thought
in 1867 that the star showed periodic changes in *color*,
from reddish to yellow; his observations were seemingly
confirmed in 1881 by a series of measurements made with a
polarizing colorimeter by M.Kovesligethy in Hungary; an
approximate period of 54½ days was derived. But, as R.H.
Allen wrote in 1899, "this is still in doubt". Martha E.
Martin, in 1907, wrote that *"There is some difference of*

*opinion as to these changes, and it may be that the period
of change is not so regular as has been supposed. It is
certain, however, that the star seems redder at some times
than at others even to the naked eye...."*
  The computed distance of Alpha Ursae Majoris is about
105 light years, giving the total luminosity as about 145
times the Sun, and the absolute magnitude as -0.7. The star
shows an annual proper motion of 0.14"; the radial velocity
is about 5.5 miles per second in approach.

BETA Name- MERAK or MIRAK, from the Arabian *Al
Marakk*, the "Loin of the Bear"; the star was
known to the Greeks as *Helice* or *Helike*, from the city of
Callisto in Arcadia. R.H.Allen states that it appears as
a celestial sphere or Armillary sphere in Chinese charts,
with the title *Tien Seuen*; the Hindu name *Pulaha* honored
one of the legendary sages or *rishis*. The star is magnitude
2.37; Spectrum A1 V; Position 10588n5639. This is one of
the bright members of the Ursa Major moving cluster, and
lies 5.4° south from Alpha; these are the two "Pointer"
stars which indicate the direction to Polaris. Beta Ursae
is very similar to Sirius in general type, but somewhat
more luminous. The accepted distance is about 80 light
years, the actual luminosity about 65 times that of the
Sun, and the absolute magnitude about +0.5. The star shows
an annual proper motion of 0.09"; the radial velocity is
7 miles per second in approach with slight variations.
  About 1.5° to the SE the observer will find a bright
edge-on galaxy NGC 3556, and less than 1° further on the
large "Owl Nebula" M97, one of the nearest of the planetary
nebulae.(Refer to page 1996)

GAMMA Name- PHAD or PHECDA, sometimes given as Phacd
or Phekda, all from the Arabic *Al Fahdh*, "The
Thigh". Magnitude 2.44; Spectrum A0 V; Position 11512n5358.
Another bright member of the Ursa Major group, probably at
a slightly greater distance than Beta, the computed distance being about 90 light years. This gives an actual luminosity of about 75 suns and an absolute magnitude of +0.2.
The star shows an annual proper motion of 0.09"; the radial
velocity is about 7.5 miles per second in approach. This
star and Beta mark the bottom of the Bowl of the Dipper.

About 0.7° to the SE will be found the hazy spot of NGC 3992, a bright barred spiral galaxy of the 11th magnitude. This galaxy has been added to the original Messier catalogue in some observing guides, under the designation M109.

DELTA   Name- MEGREZ, from the Arabic *Al Maghrez*, the "Root of the Tail". The Chinese title *Tien Kuen* is interpreted "The Authority of Heaven". This is the faintest of the seven stars which compose the Great Dipper, and marks the junction of the Bowl with the Handle. Delta Ursae Majoris is magnitude 3.30; Spectrum A3 V; Position 12130n5719. The computed distance is about 65 light years; the actual luminosity about 20 times that of the Sun, and the absolute magnitude +1.9. The star shows an annual proper motion of 0.10"; the radial velocity is 7.5 miles per second in approach.

Megrez is one of the suspected "secular variables" as they are called; stars which have supposedly faded or brightened over the centuries. Miss Agnes Clerke, in her book *Problems in Astrophysics* (1903) considered it to be one of the stars which "have undeniably faded with the lapse of centuries". Tycho rated it as 2nd magnitude, but Ptolemy, some 1800 years ago, had it as 3rd. There appears to be no certain evidence of any real change in modern times.

EPSILON   Name- ALIOTH or ALLIOTH, of uncertain meaning but possibly from the Arabic *Alyat* which is said to refer to the "fat tail" of an animal, or from the phrase *Al Haur* or *Al Hawar*, "The White of the Eye" or the "Bright Eye". Various forms as *Aliot*, *Alhaiath*, and *Aliath* appear on medieval charts. The Chinese *Yu Kang* seems to refer to a "gemmeous transverse", a portion of an astronomical sighting instrument used in ancient times. The Hindu name *Angiras* is said to honor another of the seven sages. Epsilon is magnitude 1.79; Spectrum A0p; Position 12519n5614. This is the brightest of the seven stars which form the outline of the Great Dipper, and one of the more important members of the Ursa Major cluster. Its computed distance is about 70 light years; the actual luminosity about 85 times that of the Sun, and the absolute magnitude +0.2.

Epsilon Ursae Majoris has a peculiar spectrum displaying abnormally strong lines of chromium and europium. Certain spectral features attributed to chromium and calcium show periodic changes in intensity in a cycle of 5.089 days; these variations are accompanied by a very slight fluctuation in light with a range of about 0.02 magnitude. The variations are connected in some way with the star's regularly changing magnetic field. The standard star of the type is Alpha Canum Venaticorum.

In addition to its spectrum variability, the star also shows two longer periods of radial velocity variation which are probably due to unseen companion stars; one has a cycle of 0.95 day, and the other about 4.15 years. The whole system shows an annual proper motion of 0.11" and a radial velocity of about 5.5 miles per second in approach.

ZETA   Name— MIZAR, from the Arabic *Mi'Zar*, a girdle or waistband; the alternate forms of *Mirza* or *Mizat* appear in medieval manuscripts, but sometimes are applied to one of the other bright stars of the figure. The star also was the *Anak al Banat*, or "Necks of the Maidens"; R.H.Allen points out that some editions of Ulug Beg's star tables have *Al Inak* or *Al 'Inz* (The Goat) in place of *Anak*. Mizar is magnitude 2.40; Spectrum A2 V; Position 13219n5511 at the bend of the Great Dipper's handle.

This is probably the best known member of the Ursa Major group; the first double star to be discovered, by Riccioli in 1650, reobserved and described by G.Kirch and his wife in Germany in 1700, and measured repeatedly since the time of Bradley in 1755. For over three centuries it has remained one of the most celebrated double stars, and is often the first example of that class of object to be observed by many a modern amateur. The two stars are magnitudes 2.4 and 4.0 and are separated by 14.4", an easy and striking pair in any good small telescope. As to the colors of this pair, there is slight difference of opinion; W.T. Olcott calls them both white; T.W.Webb sees them both as greenish-white; R.H.Allen calls them "brilliant white and pale emerald". C.E.Barns designates Mizar a "pioneer star" which "never fails to inspire awe, however frequently observed". Not only was it the first known double; but was also the first one to be photographed, in 1857 by G.P.Bond

DESCRIPTIVE NOTES   (Cont'd)

at Harvard. (Fig 1)   The primary is an A2-type star with
about 70 times the luminosity of the Sun; the fainter star
has a somewhat peculiar metal-lined spectrum, currently
classed as A7. The minimum separation of the two stars is
about 380 AU, nearly 5 times the diameter of the orbit of
Pluto.

Since the measurements of Bradley in 1755, the PA of
the pair has increased from 143° to 151°, a change of 8°
in a little more than 200 years; evidently the period of
the pair must be many thousands of years. *"Many radiant
couples,"* wrote Miss Agnes Clerke in 1905, *maintain, decade
after decade, an all but absolute fixity....The nascent
displacements of ζ Ursae Majoris suggest the possible ac-
complishment of a circuit in 10,000 years....."*

Mizar also has the distinction of being the first
binary to be detected spectroscopically. In 1889, Pickering
found that the spectrum lines of the primary became doubled
at period intervals, and announced that the star was a pair
too close to be resolved with any telescope.   (Fig 2)

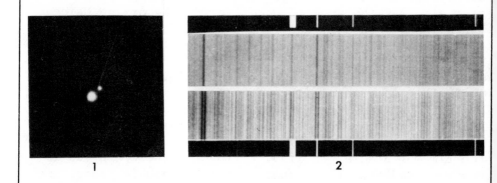

1   2

From the equal intensity of the doubled lines, it is
evident that the two stars are closely comparable in actual
luminosity, each being about 35 times as luminous as the
Sun. The revolution period is 20.5386 days, and the computed
separation of the components about 18 million miles; the
orbit has the fairly large eccentricity of 0.54.

The fainter component of the visual pair, Mizar B, is
known from later studies to be a spectroscopic double also.
Radial velocity shifts in its spectral lines were detected
as early as 1908 by E.B.Frost. A period of about half a

DESCRIPTIVE NOTES (Cont'd)

year was found by F.Gutmann of the Dominion Astrophysical
Observatory, and a value of 182.33 days was announced by
W.R.Beardsley in 1964. In addition, astrometric measurement
of the star's position indicate a third component with a
period of 1350 days. There are also rapid oscillations in
radial velocity, apparently caused by physical pulsations
in the atmosphere of one member of the close pair.

The computed distance of Mizar is 88 light years; it
shows an annual proper motion of 0.12" in PA 105° and a
radial velocity of about 5.5 miles per second in approach.

ALCOR. Immediately to the east of Mizar in the sky will be
noticed the 4th magnitude star 80 Ursa Majoris, called
Alcor, separated from Mizar by 11.8' in PA 72°. The name
itself, according to Admiral Smyth, is supposed to have
been derived from *Al Jawn* or *Al-jat*, a "courser" or "rider"
though the name *Suha*, "The Lost One" or "The Forgotten One"
was also in popular use among the Arabians. The two stars
form the pair now often called "The Horse and Rider", and
are traditionally considered a good test for the naked eye.
T.W.Webb suggested that the brightness of Alcor may have
increased since medieval times, as it is no longer a truly
difficult naked-eye object. The Arabian writer Al Firuz-
abadi, in the 14th Century, however, refers to it as *Al
Sadak*, "The Test", or "The Riddle", while the Persian Al
Kazwini, in the 13th Century, stated that "people tested
their eyesight by this star". Bayer's name *Eques* is translated "The Cavalier"; R.H.Allen tells us that a popular
name in Old England was "Jack on the Middle Horse". In
several German legends, Alcor is connected with the story
of a wagoner named Hans who chose the right to drive his
heavenly wagon across the universe as a reward for his
services to Christ. A Latin name, *Eques Stellula*, is translated "The Little Horseman of the Stars". (Photo, pg 1961)

Alcor is magnitude 4.02, spectrum A5 V, and is a true
member of the Ursa Major group, sharing the proper motion
of Mizar, and distant from it at least a quarter of a light
year. The star is a spectroscopic binary of uncertain period; the two stars have a total luminosity of about 15 suns.
Forming a flattened triangle with Mizar and Alcor is the
8th magnitude star *Sidus Ludovicianum*, so named in 1723 by
admirers of Ludwig V, under the mistaken impression that it
was a newly-appeared star or possibly a planet.

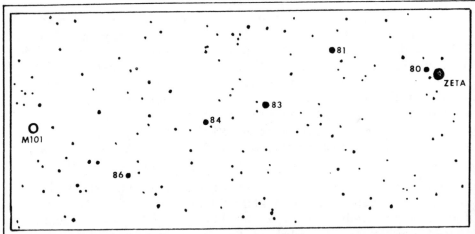

To the east of Mizar, the chain of 4th and 5th magnitude
stars labeled 80, 81, 83, 84 and 86 will aid the observer
in finding his way to the large spiral galaxy M101, about
5½° to the east. The star 83 Ursae Majoris (Mag 4.73, Spect
M2 III) is a suspected variable which is said to have shown
a single nova-like outburst in 1868. According to Miss
Agnes Clerke, "On 6th August, 1868, 83 Ursae Majoris, a
sixth-magnitude star near Mizar, was seen by Birmingham to
be the equal of Delta Ursae; though for that night only.
The next, it had visibly gone off, and before long the
whole of its added splendour had departed..." This is a
star obviously worth an occasional curious glance. On the
above chart the field is approximately 6° wide with north
at the top; stars to about 9th magnitude are shown.

ETA   Name- BENETNASCH, sometimes called ALKAID;
both names derived from the Arabic *Ka'id Banat
al Na'ash*, "The Governor of the Daughters of the Bier",
sometimes rendered "The Leader of the Daughters of the
Mourners". The name appears in various forms; Bayer has it
as *Elkeid* and also *Benenacx*, Riccioli refers to it as *Benat
Elnanschi* and *Benenatz*, while *Bennenazc* appears in the
Alfonsine Tables of 1521. The Hindu *Marici* is one of the 7
Rishis or sages of India; the Chinese *Yao Kwang*, according
to R.H.Allen, is translated "A Revolving Light".
    The star is magnitude 1.87; Spectrum B3 V; Position
13456n4934. This is the star at the end of the handle of
the Great Dipper, but is not a true cluster member, and has

a different proper motion than Beta, Gamma, Delta, Epsilon and Zeta. Sir John Herschel, in 1847, thought Eta the first star in brightness in the constellation; in 1838, however, he had rated Epsilon in first place, which agrees with the modern measurements of the Seven Stars. Miss Agnes Clerke thought that *"The immemorially observed constituents of the Plough preserve no fixed order of relative brilliancy, now one, now another of the septet having at sundry epochs assumed the primacy.."* No definite proof of clear changes in the order of brightness has been obtained in modern times.

Eta Ursae Majoris has a computed distance of about 210 light years; the actual luminosity must be close to 630 suns, and the absolute magnitude -2.1. The star shows an annual proper motion of 0.12"; the radial velocity is 6.5 miles per second in approach. The fine "Whirlpool Galaxy" M51 in Canes Venatici lies just 3½° distant, to the SW. (Refer to page 369)

THETA   (25 Ursae Majoris) Position 09296n5155; Magnitude 3.19; Spectrum F6 IV. The star has no generally used proper name. According to R.H.Allen, the term *Sarir Banat al Na'ash* was in use among the Arabs for the group including Theta, Upsilon, Phi, and several other faint stars in the neck of the Bear. The name signifies "The Throne of the Mourners". Theta has a computed distance of about 65 light years. The actual luminosity is about 16 times that of the Sun, the annual proper motion is 1.09" in PA 240°, and the radial velocity is 9 miles per second in recession.

The faint companion, of magnitude 13½, was first seen by S.W.Burnham with the 36-inch refractor at Lick Observatory in 1889. This is a common motion pair, but the orbital revolution is quite slow; between 1889 and 1958 the PA changed from 75° to 101°, and the separation decreased from 5.1" to 4.1". The projected separation at present is about 85 AU, and the faint star is a dwarf of uncertain type, some 700 times fainter than the Sun.

Approximately 1.7° WSW from Theta Ursae Majoris lies the fine spiral galaxy NGC 2841, one of the best examples of a very regular and symmetrical Sb-type system. (Refer to photograph on page 2005)

IOTA　　　　Name- TALITHA or TALITA, probably from the
　　　　　　　Arabic *Al Kafzah al Thalithah*, usually translated "The Third Leap" or "The Third Spring"; sometimes it
appears as "The Third Vertebra". The name seems to have
been used for Iota and Kappa together; the two stars marking the front feet of the Bear. Iota is Magnitude 3.12;
Spectrum A7 V; Position 08558n4814. The star is also designated 9 Ursae Majoris.

　　　　The computed distance is about 50 light years; the
annual proper motion is 0.50" in PA 241°, and the radial
velocity is about 7½ miles per second in recession. The
star has an actual luminosity of about 11 suns.

　　　　John Herschel, in 1820, seems to have been the first
to note the 10th magnitude companion which forms a common
proper motion pair with Iota. The separation at discovery
was about 10" but has slowly narrowed to about 4.5"; the
PA in the same interval increased from 280° to 360° (1909)
and then up to 16° in 1958; a measurement made in 1969 gave
4.4" in PA 21°. The period is unknown, but must be several
centuries at the very least; the present projected separation is about 70 AU. T.W.Webb quotes a statement by Buffham
that the companion appears "very dull for its size" and

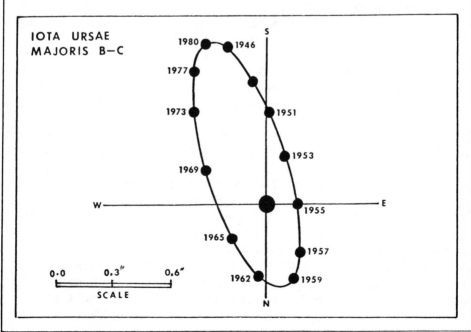

IOTA URSAE MAJORIS B—C

that E.S.Holden suggested the possibility that the companion might be shining by reflected light. This impression was evidently a result of the contrast between the bright white or yellowish blaze of the primary and the weak ruddy glow of the red dwarf companion. R.H.Allen calls the colors "topaz yellow and purple".

The faint reddish companion is itself a close binary for large telescopes; the orbit is quite rapid with a computed period of 38.7 years. According to calculations by P.Baize (1961) the semi-major axis is 0.7", the inclination is 109°, the eccentricity is 0.39, and periastron was in 1958. Both stars are red dwarfs of spectral types dM1, and differ in brightness by less than half a magnitude. Each star is roughly 1/100 the luminosity of our Sun, and the mean separation is about 11 AU, or just slightly more than the separation of Saturn from our Sun.

The star once called "10 Ursae Majoris", but now actually located in Lynx, is 6.2° to the south, but seems to show the same proper motion and parallax as Iota. It is also a binary, and a very rapid one, with a 22 year period. (For data refer to page 1125)

KAPPA Name- AL KAPRAH, but possibly used by error, as some star charts apply the name to Chi Ursae Majoris instead; the Arabian name *Talitha* was given to Iota and Kappa together. The two stars are just 1° apart and mark the Bear's front paw. Kappa is magnitude 3.68; spectrum B9 or A0; Position 09002n4721. The star lies at a computed distance of about 300 light years; the actual luminosity then must be about 250 times that of the Sun. The annual proper motion is 0.07"; the radial velocity is about 2.7 miles per second in recession.

Kappa Ursae Majoris is a very close binary system, discovered by R.G.Aitken in 1907, with the 36-inch refractor at Lick Observatory. The two stars are magnitudes 4.3 and 4.5 and the orbital period is slightly under 58 years with a mean separation of 0.3". According to computations by P.Baize (1951) the semi-major axis is 0.27"; the orbit inclination is 101° and the eccentricity is 0.30. If the computed distance is correct, the true separation must be fairly comparable to that of Uranus and the Sun. The pair is suitable only for powerful telescopes.

LAMBDA   Name- TANIA BOREALIS, from the Arabic *Kafzah al Thaniyah*, "The Second Leap" or "The Second Spring"; the star forms a bright 1.5° pair with Mu Ursae, the two stars marking the Bear's left hind foot. In the Latin translation of Ulug Beg's star tables the name is given as *Al Phikra al Thania*, but the word *Phikra*, from *Al Fikrah*, "The Vertebra", seems to be an error, says R.H. Allen, as the pair of stars has always been considered to mark the feet of the Bear.

Lambda Ursae Majoris is magnitude 3.45; Spectrum A2 IV; Position 10141n4310; the star is also known as 33 Ursae Majoris. The computed distance is about 150 light years, the actual luminosity about 75 times that of the Sun, the annual proper motion is 0.17" and the radial velocity is about 11 miles per second in recession. This star is one of a large number which show about the same space motion as the Hyades cluster in Taurus, and are considered to be outlying members of the "Taurus Stream" associated with the cluster.

MU   Name- TANIA AUSTRALIS, marking, with Lambda, one of the feet of the Bear. The star is also known as 34 Ursae Majoris. Magnitude 3.05; Spectrum M0 III; Position 10194n4145. Observe the Lambda-Mu pair with low power field glasses or small telescope to note the interesting difference in color. Mu Ursae is a red giant star of moderate luminosity, located at a distance of about 105 light years; the computed luminosity is about 50 times the brightness of our Sun. The star shows an annual proper motion of 0.09"; the radial velocity is 12.5 miles per second in approach. From spectroscopic studies the star is known to be a close binary with a period of 230.089 days; the semi-major axis of the orbit of the visible star is about 14.5 million miles with an eccentricity of 0.06.

The fine face-on spiral galaxy NGC 3184 will be seen in the field with Mu Ursae, about 0.7° distant nearly due west. In small telescopes it appears as a round glow about 4' to 5' in diameter with a slightly brighter center. Another fainter galaxy, NGC 3319, lies 2.8° distant, almost due east; this is a much-elongated barred spiral with a flattened central region; the total length is about 6'.

STAR FIELDS IN URSA MAJOR. Top left: α U.Maj. and its 7th magnitude companion. Top right: Mizar and Alcor. Below: Mu U.Maj. and the galaxy NGC 3184.   Lowell Observatory

XI   Name- ALULA AUSTRALIS, from the Arabic *Al Kafzah al Ula*, "The First Spring". The star Nu Ursae, just 1.6° to the north, is Alula Borealis. Ulug Beg's tables give *Al Fikrah al Ula* for Xi Ursae. The star is also designated 53 Ursae Majoris. Magnitude 3.74; Spectrum G0 V; Position 11155n3149 in the southern portion of the constellation, about 11° north fron Delta Leonis.

Xi Ursae Majoris is a fine close binary of great historical interest, found by Sir William Herschel in May 1780, and announced as a physically associated pair in 1804 when the change in PA since discovery amounted to 59°. This star was the first binary to have an orbit computed, by M. Savary in 1828; as Miss Agnes Clerke phrases it, the star was "the subject of the first experiment in the extension of Newtonian principles to the sidereal universe". The pair has a period of slightly under 60 years, with closest apparent separation of 0.9" in 1933 and widest separation of 3.1" in 1975. The individual magnitudes are 4.3 and 4.8 and both components closely resemble our own Sun in type, size, and brightness. This is one of the nearest binaries to the Solar System, 26 light years distant, and shows a fairly large annual proper motion of 0.73" in PA 216°. No very definite color contrast exists in this pair; to most

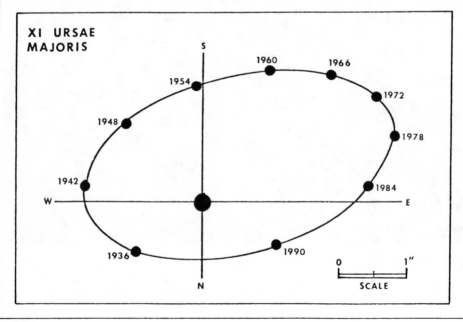

XI URSAE MAJORIS

observers both stars appear a clear pale yellow; R.H.Allen,
however, calls them "subdued white and grayish white".
    According to orbit computations by T.Rakowiecki in
1938, the period is 59.74 years with periastron in 1935;
the semi-major axis is 2.56" or about 21 AU; the eccentricity is 0.40 and the inclination 121°. From the orbtial elements, the total mass appears to be about 2½ suns. Predictions for the system are given in the following table,
according to an ephemeris by W.Wegner in Germany:

1975	3.10"	115.0°	1984	2.48"	94.5°	1993	0.85"	359.9
1976	3.09	113.0	1985	2.33	91.2	1994	0.93	335.5
1977	3.07	111.0	1986	2.15	87.4	1995	1.10	316.9
1978	3.03	109.0	1987	1.95	82.9	1996	1.29	303.6
1979	2.98	106.9	1988	1.73	77.4	1997	1.47	293.7
1980	2.92	104.8	1989	1.50	70.1	1998	1.61	285.8
1981	2.84	102.5	1990	1.27	60.2	1999	1.71	278.9
1982	2.74	100.0	1991	1.05	45.9	2000	1.77	272.7
1983	2.62	97.4	1992	0.89	25.5	2001	1.80	266.8

    Both stars of this pair are spectroscopic binaries.
In the case of the brighter star, both spectroscopic and
astrometric studies confirm the duplicity, and give the
period as 669.17 days; the semi-major axis of the orbit of
the brighter component is about 38 million miles, with an
eccentricity of 0.53. Xi-B is a much closer pair with a
period of only 3.98 days and a nearly circular orbit. This
well known double star is thus actually a quadruple system
of suns, approaching us at about 10 miles per second.

OMICRON   Name- MUSCIDA according to Bayer, from a word
             attributed to the "Barbarians" of the Middle
Ages, and is a portion of the asterism known to the
Persians as *Al Thiba*, the Gazelle, including also Sigma, Pi
and Rho. Omicron is also designated 1 Ursae Majoris. Magnitude 3.37; Spectrum G5 III; Position 08262n6053. The star
has a computed distance of about 150 light years, giving
the actual luminosity as about 85 suns. The annual proper
motion is 0.17"; the radial velocity is 12 miles per second
in recession.
    A 15th magnitude companion was discovered in 1889 at
Lick, by S.W.Burnham, using the 36-inch refractor;   the

separation at discovery was 7" in PA 191°. This small star evidently shares the proper motion of the primary, since no relative change could be detected by Burnham in a 10-year interval. Unfortunately, the star can be observed only in great telescopes, and standard double star catalogues list no measurements of the pair since 1899. The projected separation of the components is about 320 AU, and the companion is a dwarf star of uncertain type with an actual luminosity of about 1/600 that of the sun.

SIGMA-2   (13 Ursae Majoris) Position 09060n6720, forming with Sigma-1 and Rho a small group marking the ears of the Great Bear. Magnitude 4.78; Spectrum F7 IV or V. The star is a slowly rotating binary of very long period, discovered by Sir William Herschel in 1783, and measured repeatedly since the time of F.G.Struve in 1832. The stars are magnitudes 4.8 and about 8.3 with a separation of 3.7" (1974); the pair is now widening and will continue to do so for several centuries. Several orbits have been computed, ranging from about 600 years up to 1100 years. According to calculations by W.Rabe (1961) the best modern computation is about 706 years with periastron occurring in 1918; the semi-major axis of the orbit is about 4.9" and the eccentricity 0.75. Orbital motion is retrograde, with an inclination of 137°.

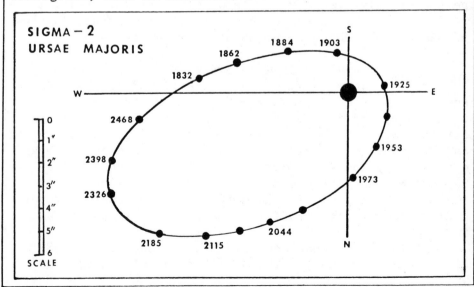

SIGMA−2
URSAE MAJORIS

T.W.Webb reports that both Sadler and Dembowski considered the smaller component to be variable, with an estimated magnitude range of 8 to 10. No strong colors are seen
is this pair; Webb thought the primary "greenish" and found
no certain contrast in the colors of the two stars.

From a direct parallax measurement the distance seems
to be about 60 light years; the actual luminosities of the
stars are then 5.0 and 0.2 suns, and the mean separation
about 90 AU. The star shows an annual proper motion of 0.09"
and the radial velocity is less than 1 mile per second in
approach.

PHI   (30 Ursae Majoris) Position 09487n5418. Magnitude 4.54; Spectrum A3. This is the pair
0Σ208, a rather close binary system discovered by F.G.W.
Struve in 1843, and suitable as a test object for a good
10-inch telescope. The two stars are very nearly equal in
magnitude, and are in slow direct revolution in a period of
about a century. According to an orbit computation by S.
Wierzbinski (1957) the period is 107 years with periastron
in 1882; the semi-major axis is 0.35", the inclination 27½°
and the eccentricity is 0.44. From a direct trigonometric

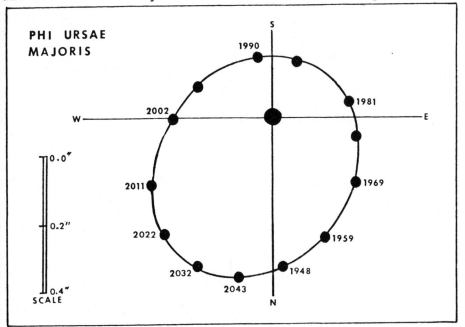

PHI URSAE
MAJORIS

parallax obtained at Sproul Observatory, the system appears
to be at a distance of about 120 light years; the resulting
total luminosity is then about 17 suns, and the true mean
separation of the components is about 13 AU. These results,
however, would give the system an unusually small total
mass, suggesting that the distance is actually much greater
than the adopted parallax seems to imply. The star shows a
small proper motion of about 0.01"; the radial velocity is
near 7 miles per second in approach.

The rapidly revolving dwarf binary system W Ursae
Majoris is located near Phi, approximately 2.2° to the NNW.
(Refer to page 1968)

PSI
(52 Ursae Majoris) Position 11069n4446, the
brightest star about 10° below the bowl of the
Great Dipper. Magnitude 3.01; Spectrum K1 III; a yellow
giant star about 85 times the luminosity of our Sun. The
computed distance is about 130 light years; the annual
proper motion is 0.07" and the radial velocity is 2½ miles
per second in approach. According to R.H.Allen, John Reeves
states that this is the Chinese *Ta Tsun*, "The Extremely
Honorable".

R
Variable. Position 10412n6902. The first of
the long-period variables in Ursa Major, found
by N.Pogson in 1853. A maximum in mid-May 1843 has been
established, however, from pre-discovery observations made
with the transit circle at Oxford. The star is located 7°
NNW from Alpha Ursae Majoris, and is easily found in the
field of the 5th magnitude star GC 14713 (Magnitude 5.22,

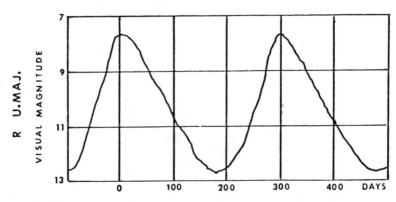

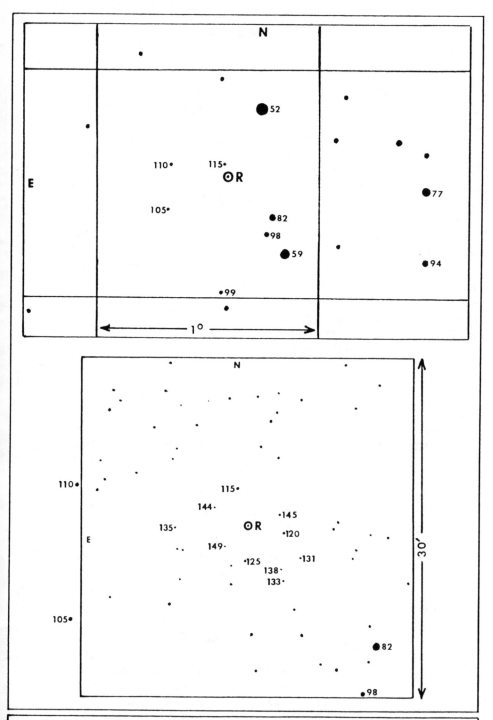

R URSAE MAJORIS Identification fields. Comparison magnitudes are given (with decimal points omitted) according to observations of the AAVSO.

1967

Spectrum K3 III) which is located at 10395n6920. This is
the brightest star on the identification chart (page 1967)
and is just 20' NNW of the variable. The field is one that
is easily identified, making this an especially suitable
variable star for the amateur observer.

R Ursae Majoris is a pulsating red giant of the Mira
class, closely comparable to Mira itself in period, range,
and spectral type. Its usual range is from magnitude 7½ to
about 13 in a cycle averaging 302 days; the star occasionally reaches 6.7 or so at a high maximum. When near the
bottom of its cycle R Ursae is one of the reddest of the
long-period variables; the color, however, weakens as the
star rises to maximum. From the difference in apparent
brightness, R Ursae appears to be about 6 times more remote
than Mira; the estimated distance is about 1350 light years
and the actual peak luminosity about 250 times that of our
Sun.

The light curve of the star on page 1966 is based on
a study of 33 cycles of the star by Leon Campbell of the
AAVSO. During these cycles the maximum magnitude of the
star ranged from 6.7 to 8.2, and at minimum from 12.3 to
about 13.4. The star brightens about twice as rapidly as it
fades; the average time from minimum to maximum is about
116 days. R Ursae shows an annual proper motion of 0.047"
according to R.E.Wilson's *General Catalogue*; the radial
velocity is about 20½ miles per second in recession. (See
also Omicron Ceti, page 631)

W   Variable. Position 09403n5611, about 2.2° NNW
             from Phi Ursae Majoris. This is a noted dwarf
eclipsing binary system, discovered by G.Muller and P.Kempf
at Potsdam in 1903. It is the typical example of a class of
binaries characterized by small diameters, low masses, and
very short periods. Each revolution of the components of W
Ursae Majoris is completed in 0.33364667 day, or just a
fraction of a minute over 8 hours. There are two eclipses
in each revolution, the primary eclipse being only 0.1 magnitude deeper than the secondary one; the photographic
range of the system is 8.3 to 9.06. Although the tilt of
the orbit is thought to be some 10° from edge-on, primary
eclipse is total or very nearly so, and lasts for about 20
minutes; the full duration of each eclipse is about 2 hours.

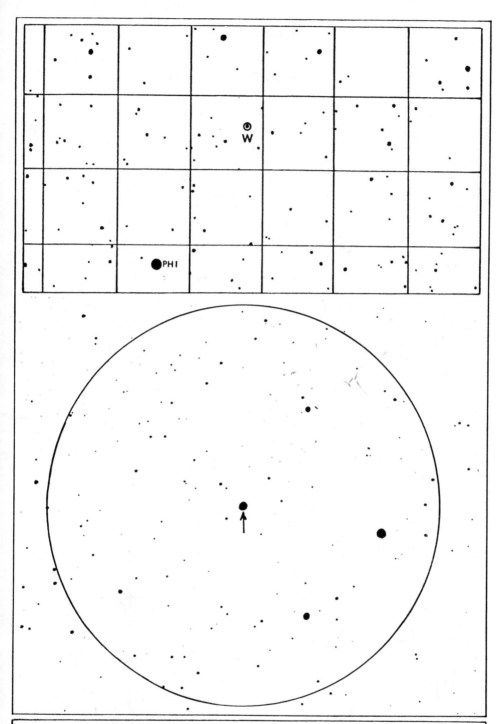

W URSAE MAJORIS Identification charts. Top: Grid squares
are 1° on a side, limiting magnitude about 9½. Below: One
degree field surrounding the star. North is at the top.

Secondary eclipse follows just 4 hours after primary minimum, and is evidently a large partial obscuration. Both of
the stars are dwarfs of class F, fairly comparable in size
and luminosity to our Sun. W Ursae has been the subject of
numerous investigations; from the combined results of seven
different studies, the table given here has been prepared:

	Diam.	Mass.	Spect.	Abs.Mag.	Lum.
A	1.14	0.99	dF8	+4.42	1.45
B	0.83	0.62	dF6	+4.75	1.0

There are substantial discrepancies in the results
obtained from the different studies, owing to the fact that
the precise inclination is uncertain, and that the radial
velocity measurements are distorted by rapidly moving gas
streams between the stars. There is a general agreement
that the primary is closely comparable to the Sun in mass
and diameter, and that the secondary has close to 60% the
mass of the primary. The difference in luminosity is at
least 0.3 magnitude, and, according to one study, may be
as great as 1.0 magnitude. W Ursae Majoris is the standard
example of a binary in which the stars are nearly or quite
in contact, and revolve in the same atmospheric envelope.
The computed separation, center to center, is close to one
million miles, which places the surfaces virtually in con-

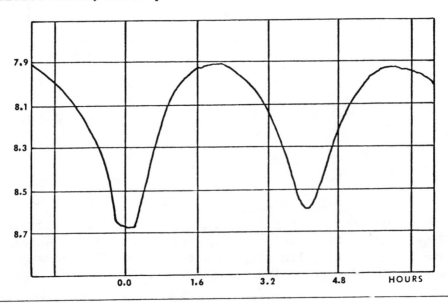

tact. Both stars are distorted into egg-shaped ellipsoids by rapid rotation and tidal effects; the best radial velocity measurements indicate that the components are moving at about 75 and 150 miles per second in their orbits. The entire system shows an approach radial velocity of about 30 miles per second. From a direct trigonometrical parallax the distance appears to be about 200 light years.

A common proper motion companion of the 13th magnitude was discovered by T.E.Espin in 1920; it is 7" distant from the primary in PA 50° and both stars share the motion of 0.06" in PA 132°. At the adopted distance the projected separation is about 435 AU. This pair is also known as ADS 7494 and Es 1825.

According to a study by Harlow Shapley, the W Ursae Majoris stars probably outnumber all other known varieties of eclipsing binary stars, by a factor of about twenty. In the Moscow *General Catalogue of Variable Stars* (1971) 374 stars of the type were listed. As the majority are faint objects, only about 30 have been observed in detail. It is thought that such "contact binaries" are probably the ancestors of the eruptive SS Cygni stars; both classes of objects have similar masses and periods, and are similarly distributed in space. (Refer also to SS Cygni, U Geminorum and U Pegasi)

SU Variable. Position 08081n6246, about 3° NW from Omicron Ursae Majoris, at the tip of the Bear's nose. SU Ursae is a faint erratic variable of the "dwarf nova" or "cataclysmic variable" class, resembling SS Cygni and U Geminorum. It is normally a 15th magnitude object, but at intervals of about 16 or 17 days it rises suddenly to magnitude 11. The light increase thus amounts to about 40 times, and is accomplished in about 24 hours. On occasion the period has been as short as 8 days; as of

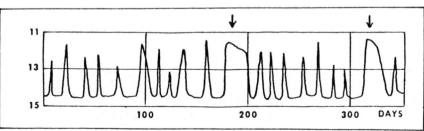

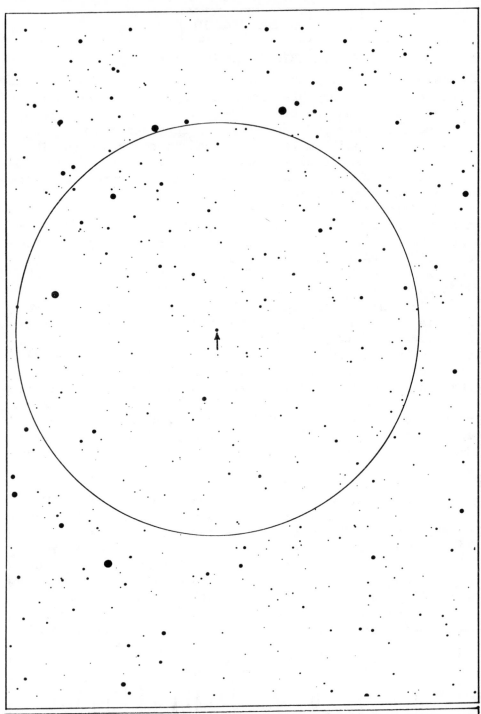

SU URSAE MAJORIS Identification field, from a Lowell Observatory 13-inch telescope plate. The circle is 1° in diameter with north at the top; limiting magnitude about 15.

1975 this remains the shortest interval yet observed for
stars of the type. The light variations were first detected
by L.Ceraski at Moscow in 1908.

SU Ursae has several notable peculiarities. It shows
frequent periods of accelerated activity when the maxima
may follow each other at about half the normal interval;
the star SS Aurigae shows the same sort of restless sput-
tering, as illustrated by the light curve on page 283. SU
Ursae also shows occasional "supermaxima" when the star
remains bright for a time equal to a whole average period;
the brightness at these times is also about a magnitude
higher than a normal maximum. Two such supermaxima are indicated by the arrows on the light curve on page 1971; the
average interval between supermaxima is about 6 months. A
similar activity is shown by the star AY Lyrae.

SU Ursae has a peculiar spectrum, appearing nearly
continuous at maximum, and with an energy distribution corresponding to spectral type late B or early A. Bright lines
of hydrogen appear as the star fades, and there are faint
indications of the superimposed spectrum of a yellow dwarf

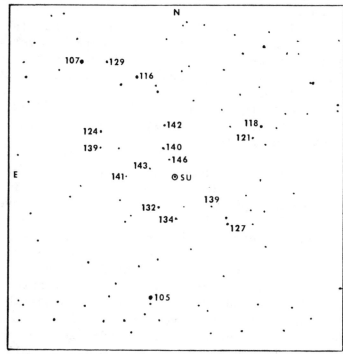

SU
U. MAJ.

30' FIELD

of type G. According to current theory, all the stars of
the type are believed to be extremely close binaries in
which one of the stars is a white dwarf or semi-degenerate;
a typical period for systems of this type is about 5 hours.
SU Ursae, however, is probably not an eclipsing system, as
no periodic dips have been detected in the light curve. The
star shows constant "flickering" at minimum, with rapid
oscillations on the order of 0.3 to 0.4 magnitude, with a
rough periodicity of about 5 minutes. The flickering often
remains detectable while the star is rising to a maximum.

According to a study by R.P.Kraft (1962) stars of the
class have absolute magnitudes near +9 at minimum; during
an outburst the luminosity rises to about that of the Sun.
In the case of SU Ursae, the blue component appears to be
about 1 magnitude fainter than the cooler G-star, so the
actual range of the nova-like component may be about 4½
magnitudes. No parallax has been detected for the star, but
if SU Ursae is comparable in actual luminosity to SS Cygni
the computed distance must be about 1200 light years. W.J.
Luyten has measured an annual proper motion of about 0.02"
for the star.

The field lies about 1° west of the star labeled 57
Camelopardalis on star atlases; the star, however, is now
actually in the constellation Ursa Major. In the chart on
page 1973, comparison magnitudes are given (with decimal
points omitted) according to observations of the AAVSO.

(Refer also to SS Cygni, U Geminorum, SS Aurigae, AE
Aquarii, and WZ Sagittae. For an account of novae in general, see Nova Aquilae 1918)

UX   Variable. Position 13374n5210, almost exactly
between Zeta and Eta Ursae Majoris in the handle of the Great Dipper. The star lies slightly over 1° SSW
from 82 Ursae Majoris, which appears in the upper left corner of the chart opposite. UX Ursae Majoris is a remarkable
dwarf eclipsing binary star, discovered by S.Beljawsky in
1933, and famous at the time for having the shortest period
of revolution known, 4 hours and 43 minutes. Several still
shorter periods are known at present, including the former
nova DQ Herculis, the eruptive variable stars U Geminorum
and SS Aurigae, and the present record-holder WZ Sagittae,
which has an eclipsing cycle of 81.6 minutes. All of these

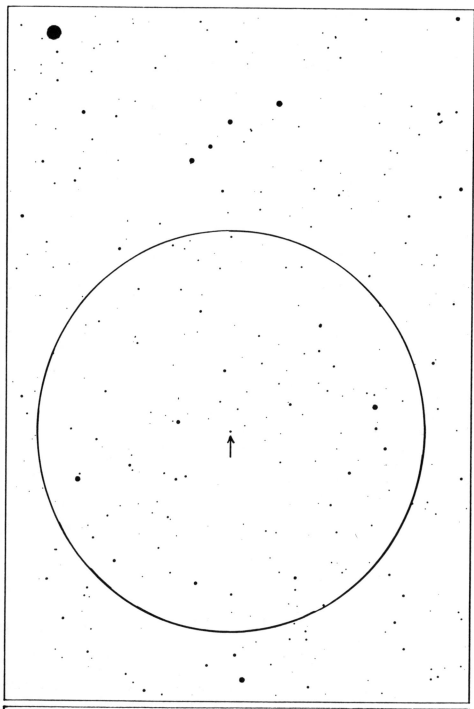

UX URSAE MAJORIS Identification field, from a Lowell Observatory 13-inch telescope plate. The circle is 1° in diameter with north at the top; limiting magnitude about 15.

DESCRIPTIVE NOTES   (Cont'd)

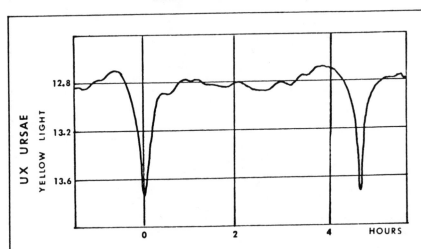

stars appear to be very similar systems, which raises some interesting speculations about the past or future history of UX Ursae.

   The star has a magnitude range of 12.7 to 13.8 in yellow light, and a primary eclipse which lasts about 40 minutes. The orbit of the two stars is oriented about 5° from the edge-on position, and the primary eclipse appears to be nearly, but not completely, total. No secondary minimum has been definitely recorded. The extremely short period implies very small diameters, and suggests that the two stars must be very nearly in contact, making the system an object of unusual interest. Both components are dwarfs, and must be intermediate between the main sequence and the white dwarf state. A spectral type of about B3 has been obtained for the primary; the spectral class of the companion is not definitely known.

   In an analysis of the system in 1949, P.Parenago found that the stars are separated by about 1.2 million miles, and that each star is about half the diameter of the Sun. The computed masses are close to 3.0 and 2.0 suns, giving the densities as 25 and 15 times the solar density. This is at least 100 times denser than a normal main sequence star of the same spectral class. With an absolute magnitude of about +3, the primary star lies about 5 magnitudes below the main sequence. The star is underluminous by a factor of about 100.

   The light curve is distinctly peculiar. There is a rising hump of about 0.2 magnitude just before the descent

DESCRIPTIVE NOTES   (Cont'd)

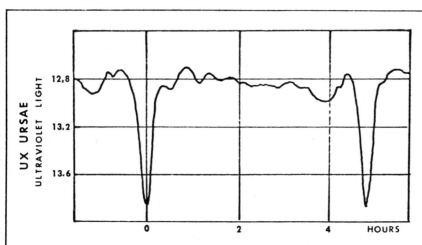

into primary eclipse, and frequently a slight "standstill"
in brightness just before the end of the eclipse. The light
curve, however, does not always repeat exactly, and the
differences are sometimes as great as half a magnitude in
ultraviolet light, somewhat less visually. In addition to
the light variations caused by the eclipses, the system is
evidently somewhat variable on its own account.

Intrinsic fluctuations outside of eclipse were first
detected at Harvard's Oak Ridge Station in 1949. They are
of very short duration, sometimes showing a periodicity of
about 20 minutes, and an amplitude generally under 0.2 magnitude. The cause of these variations is uncertain, but may
be the result of rotating gas streams between the stars;
there appears to be a bright gas cloud associated with the
primary component, and sharing its orbital revolution.

A striking feature of UX Ursae Majoris is its great
similarity to DQ Herculis, the famous eclipsing system
(period = $4^h 39^m$) which flared up as a 1st magnitude nova
in 1934. Is UX Ursae scheduled to become a nova at some time
in the future, or has it possibly had such an outburst already in the past? Although no definite answer can be given
at present, this star seems to be one of the most likely
candidates for novahood known in the heavens. Although a
rather faint object for small telescopes, the star may be
observed in a good 8-inch or 10-inch reflector; any sign of
an increase to unusual brightness should be reported to a
major observatory immediately. (Refer also to DQ Herculis,
GK Persei, and Nova Aquilae 1918)

GROOMBRIDGE 1830 (LFT 855) (GC 16253) (BD+38°2285)
Position 11501n3805, about 16°
due south from Gamma Ursae Majoris, and 7.8° NE from Nu.
The star is plotted on the Skalnate Pleso *Atlas of the
Heavens*, but is not labeled. Groombridge 1830 is noted for
its exceptionally large proper motion of 7.04" per year,
the third fastest known. It was first noted by F.W.A.Arge-
lander in 1842, and remained for over 50 years the largest
proper motion known. Kapteyn's Star (8.70" per year) was
discovered in 1897, and Barnard's Star (10.29" per year) in
1916. According to W.J.Luyten, the precise annual proper
motion of Groombridge 1830 is 7".042 in PA 145.5° and the
star requires 511 years to change its position by 1° in the
sky, a rate of progress which, as Miss Agnes Clerke stated,
"would carry it in 185,000 years round the entire [celesti-
al] sphere, or in 265 [years] over as much of it as the
sun's diameter covers". Obviously, if the majority of bright
stars had proper motions of this size, the constellation
patterns would become unrecognizable in a few centuries.
    The apparent motion of the star is illustrated below
as it appears on the blink-field of the Lowell Observatory
blink-comparator; the two plates were made in 1929 and 1961
and the scale is approximately 19" per mm. The total motion
in 32 years amounts to 225".
    Groombridge 1830 is a G8 yellow star of visual magni-
tude 6.5, slightly underluminous for its spectral class and

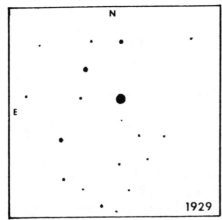 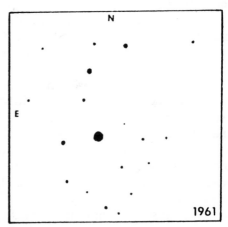

PROPER MOTION OF GROOMBRIDGE 1830, from Lowell Observatory
13-inch telescope plates. Scale approximately 19"/mm.

1978

DESCRIPTIVE NOTES   (Cont'd)

thus defined as a subdwarf. The actual luminosity is about
1/7 that of the Sun, and the absolute magnitude about +6.8.
From direct parallaxes, the distance is close to 28 light
years. As this is not an unusually small distance (Barnard's Star is only 6 light years away) the large apparent
motion must indicate an abnormally high true space velocity
of about 216 miles per second. This is a remarkable example
of a "high velocity star" and one of the nearest known members of the Pop.II galactic halo. The star has a peculiar
metal-deficient spectrum indicating that the content of the
metallic atoms is only about 1/30 that of the Sun. Presumably the star was formed at a very early period in the

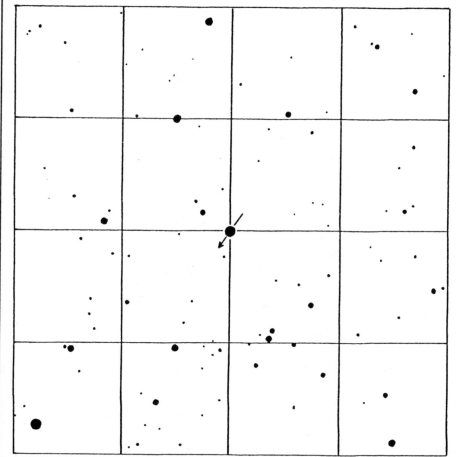

FIELD OF GROOMBRIDGE 1830. Grid squares are 1° on a side
with north at the top; limiting magnitude about 11.

history of the Galaxy, probably preceding the formation of
the Galactic disc and the pattern of the spiral arms. O.J.
Eggen has identified several stars which appear to share
the space motion of Groombridge 1830; these include the
noted variable RR Lyrae, the two faint doubles ADS 10938 in
Serpens and ADS 16655 in Cassiopeia, and BD+72°94 which is
also in Cassiopeia. With the exception of RR Lyrae, all the
members are subdwarfs of types sdF8, sdF8 and sdF2, with
spectral peculiarities resembling those of Groombridge
1830 itself.

At present the star shows a high approach radial velocity of about 61 miles per second; the approach will continue until about the year 9900 AD when the star will be
0.1 magnitude brighter than at present and the annual proper motion will have increased to 8.11". Some 100,000 years
from now the star will have faded to an apparent magnitude
of 9.1, and will have moved to the constellation Lupus.

In 1968 the discovery of a faint companion to the
star was announced by P.van de Kamp at Sproul Observatory.
The companion, two magnitudes fainter than the primary, was
at 1.7" distance in PA 166°. Since the faint star does not
appear on many other plates, it is concluded to be a rapid
variable, possibly a flare star, with a range of at least
8.5 to 12.0. Possibly its variations account for a reported
0.6 magnitude increase in the brightness of Groombridge
1830 observed in 1939. If the total mass of the system is
close to one sun, the expected orbital period is on the
order of 50 to 100 years. (Refer also to Barnard's Star in
Ophiuchus, and Kapteyn's Star in Pictor)

LALANDE 21185 (LFT 756) (GC 15183) (BD+36°2147)
Position 11007n3618, about 2.2° due
south of the double star 51 Ursae Majoris. Lalande 21185 is
one of the nearest known stars, ranking presently in fourth
place, immediately after the Alpha-Proxima Centauri system,
Barnard's Star in Ophiuchus and Wolf 359 in Leo. With a
parallax of 0.395" the star is 8.3 light years distant, a
little farther than W359, but about half a light year closer than Sirius. The star shows an unusually large annual
proper motion of 4.78" in PA 187°, the eighth largest known
proper motion. (Refer to list on page 1257) Observers will
find the star plotted, but not identified, on the Skalnate

Pleso *Atlas of the Heavens*; the field is about 4° NW from
Nu Ursae Majoris, and roughly 9° west and 1.8° south from
Groombridge 1830.

Lalande 21185 is a red dwarf star of apparent visual
magnitude 7.6, absolute magnitude about +10.5 and actual
luminosity of 0.0048 that of the Sun. It has a spectral
class of dM2 and a computed mass of about 0.3 sun. The very
large proper motion is a result of the nearness of the star
combined with a large space velocity of about 62 miles per
second. The radial velocity is 52 miles per second in
approach. Variations in the proper motion were discovered
by G.Land in 1941, and the presence of an unseen companion

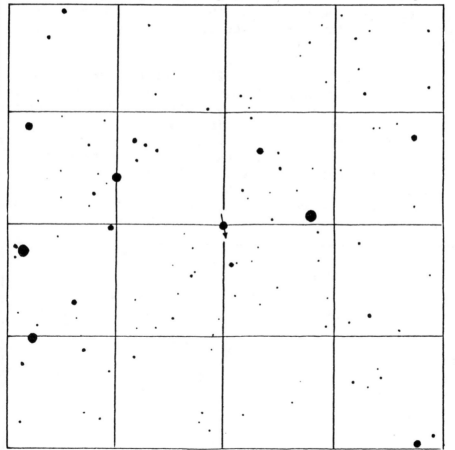

FIELD OF LALANDE 21185. Grid squares are 1° on a side with
north at the top; stars to about 11th magnitude are shown.

seems well established. An analysis of measurements from 1912 to 1959 has been made by S.Lippincott; the resulting period is about 8 years and the computed mass of the unseen companion is about 1/100 the solar mass. From the computed orbit the maximum separation is expected to be about 0.04" (1953, 1961, 1969, etc) and it is unlikely that the faint star will ever be detected visually. The computed semi-major axis is 0.03"; the eccentricity is 0.30 and the inclination is 79°. Greatest separation occurs near western elongation (PA near 260°) and closest approach occurs near PA=0° when the separation is about 0.005". As the computed mass of the unseen star is less than that of any known visible star, it has been suggested that we may have detected here a very massive planet instead. (Refer also to 61 Cygni and Barnard's Star in Ophiuchus)

M40   Position 12198n5822, about 1.5° NE from Delta Ursae Majoris. M40 is one of the few real mistakes in the Messier catalogue, and is listed here only for the benefit of members of Messier clubs who wish to complete their "set". No nebula or cluster exists at the spot; the object is merely a close pair of faint stars which, of course, might appear fuzzy under inferior seeing conditions and was, in fact, considered by J.E.Bode to be slightly nebulous. The original authority for a nebula at this position seems to have been Hevelius in 1660; he listed as

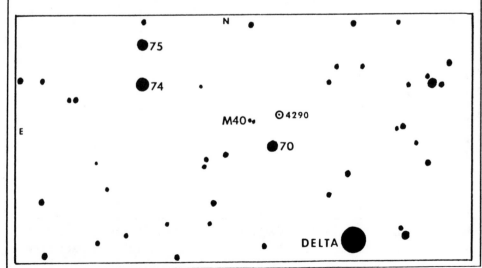

No.1496 in his *Prodromus Astronomiae* a "nebula above the
back [of Ursa Major]". Wm.Derham and Maupertius subsequently
included it in their catalogues. Messier, in 1764 "looked
for the nebula which is above the back of the Great Bear...
I found by means of this position, two stars, very close
together and of equal brightness, about 9th magnitude, situated at the root of the tail of the Great Bear... They are
difficult to distinguish with an ordinary telescope of 6
feet.... It is presumed that Hevelius mistook these two
stars for a nebula..."
       Messier, nevertheless, included the object in his
catalogue as M40. It is identical to the double star Wcn 4
recorded by A.Winnecke in 1863; the components are 50.1"
apart in PA 83° with visual magnitudes of about 9.0 and 9.3
according to recent measurements; the PA has possibly decreased somewhat since Winnecke's original measurement of
88°. The Lick Observatory *Index Catalogue* gives the spectrum of the primary as G0. The pair lies just 16' NE from
70 Ursae Majoris; there is also a faint galaxy, NGC 4290,
in the field, but this is definitely not the object seen by
Messier, and it seems very unlikely that it could have been
noticed by Hevelius. It is a faint barred spiral of total
magnitude 12.7, about 1.5' in extent.

M81   (NGC 3031) Position 09515n6918, about 10° NW
             from Alpha Ursae Majoris and 2° east from 24
Ursae Majoris, just behind the ears of the Great Bear. The
pair M81-M82 is one of the finest pairs of galaxies available to small telescopes; they are 38' apart and form a
striking duo even in 20X70 binoculars. The two systems form
the nucleus of a small group which may be the nearest group
of galaxies beyond our own Local Group. Published distances
range from 6½ to about 9 million light years, and it is not
quite certain whether this aggregation or the Sculptor
Group is the closest to us. S.van den Bergh, in a list published in 1977, gives 6.5 million light years as the distance of M81; J.L.Sersic has 7.4 million, while A.Sandage
in the *Hubble Atlas of Galaxies* (1961) reports a modulus of
27.1 magnitudes, corresponding to a distance of about 8.5
million light years. A "compromise figure" of about seven
million light years appears in many modern textbooks. The
M81 group possibly contains about a dozen members.

GALAXIES M81 and M82 in URSA MAJOR. The upper object is M82 and the bright star near the bottom is the double Σ1386. Lowell Observatory 13-inch telescope photograph.

1984

M81 and M82 were discovered by J.E.Bode at Berlin in December 1774; M81 was recorded as "a nebulous patch, more or less round, with a dense nucleus in the middle". Messier added it to his catalogue in February 1781: "This nebula is a little oval, the centre clear and can be seen well in an ordinary telescope of 3½ feet..." T.W.Webb saw it as bright "with vivid nucleus, finely grouped with small stars, two of which are projected on the haze to which J.H. gives nearly 15' length. Two little pairs S.p." Admiral Smyth also mentions the two double stars in the field, which he identifies as Σ1386 and Σ1387; they are 144" apart and may be seen on the photograph opposite as the two brightest stars below the image of M81. The brighter and more southerly one is Σ1386.

| Σ1386 | 2.1" | 112° | Mags 9 + 9 | Spect F5 | 09510n6908 |
| Σ1387 | 8.9 | 271 | 10 + 10 | G | 09510n6911 |

Smyth described M81 itself as "A fine bright oval nebula of white colour....Major axis lies N.p. and S.f. and certainly brighter in the middle..." Isaac Roberts, on an early photograph made in March 1889, found it to be "spiral with a nucleus which is not well defined at its boundary and is surrounded by rings of nebulous matter". On modern photographs, M81 is one of the most magnificent spirals in the heavens, remarkable for its strikingly symmetrical structure and dynamic appearance. The spiral whorls enclose a bright ovoid central mass which undoubtedly is composed of millions of faint stars, although resolution has not been achieved, and may be somewhat beyond the capabilities of any present telescope. According to A.Sandage the brightest stars of the central hub (red giants of absolute magnitude -3.0) would appear at about apparent magnitude +24, if the computed distance is correct. This bright central region surrounds an almost stellar nuclear condensation, usually lost on photographs owing to strong over-exposure. The well defined spiral arms are bordered on their inside edges by narrow dust lanes which can be traced to within 35" of the nucleus on short exposure photographs; the arms themselves are well resolved into star clouds and bright patches of nebulosity. In stellar content M81 seems to resemble the great galaxy M31 in Andromeda.

SPIRAL GALAXY M81 in URSA MAJOR. An exceptionally fine
example of a large symmetrical Sb-type spiral. Palomar
Observatory 200-inch telescope photograph.

1986

On modern photographs the apparent dimensions are about 18' x 10' and the major axis is oriented toward PA 149°; the long dimension corresponds to about 36,000 light years at the accepted distance, and the actual luminosity is equal to about 20 billion suns. M81 was studied by M. Wolf in 1914 in an early attempt to measure the rotation of a "spiral nebula"; from the inclination of its spectral lines he found that the outer portions had a rotational velocity of close to 180 miles per second. In a similar study by E.Holmberg a total mass of about 250 billion suns was derived, somewhat exceeding the computed mass of our own Milky Way system. M81 is among the densest galaxies known, containing about 0.1 solar mass per cubic parsec; it has possibly about twice the density of a large typical Sb spiral like M31. Well over a third of the mass appears to be concentrated in the amorphous central hub. Because of the vast numbers of red and yellow giants, the light of M81 appears somewhat yellower than usual for large spirals; the integrated spectral type is about G3.

According to A.Sandage in the *Hubble Atlas of Galaxies*, individual stars detected in M81 are brighter than absolute magnitude -4.5; among the objects identified are seven irregular blue variables of high luminosity, three cepheids, fifteen other suspected cepheids, a number of irregular red variables, and 25 novae. Up to 1977 no supernova had been detected in this galaxy. As one of the nearest galaxies beyond the Local Group, M81 shows only a very moderate red shift of about 48 miles per second.

In addition to M81 and M82, this galaxy group also contains the peculiar galaxies NGC 3077 and 2976, the faint irregular systems NGC 2366, IC 2574 and Ho II, and possibly a few other faint members. The large spiral NGC 2403 in Camelopardalis also appears to be a dynamical member of this group. (See further notes on page 1993)

M82   (NGC 3034) Position 09519n6956, just 38' to the north of M81. A bright spindle-shaped galaxy of unusual type, forming a striking pair with M81. Both galaxies were discovered by J.E.Bode in December 1774; he found M82 "a nebulous patch, very pale, elongated"; Messier in 1781 thought it "less distinct than the preceding; the light is faint and elongated with a telescopic

DESCRIPTIVE NOTES (Cont'd)

star at its extremity..." Admiral Smyth saw it as "very
long, narrow, bright, especially in northern limb, but
paler than M81". Lord Rosse with his giant reflector, saw
some of the bright and dark detail in the galaxy in 1871:
"A most extraordinary object, at least 10' in length and
crossed by several dark bands". To H.L.D'Arrest it was a
striking and curious spectacle: "Vividly luminous and
sparkling; 7' long and 100" wide with two nuclei eccentrically disposed in the major axis. It scintillates as if with
innumerable brilliant points..." In Vol.XIII of the Lick
Publications, M82 is described as "an irregular, elongated
mass, 7' x 1.5' in PA 65° showing numerous rifts; an irregular lane divides it approximately along the shorter axis.
It is possibly a very irregular spiral seen edge-wise.....
Exceedingly bright, the brighter condensations show easily
on a 5 min. exposure". J.Mullaney and W.McCall, in their
publication *The Finest Deep-Sky Objects*, report that some
of the dark lanes crossing M82 are visible with an aperture
of 10 inches. In small telescopes it often appears more
eye-catching than M81, owing to its distinctive spindle-
shaped outline and high surface brightness.

M82 is a peculiar object which could easily be taken
for an edge-on spiral in the small telescope. Photographs
made with great reflectors, however, show an elongated and
amorphous mass of rather nebulous appearance, fading away
into hazy wisps and filaments around the edges, and heavily
spotted with irregular streaks and masses of absorbing
material. There is no indication of any spiral pattern and
no sign of resolution into stars. The integrated spectral
type of the system is about A5, but color measurements show
that the outer parts are bluer than the central regions,
and that the over-all color index is equivalent to class
G0. Another strange feature is the fact that the light of
M82 is strongly polarized, apparently indicating the existence of a strong magnetic field. M82 has long presented a
number of unsolved problems.

In 1962 this peculiar galaxy was made the subject of
a thorough study by A.Sandage and C.R.Lynds, and the results were reported in the *Astrophysical Journal* in May
1963. The galaxy was identified as a strong radio source,
and the polarization of light was found by A.Elvius and
J.S.Hall to be from 10 to 15% in the outer filaments, as

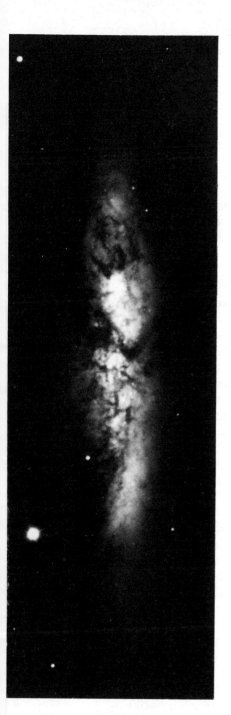 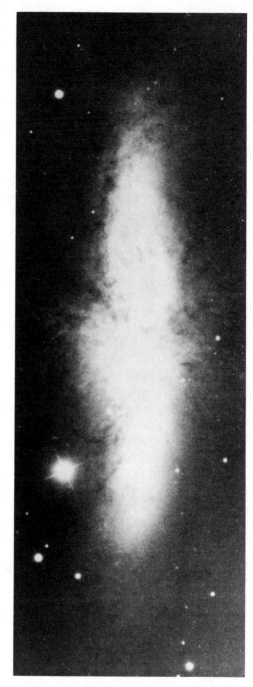

THE UNUSUAL GALAXY M82 in URSA MAJOR. (Left) A photograph
obtained with the 61-inch reflector at the U.S.Naval Observatory. (Right) A Palomar 200-inch telescope photograph.

1989

DESCRIPTIVE NOTES (Cont'd)

compared to about 3% in the main body of the galaxy. The discrepancy between the integrated spectral class and the color index has now been attributed to reddening of the light caused by large amounts of dust.

The outer filaments were found to record strongly on red-sensitive plates, and evidently radiate chiefly in the familiar red light of hydrogen, the frequency called H-Alpha at 6563 angstroms. A direct photograph of M82, taken in the light of H-Alpha with the 200-inch telescope, is reproduced on the facing page (top) and a composite photograph made with yellow-sensitive emulsion through a polaroid filter is shown (below). These techniques reveal an extensive system of filaments reaching out from the central region, at right angles to the plane of the galaxy; they may be traced out to approximately 10,000 light years on either side of the central hub, and appear to follow magnetic lines of force. Spectroscopic measurements reveal that the material of the filaments is expanding outward from the center of M82 at a velocity of about 600 miles per second. The total mass of the expanding material is computed to be about 5 million times the mass of the Sun.

These discoveries seem to suggest that a violent outburst of some sort occurred in the central region of M82 some 1.5 million years ago. Such an explosion, involving a mass of several million suns, would be on a scale far exceeding anything previously known, but a few other examples of similar activity in galaxies are now known, as NGC 1275 in Perseus, M87 in Virgo, and possibly NGC 5128 in Centaurus. It is suggested that in the center of a galaxy which still contains much dust and gas, vast contracting "pseudo stars" or "hyper-stars" might be formed, containing many thousands or millions of solar masses. Depending upon the exact conditions, such a contracting body might collapse directly into the "black hole" state, or destroy itself in a huge outburst when pressures and temperatures reached a critical stage. It now appears that some of the enigmatic "radio galaxies" may be objects of this sort.

Such odd galaxies as Cygnus A, M87, NGC 1275, and NGC 5128 all combine structural peculiarities with very strong radio emission; some of these were thought at one time to be colliding galaxy systems, but it is now known that the radio energy emitted is in many cases far too intense to be

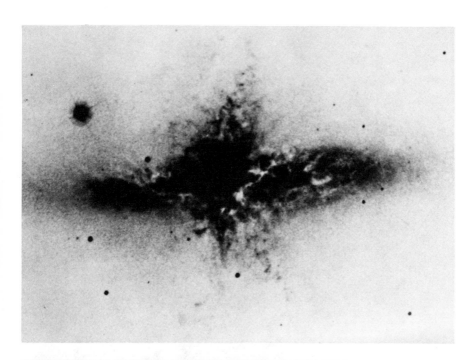

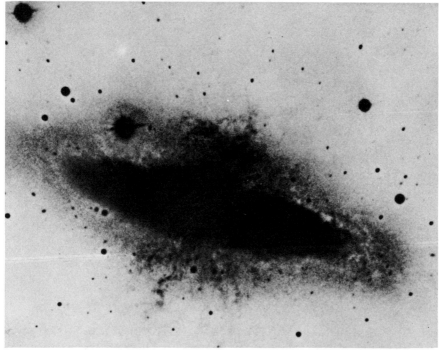

FILAMENTARY STRUCTURE IN M82. Top: A red light (H-alpha) photograph made with the 200-inch reflector. Below: Yellow-sensitive exposure made through a polaroid filter oriented along the major axis.   Palomar Observatory

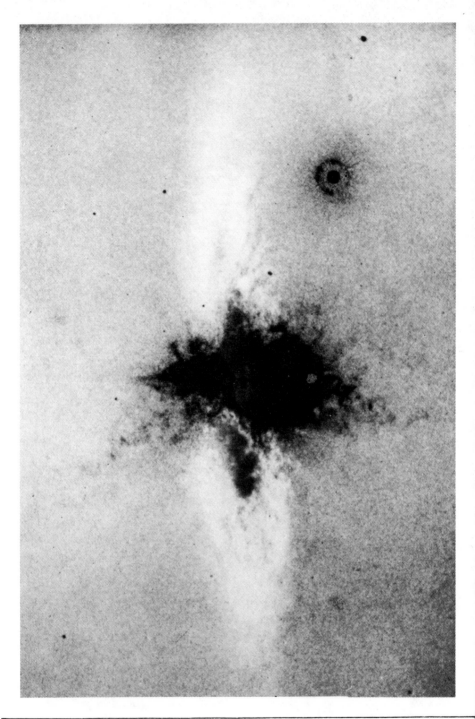

COMPOSITE PHOTOGRAPH OF M82. The filamentary detail is intensified by superimposing an H-alpha positive upon a yellow-sensitive negative.   Palomar Observatory

1992

DESCRIPTIVE NOTES   (Cont'd)

explained by such a mechanism. A new variety of collision hypothesis was, however, suggested in 1977, attributing the unusual features and radio emission of M82 to its collision with a vast intergalactic dust cloud, rather than with another galaxy. Although this theory seems inadequate to account for all the unusual features of M82, there is good evidence that much intergalactic dust is present in the region. In the photograph on page 1986 we may notice long narrow streaks crossing the upper portion of M81, at nearly right angles to its major axis; these are evidently seen *in front* of the galaxy, and very possibly mark the "front" of an advancing shock wave from M82 which lies some 30' distant. The light of the M82 filaments, in any case, seems to be the same "synchrotron radiation" which was first discovered in the supernova remnant called the "Crab Nebula" (M1) in Taurus. In this odd object, a gas cloud some 6 light years in diameter seems to be emitting this strange radiation, but in M82 we are witnessing a similar phenomenon on an enormous scale, affecting the entire central region of a galaxy.

Intrinsically, M82 is among the smaller galaxies, with a computed diameter of about 16,000 light years, and a mass of about 50 billion solar masses, roughly a fifth of the mass of M81. According to R.Minkowski at Palomar, it shows a much larger red shift than M81; the corrected measurement is about 240 miles per second.

OTHER MEMBERS OF THE M81 GROUP. At least one of the other smaller members of the group shows a structure somewhat resembling that of M82. NGC 3077, which looks like a small elliptical galaxy in the telescope, has the same wispy edges and scattering of dust clouds which characterizes M82. It is located 45' ESE of M81 and is easily found in a 6-inch telescope, the apparent magnitude being about 11.0, and the longer diameter about 2.6'. This is a dwarf galaxy only about 6000 light years in diameter, usually classed as type E2 with the added notation "peculiar".

The fourth member of the M81 group is NGC 2976, located about 1.4° from M81 toward the SSW. This is another odd galaxy, usually classed as a spiral of type Sd although very little evidence of spiral structure appears on photographs. The outline is elliptical; there does not appear

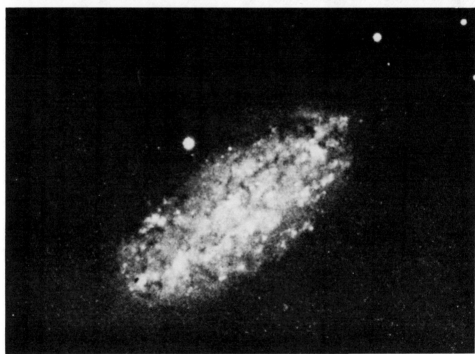

COMPANION GALAXIES TO M81. Top: NGC 3077, an elliptical
system with dusky spots. Below: The odd galaxy NGC 2976,
a "spiral with no definite spiral structure".

1994

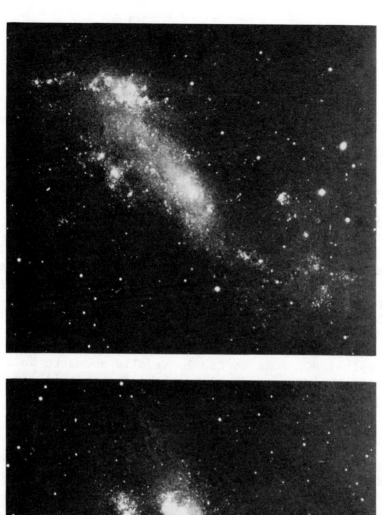

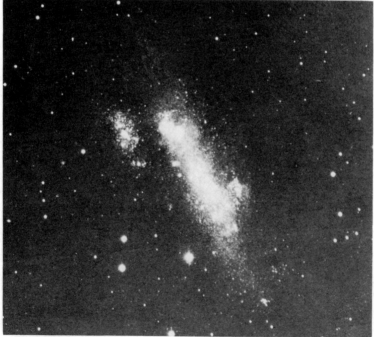

IRREGULAR GALAXIES OF THE M81 GROUP. Top: IC 2574 in URSA MAJOR. Below: NGC 2366 in Camelopardalis. The photographs on these two pages were made with the 200-inch reflector.

to be any definite nucleus, and the entire surface of the
galaxy is resolved into a mottled conglomeration of star
clouds and dust patches. This is another dwarf system with
a computed diameter of about 7000 light years.

Three faint irregular galaxies are known to be members of the M81 group. NGC 2366 is about 13° distant in
Camelopardalis, while IC 2574 is located some 3° east of
M81 and about 0.8° south. Some faint indications of a true
spiral structure appear on long exposure photographs of IC
2574, and there is a large stellar association resembling
a giant cluster on the north edge. This is a large system,
about 9' x 4', but difficult visually as the total light is
only magnitude 13. The true diameter across the longer axis
must be about 18,000 light years. Another much smaller
system called Ho II was discovered by E.Holmberg during a
study of the group in 1950. Finally, the very impressive
large spiral NGC 2403 in Camelopardalis seems to be a true
member of the M81 aggregation, although about 14° from the
main group. (Refer to page 333)

M97   (NGC 3587) Position 11120n5518, about 2.4° SE
from Beta Ursae Majoris, under the front of
the Great Dipper's bowl. This is the well known Owl Nebula,
one of the largest of the planetary nebulae, but rather a
featureless object in small telescopes owing to its fairly
low surface brightness. It was discovered by P.Mechain in
1781; he reported it "difficult to see, especially when one
illuminates the micrometer wires; its light is faint, without a star". John Herschel saw it as "a large, uniform,
nebulous disc, quite round, very bright, not sharply defined but yet very suddenly fading away to darkness"; he
found the apparent size to be about 2'40" while Lord Rosse
gave the dimensions as 163" x 147". It was Rosse, also, who
found in 1848 a striking resemblance to the face of an owl,
with two dark circular perforations and "a star in each
cavity" giving the impression of two gleaming eyes. From
the reports of Rosse it is claimed that both stars were
visible up through March 1850, but "five weeks later the
fainter one had vanished, nor could it ever again be found,
though looked for about forty times during the ensuing
quarter of a century." The central star, about 14th magnitude visually, lies on the central "bridge" of luminosity

FIELD OF THE OWL NEBULA. M97 is the circular object at lower left; Galaxy NGC 3556 appears above. Lowell Observatory
photograph made with the 13-inch telescope.

which separates the two dark vacancies; the faint swirls of
nebulosity surrounding the two "eyes" gave a vague impression of "spiral arcs" to Rosse and to Dr.T.R.Robinson, who
observed the nebula in 1848 through Rosse's 6-foot reflect-
or at Birr Castle. According to Lick Observatory *Publications* Vol.XIII, "the brighter central oval lies in PA 12°
and the diameter along this line is 199". At right angles
to this direction it is 203" in diameter to the outside of
the whorls. Aside from the outer whorls, all structural details are very vague and indistinct". Admiral Smyth called
it "This very singular object...circular and uniform after
long inspection, looks like a condensed mass of attenuated
light seemingly the size of Jupiter". A very faint, nearly
circular outer shell encloses the somewhat elliptical main
mass of the nebula. K.G.Jones thinks that both the central
star and the "owl" perforations might be seen in good conditions with an aperture of about 12 inches. Even in the
great reflectors of William Herschel, however, it appeared
merely as "a globular body of equal light throughout", and
to most visual observers today it is merely a large pale
disc of light. With great telescopes there is a distinct
impression of a spectral bluish-green tinge.

      Although this is presumably one of the nearest of the
planetary nebulae, the exact distance is uncertain. In a
study made in 1961, L.Kohoutek obtained a distance of about
1630 light years; C.R.O'Dell in 1963 derived a very similar

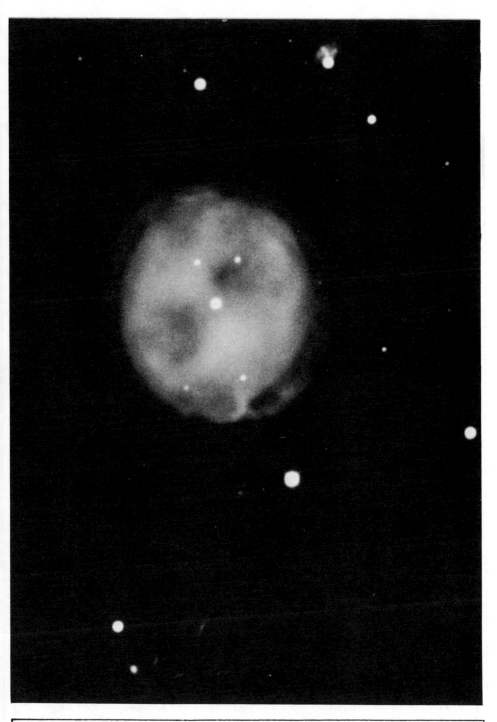

THE OWL NEBULA. M97 is one of the largest of the planetary nebulae. The photograph was made at Mt.Wilson with the 60-inch reflecting telescope.

DESCRIPTIVE NOTES   (Cont'd)

figure of 1600 light years; K.M.Cudworth (1974) has 2580 light years, I.S.Shklovsky has 1430, and the Skalnate Pleso *Atlas Coeli* Catalogue (1960) has 7460. Voroncov-Vel'jaminov has published a larger figure than any of these, 8150 light years, while K.G.Jones in his book *Messier's Nebulae and Star Clusters* gives the distance as 10,000 light years. If a working value of about 3000 light years is accepted, the actual diameter of the Owl is about 3 light years, and the central star is about half a magnitude brighter than our Sun in true luminosity. C.R.O'Dell finds that the star has an effective surface temperature of about 85,000°K, one of the hottest known, and a radius of about 4% that of the Sun, clearly implying a density near that of a white dwarf star. A total mass of 10 or 15% of the Sun has been computed for the Owl, fairly large for a planetary nebula. Objects of this sort were once regarded as possibly the debris clouds of ancient novae, but it now seems more likely that they are the products of a much quieter type of emission activity which affects stars in their old age. *"They are wreaths placed by Nature"*, says L.H.Aller, *"around dying stars."*

Just 48' to the NW the observer will notice the pale glow of the nearly edge-on spiral NGC 3556, looking like a little ghostly cigar about 7.8' in length, oriented almost due E-W. This was apparently first noticed by P.Mechain in 1781, and has been added to the original Messier catalogue in some observers' books, under the designation M108. (See photographs on pages 1997 and 2004)

M101   (NGC 5457) Position 14014n5435, about 5.5° east from Mizar (Zeta U.Maj) and about the same distance NNE from Eta. (Refer to chart on page 1956) M101 is one of the finest examples of a large face-on Sc type spiral, and a beautiful object on long-exposure photographs. It was discovered by P.Mechain in 1781 and described in the same year by Messier as "A nebula without star, very obscure and pretty large, 6' or 7' diam. between the left hand of Bootes and the tail of the Great Bear........ Difficult to distinguish when graticule lit..." William Herchel could distinguish little in it but "a mottled nebulosity" but Admiral Smyth, in 1844, spoke of it in a paragraph which reveals a remarkable prophetic insight: *"It is one of those globular nebulae that seem to be caused by a*

"Across the seas of space lie the new raw materials of the imagination... Strangeness, wonder, mystery, adventure, magic - these things, which not long ago seemed lost forever, will soon return to the world..."   ARTHUR C. CLARKE

2001

DESCRIPTIVE NOTES   (Cont'd)

*vast agglomeration of stars rather than by a mass of dif-
fused, luminous matter... the paleness tells of inconceiv-
able distance...*" Lord Rosse thought the extreme diameter
to be at least 14' and found it to be "A large spiral.....
faintish; several arms and knots.." D'Arrest found the
structure complex "with two intertwining nuclei, though not
well distinguished". The central mass is dominated by a
bright, nearly stellar nucleus, while the far extending
outer whorls contain a multitude of nebulous condensations
which found their way into the N.G.C. catalogue as separate
numbers. According to Vol.XIII of the Lick *Publications,*
these include NGC 5449, 5450, 5451, 5453, 5455, 5458, 5461
and 5462. The large galaxy, however, does have a few genuine companions, chiefly NGC 5474 to the SSE and NGC 5485 to
the NE. Probably at least 8 other smaller members belong to
the group. A.Sandage amd M.Humason report a red shift of
about 245 miles per second for M101. The distance is thought
to be close to 15 million light years; the apparent size of
about 20' then corresponds to about 90,000 light years, one
of the larger examples of an Sc spiral.

    According to E.Holmberg, the computed mass of this
system is about 16 billion suns, which is only about 10%
the estimated mass of the Milky Way; the resulting ratio of
mass to density is therefore unusually low, about 1 solar
mass per 160 cubic parsecs. M101 is also one of the bluest
galaxies known, as much of the light comes from the hot
Population I stars of the spiral arms; the integrated spectral type is about F8 and the measured color index close to
+0.16. At the derived distance, the computed absolute magnitude is near -19½, or about 5 billion times the light of
the Sun.

    Three supernovae have been recorded in M101. The first
was discovered by Max Wolf in 1909 as a 12th magnitude star
and now known as "SS Ursae Majoris". A much fainter nova,
probably discovered well past maximum, appeared in February
1951 in the brightest star cloud near the top of the print
on page 2001. In July 1970 the third known example was discovered by M.Lovas at Budapest; it attained magnitude 11
and appeared in the compact star cloud which looks like a
bright star on the outer left edge of the galaxy in the
photograph. (Refer also to M83 in Hydra and NGC 6946 in
Cepheus)

GALAXY M101 in URSA MAJOR. An impressive system with well
defined spiral arms. Palomar Observatory photograph made
with the 200-inch telescope.

UNUSUAL GALAXIES in URSA MAJOR. Top: NGC 3556 shows many obscuring dust masses. Below: The odd system NGC 2685 with its encircling ring.   Palomar Observatory photographs.

SPIRAL GALAXY NGC 2841 in URSA MAJOR. One of the finest examples of a bright symmetrical Sb spiral. U.S. Naval Observatory photograph made with the 61-inch reflector.

URSA MINOR

LIST OF DOUBLE AND MULTIPLE STARS

NAME	DIST	PA	YR	MAGS	NOTES	RA & DEC
α	18.4	218	55	2 - 9	POLARIS. (Σ93)	01488n8902
	44.7	83	00	-13	PA slow inc, spect	
	82.7	172	00	-12	F8; Primary cepheid	
					variable (*)	
Σ1583	11.1	285	25	7½- 8½	(OΣ238) relfix,	11578n8718
					spect A2	
β799	1.0	250	67	6½- 8½	PA & dist inc,	13033n7318
					spect A5	
OΣ267	0.2	332	60	9 - 9	PA measures dis-	13244n7615
					cordant, spect F5	
Σ1771	1.7	78	67	8 - 8½	PA inc, spect F5	13354n7002
h2682	26.1	280	56	7 - 10	spect A5	13402n7705
	45.9	316	56	- 9½		
Σ1798	7.3	12	25	7½- 9½	relfix, spect F2	13551n7839
Σ1840	27.0	222	24	6½ - 9	relfix, spect A0	14189n6801
Σ1849	1.1	4	57	8½- 9	Slight dist dec,	14199n7656
					spect F5	
Σ1915	2.5	321	35	7 - 10	PA slow dec, spect	14409n8610
					K0	
Hu 908	1.2	258	35	6½ - 10	PA dec, spect K0	14542n7823
	113	142	14	- 9		
Σ1905	3.7	160	35	8½- 8½	relfix, spect F8	14565n7102
Σ1928	6.8	274	37	9 - 9½	cpm, relfix, spect	15093n7239
					G5	
Σ3125	2.2	269	58	9 - 9½	relfix, spect K2	15250n6714
π^1	31.1	80	59	6½- 8	(Σ1972) cpm, slight	15320n8037
	135	104	25	- 11½	PA dec, spect G0,	
					G8	
Σ1980	10.0	50	18	9 - 9½	relfix, spect F0	15356n8113
π^2	0.4	46	25	7 - 8	(Σ1989) Binary,	15423n8008
					about 150 yrs; PA	
					dec, spect dF3	
Σ2034	1.2	117	56	7½- 8	relfix, spect A3	15547n8346
A1134	2.2	38	47	7 - 12	cpm, PA dec,	15586n7102
					spect G5	
Hu 917	2.9	189	58	6 - 9½	(Ku 1) AB cpm; AC	16453n7736
	115	14	59	- 10	dist dec, spect	
					dF2	

URSA MINOR

LIST OF VARIABLE STARS

NAME	MagVar	PER	NOTES	RA & DEC
α	1.95— 2.05	3.9698	POLARIS (North Star) Cepheid; also visual double; Spect F8 (*)	01488n8902
γ	3.1----	---	Irr; Spect A3 (*)	15208n7201
ε	4.2--4.3	39.481	Ecl.Bin; Spect G5 (22 Ursae Minoris)	16510n8207
R	8.7--11.0	324	LPV. Spect M7e	16306n7223
S	7.8--12.7	327	LPV. Spect M7e--M9e	15314n7848
T	8.5--15.0	314	LPV. Spect M5e	13336n7341
U	7.5--13.0	326	LPV. Spect M6e	14162n6701
V	7.4--8.8	72	Semi-reg; Spect M5	13378n7434
W	8.6--9.7	1.7012	Ecl.Bin; Spect A3	16208n8619
RR	4.7-- 5.1	40:	Semi-reg; Spect M5	14568n6608
RW	6.0----21	---	Nova 1956	16498n7707

LIST OF STAR CLUSTERS, NEBULAE AND GALAXIES

NGC	OTH	TYPE	SUMMARY DESCRIPTION	RA & DEC
6217	280[1]	⌀	Sc; 12.0; 1.8' x 1.2' B,cL,1E,1bM	16348n7818

2007

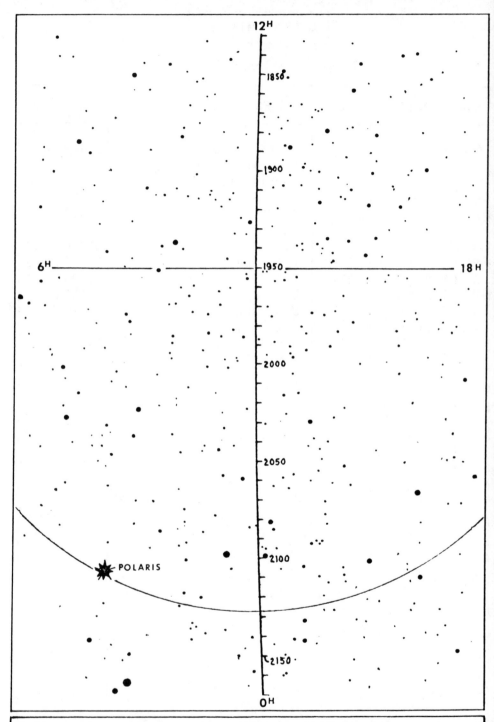

PATH OF THE NORTH POLE from 1850 to 2150 AD. The portion of the circle is 1° in radius, centered on the position for 1950.

2008

URSA MINOR

ALPHA Name- POLARIS, or "Stella Polaris", the NORTH
STAR or POLE STAR. Magnitude 1.99 (slightly
variable); Spectrum F8 Ib; Position 01488n8902. Polaris
is unquestionably one of the most famous stars in the sky,
although it ranks only 49th in brightness among all the
stars. To navigators and travelers it is one of the most
important of the fixed stars, owing to its location near
the north pole of the heavens. The star is not located precisely at the true Pole, however, but is a little less than
1° distant; during the next 150 years this distance will
diminish and the star will be closest to the exact pole
(27' 31") in the year 2102 AD. The change in distance is,
of course, due to the gradual change in the direction of
the Earth's axis in space, in a period of about 25,800
years, the so-called "precessional cycle". During this time
the polar point moves in a great circle some 47° in diameter. The star Thuban (Alpha Draconis) was the Pole Star
some 4600 years ago, during the age of the Pyramid builders
of ancient Egypt, while 12,000 years ago the bright star
Vega was near the position of honor. At the present time
Polaris is much closer to the true Pole than Vega can ever
be, and is a much more accurate guide to true north than
the compass. If a greater degree of accuracy is required,
the true Pole lies very nearly on a line drawn from Polaris
toward Eta Ursae Majoris, the last star in the handle of
the Great Dipper. For precise alignment of the polar axis
of a telescope, the observer should refer to the chart on
page 2008, where the scale is 40" per mm.

Observing with binoculars, one may notice the little
circlet of 7th and 8th magnitude stars, about 35' in size,
called the "engagement ring" of Polaris, just to the south
of the star; Polaris itself sparkling as the celestial
solitaire of the ring.

For the small telescope, Polaris is an interesting
double star, having a small companion which seems of a pale
bluish tint, about 18½" distant. The companion, of the 9th
magnitude, was first seen by Sir William Herschel in 1780,
and is an excellent test object for small telescopes. S.W.
Burnham found it detectable in a 1.6 inch refractor, but
most observers will probably require at least a 2.5 or 3-
inch glass. The two stars form a common motion pair, with
a change in PA of about 2° per century; the orbital period

must be many thousands of years, and the projected separation of the pair is close to 2000 AU.

The distance of Polaris is too great to be measured
accurately by direct parallax, and is certainly over 300
light years. The apparent magnitude of the companion, which
is an F3 main sequence star, implies a distance of about
360 light years. The absolute magnitudes are then -3.2 and
+3.2; the primary star is about 1600 times brighter than
our Sun. Polaris shows an annual proper motion of 0.046";
the radial velocity is somewhat variable, averaging about
10.4 miles per second in approach.

The bright star is a cepheid pulsating variable of
type II (the W Virginis class) with a period of 3.9696 days
and the unusually small range of about 0.1 magnitude in the
visual; the range is 0.17 magnitude in the ultraviolet and
only 0.04 in the infrared. In addition, variations in the
radial velocity indicate the presence of an unseen companion with a period of 30.5 years; orbit computations by J.H.
Moore (1929) and E.Roemer (1955) agree in placing periastron in 1928. The orbital eccentricity is 0.64 and the
semi-major axis of the orbit of the bright star is about
290 million miles. No evidence of the light of the unseen
companion is detected in the spectrum; presumably the star
is some 5 or 6 magnitudes fainter than Polaris itself.

Polaris was the *Cynosura* of the Latins, a name in
common use in Europe some two centuries ago, but, according
to R.H.Allen "this was one of the ancients' titles for the
whole of Ursa Minor, and never, by them, limited to the
lucida". Marlowe, in his *History of Doctor Faustus,* tells
us that the stars *"All jointly move upon one axletree*
              Whose terminine is term'd
              The world's wide pole...."
For many centuries Polaris has been the *Lodestar* or
Steering Star of navigators, the *Pivot Star* of seamen and
the Latin *Navigatoria* of voyagers. Anglo-Saxon tribes, a
thousand years ago, called it the *Scip-Steorra,* or "Ship-
Star"; an earlier Greek name, *Phoenice,* is said to have
originated from the legend that the wise man Thales learned
of its use in navigation by the Phoenicians; hence the name
Ursae Phoenicia for the constellation. Arabian astronomers
knew the star as *Al Kaukab al Shamaliyy,* the "Star of the

THE ETERNAL CIRCLING OF THE STARS is revealed by a time exposure centered near Polaris. This photograph was made by the author with a 1.7-inch Xenar lens.

North", but R.H.Allen suggests that it may have been Beta
Ursae Minoris which had the name in ancient times, when the
actual Pole was about equally distant from both stars. The
Arabian name is evidently the source of the *Alrucaba* of the
Alfonsine Tables, and Bayer's title *Alruccabah*. Another
name in use in Moslem lands was *Al Kiblah*, referring to a
concept of determination of true compass points so that the
traveler might know the accurate direction to Mecca. The
name *Al Kutb al Shamaliyy* has been translated "The Northern
Axis" or the "Northern Spindle".

Bryant refers to the traditional use of Polaris as
a guide to seamen in his *Hymn to the North Star*:

> *"Constellations come, and climb the heavens, and go.*
> *Star of the Pole! And thou dost see them set.*
> *Alone in thy cold skies,*
> *Thou keepest thy old unmoving station yet,*
> *Nor join'st the dances of that glittering train,*
> *Nor dipp'st thy virgin orb in the blue western main.*
> > *On thy unaltering blaze*
> *The half wrecked mariner, his compass lost,*
> > *Fixes his steady gaze,*
> *And steers, undoubting, to the friendly coast;*
> *And they who stray in perilous wastes by night,*
> *Are glad when thou dost shine to guide*
> > *their footsteps right.*
> *A beauteous type of that unchanging good,*
> *That bright eternal beacon by whose ray*
> *The voyager of time should shape his heedful way."*

Christina Rossetti saw in the Pole Star a celestial
monarch of the North:

> *".....one unchangeable upon a throne*
> *Broods o'er the frozen heart of earth alone,*
> *Content to reign the bright particular star*
> *Of some who wander...."*

And Dante, in the *Paradiso*, saw it as the central
axis of the world:

> *".......the mouth imagine of the horn*
> *That in the point beginneth of the axis*
> *Round about which the primal wheel revolves.."*

POLARIS. The North Star, as photographed with the 13-inch telescope at Lowell Observatory. The brighter stars in the field form the "engagement Ring" of Polaris.

DESCRIPTIVE NOTES   (Cont'd)

Wordsworth, looking down the long corridors of time, imagines that the men of earliest times paid reverence to their guiding star:

> *"Chaldaean shepherds, ranging trackless fields,*
> *Beneath the conclave of unclouded skies,*
> *Spread like a sea, in boundless solitude,*
> *Looked on the polar star, as on a guide*
> *And guardian of their course, that never closed*
> *His steadfast eye...."*

In many lands and ages, the North Star has been seen as a symbol of constancy and faithfulness. Sir James Frazer in *The Golden Bough* describes an ancient Hindu rite in which a newly married couple honored the Pole Star as the symbol of fidelity, and vowed in its light to remain faithful to each other for "a hundred autumns". Confucius, in the 5th Century BC, had expressed the same concept: *"He who rules by moral force is like the Pole star, which remains in its place while all the lesser stars do homage to it.."* John Keats, in his *Last Sonnet*, hails the Pole Star:

> *"Bright star! would I were steadfast as thou art-*
> *Not in lone splendour hung aloft the night...."*

And Shakespeare has Julius Caesar say, only a few moments before his assassination:

> *"If I could pray to move, prayer would move me;*
> *But I am constant as the northern star,*
> *Of whose true-fix'd and resting quality*
> *There is no fellow in the firmament.*
> *The skies are painted with unnumb'red sparks,*
> *They are all fire and everyone doth shine;*
> *But there's but one in all doth hold his place.*
> *So in the world....."*

Again, in *Othello*, the Pole star appears as a symbol of what T.S.Eliot called *"the still point of the turning world"*:

> *"The wind-shak'd surge, with high and monstrous mane*
> *Seems to cast water on the burning Bear,*
> *And quench the guards of th' ever fixed pole..."*

The "guards" in this case being the stars Beta and Gamma Ursae Minoris, long known as the traditional "Guards"

or "Guardians of the Pole". About 3000 years ago the true
Pole was closer to Beta than to the present Polaris, so it
is likely that some of the very early references to the
"Star of the North" actually refer to Beta Ursae Minoris
instead. Aeschylus, in the dramatic opening scene of the
Agamemnon, shows us the watchman on the palace roof at
Mycenae, hopefully awaiting the appearance of the beacon-
fires that shall tell of the fall of far-off Troy:

> *"I pray the gods a respite from these toils;*
> *This long year's watch that, dog-like, I have kept,*
> *High on the battlements of the Atridae, beholding*
> *The nightly council of the stars,*
> *The eternal circling of the celestial signs,*
> *And those bright regents,*
> *High-swung above earth, that to mortal men*
> *Bring summer and winter..... Here I watch,*
> *Longing for the beacon flame that shall*
> *   wing its voice*
> *From far Troy, telling of victory....."*

If this memorable scene took place in the traditional year of the conquest of Troy, 1184 BC, the closest star
of any prominence to the true Pole was Beta Ursae Minoris,
then about 6° distant. A thousand years after, in the time
of Caesar, Alpha and Beta were both about equally distant
from the actual Pole. The Jewish shekels struck during the
period of the Second Revolt, about 132 AD, bear a design of
a star over the temple at Jerusalem, often thought to be a
symbol of the Star of the North, since the name of the
great leader Bar Kochabh obviously comes from the same root
as the Arabic *Al Kaukab*. The leader's name apparently means
"Son of the Star".

Polaris, in legends found virtually throughout all of
Asia, holds a position of great importance, as the star is
regarded as the spire or pinnacle of the cosmic *Mountain
of the World* or "Axis of the Universe", the fabled *Mt.Meru*
or *Sumeru* of Hindu and Buddhist lore. On the summit of
Meru, "84,000 leagues above the Earth", the sky god Indra
has his heavenly palace; in various myths we find Surya,
Vishnu, and other deities dwelling on Meru as on the Greek
Olympus. Virtually all of the great temples of India are
symbolic representations of the cosmic mountain; especially

THE COSMIC MOUNTAIN. The Celestial Axis of the World is symbolized by these structures. (1) The Kandarya Mahadevi Temple at Khajuraho. (2) The Great Temple at Tanjore. (3) The huge Buddhist shrine of Borobudur in Java.

striking examples are the Kandarya Mahadevi temple at
Khajuraho (about 1000 AD), the Lingaraja Temple at Bhuvan-
eshvara (about 1000 AD), and the vast but ruined *Surya Deul*
or "Black Pagoda" at Konarak (about 1240 AD) where the
whole enormous temple-mountain is shown supported by great
stone wheels carved at its base. The wheels, some 10 feet
in diameter, transform the entire temple into the symbolic
representation of the Chariot of the Sun, but also, it is
thought, symbolize the axis of the cosmic mountain on which
the heavens turn. At Tanjore, the great temple-tower of the
Chola king Rajaraja (about 1000 AD) offers an especially
intriguing concept of the cosmic mountain. The huge cap-
stone, a single ornately carved mass of rock weighing some
20 tons, was raised to the temple summit, 180 feet above
the ground, by means of a ramp two miles long, and 10 years
of dedicated labor. Sherman E.Lee of the Cleveland Museum
of Art speaks of this achievement as "one of the most extraordinary and in its implications, frightening feats of
engineering in India". The temple pinnacle, we may imagine,
represents the Pole star itself, and its role as the cosmic
axis of the world.
    At Budh Gaya in northern India, a similar temple-
pinnacle marks the spot where the young prince Siddhartha
attained ultimate enlightenment (about 528 BC) and became
the Buddha.
    In the fantastic *Kailasa* or "Kailasanatha" temple at
Ellora in Hyderbad we find another interpretation of the
cosmic mountain - in this case the celestial palace of the
great god Siva, who was believed to dwell on Mt.Kailasa, an
actual peak in the Himalaya. The Kailasa Temple, completed
about 785 AD, is a complete celestial palace with pillars,
courtyards, balconies, corridors, chambers, towers, and
statuary, wonderfully decorated with intricate ornamenta-
tion, the whole gigantic structure *carved in one single
piece* from the great rock cliffs of Ellora. The temple cannot be called a "building", it is *sculpture,* on a scale
never attempted anywhere else on earth. The builder himself, King Krishna I, seems to have been as astounded by
his completed dream palace as any modern traveler. "How is
it possible," he wrote, "that I built this other than by
magic?" It was with the same helpless feeling of incredu-
lous wonder that the simple people of Cambodia, a century

THE COSMIC MOUNTAIN. (1) The Lingaraja Temple at Bhuvanesh-
vara. (2) The Budh Gaya Shrine near Benares. (3) The Great
Temple of Angkor Vat, vast Khymer shrine in Cambodia.

ago, looked upon the stunning immensity of Angkor Vat and refused to believe that such a structure could be anything other than the work of the gods.

The great temple at Angkor is indeed the ultimate representation of the cosmic mountain brought to earth, and one of the world's supreme achievements in architecture. That unquenchable globetrotter Richard Halliburton called it the greatest man-made wonder of the world and "the greatest mystery in history". Some of the mystery has possibly been dispelled by dedicated modern archeologists, but the wonder remains. Not only is Angkor Vat one of the most colossal and exquisitely ornamented structures ever erected on earth, but all the major dimensions and proportions of the whole vast complex are known to express fundamental concepts of Hindu cosmology. Completed during the reign of the Khymer king Suryavarman II, about 1150 AD, Angkor was abandoned to the ravages of the jungle in 1431, and remained almost totally unknown to the outside world for over 400 years. With its miles of richly sculptured corridors and cloistered galleries, its acres of intricate stone carvings and soaring towers of stone lace, Angkor lies in the jungle in its brooding magnificence, "powerful and beautiful and desolate beyond belief...."

One wonders if, in the great days of Angkor, a sacred flame might have burned on the Temple summit, in honor of the Pole Star, axis of the world....

It is interesting to find in Hebrew scripture some reference to a holy mountain which seems to be connected in some way with the northern sky or northern stars. In the 48th Psalm occur the lines:

> *"Great is the Lord, and greatly to be praised*
> *In the city of our God, in the mountain of*
> *   his holiness.*
> *Beautiful for situation, the joy of the whole earth*
> *Is Mount Zion, on the sides of the North,*
> *The city of the great king. "*

Again, in the 14th chapter of *Isaiah*, we find the following lines in which the prophet, addressing the degenerate monarch of Babylon, satirically compares him to the fallen angel Lucifer:

DESCRIPTIVE NOTES   (Cont'd)

> *"How art thou fallen from heaven. O Lucifer,*
> *Son of the morning!*
> *How art thou cut down to the ground,*
> *Which did weaken the nations!*
> *For thou hast said in thy heart,*
> *I will ascend into heaven,*
> *I will exalt my throne above the stars of God;*
> *I will sit also upon the mount of the congregation,*
> *In the sides of the north;*
> *I will ascend above the heights of the clouds;*
> *I will be like the most High.."*

The phrase "sides of the north" seems somewhat ambig-
uous, but appears in some other translations as "the utter-
most parts of the north". Sumerian inscriptions mention a
great peak called "the mountain of the world" in the far
north or northeast, the home of the gods, and the supporting pillar of heaven. In many cultures we find this concept symbolized on Earth by some sort of "temple-mountain"-
the *ziggurat* in Babylonia and Sumer, the *pyramid* in Egypt,
the huge dome-shaped *dagoba* in Ceylon, the *teocalli* or
pyramid-temple in ancient Mexico, and the *sikhara* or temple
spire in India. In Norse and Teutonic mythology too, the
same concept appears in the legend of the *Himinbiorg,* the
Sky-Mountain, where dwell the guardians of the *Bifrost*
Bridge to Asgard and Wotan's stronghold of Valhalla.
    Again, in the ancient Horse Sacrifice of the Altai
tribes of Siberia, we find the concept of the Cosmic Axis
of the world, represented in the ceremony by a birch tree;
at the climax of the ritual the celestial power is invoked
by the shaman with the words *"Thou who the starry heavens*
hast turned a thousand thousand times..." The symbolism
of a tree as the cosmic axis suggests some connection with
the Norse myth of the great World Ash *Yggdrasil,* which was
said to stand in the far north and unite heaven and earth.
    Glastonbury in Somerset, with its indefinable aura
of mysterious sanctity, is the spot in England most often
connected with the concept of the cosmic world-axis. The
"holiest spot in Britain", Glastonbury has been identified
with the legendary lost land of Avalon where the sword
Excalibur was forged. Here was the site of the first Christian church in England, established possibly as early as

DESCRIPTIVE NOTES   (Cont'd)

the 1st or 2nd Century AD; here, at Glastonbury Abbey, so
legend said, was enshrined the Holy Grail itself, brought
here by Joseph of Arimathea; here in the year 1190 AD was
discovered a tomb believed, from its inscription on a lead
cross, to be that of the semi-mythical King Arthur himself:
*"HIC JACET SEPULTUS INCLYTUS REX ARTURUS IN INSULA AVAL-
LONIA"*. The Glastonbury fables range from the historically
plausible to the wildly bizarre; any factual truth underlying the legends will probably remain forever unknown as
virtually all the physical evidence was destroyed in the
heart-breaking time of the dissolution of the monasteries
under Henry VIII.

Dominating the haunted ruins of Glastonbury Abbey
stands the huge mound of Glastonbury Tor, 500 feet high,
and topped by the medieval tower of the chapel of St.Mich-
ael; here, according to some modern scholars, is the central axis of a huge astronomical calendar or sky map, the
"Glastonbury Zodiac". This idea was probably first advanced
by the last of England's royal magicians, John Dee, in the
days of Elizabeth I, but was revived in modern times by the
British writer Mrs.K.Maltwood (1924) who used English topo-
graphical survey maps to demonstrate the existence of a
vast ring of earthworks representing the constellations of
the zodiac, and centered on Glastonbury Tor. Many of these
are so badly eroded that the reality of the whole pattern
is open to serious question, but the discoveries concerning the astronomical significance of nearby Stonehenge at
least make the idea plausible. There can be little doubt
that Stonehenge, in addition to its religious significance,
was used as a celestial calendar and astronomical computer.

In the texts of ancient China, the concept of the
"cosmic mountain" appears under the name *T'ai Chi*, variously translated as "The Axis of the Universe", "The Center",
"The Ridgepole", the "Great Root", the "Wheel of the Universe", "The Pole Star", or "The Wheel of Life". In early
Buddhist thought the endlessly turning heavens visibly
express the concept of the *Wheel of the Law*. To Taoist
sages the cosmic axis symbolizes the absolute, the infinite
intelligence of the Universe, the indefinable *Tao*, the "Way
of Nature". In the opening sentence of the great Buddhist
scripture called the *Dhammapada*, often attributed to Buddha

himself, the turning wheel appears as a metaphorical symbol of the inviolable principle of universal *karma:*

> *"All that we are is the result of what we have thought... It is founded on our thoughts, it is made up of our thoughts..... If a man thinks or acts with an evil thought, pain will follow him as surely as the wheel follows the foot of the ox that draws the carriage...."*

In the popular mythology of China, the North Pole of the sky was the home of *Huan-T'ien Shang-Ti,* the "Supreme Lord of the Dark Heavens", or the "Supreme Prince of the North Pole". Great accuracy is claimed for the traditional representations of this deity, as they are all based upon a famous painting made by the Sung Dynasty artist-emperor Hui-tsung, after the God had honored him with a personal visit in 1118 AD. The "cosmic mountain" was also identified in China with various actual peaks, usually with the *T'ai Shan* in eastern Shantung, sacred since at least the 7th Century BC, and from which, according to legend, the Sun began his daily journey across the sky. In the annals of the Ch'in Dynasty we find an interesting statement that the arrogant Shi Huang Ti, builder of the Great Wall and first ruler of a united China, attempted an ascent of T'ai Shan, to offer sacrifice on the summit in celebration of his victories, but was repeatedly driven back by fierce storms. *"Only sages of supreme merit may perform such a sacrifice,"* explained later historians of the Han Dynasty, after their own emperor Wu reached the summit in 110 BC.

After the introduction of popular Buddhism into China, however, the location of the "cosmic mountain" was gradually shifted in popular belief to the legendary peak of *Khun-Lun* or *Tien Shan,* immeasurably far to the west. Here was the *Pure Land* or *Western Paradise,* the *"Sukhavati"* of Indian scripture, presided over by the great Buddhist saint Amida or *Amitabha,* the "Buddha of Infinite Light". This same far land, it would seem, appears in the lore of many lands and cultures under various names, it is the kingdom of *Peng-Lai,* the mysterious island of *Dériyabar* of the Rainbow Mists, the Heavenly Kingdom of *Amel-Ric,* and, undoubtedly, Lovecraft's *Unknown Kadath,* and the *Shangri-La* of James Hilton's *Lost Horizon.* In Taoist legends it is

DESCRIPTIVE NOTES   (Cont'd)

Shou Shan, the "Jade Mountain of Immortality" or the famed "Island of the Eight Immortals", where those sages who have attained ultimate wisdom dwell on the shores of the "Green Jade Lake", living on celestial dew, the essence of pure jade, and the peaches of immortality. The association of jade with the cosmic mountain and celestial matters was a universally held concept in China. The traditional Chinese jade, incidentally, was the mineral known today as nephrite, regarded from the most ancient times as the "stone of heaven". More highly valued than gold, it was revered for its unique power of placing Man into contact with cosmic forces and "open the doors of wisdom" to higher states of consciousness. *"When I think of a wise man,"* states Confucius, *"his merits appear to be like jade.."* It seems particularly appropriate, therefore, that a number of jade ritual objects are known from ancient China which seem to be astronomical instruments of some sort.

The type most often found is a round flat disc up to a foot in diameter, with a central hole, often decorated with ornamental carvings. A number of these are known which date back to the earliest historical period of China, the almost legendary Shang Dynasty of the 16th to 11th centuries BC. There seems to be little doubt that such a jade disc, called a *pi,* was some sort of a symbol of the heavens; a number of ancient texts support this interpretation, as in the *Li Chi* or Book of Rites where we read:

"With a blue-green pi reverence is paid to Heaven...
With a yellow ts'ung homage is rendered to earth...

Especially interesting is the variety called the *hsun-chi,* characterized by a number of irregularly spaced notches on the outer rim. A number of authorities on the jade-lore of China believe that these discs were some type of astronomical sighting device, most probably used to locate the exact position of the true celestial Pole. The curator of the Art Museum of the University of Singapore, William Willetts, shows an interesting example of a *hsun-chi* in his book, *Foundations of Chinese Art;* the notches of his specimen, when held against the sky, may be aligned with the stars Alpha, Delta, Epsilon and Zeta in the Great Dipper, Chi Cephei, and Zeta and Phi in Draco. The true pole, as it would have been seen about 3000 years ago, is

then seen neatly positioned in the central aperture. The
nearest bright star to the Pole at that time was Beta Ursae
Minoris, about 6° distant.

In an obscure passage in the ancient Chinese classic,
the *Shu Ching* or "Book of History", of the 6th Century BC,
we find a cryptic reference to a "turning sphere" and a
"jade transverse" with which the emperor performed a ritual
to "regulate the Seven Directors". S.Howard Hansford, in
his book *Chinese Carved Jades* suggests that the passage refers to the use of the *hsun chi* to determine the direction
of true north, which was necessary "not only for ritual
purposes and the auspicious orientation of buildings, but
for accurate calibration of the calendar, believed to be a
prime necessity for determining the days upon which agricultural tasks, ploughing, sowing, etc., should be put in
hand".

PI

HSUN CHI

TS'UNG

Possibly connected with the use of the *hsun chi* was
another type of artifact called a *ts'ung*, essentially a
square hollow tube with cylindrical ends. These range in
shape from squat thick forms to long tubes exceeding twenty
inches in length. A statement in the *Book of Rites* implies
that these objects were used in some sort of ritual connected with worship of the earth, while the name seems to
derive from the Chinese ideogram for "shrine" or "sacred".
Some students of Chinese lore believe that the *ts'ung* was
originally a container for inscribed ancestral tablets or
rolled up scrolls, though in later times it was almost certainly used as some sort of sighting device. It seems more
than likely that the *ts'ung* was, in fact, the "jade transverse" mentioned in Chou Dynasty annals, and that one of

DESCRIPTIVE NOTES (Cont'd)

these tubes, of the proper size, was fitted into the central hole of the disc of the *hsun chi* to make the complete astronomical sighting instrument. Joseph Needham, in his authoritative study *Science and Civilization in China,* makes the statment that the use of such instruments was so well known by late Chou times (3rd Century BC) as to lead to a proverbial expression: *i kuan k'uei t'ien* - "to look at the heavens through a tube". Hansford carries this speculation one step further when he makes the intriguing suggestion that such a device *could* have been equipped with simple lenses and used as a primitive telescope, more than a thousand years before Galileo. "There is no doubt that such lenses could have been made and finely polished by Chinese lapidaries from the fifth or fourth centuries BC onward". Needless to say, however, no such lenses have ever been found in any ancient site in China; neither is there any mention in Chinese writings of the wondrous discoveries that would certainly have been made in the heavens had a telescope actually been in use.

BETA   Name- KOCHAB, from the Arabic *Al Kaukab* which was also a portion of the title of Polaris. Magnitude 2.06; Spectrum K4 III; Position 14508n7422. This star and Gamma mark the front of the bowl of the Little Dipper, and have for centuries been designated "The Guardians of the Pole". Some 3000 years ago it was the closest bright star to the true Pole, and is probably the Greek "Polos" and other objects referred to in very early writings as the "Polar Light" or "Polar Star". As the Arabian *Al Farkad* it was a symbol of faithfulness; as *Al Na'ir al Farkadain*, it was the "Bright One of the Two Calves"; the name originating from the tradition that Alpha, Beta, Gamma and Delta comprised a herd of animals closely circling about their mother. The Latin names *Circitores, Saltatores,* and *Ludentes* refer to Circlers, Leapers, or Dancers around the Pole.

Beta Ursae Minoris lies at a computed distance of about 100 light years, and has an actual luminosity of 130 times that of the Sun. (absolute magnitude about -0.5) The star shows an annual proper motion of about 0.03"; the radial velocity is 11 miles per second in recession.

DESCRIPTIVE NOTES   (Cont'd)

GAMMA   Name- PHERKAD, from the Arabic *Alifa al Farka-dain*, the "Dim One of the Two Calves". This star and Beta form the two "Wardens" or "Guardians of the Pole"; Riccioli calls them the *Vigiles*. Gamma is magnitude 3.08 (slightly variable); Spectrum A3 II or III; Position 15208n7201, at the present time just 18° from the true Pole. Gamma is 13 Ursae Minoris, and forms a wide pair with 11 Ursae Minoris (Mag 5.01, Spectrum K4 III) just 17' almost due west. The color contrast is noticeable in the small telescope, but the two stars do not form a true pair. The computed distance of Gamma is about 270 light years, and the actual luminosity about 330 times that of the Sun, with an absolute magnitude of about -1.5. The star shows an annual proper motion of 0.03"; the radial velocity is about 2½ miles per second in approach.

Gamma Ursae Minoris is a peculiar variable star of uncertain class. The light variations, discovered by P. Guthnick in 1914, are irregular and of small amplitude, not exceeding 0.1 magnitude, but the changes often occur in intervals as short as 2 or 3 hours. No regular periodicity has been determined. O.Struve in 1923 found rapid changes in radial velocity, sometimes in a period of about 2.6 hrs, but on other occasions the measurements showed constant velocities for long intervals. The star is a puzzling object, probably pulsating in several superimposed periods, and has been assigned by different investigators to the Beta Cephei class, the dwarf cepheid class, the Alpha Canum type, and the rapid spectroscopic binary category. In the Yale *Catalogue of Bright Stars* (1964) it is identified as a spectroscopic binary with a period of 0.108449 day, but with a note that the elements are variable. From recent studies made at the Uttar Pradesh State Observatory in India, it seems that the star is most likely a dwarf cepheid oscillating in two or more superimposed cycles; the light curve repeats itself very closely after every 21 cycles, and the best period determination for an individual cycle now appears to be 0.143009 day or about 3^h26^m. Owing to the "beat" phenomenon, the star at times shows virtually no measurable variations, while the shape of the light curve varies from cycle to cycle. As the derived luminosity seems unusually high for a dwarf cepheid, the adopted distance may be overestimated. (Refer also to page 1158)

LIST OF DOUBLE AND MULTIPLE STARS

NAME	DIST	PA	YR	MAGS	NOTES	RA & DEC
Jc 12	27.1	16	13	8½- 8½	Spect A	08036s4516
Rst 5285	0.9	278	49	8- 11½	spect A0	08056s4802
γ	41.0	220	51	2½- 4	(Δ65) Beautiful	08079s4712
	62.3	151	07	- 8½	easy pair; typical	
	93.5	141	02	- 9½	Wolf-Rayet star;	
					Spect WC7, B3 (*)	
γd	1.8	146	28	9½- 13	(I 1175)	
λ96	0.3	291	42	6½- 7	spect B4	08109s4607
h4069	0.4	293	36	6½- 8	(φ113) Spect B3;	08128s4541
	33.3	250	59	- 8½	AC PA slight dec	
Brs 3	5.2	327	40	8½- 9	spect A0	08173s4453
	77.6	143	00	- 9		
Rst 4884	7.8	136	47	7- 13½	spect K0	08175s4602
h4081	42.5	185	20	6½- 11	spect B3	08176s4802
I 67	0.9	138	46	5 - 6½	(B Vel) Spect B1	08210s4820
Δ69	25.7	220	31	5 - 8½	(Gls 95) Spect B3	08241s5134
Cor 71	7.0	105	28	8 - 11	spect Ao	08262s4105
h4102	68.5	281	13	6½- 9½	Spect B3, G0	08271s4225
h4104	0.1	120	59	6½- 6½	(φ315) AB PA inc,	08275s4746
	3.7	245	51	- 8	Spect B5; AC PA inc	
	18.8	39	34	- 9½	(A Vel)	
Δ70	5.0	350	51	5- 6½	relfix, spect B3	08278s4433
I 168	3.3	74	34	6½- 11	PA dec, spect B5	08290s4434
I 313	3.8	216	37	7 - 9	spect K2	08290s4121
h4107	4.5	329	20	6½- 8½	relfix, spect B3	08294s3854
	30.8	100	20	- 9½		
Slr 8	0.8	292	47	6 - 7½	PA dec, spect G5	08307s5302
Cor 72	3.6	98	35	8½- 10	spect F8	08308s5048
Hd 202	15.0	225	00	6½-10½	spect B5	08314s4535
I 195	1.9	42	54	6½- 8½	relfix, spect K5	08326s3726
h4119	10.1	226	13	7½-10½	spect A0	08356s4915
Cor 74	4.0	64	35	5½- 9	PA slight inc;	08384s4004
					spect B9; labeled	
					"Cor 18" in Norton	
h4126	16.7	30	33	5½- 9½	relfix, spect B5;	08385s5253
					cluster IC 2391 in	
					field	
I 815	4.4	2	37	7 - 11	spect B5; D= 10"	08428s4106
	8.3	130	38	- 13½	pair, mags 11, 12½	
	35.0	225	37	- 11	in PA 325°	

LIST OF DOUBLE AND MULTIPLE STARS   (Cont'd)

NAME	DIST	PA	YR	MAGS	NOTES	RA & DEC
δ	2.6	153	52	2 - 6½	(I 10) (h4136) AB	08433s5431
	69.2	61	13	- 10½	cpm, PA dec, spect AO (*)	
δc	6.2	102	35	10½-13		
Jc 13	2.1	313	41	7½- 8½	relfix, spect B9	08448s4223
Hu 1590	0.3	338	33	8½- 9	spect A3; in star cluster H3	08449s5239
I 70	1.3	113	53	7 - 9	relfix, spect AO	08459s3845
I 1181	3.8	338	33	7½- 12	spect FO	08486s4944
Hd 205	3.4	84	59	5 - 10	(f Vel) relfix, spect BO	08489s4620
h4148	5.8	111	17	8 - 10½	spect A2	08494s5355
Cp 9	3.0	79	37	7 - 8½	relfix, cpm, spect AO	08512s5156
h4150	17.5	265	20	7½- 10	spect B9	08521s4138
H	2.7	339	38	5 - 7½	(R87) PA slight dec? spect B5	08548s5232
△73	65.2	359	36	7½- 8	PA slight inc, both	08548s5519
	27.5	240	13	- 10½	spect KO	
Cor 78	10.9	144	19	8 - 8½	spect AO	08550s5336
λ108	3.1	46	35	7½- 10	(Gls 102) spect B3	08553s4304
	43.1	2	00	- 10		
	48.1	240	00	- 9		
I 71	9.1	307	34	7½- 10	(Gls 104) spect KO	08584s4922
h4165	0.9	118	54	5½- 7	PA inc, spect B9	09002s5159
h4177	0.6	126	38	7 - 8½	(Rst 3620) spect	09031s5608
	12.4	241	59	- 9½	B9; AC PA dec	
	35.5	296	13	- 9½		
I 492	0.3	147	45	7½- 8	spect B5	09059s4425
h4188	0.1	130	59	6½- 7	(φ317) Spect B8;	09107s4324
	2.7	282	51	- 7	AC PA slight dec	
h4191	5.6	15	35	5 - 9½	(z Vel) PA slight inc, spect B5	09125s4301
I 11	0.9	284	60	6½- 7½	PA inc, spect AO	09134s4521
Rst 4907	3.5	319	44	6½- 14	spect KO	09185s4458
Rst 4909	1.6	21	49	7½- 13	spect G5	09202s4633
Cor 83	19.4	152	20	7 - 10	spect F8	09240s5302
λ112	9.3	270	33	7½- 12	(△76) spect B5	09267s4517
	60.7	98	13	- 8		

VELA

LIST OF DOUBLE AND MULTIPLE STARS   (Cont'd)

NAME	DIST	PA	YR	MAGS	NOTES	RA & DEC
ψ	0.5	202	61	4 - 5	Binary, 34 yrs; PA inc, spect F2	09287s4015
Syd 1	10.8	19	37	7½-10½	spect B5	09312s5652
h4220	2.0	214	55	6 - 6½	PA inc, spect B4	09319s4847
R125	3.2	179	48	6½- 9½	PA inc, spect F0	09346s4832
λ115	0.5	180	49	6 - 6	PA inc, spect A3; cluster NGC 2925 in field	09356s5326
Hu 1465	4.5	195	33	7- 13½	spect G0	09357s4946
	13.2		00	- 11		
Hu 1467	6.3	310	32	8- 12½	spect A2	09370s5017
R129	3.2	294	59	7½- 7½	Slow PA inc, spect G5	09410s5537
B1658	2.1	200	44	6½- 11	spect B8	09417s5100
h4240	12.4	57	17	8 - 10	spect B8	09418s5948
△80	18.6	70	19	8½- 8½	relfix, cpm, spect both F8	09432s4916
h4242	7.7	358	20	7½- 9½	spect B9	09437s4126
h4245	9.4	216	39	7 - 9½	spect G5	09442s4541
φ33	8.6	191	28	7½- 13	spect K0	09465s5410
I 1520	3.1	250	42	6½- 11	spect K0	09472s4315
Cor 92	5.8	22	55	8½- 9½	spect F8	09483s4923
△81	5.3	241	52	6 - 8	relfix, spect B4	09523s4503
Hu 1472	1.6	8	32	7½-12½	spect B8	09553s4938
h4269	14.1	321	33	6½- 11	spect B3	09558s4810
h4273	15.5	135	29	7 - 11	spect A0	09574s4443
Hd 209	15.0	55	01	6½- 10	spect B8	09594s5551
h4282	47.7	199	17	7½- 8½	spect A2, A0	10013s5148
h4283	8.0	181	41	7½- 9	relfix, spect B9	10026s5134
Hu 1594	0.4	278	60	7 - 7½	PA inc, spect B9	10031s5104
h4284	6.6	66	35	7½- 9½	spect K0	10031s4539
I 173	0.2	144	26	5½- 7	binary, about 230 yrs; PA inc, spect K0	10042s4707
I 499	1.7	301	59	7½- 9½	Slight PA dec, spect B9	10053s5420
R140	3.2	281	54	7½- 8½	Slight PA inc? spect A2	10172s5546
h4307	0.8	40	48	7 - 10	(Rst 5518) spect F2	10179s5119
	13.5	262	40	- 10½	NGC 3228 in field	

LIST OF DOUBLE AND MULTIPLE STARS   (Cont'd)

NAME	DIST	PA	YR	MAGS	NOTES	RA & DEC
Rmk 13	7.2	102	52	4½- 8	relfix, spect B3; R140 in field	10191s5547
I 208	0.7	23	60	7½- 9	PA slight dec, spect A0	10217s4359
	25.4	24	60	-14		
h4319	12.1	123	34	7½-11½	spect K5	10243s5338
Rst 4928	0.5	314	49	6½- 8½	spect A5	10247s4725
h4329	46.4	94	38	5 - 8½	optical, PA & dist inc, spect F6, K5	10295s5328
Cp 10	2.2	345	40	7½- 9	spect B8	10298s5158
△ 88	13.4	218	51	6 - 6½	(s Vel) relfix, spect B8	10298s4449
h4330	40.5	162	60	5 - 8½	(t Vel) relfix, spect K4+A; h4332 in field	10308s4645
△89	26.0	30	38	6½- 8½	(Hld 106) relfix, AB spect G5, A0	10313s5507
△89b	1.5	250	54	8½- 9		
h4332	28.4	162	33	7 - 9½	relfix, spect B9	10314s4643
I 175	1.9	156	60	7½-10½	relfix, spect G0	10342s4735
λ119	0.4	286	58	4½- 5	(p Vel) Binary, 16 yrs; PA dec, spect F2, A3	10352s4758
△95	51.9	105	38	4½- 6½	(h4341) (x Vel) spect G2	10374s5521
△95b	20.2	176	34	6½- 11		
μ	0.7	90	42	3 - 7	(R155) cpm, PA inc, dist dec, spect G5	10446s4909
h4388	34.7	206	30	7- 11½	spect K0	10554s4536
Hu 1601	0.2	4	60	8½- 9	spect K0	10559s5404
h4394	28.6	261	13	8½- 9	spect G5	10571s4252
Rst 4933	0.7	312	44	7½- 10	spect K5	10580s4438

VELA

LIST OF VARIABLE STARS

NAME	MagVar	PER	NOTES	RA & DEC
λ	2.14--2.22	Irr	Spect K4	09062s4314
O	3.56--3.67	.13198	β Canis type; Spect B3	08389s5245
N	3.1---	Irr?	Variability uncertain; Spect K5	09297s5649
S	7.7--9.5	5.9337	Ecl.Bin; Spect A5e + K5e	09314s4459
T	7.8--8.3	4.6397	Cepheid; Spect F6--F9	08361s4711
U	7.9--8.6	40:	Semi-reg; Spect M5	09313s4517
V	7.2---7.9	4.3710	Cepheid; Spect F6--F9	09208s5545
W	8.3--14..	394	LPV. Spect M8e	10134s5414
X	8.4--10.7	140:	Semi-reg; Spect N	09534s4121
Y	8.2--14..	445	LPV. Spect M8e--M9e	09274s5158
Z	7.9--14..	422	LPV. Spect M9e	09512s5357
RS	8.0--11..	410	LPV. Spect M7e	09221s4839
RT	9.2--11..	141	LPV. Spect M7e	10281s4627
RU	9.4--11..	125	Semi-reg; Spect M3e	10515s5255
RV	10.0-10.8	4.8211	Ecl.Bin; Spect G0	09173s5021
RW	7.8--12..	452	LPV. Spect M7e	09186s4919
RY	7.9--8.7	28.127	Cepheid; Spect F6--G3	10188s5504
RZ	6.4--7.6	20.397	Cepheid; Spect G8	08353s4356
ST	9.4--10.0	5.8584	Cepheid; Spect K	08434s5022
SU	8.8--11..	150:	Semi-reg; Spect M7?	09480s4147
SV	7.9--9.1	14.097	Cepheid; Spect F6--G5	10429s5602
SW	7.5--9.0	23.474	Cepheid; Spect K2	08420s4713
SX	7.8--8.6	9.5499	Cepheid; Spect G5	08432s4610
SY	7.6--8..	63:	Semi-reg; Spect M5	09106s4334
SZ	8.2--9.5	150:	Semi-reg; Spect M5e	09482s4425
UX	8.7--15..	226	LPV. Spect M0e	10382s5427
VZ	9.5--11..	317	Semi-reg; Spect M6e	10262s5055
WW	9.4--13..	187	LPV. Spect M5e	10482s4824
WY	7.6--9..	Irr	Z Andromedae type; Spect M3e + B	09203s5221
WZ	7.7--9..	130:	Semi-reg; Spect M3	10155s4742
ZZ	9.8--10.4	2.8762	Ecl.Bin; Spect A0	10359s5541
AC	8.7--9.2	4.5622	Ecl.Bin; Lyrid, Spect B6	10443s5634
AH	5.5---5.9	4.2272	Cepheid; Spect F8p	08104s4630
AI	6.4--7.1	.11157	Dwarf Cepheid; Spect A2 ---F2p (*)	08124s4425
AL	8.6--9.1	96.112	Ecl.Bin; Spect K0 + A3	08296s4730
AO	9.6--10.0	1.5846	Ecl.Bin; Spect B9	08104s4836
AP	9.5--10.4	3.1278	Cepheid; Spect F9	08380s4341

VELA

LIST OF VARIABLE STARS (Cont'd)

NAME	MagVar	PER	NOTES	RA & DEC
AS	8.7--9.3	1.5579	Ecl.Bin; Spect A3	08264s3848
AX	7.9--8.5	2.5928	Cepheid; Spect F8	08093s4733
AY	9.1--9.8	1.6177	Ecl.Bin; Lyrid, Spect B9	08186s4343
BG	7.4--8.0	6.9236	Cepheid; Spect G5	09067s5114
CF	9.3--14..	245	LPV.	10090s5016
CH	9.0--12..	327	LPV.	10461s4114
CM	7.4--10..	780	Semi-reg; Spect M0--M2	10057s5301
CN	9.7---16	---	Nova 1905	11005s5407
CQ	9.0--16.5	---	Nova 1940	08574s5308
CV	6.6--7.3	6.892	Ecl.Bin; Spect B2 + B2	08591s5122
CW	9.5--10.6	2.3609	Ecl.Bin; Spect B9	09008s5239
DN	9.6--10.1	12.898	Ecl.Bin; Spect B9	09178s4528
DR	9.3--9.9	11.200	Cepheid	09299s4926
DW	8.8--15..	476`	LPV.	09483s5146
DZ	9.7--10.2	2.8104	Ecl.Bin; Spect G0	09527s4946
EP	8.4--10.1	240	Semi-reg; Spect M6	08401s4756
FW	9.4--10.1	2.3841	Ecl.Bin; Spect A2	10438s5209
FX	9.2--10.4	1.0526	Ecl.Bin; Lyrid, Spect Be	08307s3749
FY	6.84-7.06	33.72	Ecl.Bin; Lyrid? Spect B2	08309s4926
FZ	5.15-5.17	.065	δ Scuti type; Spect F0	08572s4702
GG	8.0--8.5	1.4752	Ecl.Bin; Spect A0	09110s4317
GH	9.0--9.3	75:	Semi-reg; Spect M3	09120s4827
GI	7.9--8.1	120:	Semi-reg; Spect M3	09232s4510
GK	6.2--6.5	120:	Semi-reg; Spect M5	09233s4346
GL	7.3--8.0	117	Semi-reg; Spect M5?	09251s5318
GM	8.8--9.1	120	Semi-reg; Spect M6	09458s4625
GO	6.6---6.9	Irr	Spect M5	08358s4016
GP	6.7---6.9	8.962	Ecl.Bin; Spect B0	09002s4021
GS	9.2---9.5	Irr	Spect M2	10436s5619

VELA

LIST OF STAR CLUSTERS, NEBULAE AND GALAXIES

NGC	OTH	TYPE	SUMMARY DESCRIPTION	RA & DEC
2547	△ 411		Mag 5.5; Diam 15'; B,L,cC; about 50 stars mags 7....15; class D; 2° south of Gamma	08089s4907
2626		□	10ᵐ B2 star in Small neby, diam about 2'	08338s4028
2659			Mag 9.5; Diam 10'; Class D; about 50 stars mags 11.... L,pRi,pmE	08409s4446
2660			Mag 11; Diam 2'; Class D; pS,mC,R, about 25 stars mags 12....	08410s4702
----	I.2391		vL,B, scattered group of 10 stars incl. Omicron Velorum; class C	08412s5245
----	I.2395		Mag 6; Diam 10'; Class E; about 16 stars mags 9.... cluster 2670 in field 30' S	08434s4800
2671			Mag 10; Diam 3'; Class E; pL,1cM, about 25 stars mags 11....	08444s4142
2670			Mag 9; Diam 15'; Class D; pL,P,1C, about 20 stars mags 11...13; cluster I.2395 in field	08446s4836
----	H3		Mag 6; Diam 7'; Class E; about 35 stars mags 10..... Incl double star Hu 1590	08446s5236
2792		◎	vS,F,disc; Mag 13, Diam 12"; 1.1° NE from Lambda Vel.	09106s4214
2899			pL,pF,R, Diam 1.5' in rich field 1.3° SE from Kappa Vel.	09255s5554
----	I.2488		Mag 7; Diam 20'; Class D; about 50 stars mags 11....	09257s5645
2910			Mag 8; Diam 6'; Class F; cL,pRi,pC; about 30 stars mags 10....	09284s5241
2925			Mag 8; Diam 11'; Class D; pL,pC,pRi; about 30 stars	09319s5313
3105			Mag 11; Diam 2'; Class F; S, C,1E; 15 stars mags 12....	09589s5432

LIST OF STAR CLUSTERS, NEBULAE AND GALAXIES   (Cont'd)

NGC	OTH	TYPE	SUMMARY DESCRIPTION	RA & DEC
3132		◎	! vB,L,1E; Mag 8.2; Diam 84" x 52"; 10 mag star in center; "Eight-Burst" planetary nebula; on Vela-Antlia border (*)	10049s4011
3201	Δ445	⊕	Mag 8.5; Diam 10'; Class X; vL,B,Ri,1cM; stars mags 13.. loose-structured globular	10155s4609
3228	Δ386	⦂	Mag 6.5; Diam 20'; Class F; B,1C; about 25 stars mags 9....	10197s5129
3256		⊖	S/pec?; 12.1; 2.0' x 1.5' cB,S,E, double nucleus; Two interacting galaxies?	10257s4338
3261		⊖	SBb; 12.8; 2.5' x 1.8' F,S,1E	10268s4423
3318		⊖	SBb; 12.6; 2.0' x 1.2' cF,pL,bM, BN	10351s4122

DESCRIPTIVE NOTES

GAMMA   Name- Sometimes known as SUHAIL, from the Arabic *Al Suhail al Muhlif,* thought to be a form of *Al Sahl,* "The Plain", and later made a term of admiration, implying "brilliant" or "glorious". The name, however, in various forms, was applied also to Lambda and Zeta Puppis. Magnitude 1.88; Spectrum WC7 or possibly composite WC7+07; Position 08079s4712, near the western edge of the constellation. Gamma Velorum is a splendid and easy double star for any small telescope, resolvable in a good pair of binoculars, although rather low in the sky for observers in the U.S. The companion, magnitude 4½,is located 41" from the brilliant primary in PA 220°; there has been no definite change in separation or angle since the early measurements of John Herschel in 1835. The two stars probably form a true physical pair, although the

apparent motion is so slight that it is difficult to be
certain about a common proper motion. The projected separation is about 6800 AU. E.J.Hartung finds the companion
visible by day with an aperture of 30 cm; he calls both
stars "white". A third fainter star, mag 8½, may be noticed
at 62" in PA 151°; R.H.Allen calls the colors of the three
objects as "white, greenish-white, and purple".

The computed distance of Gamma Velorum is about 520
light years; the actual luminosity of the primary is then
about 3900 times that of the Sun and the absolute magnitude
about -4.1. The annual proper motion is only 0.01" and the
radial velocity is about 21 miles per second in recession.
If at the same distance as the primary, the 4th magnitude
companion is about 400 times the solar luminosity, with an
absolute magnitude of about -1.7; the spectral type is B3.

The bright star is of special interest, as it is the
classic example of a "Wolf-Rayet" star. Once included in
spectral class O, these stars are now usually classed by
themselves in a separate class "W". The first stars of the
type were discovered in 1867 by C.Wolf and G.Rayet at the
Paris Observatory, with a visual spectroscope. Wolf-Rayet
stars are extremely hot and luminous giants like the stars
of the "helium giant" type O, but show strong continuous
spectra on which are superimposed numerous broad bright
lines. Miss Agnes Clerke, writing in 1905, described the
"extraordinary beauty" of the spectroscopic appearance of
this star which was "studied with some care and much de-
light" by Dr.Copeland in the Andes, in April 1883:

*"An intensely bright line in the blue, and the gorg-
eous group of three bright lines in the yellow and orange,
render the spectrum.... incomparably the most brilliant and
striking in the whole heavens"*. Miss Clerke adds that *" A
vivid continuous spectrum extends into the violet as far as
the eye has power to follow it, and accounts for the brilliant whiteness of the star"*.

The bright emission lines are frequently up to 100
angstroms wide, and 10 to 20 times brighter than the background spectrum. Occasionally the bands are bordered by
weak absorption lines on their violet edges. Hydrogen lines
are unusually weak in the star; the bands are produced
chiefly by ionized helium, nitrogen, oxygen, carbon, and
silicon. Spectra of the Wolf-Rayet stars appear to fall

into two subclasses called "WC" and "WN"; one group having
strong carbon lines and the other showing strong nitrogen
features.

The peculiar spectral features of the Wolf-Rayet type
stars evidently originate in a continually expanding gaseous envelope, supplied by material streaming out of the
star at tremendous velocities. Expansion velocities of up
to 1800 miles per second seem to indicate that the outer
layers of the star are being blown away into space in an
almost nova-like fashion. This interpretation is supported
also by the absorption lines on the violet edges of the
bright bands; these are shifted to the violet by the Doppler effect, since they are produced by material lying
directly in front of the main body of the star, where the
approach velocity is naturally greatest.

Wolf-Rayet stars are objects of very high intrinsic
luminosity, and, as á class, are also very distant objects.
Gamma Velorum itself appears to be the nearest star of the
type. Much of our knowledge of these stars has been derived
from studies of numerous examples in the Magellanic Clouds,
the two irregular galaxies which are the nearest optically
visible members of the Local Group. The absolute magnitudes
are then found to range from -4 to about -8, or up to about
100,000 times the luminosity of the Sun. They are also
among the hottest stars known, with temperatures ranging
from 50,000° to 100,000°K. It is a curious fact that many
of the central stars of the planetary nebulae show spectra
of the Wolf-Rayet type, but the actual luminosities are
very much lower, often less than the luminosity of the Sun.

A number of Wolf-Rayet stars are known to be close
spectroscopic binaries, and a widely supported theory regards all stars of the class as close doubles. The development of a Wolf-Rayet star possibly depends upon the effect
of a close and very massive companion. Gamma Velorum itself
shows periodic radial velocity shifts in a cycle of about
78½ days, almost certainly indicating binary motion. From a
study in 1970 with a combination of spectroscopic and interferometric techniques, the WC7 star now appears to have
a mass of about 15 suns, a diameter of 17 suns, and an
absolute magnitude close to -5.6; these results also imply
that the distance is somewhat greater than was previously
thought, probably close to 800 light years. The fainter

component of the binary pair is approximately a magnitude
less luminous than the primary, and appears to be a giant
star of type O7.

Gamma Velorum is located in a fine starry field in
the Milky Way, near the Puppis-Vela border. The bright
eclipsing variable V Puppis lies about 2½° to the SW (see
page 1502) and the galactic star cluster NGC 2547 is just
2° south. The odd short-period variable AI Velorum will be
found about 2.8° to the north. (Refer to page 2039)

DELTA   Magnitude 1.95; Spectrum A0 V; Positon 08433s
5431, about 9° SE from Gamma Velorum, near the
Vela-Carina border. This is one of the stars which seems to
share the space motion of the Ursa Major stream, along with
Sirius and a number of other bright stars. The computed
distance is 75 light years, and the actual luminosity about
70 times that of the Sun (absolute magnitude about +0.2).
Delta Velorum shows an annual proper motion of 0.09"; the
radial velocity is 1.3 miles per second in recession.

The 6th magnitude companion at 2.6" was discovered by
R.T.Innes in 1894 and shares the proper motion of the primary. There has been no definite change in the separation in
the last 80 years, but the PA has slowly decreased from
175° to about 150°. In addition, the 10th magnitude star at
69" appears to be a physical member of the system, and is
itself a double of 6" separation, magnitudes about 10½ and
13. Projected separations for this multiple star family are:
AB= 58 AU; AC= 1580 AU; CD= 140 AU. E.J.Hartung refers to
the AB components as "a pale yellow pair" and finds them
resolvable with an aperture of 10.5 cm.

KAPPA   Name- MARKAB, but very seldom used, as Alpha
Pegasi in the northern sky has the same title.
Magnitude 2.45; Spectrum B2 IV; Position 09206s5448, about
5° east of Delta. The computed distance is about 470 light
years; the actual luminosity about 1900 times that of the
Sun, and the absolute magnitude about -3.4. Kappa Velorum
shows an annual proper motion of 0.01"; the radial velocity
is 13 miles per second in recession. The star is a spectroscopic binary with a period of 116.65 days. Several galactic star clusters will be found near Kappa: NGC 2910 and
2925 lie about 2.2° to the NE.

LAMBDA   Name- SUHAIL, from the Arabic *Al Suhail al Wazn*, "The Suhail of the Weight". Several of the bright stars in the region, however, have been given some variation of the name, as Gamma Velorum. Lambda is magnitude 2.24; Spectrum K5 Ib; Position 09062s4314, about 10° ENE from Gamma. Lambda is a bright yellow-orange super-giant with a computed absolute magnitude of -4.6 (luminosity = 5800 suns); it lies some 750 light years distant. The star shows an annual proper motion of 0.03"; the radial velocity is 11 miles per second in recession. An extremely faint companion to the star was detected by T.J.J.See in 1897; the pair was entered in his catalogue as λ109. Said to be magnitude 14.8, the companion is 18" distant from the bright star in PA 137°.

Two other interesting pairs lie in the field. About 0.7° to the east and slightly south is the triple h4188; just 1° west and slightly north is the double star h4191, sometimes marked "z Velorum" on atlases, but not to be confused with the variable "Z Velorum" which is a different object. (Refer to data on page 2029). Also near Lambda is the faint pale greyish disc of planetary nebula NGC 2792; look for it about 1.1° NE of the star.

MU   Magnitude 2.67; Spectrum G5 III; Position 10446s4909, near the eastern boundary of the constellation. Mu Velorum is a rather difficult binary, first measured by H.C.Russell in 1880 when the separation was 2.8" in PA 55°; it has since closed down to about 0.7" in PA 90° (1942) and, according to a note in the Lick *Index Catalogue of Visual Double Stars* in 1961, has not been seen with certainty since.

The computed distance of the star is about 110 light years, and the total luminosity about 75 times that of the Sun. The star shows an annual proper motion of 0.08"; the radial velocity is 4 miles per second in recession.

N   (Suspected variable) Magnitude 3.15; Spectrum K5 III; Position 09297s5649, about 2.3° SSE from Kappa Velorum. Suspected of variability by various observers, the star seems to be constant in light at the present time. R.T.Innes in 1895, however, reported a range of magnitude 3.2 to 3.8 with no regular period. Isaac

Roberts thought that the star varies in color, though possibly not in light. The case is rather similar to that of Alpha Cassiopeiae, which also seems constant in light although variability was reported in the past.

Direct parallaxes indicate a distance of about 170 light years; this gives the star an actual luminosity of 120 suns, and an absolute magnitude of -0.4. The annual proper motion is 0.04"; the radial velocity about 8 miles per second in approach.

AI   Variable. Position 08124s4425. A peculiar short-period pulsating variable star, located in the Milky Way some 2.8° north and slightly east from Gamma Velorum. The star is usually classed as a dwarf cepheid, though it has a rather strange light curve. It was discovered by E.Hertzsprung in 1931, from photographs made at Johannesburg in South Africa. At first thought to be an RR Lyrae type star, AI Velorum was found to be a pulsating variable with the very rapid fundamental period of 2^h 41^m,

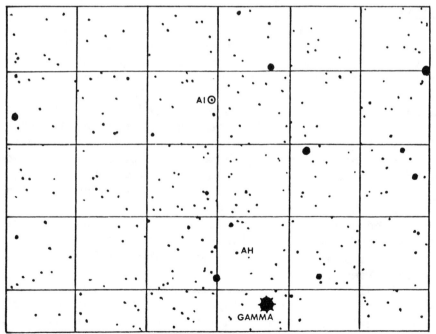

AI VELORUM FIELD, showing stars to about 8th magnitude. Grid squares are 1° on a side with north at the top.

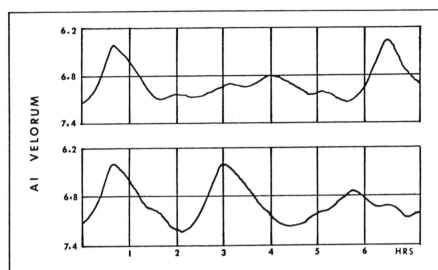

with a magnitude range of 6.4 to 7.1. The star is a white
type A2 at maximum, fading to a somewhat yellowish F2 at
minimum. According to the Moscow *General Catalogue of Variable Stars* (3rd Supplement, 1976) the precise period is
0.11157396 day.

The light curve of the star shows many puzzling and
curious irregularities which result in some maxima being
over twice the amplitude of others. A high maximum may be
followed immediately by one which shows an amplitude of
only a small fraction of a magnitude. The radial velocity,
which averages about 9 miles per second in recession, is
likewise variable, at times decreasing to zero, at other
times rising to nearly 25 miles per second. This is evidently a case of pulsation with several superimposed periods.
In addition to the primary cycle of 161 minutes, the star
appears to be oscillating in periods of 46 minutes, 64 minutes, and 124 minutes. When these multiple oscillations re-
inforce each other the total range in brightness is nearly
one magnitude; when they interfere the range is only a
fraction of a magnitude. A "beat period" of 0.379188 day
has been identified.

In discussing this peculiar star, O.Struve pointed
out that AI Velorum is analogous to a vibrating organ pipe;
it may oscillate not only in its fundamental frequency, but
also in various "overtones". For a fundamental period of
161 minutes, overtones of 70 and 46 minutes are predicted;

DESCRIPTIVE NOTES (Cont'd)

the first of these has not been detected, but the 46-minute oscillation was found in the analysis of T.Walraven in his study at Johannesburg. The interaction of the primary period with the two overtones produces the secondary cycles of 64 and 124 minutes. This is a complex situation, and the physical processes at work in the star are largely unknown. Pulsations of this type probably originate in the deeper layers of the star's photosphere, and not in the central body or "core".

The distance of AI Velorum is not accurately known. If it is roughly a magnitude fainter in actual luminosity than typical RR Lyrae stars, the absolute magnitude may be about +1.8 and the distance would then be close to 300 light years. An annual proper motion of 0.06" has been measured, consistent with the assumed distance. (Refer also to RR Lyrae, CY Aquarii, and SX Phoenicis)

NGC 3132 Position 10049s4011. Fine planetary nebula, located squarely on the Vela-Antlia border, and included in some observing guides under Antlia. More or less comparable in size to the Ring Nebula in Lyra, NGC 3132 appears more conspicuous visually than the Ring, owing to the presence of the unusually bright central star, listed as either 9th or 10th magnitude in standard catalogues. The disc of the nebula is noticeably elliptical, measuring about 84" x 52" on photographs, with much diaphanous detail and a structure suggesting the appearance of several oval rings superimposed and tilted at different angles. From the complex structure on photographs, NGC 3132 has been called the "Eight-Burst" Nebula. E.J.Hartung reports his impression that the light of this nebula appears "bright white" and without a sign of the usual bluish-green tint common in the planetaries. The total integrated magnitude is 8.2.

C.R.O'Dell (1963) derives a distance of about 2800 light years for this nebula; I.S.Shklovsky obtains 2150 and L.Kohoutek about 1955. The actual diameter is close to 0.5 light year. In a recent study (1977) it was found that the "central star"(HD 87892, Mag 10, Spectrum A0) is not truly the illuminating star of the nebula; the radiation instead is supplied by a 16th magnitude dwarf companion 1.65" distant, with a surface temperature of possibly 100,000°K. (See photograph on page 1175)

LIST OF DOUBLE AND MULTIPLE STARS

NAME	DIST	PA	YR	MAGS	NOTES	RA & DEC
Σ1560	4.9	278	58	6 - 10	cpm, relfix; spect gK1	11358s0210
A2578	0.9	140	33	7½- 12	spect A3	11378n0114
Σ1568	9.6	222	30	9 -9	relfix, spect F5	11407n0102
Σ1575	30.4	210	40	7½- 8	relfix, spect G0+F5	11494n0907
Σ1593	1.3	17	56	8½- 8½	relfix, spect F0	12010s0210
	49.7	3	41	- 11½		
A2982	142	195	17	8 - 11	spect K2	12047s0548
A2982b	0.4	76	57	11-11½		
Σ1605	23.7	278	31	8 - 8½	relfix, spect F2	12079s0158
Σ1616	23.4	295	53	7½- 10	AB cpm spect G0+K2;	12119n0905
	157	293	20	- 9½	AC dist inc	
Σ1618	26.3	245	27	8½- 8½	relfix, spect F5	12125n1016
Σ1619	7.2	273	55	7½- 8	PA dec, spect G5	12126s0658
	98.7	176	25	-10½		
Σ1627	19.9	196	58	6½- 7	cpm, relfix, spect dF4, dF5	12156s0341
Σ1628	9.8	239	38	8½- 9	cpm, relfix, spect G0	12162n1204
	47.5	349	20	- 11		
14	3.3	165	39	7- 14½	(A145) spect K0	12168s0838
Σ1635	13.3	173	38	7½- 8½	cpm, relfix, spect A0	12186s1111
17	19.6	337	58	6½- 9	(Σ1636) relfix, cpm spect dF8, dK5	12200n0536
Hn 13	1.2	145	57	9 - 9	AB PA dec, spect F5	12239s0137
	134	74	09	-12		
Hn 13c	15.1	350	09	12- 14		
AGC 4	0.7	200	56	7½- 11	PA dec, spect A5	12242n0006
A78	0.3	113	66	8 - 8½	PA inc, spect F2	12242s0519
β923	2.6	60	25	7- 13½	spect A0	12258n0440
Σ1647	1.2	236	67	7½- 8	cpm, PA slow inc, spect F2	12280n1000
Σ1648	7.8	43	34	7½- 9½	slight PA inc? spect K0; galaxies NGC 4457 & 4496 in field	12280n0347
AG 178	1.5	288	60	8½- 9	relfix, spect F0; Quasar 3C 273 in field	12285n0223
A2583	50.0	290	21	8 - 8½	(Sh 146) spect A5	12286n0136

LIST OF DOUBLE AND MULTIPLE STARS (Cont'd)

NAME	DIST	PA	YR	MAGS	NOTES	RA & DEC
A2583b	5.0	336	52	$8\frac{1}{2}$- 13	(Sh 146b)	12286n0136
Σ1649	15.4	194	55	7 - 8	relfix, cpm, spect A5, F2	12290s1048
Σ1658	2.5	9	58	8 - 10	AB cpm, PA inc,	12326n0743
	112	261	23	- $9\frac{1}{2}$	spect F8; AC optical, dist inc. Galaxy NGC 4526 in field.	
Σ1661	2.4	247	59	$8\frac{1}{2}$- $8\frac{1}{2}$	PA inc, spect G5	12335n1141
Σ1665	8.9	98	38	$8\frac{1}{2}$- 9	relfix, spect K0; galaxy NGC 4593 in field	12361s0503
Σ1668	1.4	190	58	$7\frac{1}{2}$- 8	PA slow dec, spect F2	12384n0906
γ	4.7	306	66	$3\frac{1}{2}$- $3\frac{1}{2}$	Fine binary; 171 yrs; PA dec, both spect F0 (Σ1670) dist dec (*)	12391s0111
31	4.0	37	58	$5\frac{1}{2}$-$11\frac{1}{2}$	(β924) cpm, PA inc, spect B9	12394n0705
Rst 4965	7.6	97	43	8 - 11	spect K0	12424s0816
Σ1677	16.2	348	38	7 - 8	relfix, spect A3	12427s0337
OΣ255	17.9	343	29	7 - 12	dist dec, PA slight inc, spect A5	12437n0244
Σ1681	8.7	197	40	$8\frac{1}{2}$- $8\frac{1}{2}$	cpm, slight PA inc; spect G0	12470n0406
Σ1682	30.8	301	59	$6\frac{1}{2}$- 9	optical, PA slow	12488s1004
	144	201	11	- 11	dec, spect gG8	
Σ1689	28.6	211	27	$6\frac{1}{2}$- 9	PA inc, spect M	12530n1146
Σ1690	5.8	148	43	$7\frac{1}{2}$- 9	relfix, spect A0	12537s0435
OΣ256	0.8	93	67	7 - $7\frac{1}{2}$	PA inc, spect F5	12539s0041
A146	2.0	296	60	7 - 10	cpm, PA dec, spect F5	12549s0929
Σ1701	21.6	306	27	$7\frac{1}{2}$- $9\frac{1}{2}$	cpm, relfix, spect G5	12568n0647
44	20.9	54	58	6 - 11	(Σ1704) cpm, relfix spect A2	12571s0333
46	1.1	165	58	6 - 11	(AGC 5) PA inc,	12580s0306
	3.6	122	12	- 13	spect gK2; AC slow PA inc	

LIST OF DOUBLE AND MULTIPLE STARS   (Cont'd)

NAME	DIST	PA	YR	MAGS	NOTES	RA & DEC
Σ1711	0.6	341	67	8½- 9	PA dec, spect F5	13004n1344
β928	2.1	316	67	8 - 9	Spect F0; Galaxy NGC 4951 in field	13008s0610
Σ1712	8.7	332	27	9 - 9½	cpm, relfix, spect G5	13010n0944
β341	0.8	312	47	6 - 6½	relfix, spect dF8	13011s2019
48	0.7	202	67	7 - 7	(β929) PA dec, spect A7	13013s0324
I 915	0.3	65	60	8½- 8½	spect F5	13029s2148
Σ1719	7.5	1	50	7½- 8	relfix, spect F5	13048n0051
θ	7.2	343	58	4½- 9	(Σ1724) cpm, relfix	13074s0516
	71.4	299	34	- 10	spect A1	
β931	4.9	205	28	6½- 11½	cpm, spect K0	13083n1334
54	5.3	34	58	6 - 7	(Hh 412) (Sh 151) relfix, spect A0	13108s1834
Hu 740	3.9	271	44	7½- 13	spect A2	13171s1125
Σ1734	1.0	179	68	6½- 7½	PA dec, slight dist inc, spect A0	13182n0312
HV 119	19.4	309	24	7½- 10	cpm, spect A2	13201s1256
Σ1738	3.9	281	55	8 - 8	relfix, spect F8	13206s1440
Σ1740	26.5	75	52	7 - 7	easy relfix cpm pair; both spect G5	13211n0259
β610	3.9	16	33	7 - 10½	relfix, spect K0; Galaxy NGC 5134 in field	13211s2040
Σ1742	1.0	353	68	7½- 8	relfix, spect A2	13218n0140
β460	2.3	34	37	8- 10½	relfix, spect A2	13224s1522
Σ1746	25.0	247	58	7½-10½	dist slow dec, spect K0	13257n0943
A2490	1.1	97	37	8 - 11	spect K0	13257n0230
72	29.6	16	38	6- 11½	(Σ1750) cpm, relfix spect A5	13278s0613
Σ1751	5.7	60	35	7½-10½	relfix, spect K0	13282n0934
73	0.1	183	59	7 - 7	(B2542) PA dec, spect A4m	13293s1828
Σ1757	2.6	104	62	7½- 8½	binary, about 240 yrs; PA inc, spect dK1	13317s0004
	45.0	153	21	- 12		
β114	1.2	155	67	8 - 8½	PA inc, spect F8	13317s0821

LIST OF DOUBLE AND MULTIPLE STARS   (Cont'd)

NAME	DIST	PA	YR	MAGS	NOTES	RA & DEC
β932	0.2	8	60	$6\frac{1}{2}$ - 7	Binary, about 200	13320s1258
	25.7	149	59	- $12\frac{1}{2}$	yrs; PA inc, spect A1; AC PA dec	
A1611	0.7	128	68	$8\frac{1}{2}$- $8\frac{1}{2}$	PA dec, spect A5	13342n0705
81	2.6	40	68	$7\frac{1}{2}$- $7\frac{1}{2}$	(Σ1763) relfix;	13350s0737
	14.1	328	12	- 11	spect K2	
Σ1764	15.7	32	33	7 - $8\frac{1}{2}$	relfix, cpm, spect K0, nice colors	13352n0238
84	3.0	229	58	$5\frac{1}{2}$- 8	(Σ1777) dist slight dec, PA slow dec, spect K2, dG5	13405n0347
Σ1775	27.7	336	38	7 - $9\frac{1}{2}$	relfix, cpm,	13409s0402
	38.5	178	00	-$13\frac{1}{2}$	spect K2	
86	1.3	303	58	6- $10\frac{1}{2}$	(Σ1780) (β935) PA	13433s1211
	27.1	164	58	- 11	inc, spect gG7;	
86c	2.4	272	58	$11\frac{1}{2}$-13	CD dist inc	
Σ1781	0.4	11	68	$7\frac{1}{2}$- 8	binary, about 360 yrs; PA inc, spect G0	13436n0522
Kui 65	0.3	269	66	$6\frac{1}{2}$- 8	spect gK5	13446s0928
Hu 1262	1.0	295	59	$6\frac{1}{2}$- 10	PA inc, spect K0	13518s2200
Σ1788	3.3	91	59	$6\frac{1}{2}$- $7\frac{1}{2}$	AB cpm, dist inc,	13524s0749
	126	293	22	- $10\frac{1}{2}$	PA inc, spect dF8,	
	156	215	24	- 11	dG1	
Ho 261	7.1	180	24	$7\frac{1}{2}$- 12	spect F0	13529s0917
Σ1790	5.4	241	45	$8\frac{1}{2}$- $8\frac{1}{2}$	relfix, spect F8	13535s0423
OΣ273	0.9	112	60	8 - $8\frac{1}{2}$	PA inc, spect F5; Galaxies NGC 5363, 5364 in field	13538n0532
τ	80.1	290	27	$4\frac{1}{2}$- 9	(93 Virg) (Hh 432) (Sh 171) optical, spect A3	13591n0147
Σ1799	4.2	293	44	8 - 9	relfix, spect F0	14022s0618
Σ1802	5.6	279	59	$7\frac{1}{2}$- 9	cpm, slight dist inc, slow PA dec, spect G0	14054s1241
Σ1805	4.8	33	52	$8\frac{1}{2}$- $8\frac{1}{2}$	relfix, spect G5	14075n0415
Σ1807	6.9	29	55	$7\frac{1}{2}$- 8	cpm, slight PA inc, spect F8	14087s0306
Σ1813	4.8	192	55	8 - 8	relfix, spect A3	14109n0538

LIST OF DOUBLE AND MULTIPLE STARS   (Cont'd)

NAME	DIST	PA	YR	MAGS	NOTES	RA & DEC
β939	0.4	87	66	8 - 8	PA dec, spect F2	14115s0817
	87.6	281	14	- 9		
Σ1819	0.7	268	68	7½- 8	binary, about 360 yrs; PA dec, spect dF8	14128n0322
Hn 18	3.7	354	60	7½- 11	cpm, relfix, spect A2	14167s1817
	134	336	09	- 12		
β116	3.2	276	50	8 - 8½	cpm, relfix, spect G0; in field of λ Virginis	14168s1329
Σ1833	5.7	172	54	7 - 7	PA inc, spect G0	14200s0733
Σ1842	2.7	16	49	8½- 8½	relfix, spect G0	14245n0355
φ	4.7	110	58	5 - 9½	(105 Virg) (Σ1846)	14256s0200
	92.8	209	60	- 12½	AB cpm, relfix, spect G2, dK0	
Σ1852	25.0	267	37	7 - 10	relfix, cpm, spect F2	14274s0401
Rst 4529	0.3	168	62	8½- 8½	binary, 42½ yrs; PA inc, spect dG5	14284s0535
A2227	2.0	139	56	7 - 11	cpm, spect F8; Galaxy NGC 5690 in field	14350n0230
Rst 5004	1.6	210	43	7½- 13	spect A2	14389s0150
A1109	1.3	74	60	7½- 10	PA & dist inc, spect F8	14403n0648
Σ1876	1.2	98	68	8 - 8½	PA inc, spect G0	14437s0711
Σ1881	3.5	359	48	7 - 9½	cpm, relfix, spect A0	14446n0111
β1113	4.0	137	24	6- 11½	spect A2; in field of 109 Virginis	14449n0215
Σ1883	0.2	141	41	7½- 7½	binary, about 300 yrs; PA dec, spect dF5	14464n0610
β348	0.4	111	58	5 - 7½	slight PA dec, spect gM2; AC optical	14592n0003
	31.7	217	60	- 14		
Σ1904	9.9	347	58	7 - 7	relfix, spect F0	15017n0541

VIRGO

LIST OF VARIABLE STARS

NAME	MagVar	PER	NOTES	RA & DEC
α	0.97--1.04	4.0142	Ecl.Bin or Ell.Bin; SPICA; Spect B1 (*)	13226s1054
ψ	4.7-- 4.8	Irr	(40 Virginis) Spect M3	12518s0916
ω	5.23--5.37	Irr	(1 Virginis) Spect M4	11359n0825
R	6.2--12.1	146	LPV. Spect M4e--M8e (*)	12360n0716
S	6.2--13..	378	LPV. Spect M6e--M7e	13304s0656
T	8.9--14..	339	LPV. Spect M6e	12121s0545
U	7.5--13.5	207	LPV. Spect M2e--M5e	12486n0550
V	8.1--14.5	250	LPV. Spect M3e--M5e	13252s0255
W	9.5--10.7	17.274	Cepheid; Typical Pop II type; Spect F0--G0 (*)	13235s0307
X	7.3--11.2	?	Class uncertain; spect sgFp	11593n0921
Y	8.4--15..	219	LPV. Spect M3e--M5e	12313s0409
Z	9.8--15..	307	LPV. Spect M5e	14077s1304
RS	7.2--14..	353	LPV. Spect M6e--M7e	14248n0454
RT	7.6--9..	155:	Semi-reg; Spect M8	13001n0527
RU	9.0--14..	437	LPV. Spect R3e p	12448n0425
RW	7.1--8...	Irr	Spect M5	12047s0629
RX	7.7--8...	200:	Semi-reg; Spect K0	12022s0530
RY	9.0--10..	140	Semi-reg; Spect M3	13390s1853
SS	6.0--9.6	355	LPV. Spect Ne	12227n0103
SU	8.4--14.5	210	LPV. Spect M2e--M4e	12027n1238
SV	8.0--11..	296	LPV. Spect M4e	11578s0956
SW	6.9--7.8	150:	Semi-reg; Spect M7	13115s0233
SX	8.8--12..	106	Semi-reg; Spect Me	13338s1920
SY	9.6--13.7	237	LPV.	13560s0420
TY	8.0--8.3	30:	Semi-reg; Spect G3p	11493s0529
TZ	7.8--9.4	134	Semi-reg; Spect M5	12020n0254
UU	9.9--11.1	.47561	Cl.Var; Spect A9--F5	12060s0013
UW	8.9--12.2	1.8107	Ecl.Bin; Spect A2	13127s1713
UY	8.0--8.8	1.9945	Ecl.Bin; Spect A7	12592s1930
VY	9.0--13..	280	LPV. Spect M3e p	13159s0425
AG	8.4--9.0	.64265	Ecl.Bin; Lyrid; Spect A2 + A2	11585n1317
AH	9.2--9.7	.40752	Ecl.Bin; W U.Maj type; Spect K0 + K0	12118n1206
AL	9.1--9.9	10.303	Cepheid; W Virg type; Spect F3--G0	14084s1305

LIST OF VARIABLE STARS   (Cont'd)

NAME	MagVar	PER	NOTES	RA & DEC
AN	7.5---8.5	100:	Semi-reg; Spect M7?	14153s1429
AO	8.6--11..	256	LPV. Spect M2e--M4e	14193n0408
AQ	9.8--14..	293	LPV. Spect M5e	14439s0704
AY	8.5--9.7	113	Semi-reg; Spect M6	13493s0326
BD	9.9--11.2	2.5485	Ecl.Bin; Spect A5	13240s1552
BG	8.0--9...	Irr	Spect M5	14446n0506
BH	9.6--10.6	.81687	Ecl.Bin; Spect G0 + G2	13559s0125
BK	7.4--8.7	Irr	Spect M7	12278n0442
BZ	8.0--11..	151	LPV. Spect M5e	12581s1723
CE	8.5--9.6	70:	Semi-reg; Spect K2	13467s0141
CF	9.5--13..	228	LPV. Spect M5e	14131s0538
CH	8.0--9.5	Irr	Spect M7?	12178s0843
CI	8.5--9.3	Irr	Spect M6	12306n0731
CN	8.2--9.0	60:	Semi-reg; Spect M3	12563n0829
CO	8.6--9.6	70:	Semi-reg; Spect M5	13014n0720
CP	8.4--9.3	70:	Semi-reg; Spect M7	13545n0649
CQ	9.0--9.7	Irr	Spect M3	14189n0640
CR	8.5--9.0	Irr	Spect M5	14354n0343
CS	5.73--5.93	9.2954	α Canes Ven type; Spect A3p	14159s1829
CT	9.3--10..	Irr	Spect M1	12496n0614
CU	4.9--4.95	.52068	α Canes Ven type; Spect B9p	14098n0238
CW	4.9--4.99	3.722	α Canes Ven type; Spect A2p	13316n0355
CX	9.2--9.7	2.9621	Ecl.Bin; Lyrid, Spect F5	14067s1521
DK	6.67-6.72	.121:	δ Scuti type; Spect F1	13139s0108
DL	7.0--7.5	1.3155	Ecl.Bin; Spect F2 + A2	13499s1828
DM	8.5--9.2	4.6694	Ecl.Bin; Spect F6 + F6	14052s1055
DN	7.7---8.5	Irr	Spect M4?	14087s1014
EP	6.3--6.4	16.31	α Canes Ven type; A4p	12446n0613
EQ	9.2--9.4	3.9140	Ecl.Bin with flares ? Spect K7e	13321s0805
ER	6.45--6.6	55:	Semi-reg; Spect M5	14040s1358
ES	8.1--8.3	Irr	Spect M	14060s0838
ET	4.8--5.0	80:	Semi-reg; Spect gM3	14081s1604
EU	9.0--9.2	38:	Semi-reg; spect M3?	14096s1831
EV	6.7--7.1	120	Semi-reg; Spect M3	14104s1338
EW	9.2--9.3	100:	Semi-reg; Spect M4	14143s1612

LIST OF STAR CLUSTERS, NEBULAE AND GALAXIES

NGC	OTH	TYPE	SUMMARY DESCRIPTION	RA & DEC
3818	284[3]	⊖	E4/E5; 12.9; 0.8' x 0.5' F.pS,E, psbM	11394s0553
3952	612[3]	⊖	Pec; 13.0; 1.0' x 0.4' cF,cS,1E,bM	11511s0343
3976	132[2]	⊖	Sb; 12.4; 3.6' x 0.8' B,pL,mE,vsmbMN, nearly edge-on spiral	11534n0702
4030	121[1]	⊖	Sb/Sc; 11.2; 3.3' x 2.4' cB,L,1E,psmbM; fine nearly face-on spiral	11578s0049
4037	77[3]	⊖	SBb; 12.8; 1.5' x 1.1' cF,pL,E	11588n1341
4045	276[2]	⊖	Sa; 12.8; 1.2' x 1.0' pF,L,E,sbM; with faint outer arms	12002n0215
4073	277[2]	⊖	E1/E2; 13.2; 0.7' x 0.6' F,pS,R,sbM; field of 6 small galaxies 5' to south	12019n0211
4116		⊖	SBc; 12.3; 3.3' x 1.4' F,L,E; flattened bar-shaped central mass; 13.4' pair with 4123	12051n0258
4123	4[5]	⊖	SBb; 12.0; 3.5' x 2.4' cF,vL,E,bMN; pair with 4116	12056n0309
4124	33[1]	⊖	S0/Sa; 12.5; 3.5' x 1.2' pB,pL,mE,bM; flattened oval outline	12056n1040
4129	548[2]	⊖	SBb; 12.9; 1.9' x 0.3' F,pL,pmE,vglbM; nearly edge-on barred spiral	12063s0845
4168	105[2]	⊖	E0/E1; 12.6; 1.2' x 1.2' pB,pL,R,psbM	12098n1329
4178		⊖	SBc; 12.0; 4.5' x 1.5' vF,vL,mE; nearly edge-on; central mass is long narrow streak; 37' NNW from 12 Virg	12102n1109
4179	9[1]	⊖	E8/S0; 11.7; 2.7' x 0.6' pB,pS,pmE,bMN; edge-on; spindle-shaped	12103n0135

LIST OF STAR CLUSTERS, NEBULAE AND GALAXIES   (Cont'd)

NGC	OTH	TYPE	SUMMARY DESCRIPTION	RA & DEC
4189	106[2]	⊖	Sc; 12.7; 1.7' x 1.5' F,L,1E,vglbM; compact nearly face-on spiral	12112n1342
4215	135[2]	⊖	S0/Sb; 12.8; 1.3' x 0.4' B,S,E,sbM; edge-on; lens-shaped system	12134n0641
4216	35[1]	⊖	Sb; 10.9; 7.2' x 1.0' vB,vL,vmE,sbMN; long streak, fine nearly edge-on spiral; two others in field (*)	12134n1325
4224	136[2]	⊖	Sa/Sb?; 12.8; 1.5' x 0.6' pB,pS,1E,gbM; nearly edge-on spiral with equatorial dust lane; 4233 & 4235 in field	12140n0744
4233	496[2]	⊖	Ep?; 13.0; 1.0' x 0.4' pF,R,vsbM,SN; with faint extensions north & south	12146n0754
4234		⊖	SBc/Irr; 13.0; 0.8' x 0.7' pB,S,R,gbM; seems to be a distorted spiral	12146n0358
4235	17[2]	⊖	Sa; 12.7; 3.0' x 0.6' pB,pL,mE,bM; nearly edge-on; in field with 4224 & 4233	12146n0728
4260	138[2]	⊖	SBc; 12.7; 2.3' x 0.9' pB,E,psbM	12168n0623
4261	139[2]	⊖	E2; 11.7; 2.0' x 1.7' pB,pS,1E,gbM; 4260 is 16.7' to north	12168n0606
4267	166[2]	⊖	E2/S0; 12.0; 2.2' x 2.2' pB,vS,R,vsmbM	12172n1303
4270	568[2]	⊖	S0/Sa; 13.0; 1.4' x 0.5' pB,S,E; group with 4273 & 4281	12173n0544
4273	569[2]	⊖	Sc; 12.3; 1.7' x 1.2' pB,L,E,gbM; asymmetric; 4270 is 7.5' north; 4281 is 6.7' ENE	12174n0537
4281	573[2]	⊖	E5/S0; 12.2; 1.5' x 0.8' B,vL,E,pgbM; group with 4270 & 4273	12178n0540

LIST OF STAR CLUSTERS, NEBULAE AND GALAXIES   (Cont'd)

NGC	OTH	TYPE	SUMMARY DESCRIPTION	RA & DEC
4294	61[2]	⊖	SBc; 12.6; 2.4' x 0.9' F,L,mE; followed by 4299, 5.6' to east	12187n1147
4299	62[2]	⊖	SBc; 12.9; 1.1' x 0.9' F,L,1E,vgbM; pair with 4294	12192n1147
4303	M61	⊖	Sc; 10.2; 5.7' x 5.5' vB,vL,vsbM; large face-on spiral (*)	12194n0445
4307		⊖	Sa/Sb; 13.0; 2.8' x 0.4' pF,L,mE; spindle-shaped edge-on spiral	12195n0920
4313	63[2]	⊖	Sa/S0; 13.2; 1.4' x 0.4' pS,F,mE,bM; spindle-shaped edge-on galaxy	12201n1204
4324		⊖	S0/SB; 12.5; 1.6' x 0.7' pB,E,bM; 9' ESE from 17 Virg.	12206n0531
4339	143[2]	⊖	E0; 12.6; 1.0' x 1.0' pB,pL,R,bM	12210n0622
4342	96[3]	⊖	S0/Sa; 13.0; 2.0' x 0.5' cF,S,1E; edge-on lenticular; 4 fainter galaxies to north between 4342 & 4365	12211n0722
4348	625[2]	⊖	Sb; 13.1; 2.8' x 0.4' F,pL,mE,vlbM; spindle-shaped edge-on galaxy	12213s0310
4365	30[1]	⊖	E2/E3; 11.0; 2.0' x 1.3' cB,pL,E,gsmbM	12220n0736
4371	22[1]	⊖	S0/SBa; 11.8; 2.0' x 1.2' B,1E,1E.gbM	12224n1159
4374	M84	⊖	E1; 10.5; 2.0' x 1.8' vB,pL,R,psbM; central core of Virgo Galaxy cluster (*)	12226n1310
4378	123[1]	⊖	Sa; 12.8; 3.0' x 2.7' B,S,R, with outer ring	12228n0512
4380		⊖	Sb?; 12.8; 3.0' x 1.5' vF,pL,R,1bM; spiral pattern ill-defined & amorphous	12229n1017
4385		⊖	S0/SBa; 12.9; 1.2' x 0.6' vF,vS,1E; SS Virg in field	12231n0050

NGC	OTH	TYPE	SUMMARY DESCRIPTION	RA & DEC
4388	168[2]	⊖	SBc; 12.0; 5.0' x 1.0' vF,vmE,bM; edge-on lens-shaped galaxy, 16.7' SE from M84 in central portion of Virgo Galaxy Cluster	12233n1256
4402		⊖	Sb; 13.0; 2.0' x 0.8' F,L,mE; edge-on spiral with equatorial dust lane; 10' north from M86; Virgo Cluster	12236n1324
4406	M86	⊖	E3; 10.5; 3.0' x 2.0' vB,L,R,gbMN; In central core of Virgo Galaxy Cluster (*)	12237n1313
4412	34[2]	⊖	SBp; 12.8; 1.0' x 0.8' F,pL,lE,gbM; compact non-symmetrical spiral	12240n0414
4413	169[2]	⊖	SBa; 13.2; 1.1' x 0.7' cF,S,lE,gbM; 11.7' ESE from 4388 in Virgo Galaxy cluster	12240n1253
4417	155[2]	⊖	E7/S0; 12.2; 2.2' x 0.8' F,pL,E, edge-on lens-shaped system; 4424 in field 11' SSE	12243n0952
4420	23[2]	⊖	SBc; 12.5; 1.8' x 0.7' F,pL,lE; Quasar 3C273 about 1° ESE	12244n0246
4424		⊖	SBa; 12.5; 2.5' x 1.3' F,pL,E,bM; spiral pattern hazy & amorphous; 4417 in field 11' to NNW	12246n0942
4425	170[2]	⊖	S0/Sb; 12.9; 2.0' x 0.5' F,S,E,bM; lens-shaped edge-on system; 19' SE from M86	12247n1301
4428		⊖	Sc; 13.1; 1.5' x 0.6' vF,pL,E; compact spiral, 7' pair with 4433	12249s0754
4429	65[2]	⊖	S0/Sa; 11.3; 3.0' x 1.0' B,L,cE,psbM; oval with large Saturn-like outer ring	12249n1123
4433		⊖	Sc; 12.9; 1.4' x 0.6' pF,pL,lE; compact spiral, 4428 is 7' to north	12250s0801

LIST OF STAR CLUSTERS, NEBULAE AND GALAXIES   (Cont'd)

NGC	OTH	TYPE	SUMMARY DESCRIPTION	RA & DEC
4435	28[1]	⊖	E4/S0; 11.8; 1.4' x 0.9' vB,cL,E,; 4438 is 4.5' SSE	12252n1321
4438		⊖	Sa/Pec; 11.0; 4.0' x 1.5' B,cL,vlE, with long extending angular filaments; pair with 4435; in Virgo Galaxy Cluster	12253n1317
4442	156[2]	⊖	E5/S0; 11.6; 2.0' x 1.0' vB,pL,lE,smbM	12256n1005
4452	23[1]	⊖	Sb?; 13.2; 1.3' x 0.3' pB,S,vmE; cigar-shaped edge-on spiral	12262n1202
4454	180[2]	⊖	SBa; 12.8; 1.2' x 1.0' F,L,lE,gbM	12263s0140
4457	35[2]	⊖	SBa; 11.7; 2.0' x 1.6' cB,pS,lE,smbMN; with faint outer arms	12264n0351
4461	122[2]	⊖	S0/Sa; 12.2; 2.0' x 1.0' pF,S,E,bM	12266n1328
----	3C273	QSO	Brightest known quasar; Mag 12.8; appearance stellar (*)	12266n0219
4469	157[2]	⊖	SBp; 12.5; 3.0' x 0.9' pF,pL,mE,bM; edge-on system with box-shaped central mass; resembles NGC 128 in Pisces	12270n0902
4472	M49	⊖	E3/E4; 10.1; 4.0' x 3.4' vB,L,R,mbM (*)	12273n0816
4473	114[2]	⊖	E4/E5; 11.3; 2.0' x 1.0' pB,pL,E,bM	12273n1342
4476	123[2]	⊖	E4/S0; 13.3; 0.7' x 0.4' F,S,lE,bM; in field of M87, 12' to west	12275n1237
4478	124[2]	⊖	E1/E2; 12.4; 1.0' x 0.8' pB,S,lE,psbM; in field of M87, 9' to WSW	12278n1236
4483		⊖	Sa; 13.3; 0.8' x 0.5' pB,pS,R,bM	12282n0917
4486	M87	⊖	Elp; 10.1; 3.0' x 3.0' vB,vL,R,mbM; radio source; Two faint ellipticals 4476 & 4478 in field (*)	12283n1240

LIST OF STAR CLUSTERS, NEBULAE AND GALAXIES   (Cont'd)

NGC	OTH	TYPE	SUMMARY DESCRIPTION	RA & DEC
4487	776[2]	⊖	Sc; 12.0; 3.5' x 2.5' F,vL,1E; coarse structured spiral	12283s0748
4496	36[2]	⊖	SBc; 12.0; 3.0' x 2.0' F.cL, coarse structure	12291n0412
4503	66[2]	⊖	E2/S0; 12.8; 1.0' x 0.7' pB,S,E,gbM	12296n1127
4504	771[2]	⊖	Sc; 12.3; 4.0' x 2.0' pB,cL,E,gvlbM	12297s0717
----	R80	⊖	Sc?; 12.6; 3.5' x 1.5' S,E,BN; spiral pattern weak; 17' north from 4517	12299n0038
4517	5[4]	⊖	Sc; 11.4; 9.0' x 1.0' cB,vL,vmE, long narrow streak east-west; edge-on spiral with equatorial dust patches	12302n0023
4519	158[2]	⊖	SBc; 12.4; 2.5' x 1.8' F,pL,1E,bM; spiral pattern coarse	12310n0856
4522		⊖	Sc; 12.9; 3.0' x 0.8' F,pL,E,vlbM; edge-on spiral	12312n0927
4526	31[1]	⊖	E7/S0; 10.7; 4.0' x 1.0' vB,vL,mE,psmbM; lenticular, nearly edge-on; between two 7th mag stars	12316n0758
4527	37[2]	⊖	Sb/Sc; 11.3; 5.1' x 1.1' pB,L,mE,mbM; nearly edge-on spiral	12316n0256
4532	147[2]	⊖	I; 12.1; 2.3' x 0.8' pB,pL,pmE,vgbM; tadpole-shaped, brighter portion to north	12318n0644
4535	500[2]	⊖	SBc; 10.7; 6.0' x 4.0' pF,vL,SBN; fine S-shaped spiral, delicate arms	12318n0828
4536	2[5]	⊖	Sc/SBc; 11.0; 7.0' x 2.0' B,vL,vmE,sbM; highly tilted spiral with long curving arms	12319n0228
4546	160[1]	⊖	E6/S0; 11.4; 1.6' x 0.7' vB,cL,pmE,vsmbMN	12329s0331

LIST OF STAR CLUSTERS, NEBULAE AND GALAXIES   (Cont'd)

NGC	OTH	TYPE	SUMMARY DESCRIPTION	RA & DEC
4550	36[1]	⊖	E7/S0; 12.6; 2.0' x 0.7' pB,S,E, Small E3 galaxy in field 3.3' NNE = 4551	12329n1230
4552	M89	⊖	E0; 11.0; 2.0' x 2.0' pB,pL,R,gmbM (*)	12331n1250
4564	68[2]	⊖	E6; 12.1; 1.8' x 0.6' pB,S,E,psbM; 11' north from pair 4567 & 4568	12340n1143
4567	8[4]	⊖	Sb/Sc; 12.0; 2.4' x 1.6' vF,pL,BN; contact pair with 4568 = "The Siamese Twins"(*)	12340n1132
4568	9[4]	⊖	Sb/Sc; 11.9; 3.6' x 1.8' vF,pL,mE, 1.6' pair with 4567	12341n1131
4569	M90	⊖	Sb; 11.1; 7.0' x 2.5' pL,E,bMN; fine well-formed oval spiral (*)	12343n1326
4570	32[1]	⊖	E8/S0; 11.8; 2.5' x 0.7' cB,pS,mE,sbMN; edge-on lens-shaped system, in field of R Virginis	12344n0731
4578	15[2]	⊖	S0/Sa; 12.5; 2.0' x 1.4' pF,pS,1E,smbMN	12350n0950
4579	M58	⊖	Sb; 10.5; 4.0' x 3.5' B,L,E,vmbM; fine compact spiral (*)	12351n1205
4580	124[1]	⊖	Sa/Sb; 12.8; 1.4' x 1.1' pB,L,1E,vgbM	12353n0538
4586		⊖	Sa/Sb; 12.7; 2.6' x 1.1' vF,S,mE,bM; spiral pattern indistinct	12359n0435
4592	31[2]	⊖	Sb/Sc; 12.4; 3.0'x 0.7' F,L,mE,vgbM; 1° NNW from fine double star Gamma Virginis	12367s0016
4593	183[2]	⊖	SBb; 12.1; 3.2' x 2.5' pB,cL,E,sbMN; saturn-like central mass with extensive outer arms; 4602 in field	12370s0504
4594	M104 43[1]	⊖	Sa/Sb; 8.2; 7.0' x 1.5' ! vB,vL,vmE,vsmbMN, with dark equatorial lane (*)	12373s1121

LIST OF STAR CLUSTERS, NEBULAE AND GALAXIES (Cont'd)

NGC	OTH	TYPE	SUMMARY DESCRIPTION	RA & DEC
4596	24[1]	⊖	S0/SBa; 12.0; 2.4' x 0.9' B,pS,E,gmbM; Saturn-like disc with projecting ansae; ½° W from Rho Virg; 4608 in field	12374n1027
4597	636[2]	⊖	SBc; 12.9; 3.8' x 1.2' F,vL,E,bM; reversed "Z" structure	12375s0532
4602	184[2]	⊖	Sb/Sc; 12.4; 3.0' x 1.5' F,L,cE,vglbM; in field with 4593, 19' to NE	12380s0452
4608	69[2]	⊖	S0/SBa; 12.1; 2.0' x 1.4' F,L,E,vglbM; 11' SW from Rho; in field with 4596	12387n1026
4612	20[2]	⊖	S0/Ep; 12.6; 1.0' x 0.9' pB,1E,psmbM	12390n0735
4621	M59	⊖	E3/E4; 11.0; 2.0' x 1.5' B,pL,1E,vsvmbM; M60 is 25' to east (*)	12395n1155
4623	149[2]	⊖	E5; 13.2; 1.1' x 0.6' cF,pL,E,pslbM	12396n0756
4630	532[2]	⊖	S/pec?; 13.1; 1.1' x 0.8' cF,S,R,1bM; possibly compact ill-defined spiral	12400n0414
4632	14[1]	⊖	Sc; 12.1; 2.7' x 0.9' pB,L,E,bM	12400n0011
4636	38[2]	⊖	E1; 10.8; 1.3' x 1.2' B,L,R,vgvmbM	12403n0257
4638	70[2]	⊖	E5/S0; 12.2; 1.1' x 0.5' F,E,gbM; stubby lens-shape; in field with M59 + M60	12402n1143
4639	125[2]	⊖	SBb; 12.2; 2.0' x 1.3' pB,S,E, in field with 4654	12403n1331
4643	10[1]	⊖	SBa; 11.6; 1.7' x 0.8' cB,pS,1E,mbM; Saturn-like central mass, projecting ansae	12408n0215
4647	44[3]	⊖	Sc; 12.0; 2.3' x 1.8' pF,pL,1E; in field with M60, 2.5' to NW	12410n1151

LIST OF STAR CLUSTERS, NEBULAE AND GALAXIES   (Cont'd)

NGC	OTH	TYPE	SUMMARY DESCRIPTION	RA & DEC
4649	M60	⊖	E1/E2; 10.0; 3.0' x 2.5' vB,pL,R,1bM; 4647 in field, 2.5' to NW (*)	12411n1149
4653	662[3]	⊖	Sc; 12.7; 2.2' x 1.9' vF,pL,R, 20' WSW from 4666	12414s0018
4654	126[2]	⊖	Sc; 11.2; 4.5' x 2.5' F,vL,pmE; 4639 in field 18' WNW	12414n1323
4658	558[2]	⊖	SBc; 12.4; 1.4' x 0.6' vF,L,E	12421s0949
4660	71[2]	⊖	E5; 12.2; 1.3' x 0.6' vB,S,vsvmbMN; 25' SSE from M60	12420n1126
4665	142[1]	⊖	S0/SBa; 11.8; 3.0' x 2.0' B,pL,R,mbM	12426n0319
4666	15[1]	⊖	Sc; 11.4; 3.9'x 0.7' B,vL,vmE,psbM; nearly edge-on spiral; 4668 in field	12426s0012
4668	663[3]	⊖	SBc/pec; 13.4; 0.8' x 0.6' vF,S; 7.8' SE from 4666	12430s0017
4682	523[3]	⊖	Sc?; 13.1; 1.4' x 0.8' cF,L,E,v1bM	12447s0948
4684	181[2]	⊖	S0/Sa; 12.2; 1.6' x 0.5' B,pL,pmE, lenticular, nearly edge-on	12447s0228
4688	543[3]	⊖	Sc; 13.0; 2.2' x 2.2' eF,pL,R, asymmetric spiral	12453n0436
4691	182[2]	⊖	SBa; 11.8; 2.0' x 1.5' pB,pL,E,mbM	12456s0304
4694	72[2]	⊖	E5/S0; 12.6; 1.9' x 0.7' pF,S,mE	12457n1115
4697	39[1]	⊖	E5; 10.5; 2.5' x 1.3' vB,L,1E,smbMN	12460s0532
4698	8[1]	⊖	Sa/Sb; 11.8; 2.8' x 1.0' cB,pL,E,bM; 1° NE from 32 Virginis	12458n0845
4699	129[1]	⊖	Sa/Sb; 10.3; 3.0' x 2.0' vB,L,1E,vmbMN	12465s0824
4700	524[3]	⊖	SB; 12.5; 2.2' x 0.3' F,L,mE,v1bM; edge-on spiral	12465s1108

LIST OF STAR CLUSTERS, NEBULAE AND GALAXIES   (Cont'd)

NGC	OTH	TYPE	SUMMARY DESCRIPTION	RA & DEC
4701	578[2]	⊖	Sc; 12.8; 1.5' x 1.0' F,S,1E, compact center and faint spiral arms	12466n0339
----	An 3	⊖	Sc?; 12.4; 2.5' x 2.0' F,S,1E, spiral pattern ill-defined	12468s0951
4713	140[1]	⊖	Sc; 12.3; 2.3' x 1.4' pB,L,vlE,glbM; coarse spiral	12475n0535
4731	41[1]	⊖	SBc; 12.2; 5.5' x 2.5' vF,pL,E, loose, open-structured spiral, far-extending arms	12484s0608
4742	133[1]	⊖	E3/E4; 12.0; 1.0' x 0.6' pB,vS,E,vmbMN; Group with 4760 + 4781	12492s1012
4753	16[1]	⊖	Ep/I?; 10.6; 2.8' x 2.0' cB,L,vlE,vglbM	12498s0055
4754	25[1]	⊖	S0; 11.8; 2.5' x 1.0' B,pL,E,psbM; Pair with 4762, 11' to NW	12497n1135
4760		⊖	E1/E2; 12.5; 0.6' x 0.5' pB,S,R, between 4742 + 4781	12505s1013
4762	75[2]	⊖	S0/SBa; 11.5; 3.7' x 0.4' pB,vmE; edge-on; thick streak with extending tufts at ends; 11' pair with 4754	12504n1131
4765	544[3]	⊖	S?; 12.9; 0.7' x 0.5' F,cS,1E,gbM	12507n0445
4771	535[2]	⊖	Sb/Sc; 12.9; 3.0' x 0.5' F,pL,mE; edge-on lens-shaped spiral	12508n0133
4772	24[2]	⊖	Sa; 12.6; 2.7' x 1.0' pF,pS,E,mbM	12510n0227
4775	186[2]	⊖	Sc; 11.6; 1.7' x 1.6' F,cL,R,vglbM; compact face-on spiral	12511s0621
4781	134[1]	⊖	SBc; 11.7; 2.6' x 1.1' cB,vL,mE; coarse spiral, group with 4742 + 4760	12518s1016

LIST OF STAR CLUSTERS, NEBULAE AND GALAXIES (Cont'd)

NGC	OTH	TYPE	SUMMARY DESCRIPTION	RA & DEC
4786	187[2]	⊖	E2; 12.7; 0.8' x 0.7' pB,pS,mbM; in field with 4775, 18' to SE	12520s0635
4790	560[2]	⊖	SBc; 12.5; 1.4' x 1.0' pF,pS,1E, 16' NNE from 4781	12522s0958
4795	21[2]	⊖	Sa?; 13.1; 1.5' x 1.0' pF,pL,1E,bM; spiral pattern hazy	12525n0820
----	An 4	⊖	S; 12.9; 2.8' x 1.7' pS,F,E, spiral pattern faint and indistinct	12526n0023
4808	141[1]	⊖	Sc; 12.5; 2.2' x 0.8' pB,cL,mE, 1° N from Delta	12533n0435
4818	549[2]	⊖	Sa/SBa; 12.1; 3.4' x 1.0' pB,pL,pmE,gbM	12543s0815
4825	563[2]	⊖	E2/S0; 12.9; 1.0' x 0.7' pB,1E,bM	12543s1324
4845	536[2]	⊖	Sa/Sb; 12.6; 4.0' x 0.8' pF,pL,pmE,vgbM; nearly edge-on with equatorial dust lane	12555n0151
4856	68[1]	⊖	SBa; 11.4; 2.5' x 0.7' B,mE,psmbM	12567s1446
4866	162[1]	⊖	S0/Sa; 12.0; 4.0' x 0.7' B,pL,mE,sbMN; edge-on lens-shaped system tilted E-W	12570n1427
4880	83[3]	⊖	S0/Sa; 13.1; 2.2' x 1.3' cF,pL,R,vglbM; spiral pattern indistinct	12577n1245
4891		⊖	Sc; 13.0; 1.6' x 1.4' F,S,bMN; spiral arms faint	12581s1309
4899	300[2]	⊖	Sc/pec; 12.7; 1.9' x 1.0' pF,L,cE	12583s1341
4900	143[1]	⊖	Sc; 11.9; 1.7' x 1.5' cB,L,1E; compact spiral	12582n0246
4902	69[1]	⊖	SBb; 11.6; 2.0' x 2.0' pB,pL,R, θ shape center, triskelion arm pattern	12583s1415
4904	517[2]	⊖	SBc; 12.8; 2.0' x 1.0' pB,pS,bM	12584n0015

LIST OF STAR CLUSTERS, NEBULAE AND GALAXIES (Cont'd)

NGC	OTH	TYPE	SUMMARY DESCRIPTION	RA & DEC
4915	47[4]	⊖	E0/S0; 13.0; 0.7' x 0.6' pB,S,R,bM	12588s0416
4928	190[2]	⊖	Sb/pec; 12.9; 0.8' x 0.6' F,pS,vlE,bM	13003s0749
4933	191[2]	⊖	S0/Ep; 12.8; 1.2' x 0.6' pB,pL,R	13012s1114
4939	561[2]	⊖	Sb/Sc; 12.2; 5.0' x 2.0' pB,L,1E,gmbM; beautiful well defined spiral with delicate arms; 1° NW from 49 Virg.	13017s1005
4941	40[1]	⊖	Sa/SBa; 12.2; 3.0' x 1.0' pF,L,E,gbM,BN; smooth-textured spiral	13016s0517
4951	188[2]	⊖	Sc; 12.7; 3.0' x 1.0' F,pL,1E; nearly edge-on, spiral form indistinct	13025s0614
4958	130[1]	⊖	E6/S0; 11.5; 2.0' x 1.0' vB,pS,E,bM,BN; nearly edge-on lenticular	13031s0745
4981	189[2]	⊖	Sb/Sc; 12.2; 2.0' x 1.6' B,pL,1E	13061s0631
4984	301[2]	⊖	SB; 11.9; 1.3' x 1.0' B,pL,R,psmbM; dense but ill defined structure	13064s1515
4995	42[1]	⊖	Sb; 11.8; 2.2' x 1.5' pB,pL,1E,vgpmbM; dense spiral	13070s0734
4999	537[2]	⊖	SBb; 12.8; 2.0' x 1.8' cF,pL,R,1bM; θ pattern	13072n0155
5017	669[3]	⊖	E2; 13.3; 0.5' x 0.4' F,R,bM	13103s1630
5018	746[2]	⊖	E3/S0; 12.2; 1.5' x 0.8' cB,S,1E,mbM,pBN; 30' NW from 55 Virg.	13103s1915
5037	510[2]	⊖	Sa; 13.1; 1.6' x 0.5' cF,pS,vcE,bM; nearly edge-on; 13' SSW of 5044	13124s1620
5044	511[2]	⊖	E0; 12.2; 1.0' x 1.0' pB,pL,R,bM; 5037 and two other faint galaxies in field	13128s1608

NGC	OTH	TYPE	SUMMARY DESCRIPTION	RA & DEC
5054	513[2]	⊖	Sb; 11.9; 4.0' x 2.3' F,pS,1E, three-branch spiral	13143s1623
5068	312[2]	⊖	SBc; 11.6; 6.0' x 5.5' F,vL,R,bM; coarse spiral, 5087 in field	13162s2047
5077	193[2]	⊖	E3; 12.4; 1.0' x 0.7' pB,S,vlE,sbM; 5088 and two other faint galaxies in field 2° SW from Spica	13169s1224
5084	313[2]	⊖	E8/S0; 12.4; 6.0' x 1.0' cB,cS,mE; edge-on spindle with faint extensions at tips	13175s2134
5087	724[3]	⊖	E4/S0; 12.4; 0.8' x 0.5' cF,vS,E; in field with 5068, 32' to NE	13177s2021
5088		⊖	Sb/Sc; 13.2; 2.0' x 0.6' pB,pS,E,bM; 13' ENE from 5077	13177s1219
5134	314[2]	⊖	Sb; 12.4; 1.8' x 0.8' F,pS,1E,vgbM; Saturn-like oval with hazy arm pattern; double star β610 in field	13226s2051
5147	25[2]	⊖	Sc; 12.1; 1.3' x 1.0' pB,pL,vlE,vsmbM; compact spiral	13237n0222
5170	22[5]	⊖	Sc; 12.6; 7.0' x 0.7' cF,L,mE,pgbM; edge-on spiral, flat narrow streak; some obscuring lanes	13271s1742
5230	87[3]	⊖	Sc; 12.9; 1.7' x 1.6' F,L,E,vgbM; well defined face-on spiral	13330n1356
5247	297[2]	⊖	Sb/Sc; 11.9; 4.5' x 4.0' cF,vL,R,vgpsmbM,LN; very fine reversed-S double-arm spiral	13353s1738
5300	533[2]	⊖	Sc; 12.3; 3.2' x 2.0' vF,vL,1E,vgbM; multiple-arm spiral; 1.3° ENE from 84 Virg	13457n0411
5324	307[2]	⊖	Sc; 12.6; 1.6' x 1.5' cF,L,R,bM	13494s0548

LIST OF STAR CLUSTERS, NEBULAE AND GALAXIES   (Cont'd)

NGC	OTH	TYPE	SUMMARY DESCRIPTION	RA & DEC
5334	665[3]	⊖	SBc; 12.5; 3.5' x 2.2' cF,vL,1E,1bM; many-arm spiral 36' NW from Rho Virg.	13504s0053
5363	6[1]	⊖	I/Ep; 11.1; 1.7' x 1.5' B,pL,R,psbM; 14½' pair with 5364	13536n0529
5364	534[2]	⊖	Sb; 11.5; 5.0' x 4.0' cF,L,R,gbM; beautiful oval spiral; pair with 5363, 14½' to North (*)	13537n0515
5426	309[2]	⊖	Sc; 12.7; 1.5' x 1.1' pF,cL,1E,gmbM; pair with 5427	14008s0549
5427	310[2]	⊖	Sc; 12.0; 2.0' x 1.7' pF,cL,R; face-on spiral; pair with 5426, 2.8' to south	14008s0547
5468	286[3]	⊖	Sc; 12.4; 2.1' x 1.9' F,L,R,vgbM; 3-branch spiral	14040s0514
5493	46[4]	⊖	S0; 12.5; 0.9' x 0.5' pB,vS,1E,psmbM; fat lens-shaped system	14089s0449
5496		⊖	Sc/Sd; 12.8; 3.9' x 0.6' pB,vL,mE; edge-on, streaky dust lanes	14090s0056
5534		⊖	SB?; 13.0; 0.8' x 0.5' pF,S,1E; distorted spiral	14150s0711
5566	144[1]	⊖	SB; 11.4; 5.8' x 1.3' B,pL,E,psbM; compact center & faint outer arms; 2 faint galaxies in field	14178n0411
5574	145[1]	⊖	S0; 13.4; 0.9' x 0.3' pF,pS,1E; stubby oval, pair with 5576	14184n0328
5576	146[1]	⊖	E2/E3; 11.9; 0.8' x 0.6' B,S,1E,vsmbM; 5574 in field 2.8' to SW	14185n0330
5584		⊖	Sc; 12.2; 2.7' x 1.0' F,L,mE,glbM; many-arm spiral	14198s0010
5634	70[1]	⊕	Mag 10; Diam 4'; Class IV; vB,cL,R,gbM; stars mags 19... 11^m star on E edge	14270s0545

LIST OF STAR CLUSTERS, NEBULAE AND GALAXIES   (Cont'd)

NGC	OTH	TYPE	SUMMARY DESCRIPTION	RA & DEC
5638	581[2]	⊖	E1; 12.5; 1.0' x 0.9' cB,pL,R; Faint barred spiral in field 2' to north	14271n0327
5645	150[2]	⊖	SBc; 12.9; 1.6' x 1.0' cF,pL,1E,gbM	14281n0729
5668	574[2]	⊖	Sc; 12.3; 2.0' x 1.6' F,pS,vlE, nearly face-on spiral, arms weak	14309n0440
5690	582[2]	⊖	Sb/Sc; 12.9; 2.7' x 0.6' vF,mE; nearly edge-on spiral; Double star A2227 is 3' west	14352n0230
5691	681[2]	⊖	S/pec; 13.0; 1.1' x 0.8' pB,pS,1E,gbM; distorted structure	14353s0011
5701	575[2]	⊖	SBb; 12.8; 3.0' x 2.0' cB,pS,1E,mbM; outer halo- like ring	14367n0534
5713	182[1]	⊖	Sb/Sc/pec; 11.8; 2.1' x 1.9' pL,cB,R,psmbM; tight spiral with faint outer halo; Faint mE spiral 11' to east	14376s0005
5740	538[2]	⊖	Sb; 12.6; 2.5' x 1.5' pB,L,E,gbM; in field with 5746 & 109 Virg.	14419n0154
5746	126[1]	⊖	Sb; 11.7; 6.5' x 0.8' B,L,vmE,bM,BN; fine edge-on spiral with equatorial dust band; 5740 is 18' to SSW	14423n0210
5750	183[1]	⊖	SBc; 12.6; 1.5' x 0.9' pF,pS,vlE,mbM; with Saturn- like elliptical ring around nucleus	14436s0001
5775	554[3]	⊖	SBb; 12.3; 4.0' x 0.7' F,cL,vmE,gvlbM; edge-on; Faint round Sc system in field 4½' NW	14515n0345
5806	539[2]	⊖	Sb; 12.4; 1.9' x 0.9' cB,cL,E,sbMN; 45' WSW from 110 Virg; 5813 in field 21' to SE	14575n0205

LIST OF STAR CLUSTERS, NEBULAE AND GALAXIES (Cont'd)

NGC	OTH	TYPE	SUMMARY DESCRIPTION	RA & DEC
5813	127[1]	⊖	E1; 12.1; 1.0' x 0.9' B,pS,R,psbM; in field with 5806	14587n0154
5831	540[2]	⊖	E3p; 12.7; 0.5' x 0.5' pB,S,R,mbM	15016n0124
5838	542[2]	⊖	S0; 11.9; 3.2' x 0.7' pB,pS,E, 38' East from 110 Virginis	15029n0218
5846	128[1]	⊖	E0; 11.5; 1.0' x 1.0' vB,pL,R,psbMN; Pair with 5850, 10' ESE	15040n0148
5850	543[2]	⊖	SBb; 12.0; 2.6' x 2.1' cF,S,1E,psbM; delicate θ-shaped barred spiral; pair with 5846	15046n0144
5854	544[3]	⊖	S0/SBa; 12.6; 2.0' x 0.4' pB,S,1E,1bM	15053n0245
5864	585[2]	⊖	S0/E7p; 12.8; 1.8' x 0.4' pF,cS,cE,gbM; nearly edge-on lenticular system	15070n0314

DESCRIPTIVE NOTES

ALPHA Name- SPICA, from the Latin *Spicum*, said to signify the Ear of Wheat which Virgo holds in her left hand. Magnitude 1.00; Spectrum B1 V; Position 13226s1054. Spica is the 16th brightest star in the sky, and may be regarded as the standard example of a 1st magnitude star, having a measured mean visual magnitude of exactly 1.00, though the star varies slightly above and below the zero point. Opposition date (midnight culmination) is April 13.

The Arabian name, *Al Simak al A'zal*, seems to mean "The Unarmed One" and appears in a variety of forms in medieval writings, as does the Coptic *Khoritos*, the Solitary One. The 1515 edition of the *Almagest* has *Aschimech Inermis*; the *Alfonsine Tables* designate it *Inermis Asimec*; Riccioli has *Eltsamecti*, Bayer has it labeled *Alaazel*, and Schickard has *Huzimethon*. According to Thomas Hyde, the Persian *Chushe*, the Syrian *Shebbelta*, and the Turkish *Salkim* may all be translated "The Ear of Wheat". The Sogdian name *Shaghar* seems to honor the star as "The Point" of the constellation.

In Greek and Roman tradition the Virgin was usually identified with the goddess *Astraea*, the personification of Justice, daughter of Zeus and Themis, and the last of the deities to abandon the Earth at the end of the Golden Age. In another legend she is *Persephone*, daughter of the goddess of the harvest, Ceres, and consequently associated with agricultural matters. In many parts of the classical world she is the *Wheat-Bearing Maiden* or the *Daughter of the Harvest*. Aratus offers her his admiring praise:

> *"Her lovely tresses glow with starry light;*
> *Stars ornament the bracelet on her hand;*
> *Her robe in ample fold, glitters with stars;*
> *Beneath her snowy feet they shine; her eyes*
> *Lighten, all glorious, with the heavenly rays,*
> *But first the star which crowns the golden sheaf."*

In Egypt she was the goddess *Isis*, the Divine Wife and Mother, and her star was honored as the *Lute Bearer*. Prof.Lockyer thought that at least one of the temples at Akhenaton's royal city, now known as Tell el Amarna, was oriented to the setting of Spica. If so, it may be that this most individualistic and enigmatic of the Pharaohs,

DESCRIPTIVE NOTES (Cont'd)

often considered as the "first monotheist in history", was actually paying tribute to his own "Divine Wife", the exotically lovely Nefertiti. In other lands too, Virgo was associated with a great "Mother Goddess"; in Assyria she was the *Wife of Bel*, in Babylonian lands she was *Ishtar*, Queen of the Stars, and in India she was identified with *Kanya*, the Maiden, and mother of Krishna. Medieval Christians saw her as the Madonna, or as *Ruth of the Fields*; in Roman times the goddess *Pax* with her olive branch was seen here, as was the serene *Concordia* and the mysterious *Tyche* or *Fortuna*. R.H.Allen states that she was identified by the Venerable Bede with the Saxon goddess *Eoestre* or *Eostre*, whose name still survives in the modern observance of our Easter. To some writers of classical times, the proximity of Virgo to the Lion suggested an identification with the mother-goddess *Cybele*, in her lion-drawn carriage; she was also honored as *Diana*, *Minerva*, *Athena*, and occasionally as *Urania*, Muse of astronomy.

> *"Virgin August! Come in thy regal state,*
> *With soft majestic grace and brow serene*
> *Though the fierce Lion's reign is overpast,*
> *The summer's heat is all thy own as yet;*
> *And all untouched thy robe of living green..."*

Spica is a brilliant "helium" type star, about 2300 times more luminous than our Sun. From the results of a combined spectroscopic and interferometric study made in 1970, the distance appears to be about 275 light years. The star shows an annual proper motion of 0.05"; the radial velocity is less than 1 mile per second in recession with definite variations.

Spica is a massive spectroscopic binary star, first detected by H.C.Vogel in 1890; the period is 4.01416 days, remarkably short for stars of the giant class. About 80% of the light of the system comes from the primary star which has a computed mass of 10.9 and a diameter of 8.1 the solar values; the fainter star appears to be possibly half the size of the primary, and has a spectral type close to B7 and a computed mass of 6.8. From the orbital elements the center-to-center separation must be something close to 11 million miles, and the two stars form a grazing eclips-

ing system with a range of about 0.07 magnitude; the orbit
is inclined about 24° from the edge-on position, and the
computed eccentricity is 0.10. In addition to the small
light changes caused by the eclipses, the brighter star is
itself a pulsating variable, evidently of the Beta Canis
Majoris class, with a period of about 0.174 day; both the
amplitude and the shape of the light curve are variable.

GAMMA   Name- PORRIMA, in honor of the "Goddess of
Prophecy". The Arabian name *Zawiat al Awwa* is
translated "The Angle (or Corner) of the Barker"; from the
tradition that the space enclosed by the curve of Epsilon,
Delta, Gamma, Eta and Beta marked the *"Retreat of the Howl-
ing Dog"*. Gamma is the SE corner of the figure. Magnitude
2.76; Spectrum F0 V + F0 V; Position 12391s0111; the first
bright star NW from Spica, about 14½° away.
Gamma Virginis is one of the finest of the visual
binaries, the components being almost identical in type and

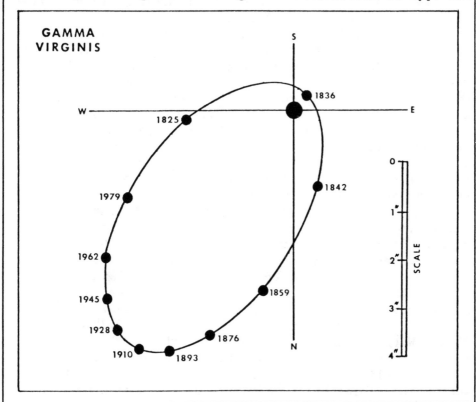

brightness; the apparent magnitude of each star is +3.65. The star was discovered as a double by Bradley and Pound in 1718, and was also observed by Cassini in 1720. Sir John Herschel calculated the orbit in 1833 and predicted that at the closest approach the two stars would be inseparable in any but the greatest telescopes. This was verified in 1836 when the apparent separation diminished to 0.3" at periastron passage, and the star appeared as single even in Herschel's telescope at the Cape of Good Hope. Within a few months the pair had widened and soon became an easy object for small telescopes.. In 1920 it attained its greatest separation of about 6.2"; the stars are now closing in and will again appear single about the year 2007.

C.E.Barns refers to this binary as "one of the finest pairs visible!"; to Mary Proctor they were "both white" while E.J.Hartung sees them as "bright yellow". To the present author they are a clear pale yellow, and look for all the world like the remote twin head-lamps of some celestial auto, approaching from deep space.

John Herschel's calculation of an orbit for a binary system was the second such case on record; the orbit of Xi Ursae Majoris had been computed by M.Savary in 1828. There are at least two interesting accounts of occultations of Gamma Virginis by the moon in 18th Century records. At the occultation of March 20, 1780, observed at Paris, it was found that the disappearances of the two stars occurred 10 seconds apart. Cassini reported having observed a similar phenomenon in 1720. It is also interesting to note that F.G.W.Struve suspected light variations in one or both of the components. In the years between 1825 and 1831 the star now regarded as the secondary appeared to be the brightest, but in 1851 the order of brightness appeared to have reversed. On many nights in 1852, Struve found the two stars to be perfectly equal in light. Modern photometric studies show that the present difference is only 0.02 magnitude.

The orbit of the pair is very well known, and three modern computations give the period as 171 years with periastron in June 1836. The semi-major axis is 3.7" or about 38 AU. With an eccentricity of 0.88, this system has a much-elongated comet-like orbit; the motion is retrograde with an inclination of 146°. At periastron passage the true separation diminishes to about 3 AU, increasing to 70 AU at

widest separation. Gamma Virginis is one of the relatively
nearby star systems, at a distance of 32 light years; both
components are F-type main sequence stars giving a total
light of 7 suns. The absolute magnitude in each case is
+3.6. The star shows an annual proper motion of 0.57" in PA
271°; the radial velocity is about 12 miles per second in
approach.
      Several distant optical companions are mentioned in
the ADS Catalogue. A 15th magnitude star at 53" in PA 159°
was noted by S.W.Burnham in 1889, and a star of the 12th
magnitude lies 124" distant in PA 88°. These stars have no
real connection with the orbiting pair, and the separation
is increasing in both cases from the proper motion of Gamma
itself. Several faint galaxies will be found in the field
of this binary: the elongated spiral NGC 4592 is about 1°
to the NNW, and the nearly edge-on spiral NGC 4666 is 1.3°
to the NE. (Refer to pages 2055 and 2057 for data)

DELTA   Magnitude 3.66; Spectrum M3 III; Position
12531n0340, about 6° NE from Gamma. The star
has no common English name; R.H.Allen states that it was
Lu Lim in Babylonian lands, the name signifying a stag or
a gazelle; the Hindu name *Apas* is translated "The Waters".
The star was called *Bellissima* by Father Secchi in recognition of its beautiful banded spectrum, resembling that of
Alpha Herculis.
      The computed distance is about 180 light years, and
the actual luminosity about 85 times that of the sun. The
star shows an annual proper motion of 0.47" in PA 263°; the
radial velocity is about 11 miles per second in approach.

EPSILON   Name- VINDEMIATRIX, from the Latin *Vindemiator*
or *Provindemiator*, "The Grape Gatherer", so
called from the rising of the star with the Sun just before
the annual vintage time. Magnitude 2.84; Spectrum G9 II or
III; Position 12597n1114, about 13½° NNE from Gamma. The
star is about 90 light years distant and has an actual
luminosity of about 50 suns (absolute magnitude about +0.6)
The annual proper motion is 0.27" in PA 273°; the radial
velocity is 8½ miles per second in approach.
      Between this star and Beta Leonis is centered the
great Virgo Cluster of Galaxies (Refer to page 2074)

ZETA   Magnitude 3.38; Spectrum A3 V; Position 13321
s0020, on the equator and about 11° north and
slightly east from Spica. Zeta Virginis is a white star
very similar in type and luminosity to Sirius. The computed
distance is about 90 light years, the actual luminosity
about 30 times that of the Sun (absolute magnitude +1.1)
and the annual proper motion is 0.29" in PA 276°, closely
matching that of Epsilon. The two stars also show about the
same radial velocity, approximately 8 miles per second in
approach.

The two variable stars V and W Virginis lie near
Zeta, about 3° to the SW. (Refer to chart on page 2073)

R   (Variable) Position 12360n0716, some 8½° N
from Gamma Virginis and 1° west of 31 Virg.
The first of the long-period variables in Virgo to be discovered, found by K.L.Harding in Germany in 1809. It varies
in brightness from magnitude 7 or so down to about 11, but
the range on occasion may be as great as 6.2 to about 12.0.
L.Campbell, in the AAVSO publication *Studies of Long Period
Variables* (1955) summarizes the results of observations of
the star from April 1921 to April 1949: the star rose above
magnitude 7.0 on 23 occasions in this interval, and the
highest observed maximum was 6.2. On only three occasions
did the magnitude fall to 12.0, and the deepest minimum
recorded was 12.1.

R Virginis is one of the Mira-type stars, but the
mean cycle of only 145½ days is unusually short for stars
of the class, less than half the period of Mira itself. At
maximum the spectral class is M3 or M4, with the usual

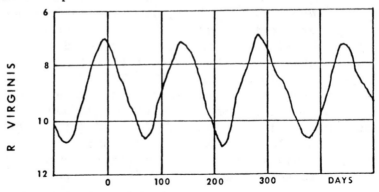

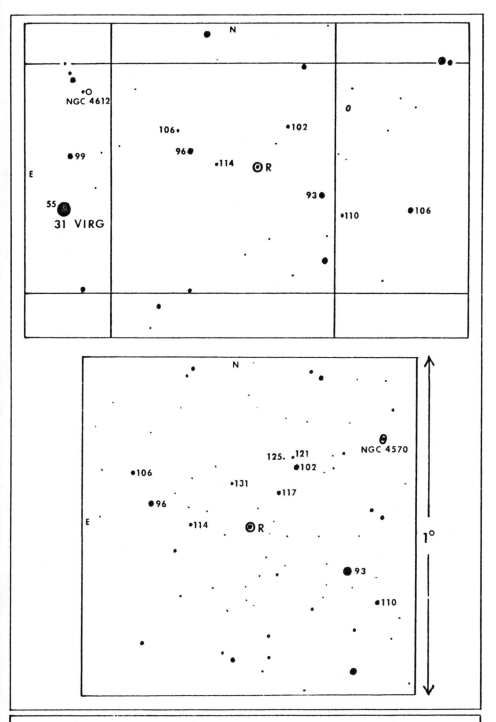

R VIRGINIS Identification fields. Comparison magnitudes are given (with decimal points omitted) according to observations of the AAVSO.

bright lines of hydrogen; the spectral type shifts to M8 at minimum and the red color deepens as the light fades, as is generally true of all the red giant variables. The star may be followed throughout its entire range with only modest telescopes; it never fades below visibility in a 6-inch reflector. From statistical studies of many stars of the class, variables of the type are believed to have maximum absolute magnitudes in the range of -1 to -2; the luminosity at an average maximum may be about 300 suns. A distance of close to 1000 light years is suggested by the spectroscopic luminosity criteria. R Virginis shows an annual proper motion of 0.03"; the radial velocity is about 15½ miles per second in approach.

Two fairly conspicuous galaxies will be found in the field, as shown on the identification charts. NGC 4570 is the brighter of the two, and is evidently a much elongated E or S0 system seen nearly edge-on. It lies about ½° from the variable, to the NW.

W (Variable) Position 13235s0307, about 3½° SW of Zeta Virginis. W Virginis is a cepheid variable star of somewhat unusual properties, discovered by Prof.E.Schonfeld in Germany in 1866, and now recognized as the typical example of a "Population II" cepheid. It exhibits several peculiarities: The position in the sky is some 57° above the galactic plane whereas the majority of cepheids lie in or near the Milky Way; the light curve is unusually broad with a square-shouldered hump on the side

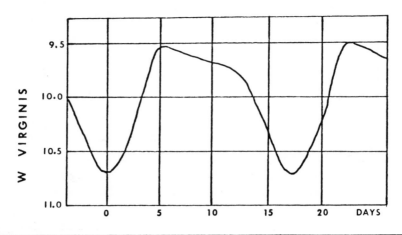

DESCRIPTIVE NOTES   (Cont'd)

following maximum light; this odd light curve cannot be fitted into the standard sequence of cepheid light curves shown on page 588; and the spectrum is abnormal in showing bright hydrogen emission lines during the rise to maximum.

Like all the cepheids, W Virginis is a pulsating giant star whose outer layers appear to be alternately expanding and contracting; the period was 17.2711 days in 1907, but has changed slightly to the present value of 17.2736 days. At the current rate of increase, the period will be about 0.16 day longer 1000 years from now. Variations of the star are accompanied by corresponding changes in the spectral class and diameter, the spectrum varying from G0 Ib at minimum at about F0 Ib at peak luminosity. Visually the range in light is about 1.2 magnitudes, but increases to 2.4 magnitudes when the star is observed in the ultraviolet. The size of the star, or at least of the light-radiating photosphere, is estimated to vary by a factor of 2 during the cycle; at maximum it may be about 30 solar diameters. The pulsations may also be followed by

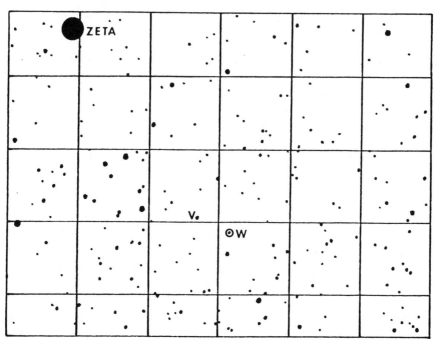

W VIRGINIS FIELD, showing stars to about 9th magnitude. Grid squares are 1° on a side with north at the top.

measuring the cyclic change in the radial velocity, which
varies from about 23 to 56 miles per second in approach,
the greatest velocity coinciding with peak brightness.

W Virginis, as a Population II star, is a member of
the great "halo" of stars which envelops the Galaxy, a much
older star than typical members of the spiral arms. Both
types of populations contain cepheids, but those of Pop.II
are scarcer and intrinsically fainter, by about 1½ magnitude, than the Pop.I cepheids. From the period-luminosity
graph (page 591) we see that the expected absolute magnitude of W Virginis at mid-range is about -2.5; the maximum
luminosity may be about 1500 times the solar value. From
this figure the computed distance is found to be about
11,000 light years; the star appear to be located a great
distance above the galactic plane, far out in the halo of
the Galaxy. A very small proper motion of about 0.006" per
year is consistent with the derived distance.

About ½° from W Virginis, to the ENE, lies the Mira-
type long-period red variable V Virginis, varying from magnitude 8½ or so to about 14, in a cycle of 250 days. (See
chart on page 2073)

THE VIRGO GALAXY CLUSTER

The Coma-Virgo section
of the sky is one of the
most remarkable areas of the heavens, occupied by the Virgo
Cluster of Galaxies, sometimes called the Coma-Virgo Cloud
of Galaxies, and often referred to in the past as " The
Field of the Nebulae". Centered near the Virgo-Coma border,
about midway between Epsilon Virginis and Beta Leonis, the
Cluster extends as far north as Canes Venatici, and south
into Corvus; the central concentration is near 12^h 24^m, and
+13°. A celestial wonderland of innumerable star cities,
the Virgo Cluster may be explored with telescopes of 6-inch
aperture or larger, and is the only great cloud of galaxies
available to the average amateur. Some 3000 members have
been identified on photographs taken with the greatest telescopes; more than 100 are within range of a good 8-inch
reflector.

The Virgo Galaxy Cluster is the nearest of the large
aggregations of galaxies, but the exact distance, in 1978,
was still being debated. G.de Vaucouleurs, in 1977, finds
a modulus of about 30.6 magnitudes, giving the distance as

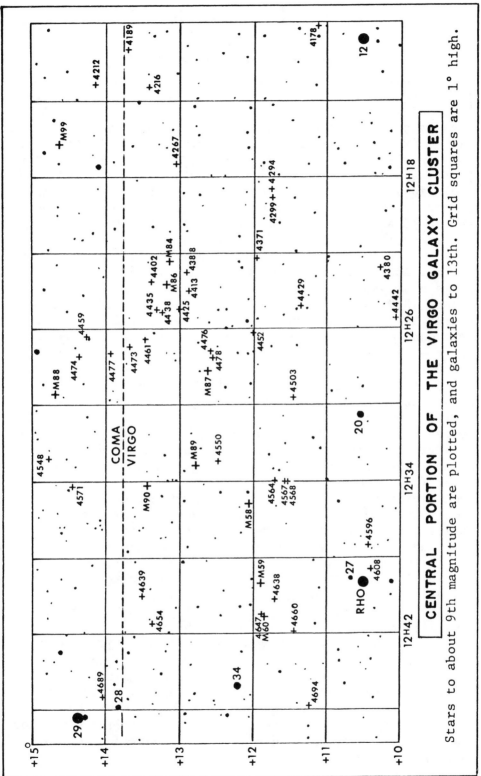

CENTRAL PORTION OF THE VIRGO GALAXY CLUSTER

Stars to about 9th magnitude are plotted, and galaxies to 13th. Grid squares are 1° high.

2075

DESCRIPTIVE NOTES   (Cont'd)

about 42 million light years. This figure is the weighted mean of six different methods of distance determination, and matches well the figure of 41 million light years which is derived from the apparent brightness of globular clusters in the group. From this result, however, de Vaucouleurs obtains a value for the Hubble Constant (red shift-distance relation) of about 16.4 miles per second per million light years, rather than the value of about 10.0 which has been accepted as the best value since the mid-1970's. The figure of about 10.0 is thought by A.Sandage and G.Tammann to be the more accurate, but it puts the Virgo Cluster at about 70 million light years. Distances given for galaxies in this book are generally based upon the Sandage figure of about 10.0 for the Hubble Constant, but in the case of the Virgo Cloud, the figure of 16.4 appears to give the more reasonable distance. One way out of this impasse is to assume that the expansion of the Universe is not necessarily uniform; both the Virgo Cloud and the Local Group may have individual space motions which either add to or subtract from the effect of the general expansion. From a red shift study of many galaxies, there seems to be some evidence that the Local Group is moving at a velocity of a few hundred kilometers per second, in the general direction of Leo, relative to the over-all background of distant galaxies. This finding suggests that all the measured radial velocities in the Virgo Cloud are systematically too low, and should possibly be increased by something like 180 miles per second. In the meantime, the distance of about 42 million light years seems reasonable for the Virgo Cloud.

There is a wide range in the radial velocities of the different galaxies, but the mean red shift for the cluster is about 680 miles per second. According to a list published by M.L.Humason, N.U.Mayall, and A.Sandage (1956) the largest red shift known among the members is that of NGC 4281, about 1545 miles per second. The object is an S0 or elliptical galaxy of the 12th magnitude, located near M61. At the other extreme, several of the galaxies (including the bright elliptical M86) show approach radial velocities, and would thus appear to be much closer than the average. It is not certain whether this peculiar situation can be completely explained by random individual motions within the cluster.

CENTER OF THE VIRGO GALAXY CLUSTER. The two bright elliptical galaxies M84 and M86 are at the top; the pair at
center is NGC 4435 & 4438. North is at the left in this
print.

CORRECTED RADIAL VELOCITIES IN MILES PER SECOND FOR MEMBERS OF THE VIRGO GALAXY CLOUD							
NGC	RV	NGC	RV	NGC	RV	NGC	RV
4179	+712	4387	+740	4477	+270	4594 (M104)	+630
4192 (M98)	-125	4394	+875	4478	+445	4621 (M59)	+210
4216	-30	4406 (M86)	+465	4479	-280	4636	+540
4254 (M99)	+1490	4421	+755	4486 (M87)	+1010	4638	+625
4261	+1300	4425	+1018	4492	+1120	4649 (M60)	+820
4267	+730	4429	+1275	4501 (M88)	+635	4660	+590
4270	+1385	4435	+220	4526	+495	4665	+425
4273	+1360	4438	+1000	4527	-65	4697	+730
4281	+1545	4442	+1140	4535	+305	4698	+590
4303 (M61)	+965	4450	+545	4546	+1235	4742	+730
4321 (M100)	+960	4458	+230	4548	+190	4754	+865
4324	+995	4459	+175	4550	+645	4762	+500
4339	+730	4461	+560	4551	+1120	4856	+680
4343	+380	4464	+130	4552 (M89)	+685	4866	+1150
4350	+695	4467	+555	4569 (M90)	+855	4941	+450
4365	+660	4472 (M49)	+1015	4570	+570	4958	+860
4374 (M84)	+545	4473	+1365	4578	+1350		
4382 (M85)	+445	4474	+1040	4579 (M58)	+905		

Radial velocities for the brighter members of the group are given in the table oppsite. The values have been corrected for the motion of the Sun in our own Galaxy.

Among the bright members, spiral galaxies are in the majority in the Virgo Cluster, numbering nearly 75% of the total. The remaining members are chiefly elliptical and S0 systems, with only a few irregulars. Only a few extreme dwarf galaxies have been identified in the group; doubtless there are many others too faint to be detected by any telescope now in existence. The majority of those known at the present time are dwarf irregulars, resembling such systems as IC 1613 in the Local Group.

The brightest spiral in the Virgo Cluster is M100 or NGC 4321, which lies across the border in Coma Berenices; it has an absolute magnitude of about -21, or about twenty billion times the luminosity of the Sun. About equally brilliant is the giant elliptical M87 (NGC 4486) which is also the strong radio source "Virgo A".

EXPLORING IN THE VIRGO CLUSTER. In small telescopes none of the members of this galaxy cluster are visually impressive; they appear mainly as pale little patches of light, round, elongated, and irregular; even the brightest members are not brilliant or striking objects. It is the knowledge of the actual nature of these glowing spots that compells the interested amateur to return again and again to the Virgo Galaxy Field, to observe and contemplate a celestial panorama surpassing the highest flights of human imagination. As a Japanese sage has said, "One must be open to the experience of the Ah! of things.." Here in the Virgo Cloud one may gaze upon the radiance of a hundred vastly remote star cities, twinkling across the millions of light years. Thomas Carlyle might have had such a panorama in mind when he wrote:

"....but is it not reckoned still a merit, proof of what we call a 'poetic nature', that we recognize how every object has a divine beauty in it; how every object still verily is 'a window through which we may look into Infinitude itself'?

For the exploration of the Virgo Cluster, at least a 6-inch telescope is recommended; a clear and moonless night is essential, and the observer should take the time to

DESCRIPTIVE NOTES   (Cont'd)

dark-adapt his eyes before attempting to locate any object.
A wide-angle low power eyepiece should be used on the telescope; 30X to 50X is ideal on 6 to 8-inch instruments. Once
located, specific objects may be examined with higher power
if desired.

The most crowded area of the field is the region
lying on the Virgo-Coma border at about $12^h\ 24^m$, the best
starting point for exploration. Two of the brighter E-type
galaxies are easy to locate here; these are M84 and M86,
about 17' apart. From these, a long curving chain of fainter objects runs eastward and then north, pointing toward a
fine bright spiral, M88, across the border in Coma, about
2½° away. When the observer has become familiar with the
region from M84 to M88, he will find it not difficult to
locate the other bright galaxies in the surrounding region.
Starting about 1° SE of the M84-M86 chain, the celestial
traveler will first note the monster elliptical M87, one of
the major sights of a tour through the Virgo Cloud. Very
similar in appearance is the E0 galaxy M89 which lies just
over 1° to the east and slightly north; from this point the
observer may move either NNE to the fine spiral M90, or SSE
to another large spiral M58. These are illustrated on page
2087. About 2° west of the M84-M86 group may be seen the
large nearly edge-on spiral NGC 4216, a bright streak about
7' in length, with two other edgewise systems in the same
field; this curious group is illustrated on page 2081.

Between the bright star Epsilon Virginis and the M84-
M86 chain lies a scattering of noble objects. The nearly
edge-on spiral NGC 4762 and its barred companion 4754 are
easily found 2½° west of Epsilon. The same distance again
brings us to the contrasting pair M60 and NGC 4647; the
former is a bright elliptical, and the companion a smaller
and unusual spiral with strangely broken and mottled spiral
arms. Another odd pair, about 2° west and slightly south,
is the intriguing double system NGC 4567 and 4568, called
the "Siamese Twins"by L.S.Copeland. Are these actually in
contact, or does one lie considerably beyond the other?
There is no sign of tidal filaments or distorted structure
as is often found in gravitationally interacting pairs.
(Photograph on page 2082)

Five degrees south of the M84-M86 chain is the fine
bright elliptical M49, a good guide to another oddly con-

GALAXIES OF THE VIRGO CLUSTER. Top: Region of M84 and M86, two bright elliptical systems. NGC 4388 is at the right. Below: NGC 4216, a nearly edge-on spiral.

Lowell Observatory

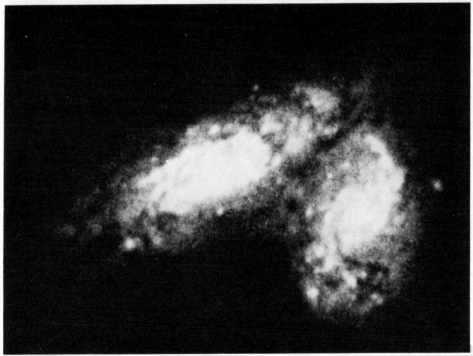

GALAXIES OF THE VIRGO CLUSTER. Top: The large elliptical system M86. Below: The "Siamese Twins" NGC 4567 & 4568. Palomar Observatory 200-inch telescope photograph.

trasting pair of objects located a little more than 1° to
the east. These are the galaxies NGC 4526 and 4535, the
former a much-elongated elliptical or SO system with a
bright center, the other a large round spiral of rather low
surface brightness, measuring 6' x 4' in size. From its
hazy phantom-like appearance in the amateur telescope this
has been christened the "Lost Galaxy" by L.S.Copeland. On
photographs an S-shaped spiral arm pattern appears, and a
small bright central nucleus. About 4° away, toward the SW,
is another, brighter spiral with strong arms and a large
central mass which shows some indication of a barred structure; this is M61, illustrated on page 2090. A degree to
the north and somewhat west is a remarkable concentration
of 9 small galaxies, including NGC 4270, 4273, 4260, and
4261. Any observer who takes the time to become really familiar with the Virgo Cluster will discover many other interesting groupings in this extraordinary region of the sky.
      Some of the bright members of the Virgo Cluster were
the first galaxies in which large recessional velocities
were found in 1913, thus introducing astronomers to the
mystery of the "red shift" and the problem of the "expanding universe".Specifically it was the galaxy NGC 4594, the
well known "Sombrero", in which V.M.Slipher in 1913 discovered a red shift of about 700 miles per second, the
highest radial velocity measured up to that time. A study
of the Virgo group has been of prime importance in the
problem of calibrating the red shift - distance relation;
the key to the scale of the Universe, as well as to its age
and its history.

THE SUPERGALAXY. The dimensions of the Virgo Cluster are
usually said to be about 12° x 10°, or something over five
million light years in actual diameter. The distribution of
neighboring galaxies, however, has suggested to many investigators that the Virgo Cloud is only the core of a much
more extensive aggregation. On the chart on page 2085 the
distribution of all galaxies brighter than 13th magnitude
is shown on the Aitoff "equal area projection". The dense
concentration near the center is the Virgo Cloud. It will
be seen that there are definite extensions to the pattern,
both north and south, increasing the total length to over
90°. This "supergalaxy" must measure at least 40 million

SPIRAL GALAXY NGC 5364. This beautiful system lies in
eastern Virgo, some 20° from the main core of the Virgo
Galaxy Cluster. Palomar Observatory 200-inch telescope.

VIRGO

DESCRIPTIVE NOTES   (Cont'd)

light years in extent, and appears to form a vast flattened
cloud of perhaps more than 10,000 individual galaxies. In
addition to the Virgo members, it appears to include many
of the bright galaxies in Coma, Canes Venatici, Ursa Major
and Leo. From the computed diameter of this gargantuan
super-system of galaxies, it seems quite possible that our
own Galaxy is located, as G.de Vaucouleurs has suggested,
on the outskirts of the Virgo aggregation, and that our
whole Local Group is merely a sub-unit, a small cluster of
perhaps 2 dozen members, riding on the edge of the Virgo
Cloud. The flattened shape of the Super-galaxy is very
evident, implying rotation in some period of many hundreds
of millions of years.

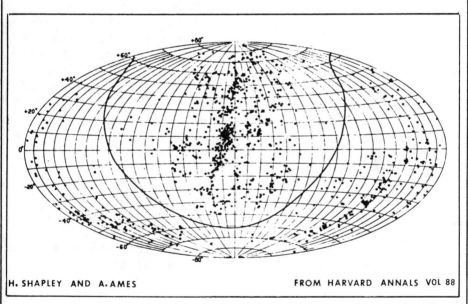

H. SHAPLEY AND A.AMES   FROM HARVARD ANNALS VOL 88

Although the Virgo Cluster is the most thoroughly
studied aggregation of galaxies known, it is not unique. A
rich cluster of galaxies in Corona Borealis appears to be
rather similar in size and type, but is located at a distance probably approaching 700 million light years, and is
therefore observable in large telescopes only. The most
luminous members are magnitude 16.5. Other large clusters
are known in Ursa Major, Hydra, and Hercules, all at mind-
chilling distances. The 48-inch Schmidt camera at Palomar

has revealed several thousand such clusters, ranging from
small families like our own Local Group up to enormous
clouds of galaxies like the Virgo system. The occurrence of
galaxies in clusters may in fact be the general rule in
space, rather than the exception. (Refer also to the Coma
Cluster on page 691, the Corona Borealis Cluster on page
714, and the Fornax Group on page 899. A portion of the
very distant Hercules Galaxy group is shown on page 996)

M49   (NGC 4472)   Position 12273n0816, about 5° S
from the center of the Virgo Galaxy Cluster.
One of the bright members of the Virgo group, discovered by
Messier in February 1771, and "seen only with difficulty in
a 3½-foot telescope". Admiral Smyth described it as "a
bright, round, and well-defined nebula" of a "pearly aspect"; T.W.Webb called it a "faint haze in a beautiful position between two 6th mag.stars". According to Lick Observatory *Publications* Vol.XIII, "the very bright nucleus is
not stellar; shows well in a 3-min.exposure. Nearly round,
2' in diameter, fading out rapidly toward the edges. No
structure discernible, though spiral character is suspected
near the centre in short exposures". M49 is slightly ovoid
on photographs, and is usually rated as about type E3; it
is one of the largest and most massive elliptical systems
known, with a computed absolute magnitude of about -21½,
and a total mass, according to E.Holmberg, of about 5 times
that of the Milky Way system. Owing to its lack of high
luminosity blue giant stars, this great system shines with
a much yellower light than do most galaxies; the integrated
spectral type is near G7. If the figure of 42 million light
years for the Virgo Cluster is accepted, M49 has a true
diameter of about 50,000 light years. The corrected radial
velocity is 570 miles per second in recession.

M58   (NGC 4579)   Position 12351n1205, about 2½° SE
of the main core of the Virgo Cluster. This
galaxy is a fine compact barred spiral, discovered in April
1779 by Messier, and recorded as "a very faint nebula.....
almost on the same parallel as Epsilon [slightly north]".
John Herschel thought it "hardly resolvable; rather mottled
as if with stars". According to K.G.Jones, the presence of
the central bar may be detected in an 8-inch telescope "as

SPIRAL GALAXIES OF THE VIRGO CLUSTER. Shown here are (top) M58, and (below) the more elongated system M90. Palomar Observatory photographs.

an extension of the central nucleus in an E-W direction".
On photographs the major diameter of 4' corresponds to an
actual size of about 50,000 light years at the accepted
distance of the Virgo Cloud. E.Holmberg finds for this system a computed mass nearly comparable to that of the Milky
Way, about 160 billion solar masses. The derived absolute
magnitude is close to -21, and the integrated spectral type
about G3.   According to M.Humason, the radial velocity is
1040 miles per second in recession.

The curious pair called the "Siamese Twins" (NGC 4567
+ 4568) lies about 0.5° distant to the SW; these are illustrated on page 2082.

M59   (NGC 4621) Position 12395n1155, about 1° east
and slightly south from M58, or 1.5° due north
from Rho Virginis. Elliptical galaxy of the 11th magnitude,
discovered in April 1779 by J.G.Koehler at Dresden, while
observing the comet of that year. It is classified by different authorities as type E3 or E4, occasionally E5; the
dimensions on photographs are about 2' x 1.5'. Messier recorded M59 on his chart of the comet of 1779, and found it
to be "of the same light as that above [M58] and as faint."
The galaxy appears to be a somewhat smaller and less massive system than either M49 or M60, but the derived figure
of about 250 billion solar masses still exceeds that of our
own Galaxy. This is a much denser system than the Milky Way
as the computed diameter is only about 24,000 light years,
about a quarter the size of our own system. At the mean
distance of the Virgo Cloud the derived absolute magnitude
is close to -19½. M59 has an integrated spectral class of
about G7, and the radial velocity is only about a third of
the mean figure for the Virgo Cloud, about 210 miles per
second in recession. A 12th magnitude supernova was recorded in the galaxy in May 1939.

M60   (NGC 4649) Position 12411n1149, about 0.5° E
from M59. Bright elliptical galaxy of about
10th magnitude, discovered by J.G.Koehler in April 1779.
Messier, observing it a few nights later, found it "a little
more distinct than the two preceding [M58 & M59]" and saw
the comet of 1779 pass through the field on April 13 and 14.
M60 forms a close pair with the fainter spiral NGC 4647,

which lies about 2.5' to the NW; Admiral Smyth refers to
these as "a double nebula, lying N.p. & S.f. about 2' or 3'
centre to centre, the preceding one being extremely faint.
The following, or brighter one, is that seen and imperfect-
ly described by Messier [M60] in 1779 and is nearly between
two telescopic stars N-S". T.W.Webb could detect only the
brighter object in his 3.7-inch refractor. H.Shapley, in
describing this unlike pair, pointed out that "there are
many similar examples of an open spiral and an elliptical
galaxy in close proximity. It is very difficult to account
for as they are thought to be very different in age and
also of development. There is the possibility of chance
encounter, but this is also remote".

M60 is one of the largest elliptical galaxies known,
comparable to M49 in mass, and only slightly smaller in
dimensions. It has a computed absolute magnitude of about
-21, an integrated spectral type of G7, and a mass of close
to a trillion suns, roughly 5 times that of the Milky Way
system. The visible diameter of 3' corresponds to about
25,000 light years at the adopted distance. The companion
spiral has about a tenth the mass of M60 itself, and 1/6
the luminosity. A red shift of 820 miles per second was
measured by E.Hubble for M60, somewhat higher than the mean
value for members of the Virgo Cluster.

M61   (NGC 4303) Position 12194n0445, about 8° NW
from Gamma Virginis. A large face-on spiral of
10th magnitude, discovered by the Italian astronomer B.
Oriani in May 1779, while observing the comet of that year;
Messier found it a few nights later and at first mistook it
for the comet. Lord Rosse resolved it into "a spiral with a
bright centre and 2 knots" but Admiral Smyth saw it as "a
large pale-white nebula...a well-defined object, but....so
feeble as to excite surprise that Messier detected it with
his 3½-foot telescope in 1779. Under the best action of my
instrument it blazes toward the middle..." In Vol.XIII of
the Lick *Publications* it is described as "Nearly round, 6'
in diameter, very bright. A beautiful spiral with a very
bright, almost stellar nucleus and many almost stellar condensations in its open, somewhat irregular whorls". E.J.
Hartung finds it "easy, but not bright" with an aperture of
7.5 cm. To K.G.Jones it appears "faintly three-lobed; it

SPIRAL GALAXY M61 in VIRGO. Arrow indicates the supernova of 1961; north is at the right in this photograph, made at Lick Observatory with the 120-inch reflector.

is almost round with a diameter of about 5'. The background is dark and the field contains several 8 - 9 mag. stars."

M61 is one of the larger galaxies in the Virgo Cloud, with a computed linear diameter of about 60,000 light years and an absolute magnitude close to -21. The integrated spectral type is about G1, and the total mass, according to E.Holmberg, may be about 50 billion solar masses, or about 25% the mass of the Milky Way system. A rather unusual effect is seen in the spiral structure of this galaxy; the arms show several sudden changes of direction at sharp angles, producing an over-all polygonal structure, and there is an exceptionally bright and thick star cloud in the arm on the north edge of the system. Supernovae were recorded in M61 in 1926, 1961, and 1964. The brightest of these was the outburst of 1961, shown on page 2090; it was located approximately 82" east of the center of the galaxy, and reached magnitude 13.0.

The easy double star 17 Virginis lies about 50' north in the field, next to the faint S0 galaxy NGC 4324; the small group consisting of NGC 4270, 4273, and 4281 is about 1° to the NNW.

M84 + M86   (NGC 4374 and 4406)   Positions 12226n1310 and 12237n1313, in the center of the Virgo Galaxy Cluster. This is the bright pair of elliptical galaxies which form the western portion of the "core" of the Virgo aggregation; they are about 17' apart, and were discovered by Messier in March 1781: "They both have the same appearance and are seen together in the field of the telescope." Each was described as a "nebula without star in Virgo... in the centre it is pretty bright, and surrounded with a pale nebulosity.." Each is rated at about magnitude 10.5, and M84 is the western member of the duo. Both are shown in the photograph on page 2081.

M84 is usually classed as an elliptical system of type E1, though on short exposures a dusky ring may be seen surrounding the bright nucleus; this suggests that the true classification shoud be changed to S0. At the accepted distance of the Virgo Cloud, the 2' apparent size corresponds to about 25,000 light years, and the derived absolute magnitude is about -20.5. E.Holmberg finds a mass of about 500 billion suns for the system, and the integrated spectral

DESCRIPTIVE NOTES   (Cont'd)

type is about G5. This galaxy has been identified as a source of radio radiation. In May 1957 a supernova appeared about 48" north of the nucleus, and reached an apparent magnitude of about 12.5.

M86, some 17' to the east, has a very similar appearance in the small telescope, though slightly more oblate; it is usually classed as type E3. D'Arrest thought that the nucleus was equivalent to a star of mag.11 or 12, but observations at Lick revealed that the nucleus was not stellar in short exposures. M86 is one of the eccentrics of the Virgo group, since it shows no red shift at all; the radial velocity is about 280 miles per second *in approach*. If it is actually a member of the Virgo aggregation, it must have an abnormally high individual space motion, and is possibly escaping from the cluster. There is the possibility, also, that M86 is actually a foreground system, merely seen in the same direction as the Virgo Cloud, but actually much closer to us; this was the solution adopted by E.Holmberg, who found a probable distance of slightly under 20 million light years, an absolute magnitude of −19.1, and a computed mass of about 130 billion solar masses. M86 is one of the redder galaxies in this region, with an integrated spectral type of about G7.

The spindle-shaped system NGC 4388 forms a nearly equilateral triangle with M84 and M86, lying about 16' to the south; this appears to be an edge-on SBc galaxy. Also in the field, about 10' north, lies another edge-on spiral with a prominent equatorial dust lane; this is NGC 4402. The group is illustrated on page 2081.

M87 (NGC 4486) Position 12283n1240, about 1.3° SE from M86. Giant elliptical galaxy, one of the largest members of the Virgo Cluster, discovered by Messier in March 1781, and described as "Nebula without star in Virgo.....below and very near an 8 mag.star.......appears to have the same light as the two nebulae M84 and M86..." D'Arrest found it "very large and bright, perfectly circular, diam 85". A little brighter towards the middle and, in the absolute centre, almost resolved to a star of 9 or 10 mag." Modern observations give the total photographic magnitude as about 9.7, and the apparent size as 3.0' with slight ellipticity; the type is usually given as E1.

M87 in VIRGO. Some of the system's many globular clusters
appear as fuzzy spots around the edges of this galaxy.
Palomar Observatory 200-inch telescope photograph.

This is undoubtedly one of the most massive and luminous of all known galaxies, with an absolute magnitude of about -21, exceeding even such systems as the Andromeda Galaxy M31. E.Holmberg derives a total mass of 790 billion solar masses and finds an integrated spectral class of about G5. The red shift for this galaxy is 755 miles per second.

M87 is remarkable for its large family of globular star clusters which appear as many small fuzzy images scattered throughout the outer parts of the galaxy. Over 500 of them can be detected on plates made with the 200-inch reflector, and a great many others may be out of sight on the far side of the galaxy, or in front of it where detection would be nearly impossible. The total number of globulars in the system may be at least 1000, and, according to a recent study (1976) may be as high as 4000. Our own Milky Way Galaxy, for comparison, possesses a known total of about 110.

M87 is a strong source of radio emission, originally detected by J.G.Bolton in 1948, and now called "Virgo A". In the Cambridge Catalogue of radio sources it bears the designation 3C 274, and ranks 5th in intensity among all the known radio sources of the sky. The strong radio energy appears to be associated with a curious optical feature first mentioned in astronomical literature by H.D.Curtis at the Lick Observatory in 1918: "The brighter central portion is about 0.5' in diameter....no spiral structure is discernible. A curious straight ray lies in a gap in the nebulosity.... apparently connected with the nucleus by a thin line of matter..." This peculiar nebulous jet appears to extend outward from the nucleus on the north-west side; it is some 20" in length and about 2" wide. Photographs made with the 200-inch reflector reveal that the jet contains 3 main condensations, and is much bluer than the light of the galaxy itself. According to M.L.Humason (1954) the spectrum of the jet is continuous, showing neither absorption nor emission lines. In 1956, W.Baade discovered that the light of the jet is strongly polarized. More recently, in 1966, it was discovered that M87 and its jet are a strong source of X-ray emission; the intensity of the X-ray energy is at least 10 times that of the combined optical and radio emission. The length of the jet is about 4100 light years, and the width about 400.

GALAXIES IN VIRGO. Top: The nuclear "jet" of M87 is shown in a short exposure with the 200-inch reflector. Below: The "Sombrero" galaxy M104, photographed at Lowell Observatory.

DESCRIPTIVE NOTES (Cont'd)

Using a technique of computer-enhancement of images in 1977, H.C.Arp at Palomar and J.Lorre of J.P.L. have resolved the M87 jet into a string of six small components, each less than 1" in size. The combined radio and optical studies suggest that the jet was formed by an ejection of material from the nucleus, and that the source of the radiation is the "synchrotron process" in which high-speed electrons are accelerated in a magnetic field. The original cause may have been some sort of gigantic outburst in the nucleus, such as now appears to be in progress in other odd galaxies such as M82 in Ursa Major and NGC 1275 in Perseus. The details of such an event are still highly speculative. Arp and Lorre point out that "The conclusion that these apparently synchrotron emitting knots are less than 1" in apparent dimension, and are being ejected so accurately aligned and spaced from the nucleus of M87, is undoubtedly very important to the understanding of galaxies and their relationship to more unusual objects".

Only one supernova is recorded for M87, a star of about magnitude 12½ which appeared in the spring of 1919 about 100" north from the nucleus and 15" west. As the star was not discovered until 1922, the true peak luminosity may have been as high as 11½, or about absolute magnitude -19.

M89   (NGC 4552) Position 12331n1250, about 1.3° E and slightly north from M87. Large elliptical galaxy of type E0, discovered by Messier in March 1781: "Its light was extremely faint and pale and it can be seen only with difficulty". In appearance it resembles M87 but is somewhat smaller and about 1 magnitude fainter. The central nucleus was thought by D'Arrest to be equivalent to a star of about 10th magnitude. E.Holmberg finds a computed mass of about 250 billion solar masses for M89, exceeding that of our own Galaxy; the total absolute magnitude is in the range of -19 or -20, depending upon the precise distance adopted. The red shift of only 130 miles per second is much smaller than the mean value for the Virgo Cluster, suggesting that M89 may be considerably closer to us than some of the other bright members. This is another of the rather "yellowish" elliptical systems, and has an integrated spectral type of about G7. As of 1977, no supernovae have been recorded in this galaxy.

M90   (NGC 4569) Position 12343n1326, about 1° NNE
from M89 and 1.7° NE from M87. Bright spiral
galaxy of the Virgo Cluster, discovered by Messier in March
1781, and recorded with the usual phrase "Nebula without
star in Virgo. Its light is as faint as the preceding, No.
89". D'Arrest, observing in 1862, thought this object to
be "splendid, immense; in the centre of an elliptical nebula, 7' x 90" is a shining and sparkling 11 mag.star... The
nucleus is a true star and no more than a point of light".
Flammarion also mentions the "brilliant nucleus at the
centre". On photographs the system is a fine much inclined
spiral of type Sb, about 7' in length, probably rather similar in type and dimensions to our own Milky Way. At the
accepted distance of the Virgo Cluster, about 42 million
light years, the actual diameter is found to be about 80
thousand light years, and the total absolute magnitude near
$-21\frac{1}{2}$. E.Holmberg reports for this galaxy an integrated
spectral type of about G0, and a total mass of near 80 billion solar masses. The corrected red shift is 555 miles per
second. (Photograph on page 2087)

M104   (NGC 4594) Position 12373s1121, on the Virgo-
Corvus border, usually accepted as a member of
the Virgo Cluster of Galaxies although it lies some 20° S
of the main concentration. This is the well known "Dark
Lane" or "Sombrero" galaxy, discovered by P.Mechain in May
1781, and added by Messier to his copy of the *Connaissance
des Temps* (1784); consequently it has been added in many
modern observing guides to the original Messier list, and
given the number M104. Other versions of the list have proposed additions up to M109, but, as K.G.Jones states, "the
trouble with sort of thing is to know where to stop". The
present author thinks it advisable to "stop" with M104,
having seen, from long experience, how things can get far
out of hand.

M104 is a fine example of a galaxy seen nearly edge-
on, and has sometimes been regarded as a transition type
between the spiral and elliptical galaxies; in most modern
lists it is classified as an "Sa" or an intermediate Sa-Sb.
It has a bright bulging main mass with a nearly stellar
nucleus, and a well-defined dark lane traversing the equatorial plane. This was probably first noticed by William

Herschel with his great reflector: "A faint diffused oval light all about it; almost positive that there is a dark interval or stratum separating the nucleus and the general mass of the nebula from the light above it..."   In Vol.XIII of the Lick *Publications* a much clearer description is given of this phenomenon: "A remarkable, slightly curved, clear-cut dark lane runs along the entire length to the south of the nucleus; probably the finest known example of this phenomenon". The dark band is not exactly easy in a 10-inch aperture, though it has been glimpsed with instruments as small as 6-inch. Much depends upon the darkness and clarity of the sky. On photographs there is some indication of spiral structure at the ends of the projecting arms, and A.Sandage suggests that the system, if seen face-on, might resemble the tightly coiled NGC 488 in Pisces. The orientation, however, is only about 6° from edge-on.

Palomar Observatory photographs reveal a rich population of globular star clusters surrounding the Sombrero; if these are comparable to the ones in our own galaxy, a distance of about 40 million light years is derived, which matches well the accepted distance of the Virgo Group. M104 is intrinsically one of the brightest and most massive of the galaxies. E.Holmberg derives an absolute magnitude of about -22, an integrated spectral class of G3, and a total mass of 1.3 trillion suns. The apparent diameter of 7' is equivalent to about 82,000 light years, but fainter outer portions increase the total size to about 130,000 light years. According to A.Sandage, the brightest globulars of the system have absolute magnitudes (pg) of about -10.6.

V.M.Slipher of the Lowell Observatory, in May 1914, announced the discovery of the rotation of M104, through a study of the inclination of the spectral lines. From later measurements made at the ends of the spiral arms, the rotation period at that distance appears to be about 25 million years.

The name of the Sombrero is forever linked with the discovery of the "red shift" and the mystery of the expanding universe. Investigation of the radial velocities of the galaxies was begun at Lowell Observatory in 1912, before the true nature of such objects was known. The Sombrero was found to show a red shift of nearly 700 miles per second. This enormous velocity, exceeding anything known at the

THE SOMBRERO GALAXY. M104 shows one of the clearest examples of an equatorial dust lane.   Palomar Observatory photograph made with the 200-inch telescope.

time, made it seem very unlikely that such a "nebula" could
be a local gas cloud in the Milky Way system, and offered
strong support for the "Island Universe" theory. The whole
question was definitely settled in 1923 by the discovery of
cepheid variables in the Andromeda Galaxy (M31); the existence of other galaxies since that time has been an established fact.

M104 lies in a very attractive field, centered in a
sparkling group of six 7th magnitude stars; several of
these, about 19' out from the galaxy to the WNW, form the
multiple star Σ1664. This is a 3.5' chain of three stars
with the westernmost member a 26" pair in PA 237°, magnitudes 8½ and 9½. About 1° ESE of M104 the observer will
find another double star, Hu 738 in Corvus; while about 85'
to the SSE is the neat 5" pair Σ1669. Data for these stars
will be found on page 716.

3C 273   Position 12266n0219, about 4.7° NW from Gamma
               Virginis. The brightest known example of a
"quasi-stellar radio source" or *quasar*, a puzzling and
enigmatic object, currently believed to be one of the most
brilliant single objects in the universe, and possibly the
most distant object visible in a 10-inch telescope. It is
located 3½° NE from Eta Virginis, and has the appearance
of a faint bluish "star" of magnitude 12.8. The position is
about 1½° W from the bright galaxy NGC 4536.

As the name implies, quasars are strong radio sources
which have the appearance of faint stars. Their spectra,
however, show enormous red shifts, which seem to indicate
that they are at distances far beyond most of the galaxies
visible in even the greatest telescopes; this in turn seems
to suggest that their actual luminosities must exceed those
of any other type of object known. They are known variously
as "quasi-stellar radio sources", "super radio galaxies",
"quasars", or "QSO's". In the following pages the term
quasar will be used. At the time of an International Symposium on these objects in Dallas, Texas, in December 1963,
nine examples had been identified; the number known at the
present time (1978) runs into the thousands. The best known
are 3C 273 in Virgo, 3C 48 in Triangulum, and 3C 147 in
Auriga. The designation "3C" indicates the number in the
Third Cambridge Catalogue of Radio Sources, published by

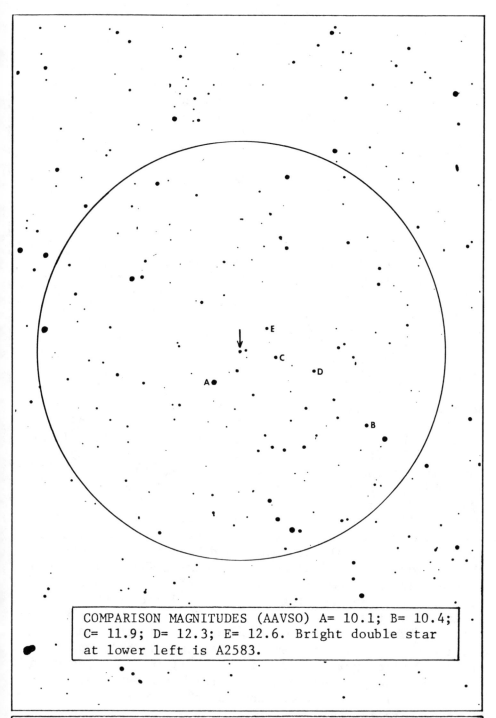

COMPARISON MAGNITUDES (AAVSO) A= 10.1; B= 10.4; C= 11.9; D= 12.3; E= 12.6. Bright double star at lower left is A2583.

3C 273 Identification field, from a Lowell Observatory 13-inch telescope plate. The circle is 1° in diameter with north at the top; limiting magnitude about 15.

the Royal Astronomical Society in London in 1959. This
Catalogue lists 471 radio sources. Many of these have been
identified with diffuse nebulae in our own galaxy, some
with supernova remnants such as the Crab Nebula (M1) in
Taurus, and others with peculiar external galaxies such as
M87 in Virgo and M82 in Ursa Major. In 1960, however, the
two strong sources 3C 48 and 3C 273 were identified with
objects which look like ordinary stars. For a time it was
thought that perhaps the first true "radio stars" had been
discovered. The first spectra obtained, however, showed a
peculiar pattern of several bright lines, resembling no
other spectra ever seen, either from stars, nebulae, or
galaxies. The spectrum lines, furthermore, did not match
those of any known element. The mystery deepened when the
spectra of several quasars were compared; no two of them
showed even a single line in common! In 1962 the key to the
puzzle was found by Maarten Schmidt at Palomar; studying
the spectrum of 3C 273 he realized that the pattern of
emission lines would match that of hydrogen, *provided that
one assumed a red shift of 16% in all the spectrum lines.*
This identification was strengthened by the fact that another of the unknown lines could then be immediately identified as the line of doubly ionized oxygen. Similarly, the
spectrum of 3C 48 could be decoded by allowing for a 37%
red shift, while 3C 147 was found to show the largest red
shift then known, about 54%.

Until this surprising discovery, the largest red
shifts measured had been those of the remote galaxies, and
upon this universal phenomenon had been built the widely-
held theory of the "Expanding Universe". Whatever the true
cause, it is an observed fact that all the distant galaxies appear to be receding, and show red shifts which increase directly with the distance. The relation between the
distance and the amount of the red shift is the quantity
called the "Hubble Constant", now generally believed to be
about 10 miles per second for each million light years of
distance. The precise value is still being debated, however, and like many other "constants" in astronomy, is open
to future refinement.

Now if the observed red shift of the quasars is in
reality this familiar "cosmological red shift", it implies
enormous distances. 3C 273 shows a red shift of about 30000

DESCRIPTIVE NOTES (Cont'd)

miles per second, and the resulting distance is close to 3 billion light years. At such a distance even an unusually bright galaxy would appear as an 18th or 19th magnitude object. Yet the apparent brightness of 3C 273 is about 12.8; this implies a luminosity of about 300 times that of a giant galaxy, or about 30 trillion times the energy output of the Sun. The absolute magnitude must be close to -27. Yet, at the same time, the quasars appear to be much smaller than any true galaxy. 3C 273 is less than 1" in apparent size, and appears stellar, but at the same distance any normal galaxy would appear as a glowing object with perceptible surface area, measuring 5" or so in diameter. Yet from this relatively small area, energy is being radiated at more than 100 times the rate of a typical galaxy. What can be the nature of such a fantastic object?

In the attack on the quasar problem, the evidence from radio studies has been of great value. With the 210-foot parabolic radio telescope near Sydney, Australia, observations of an unusual event were made in 1963, when the brightest quasar was occulted by the Moon. It is now known that 3C 273 is a double radio source, and that the two radio components can be identified with optically visible details. The 13th magnitude star-like component is the primary of the pair visually, but about 90% of the radio energy is coming from the second component which can be seen on photographs as a nebulous streak or "jet" on the SW side of the object. The jet begins about 11" from the bright core and continues out for about 10"; it is from 1" to 2" in width. If the red-shift-distance is accepted, this feature is some 300,000 light years in length.

The "stellar" component of 3C 273 has a radio "core" about 0.5" in diameter which emits about half the energy; there is also an encircling halo about 3" in size. Again, accepting the implied distance, the true size of the core must be about 7500 light years. The enormous energy output of such a relatively small object is difficult to explain. And in addition, a new problem arises from the discovery that the stellar component is variable in light; the range appears to be about 0.5 magnitude in a semi-periodicity of about 13 years, but there have also been more sudden increases of up to one magnitude lasting for a week or so. From a study of older plates, it is known that a sharp drop

in brightness occurred in 1929, followed by a steady rise
back to normal light in 1940. It is difficult to understand
how an object of galactic dimensions can show such rapid
light changes, since light itself must require centuries to
traverse the object. It is impossible to attribute such
variations to the whole body of a galaxy-sized object, and
the variations seem to prove that the actual source of the
light must be a relatively small structure perhaps no more
than a few light-months in diameter. This brings us back to
the original question: What can be the nature of such a
fantastic object?

To begin with, let us remember that the computed high
luminosity is based upon the distance of about 3 billion
light years, which in turn was derived from the observed
16% red shift. If this large red shift could be explained
in some other way it would no longer be necessary to assume
that quasars are extremely distant objects. The two known
causes for a red shift of spectral lines are (1) The well
known "Doppler effect", an increase in observed wavelength
due to velocity of recession, and (2) the "gravitational
red shift" produced by bodies of high mass and density.
This latter effect, predicted by Einstein, has actually
been observed in the case of the white dwarf stars, which
have densities of several tons to the cubic inch. In the
famous companion to Sirius, for example, a red shift of
about 16 miles per second is produced by the extremely
strong gravitational field. The star has very nearly the
mass of the Sun, but only about 2% the diameter; giving a
density of something like 125,000 times that of water.

ULTRA-DENSE STARS? As we have seen, the red shift of the
Virgo quasar is about 16% the velocity of light. To produce
this effect gravitationally, we would require a mass equal
to that of the Sun squeezed into an object about 12 miles
in diameter. Such an object would have so small a radiating
surface that it could not be seen at any great distance; to
appear as a 13th magnitude star it would be so near that it
would produce easily detected perturbations in the orbits
of the planets! It would also, of course, show an unusually
large parallax and proper motion. Aside from the fact that
an ultra-dense white dwarf would not show the characteristic quasar spectrum, the most careful measurements have
shown no detectable proper motion at all in 3C 273, at

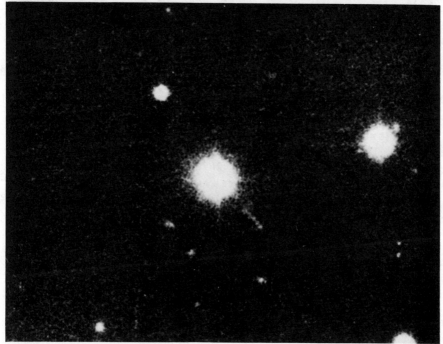

QUASAR 3C 273 in VIRGO. The photographs were made with the Lowell 13-inch camera (top) and the Palomar 200-inch telescope (below). The lower print shows the 3C 273 "jet".

least none as great as 0.001" per year. This alone indicates that the distance cannot be less than about 60,000 light years. Thus this particular quasar is at least an extragalactic object, whether or not we accept the much higher distance implied by the red shift.

HYPERSTARS? Having rejected the ultra-dense star idea, there remains an alternate possibility. The same large red shift could also be produced by an object of normal stellar size but of enormous mass, for example, an object the size of the Sun, but containing about 100,000 times the mass. The largest stellar masses actually observed are less than 100 suns, and it has been generally agreed that vastly larger masses would produce unstable "stars", which would destroy themselves in the process of formation. However, it has recently been proposed that some of the peculiar features of radio galaxies can be explained on the theory that such "hyperstars" have actually formed in the nucleus of the galaxy, and then exploded. In the peculiar galaxy M82 in Ursa Major, and in NGC 1275 in Perseus, such outbursts appear to be in progress at the present time. If the mysterious quasars are actually powered by a similar mechanism it may be that a quasar is merely an extreme case of a "radio galaxy". There are also the so-called "Seyfert galaxies" which show abnormally bright nuclei with emission-line spectra; these are possibly to be regarded as "mini-quasars". It has also been proposed that the quasar phenomenon represents a stage in the history of possibly many normal galaxies. In the majority of cases the object actually observed must be merely the intensely bright eruptive core of the galaxy; the surrounding galaxy itself is very possibly too faint to be detected.

GRAVITATIONAL COLLAPSE. At the Dallas Conference in December 1963, the central topic was the question of gravitational collapse, the mechanism considered to be the most likely explanation of the enormous energy output of the quasars. Fred Hoyle of Cambridge and W.A.Fowler of the California Institute of Technology have attempted to analyse the conditions which might exist in collapsing gas clouds with masses of a million or a billion suns. In the later stages of the collapse the release of gravitational energy could far exceed that produced by any possible

DESCRIPTIVE NOTES   (Cont'd)

nuclear reactions. The details of this theory are highly
speculative, and may possibly lead us to a quasar definition as satisfactory as Arthur Eddington's famous summary
of the nature of the electron: *"Something unknown is doing
we don't know what"*. Lewis Carroll put it into words even
more effectively:

" *'Twas brillig and the slithy toves
       Did gyre and gimble in the wabe...*"

And Richard Wagner, in the marvelous opening passage of
Das Rheingold, expressed the ultimate mystery of the cosmos
without using any words at all:

Our own scientist-poet Loren Eiseley evidently had
the far frontiers of astronomical knowledge in mind when he
wrote:

*"The one great hieroglyphic, Nature, is as unreadable
as it ever was and so is her equally wild and unpredictable
offspring, man...."*

That crusty old iconoclast, H.L.Mencken, expressed
the same skepticism concerning the nature of knowledge:

*"Nine times out of ten, in the arts as in life, there
is actually no truth to be discovered; there is only error
to be exposed..."*

Well, perhaps Mencken had a point there, since much
of scientific advance unquestionably *does* consist of the
discarding of former errors, without answering in any way
the fundamental human questions. *"See how today's achievement,"* wrote William Dean Howells, *"is only tomorrow's
confusion"*. The whole quasar controversy recalls J.B.S.
Haldane's gentle warning: *"The Universe is not only queerer
than we suppose; it is queerer than we can suppose"*. *"But
why shouldn't truth be stranger than fiction?"* queried Mark
Twain, *"fiction after all has to make sense..."*

How far, then, have we advanced toward "making sense"
out of the quasar enigma? Although large numbers of these
puzzling objects are known, our definition of a quasar may
still be said to be entirely *observational*; a quasar is an

object which exhibits certain properties; at the moment the explanation of these properties is fairly well unknown. We have a definition, then, which reminds one of Ambrose Bierce's definition of the mouse as "an animal that strews its path with fainting women"; the definition (no longer true of course) tells us more about the women than about the mouse. Assuming, however, that the concept of gravitational collapse is the key to this puzzle, and that a mass of about a billion suns can actually contract into a single object, it may be possible to explain the observed features of the quasars by the resulting release of gravitational energy through implosion, explosion, or both.

As the most plausible model at the present time, we may picture a quasar as an extremely hot cloud of gas in a state of gravitational collapse, totalling possibly about a billion solar masses, and having a diameter of not more than a few hundred light years. The light variations, as we have seen, must originate in a much smaller object, and may be attributed to rapidly changing conditions in the "core" of the quasar. There are various possible modifications of this basic hypothesis: For example there is the idea that the process of star formation might begin at a certain stage in the life of such a contracting gas cloud; thousands of high-mass stars might be formed almost simultaneously, and would evolve to the supernova stage at the same time. Further speculation leads on to ideas which must seem fantastic and bizarre, even to those acquainted in a general way with the tenets of relativity. We are now familiar, for example, with the idea that a sufficiently massive and dense object will collapse into the "black hole" state, effectively vanishing from our universe; possibly quasars represent an early stage in this collapse process. In the same way, it is theorized, a quasar may appear as a "white hole" to observers in some other space-time continuum, as it ultimately vanishes from our own. This raises the odd question of whether some of our quasars may not be the black holes of some other universe, vanishing into limbo on *their* side of the barrier, and appearing into *our* universe.

In 1972, H.C.Arp at Palomar reported finding strong evidence that some quasars are associated with galaxies and groups of galaxies, possibly implying that the quasars are

DESCRIPTIVE NOTES   (Cont'd)

physically associated with galaxies and may have been born
from explosions in galactic nuclei. If the association is
real, the quasars (in these cases at least) are not at the
very remote distances that their red shifts would seem to
imply. The red shift, then, cannot be the "cosmological"
one shown by all distant objects; it must be the result of
tremendous density in a collapsing object, or actual velocity resulting from high-speed ejection from the parent
galaxy. Yet, if the latter possibility is true, why should
all quasars show a *red shift*? Surely *some of these expelled
objects should be approaching us!*

The apparent association of a galaxy and a quasar
may occur occasionally simply by chance, but several very
convincing cases of actual relationship appear to be known.
A faint galaxy known as ESO 113-IG45 contains a 13th magnitude quasar as its nucleus; the galaxy itself is a spiral
some 800 million light years distant, and the quasar shows
a red shift of about 8400 miles per second. In this case
there can be no real doubt that the quasar is actually in
the nucleus of the galaxy, and that the derived absolute
magnitude is about -24. Even more conclusive is the case of
the E-type galaxy NGC 1199 in Eridanus, investigated by H.
Arp, and announced in the *Astrophysical Journal* in 1978.
*The galaxy itself has a red shift of about 1600 miles per
second, but a compact and nearly stellar-appearing companion, seen in silhouette in front of the galaxy, shows a
red shift of about 8250 miles per second!* There would
seem to be no other conclusion than that reached by Arp:
"...the compact object is slightly in front of, but approximately at the same distance as, the E galaxy.....most of
its redshift is of origin other than Doppler motion of
recession..." Another intriguing case is that of NGC 7603,
a faint Seyfert galaxy (Pg.mag. 14.4) in Pisces, connected
to a smaller companion by a luminous filament. NGC 7603
shows a red shift of about 5450 miles per second; that of
the companion is close to 10,500 miles per second! Neither
of the objects appears to be a quasar, but this case would
seem to prove that a large red shift very definitely can
result from some effect other than the familiar Doppler
shift. It seems safe to conclude that a quasar may not
always be at the vast distance that its red shift alone
seems to indicate. We await further revelations.

LIST OF DOUBLE AND MULTIPLE STARS

NAME	DIST	PA	YR	MAGS	NOTES	RA & DEC
I 351	10.7	334	00	6½- 10	spect K0	06417s7143
h3929	9.2	238	17	7½- 10	spect G0	06579s7158
γ	13.6	300	41	4 - 5½	($\triangle$42) Fine pair; cpm, relfix, spect G8, dF4	07092s7025
h3997	2.0	120	47	7 - 7	PA inc, spect both B9	07364s7410
ε	6.1	24	22	4½- 8	(Rmk 7) relfix, cpm, spect B5	08078s6828
I 192	2.0	173	18	7- 9½	spect A0	08089s6851
I 9	1.0	300	41	7½- 7½	PA dec, spect A5	08154s7339
h 4134	45.0	108	17	5½- 10	(Theta Vol)	08389s7013
	21.4	66	30	- 15	Spect A0	
Hrg 19	4.5	165	30	7½-10½	dist inc, spect K0	08476s6515
h4164	10.7	145	18	7½- 10	spect K0	08568s6601

LIST OF VARIABLE STARS

NAME	MagVar	PER	NOTES	RA & DEC
R	8.7--13.9	448	LPV. Spect Me	07064s7256
S	7.7--13.8	396	LPV. Spect M4e	07306s7316
T	8.5--12..	176	LPV. Spect M4e--M5e	06578s6703
X	9.1--11..	280	LPV.	07575s6509
UU	8.0---8.4		Semi-reg? Spect M4e	08158s6819

LIST OF STAR CLUSTERS, NEBULAE AND GALAXIES

NGC	OTH	TYPE	SUMMARY DESCRIPTION	RA & DEC
2397		⊖	SBa; 12.8; 1.8' x 1.0' pB,cL,cE, 1bM	07215s6854
2434		⊖	E1; 12.8; 0.8' x 0.7' pS,F,R,pmbM	07350s6910
2442		⊖	SBb; 11.8; 6.0' x 5.0' cL,vF,1E	07365s6925

VULPECULA

LIST OF DOUBLE AND MULTIPLE STARS

NAME	DIST	PA	YR	MAGS	NOTES	RA & DEC
Σ2445	12.4	263	51	7 - 8½	relfix, spect B3	19025n2315
Σ2455	6.6	40	60	7½- 8½	cpm, PA dec, dist	19048n2206
	93.5	22	21	- 12	inc, spect F0, B5	
Σ2457	10.2	200	57	7 - 8½	cpm, relfix,	19050n2230
					spect F0	
A264	3.1	287	68	8- 13½	(Ho 446) PA slow	19106n2429
	1.8	351	68	- 12½	dec; spect F5; AC	
					PA & dist dec	
2	2.0	124	53	5½- 9½	(ES Vulp) (β248)	19156n2256
					relfix, spect B0;	
					Primary variable	
Σ2499	2.6	325	49	8½- 9	relfix, spect B8	19164n2151
OΣΣ181	57.8	3	23	6 - 6½	dist inc, spect F5	19181n2633
	36.2	169	10	- 14	+ K	
Z	13.5	359	14	7 - 12	(Es 483) spect B3;	19196n2529
					Primary= Ecl.bin.	
β141	0.8	81	55	7½- 9	(h2867) Spect B5	19198n2225
	28.0	332	16	- 11½		
	50.4	90	16	- 11		
	50.0	213	16	- 12½		
β141c	5.5	181	16	11½-13		
Σ2515	4.6	68	26	8 - 9	Optical, dist dec,	19224n2125
	68.8	182	10	- 12	PA inc, spect A5	
Σ3111	2.3	117	37	9 - 9	relfix, spect A0	19230n2144
4	18.9	100	57	5½- 10	(h2871) optical, PA	19233n1942
	52.6	204	59	- 11½	& dist dec, spect	
					K0; AC PA inc; in	
					cluster Col 399	
Σ2521	26.7	35	58	6 - 10	Optical, dist inc,	19243n1948
	26.5	71	34	- 14	PA dec, spect gM0;	
	70.5	323	18	- 9½	color contrast; in	
	150	64	18	- 10	cluster Col 399	
Σ2525	1.6	294	66	7½- 7½	binary, eccentric	19245n2713
					orbit; about 475	
					yrs; PA dec, spect	
					dF9	
Σ2523	6.4	149	53	7½- 7½	relfix, spect B8	19246n2103
Σ2540	5.1	147	51	7½- 9	relfix, spect A3;	19311n2018
	147	221	12	- 12½	Primary may be 0.4"	
					pair, uncertain	

LIST OF DOUBLE AND MULTIPLE STARS   (Cont'd)

NAME	DIST	PA	YR	MAGS	NOTES	RA & DEC
9	9.3	32	34	$5\frac{1}{2}$- $13\frac{1}{2}$	(β1130) Spect B7	19324n1940
	108	318	23	- $12\frac{1}{2}$		
Σ2548	9.2	100	51	$8\frac{1}{2}$- $9\frac{1}{2}$	relfix, spect B9	19344n2453
Σ2551	6.7	42	12	9 - $9\frac{1}{2}$	relfix, spect G0	19352n2242
	44.5	320	02	- $11\frac{1}{2}$		
A164	0.4	226	53	$7\frac{1}{2}$- 9	PA inc, spect G0	19357n2241
A165	6.3	132	30	7 - 14	spect A0	19365n2256
	18.6	343	30	- $13\frac{1}{2}$		
Σ2556	0.3	52	67	$7\frac{1}{2}$- 8	binary, about 250 yrs; PA dec, spect F2	19373n2208
Σ2560	15.3	295	58	$6\frac{1}{2}$- $8\frac{1}{2}$	relfix, spect B3	19385n2336
OΣ382	0.4	333	57	7 - $7\frac{1}{2}$	PA dec, spect B8	19398n2716
β658	0.4	292	54	$6\frac{1}{2}$- 10	PA slow dec, A = composite spectrum G4+A0	19419n2701
Σ2584	1.9	295	42	$8\frac{1}{2}$- $8\frac{1}{2}$	relfix, spect F5	19462n2204
Σ2586	3.7	227	55	$7\frac{1}{2}$- 10	relfix, spect B9	19465n2450
OΣ388	3.7	139	54	8 - 8	relfix, spect A0	19502n2544
	31.3	135	50	- 9		
ΣI 48	42.2	147	00	7 - $7\frac{1}{2}$	cpm, spect both A0	19512n2012
13	0.8	243	60	5 - 8	(Dju 4) spect A0	19513n2357
Ho 584	2.3	226	24	$6\frac{1}{2}$- 12	spect K0	19582n2603
16	0.8	115	58	$5\frac{1}{2}$- 6	(OΣ395) PA inc, spect F5	19599n2448
A1200	4.9	198	29	$7\frac{1}{2}$-$13\frac{1}{2}$	spect K2	20094n2901
Σ2653	2.5	270	58	7 - 10	PA inc, spect Am	20115n2405
Σ2655	6.2	3	54	$7\frac{1}{2}$- $7\frac{1}{2}$	relfix, spect A0	20119n2204
	59.9	154	32	- 9		
OΣ402	15.4	33	07	7 - $10\frac{1}{2}$	relfix, spect B9	20124n2441
β983	0.7	166	55	6 - 10	PA inc, spect B3	20132n2526
β441	5.9	64	58	7- $11\frac{1}{2}$	cpm, relfix, spect K0	20155n2859
β984	0.6	237	65	8 - $8\frac{1}{2}$	PA inc, spect F8	20155n2613
β985	5.1	149	15	7 - 13	(h1499) spect B3	20162n2529
	21.6	356	15	- $9\frac{1}{2}$		
β985c	9.8	63	15	$9\frac{1}{2}$-$12\frac{1}{2}$		
β443	14.0	138	15	$7\frac{1}{2}$-$11\frac{1}{2}$	spect A5	20221n2850
	35.2	89	15	- $11\frac{1}{2}$		
Σ2695	0.8	87	59	$6\frac{1}{2}$- $8\frac{1}{2}$	PA inc, spect A2,G	20298n2538

LIST OF DOUBLE AND MULTIPLE STARS (Cont'd)

NAME	DIST	PA	YR	MAGS	NOTES	RA & DEC
Σ2698	4.4	302	46	8 - 9	PA slight dec, spect A0; in star cluster NGC 6940	20317n2757
A2795	0.1	295	58	8 - 8	PA inc, spect B9	20384n2145
β673	3.9	296	26	7½- 12	AB cpm, spect A3	20396n2032
	105	164	49	- 8		
Ho 138	3.0	340	58	7- 13½	PA dec, spect K5; C= 3.1" pair	20413n2525
	128	306	06	- 11		
Σ2724	2.4	327	53	8 - 8	relfix, spect G0	20423n2345
Ho 144	0.3	350	67	7½- 7½	spect F5, A0	20501n1956
OΣ417	0.8	29	59	7½- 8	PA dec, spect A0	20510n2857
	31.2	109	41	- 10		
Ho 281	13.7	300	24	7 - 13	Spect M; R Vulp in field	21013n2348
Σ2761	5.6	111	46	8½- 9	relfix, spect A2	21052n2417
Σ2769	17.9	300	54	6½- 7½	easy relfix pair, probably cpm, spect both A0	21083n2215
OΣ430	1.3	205	59	8 - 10	cpm, PA dec, spect F8	21097n2357
β447	9.3	320	34	6½- 12	PA dec, spect A2	21219n2506
	56.6	187	17	- 11½		
	56.9	82	60	- 11½		
	66.8	115	60	- 10½		
	82.4	217	60	- 10½		

VULPECULA

LIST OF VARIABLE STARS

NAME	MagVar	PER	NOTES	RA & DEC
2	5.4--5.46	.6096	(ES Vulp) β Canis Major type; Spect B0; also visual double	19156n2256
R	7.4--13.7	137	LPV. Spect M3e--M7e	21022n2337
S	9.2--10.8	69	Semi-reg; Spect G0--K2	19464n2710
T	5.4--6.1	4.4356	Cepheid; Spect F5--G0	20493n2804
U	6.8--7.5	7.9907	Cepheid; Spect F8--G2	19344n2013
V	8.1--9.4	76	RV Tauri type; Spect G4e ---K3	20344n2626
W	8.3--10.2	234	Semi-reg; Spect M5e	20080n2608
X	8.5--9.3	6.3194	Cepheid; Spect F8--G1	19554n2625
Z	7.4--9.2	2.4549	Ecl.Bin; Spect B3 + A2; also visual double	19196n2529
RR	9.8--11.3	5.0507	Ecl.Bin; Spect A2	20527n2744
RS	6.9--7.6	4.4777	Ecl.Bin; Spect B5 + A2	19155n2221
RU	8.2--12..	156	Semi-reg; Spect M3e	20367n2305
SV	6.7--7.8	45.04	Cepheid; Spect F7--K0	19459n2720
AT	9.5--10.1	3.9804	Ecl.Bin; Spect B3	19518n2326
BD	9.3--12.7	430	LPV. Spect Ne	20352n2618
BE	9.9--11.3	1.5520	Ecl.Bin; Spect A3	20235n2712
BP	9.3--10.4	1.9403	Ecl.Bin; Spect A7	20233n2052
BW	6.3--6.4	.2010	β Canis Major type; spect B2	20523n2820
CK	2.7--17..	---	Nova 1670	19455n2711
DR	8.6--9.2	2.2509	Ecl.Bin; Spect B8	20117n2636
DY	7.6--8.4	Irr	Spect M3	21013n2348
ER	7.3--7.5	.6981	Ecl.Bin; W U.Maj type; Spect G0 + G5	21003n2735
FG	9.0--9.5	80:	Semi-reg; Spect M5	20325n2806
FI	7.2--8.7	Irr	Spect M3	20466n2248
LT	6.5--6.54	.098	δ Scuti type; Spect F2	19015n2112
LU	9.2---21..	---	Nova 1968	19436n2828
LV	5.2---17..	---	Nova 1968	19459n2702
MV	9.3--9.8	Irr	Spect K0	20130n2200
MW	6.62--6.7	16.478	α Canes Ven type; spect A0p	20144n2737
NQ	6.5---18..	---	Nova 1976	19271n2022

VULPECULA

LIST OF STAR CLUSTERS, NEBULAE AND GALAXIES

NGC	OTH	TYPE	SUMMARY DESCRIPTION	RA & DEC
----	C. 399	⠂⠄⠂	vL,B, scattered naked-eye group incl 4, 5, 7 Vulp	19240n2500
6800	21[8]	⠂⠄⠂	vL,pRi, Diam 15'; about 25 stars mags 10.... Class D; 30' NW from α Vulp	19251n2505
6802	14[6]	⦂⦂	S,vC,Ri; Diam 5', Mag 11; bar shaped, N-S; about 60 stars mags 13....18 Class E	19285n2010
6815		⦂⦂	L, scattered star-field in Vulp Milky Way, probably not a true cluster	19388n2641
6820		☐	F,L,Irr; Diam 20' with bright rim features and dark clouds; Incl cluster NGC 6823	19405n2258
6823	18[7]	⦂⦂	pL,cRi,E; Mag 10, Diam 5'; 30 stars mags 11...12; Class D; Neby NGC 6820 inv.	19411n2312
6830	9[7]	⦂⦂	L,Irr, pRi,pC; Mag 9; Diam 8' 20 stars mags 11... Class D	19489n2258
6834	16[8]	⦂⦂	Ri,1C, Diam 4'; about 50 stars mags 11...15; Class D; Plan.Neb NGC 6842 is 38' ESE	19502n2917
6842		◎	F,pL,v1E; Mag 13; Diam 50" x 45"; $14\frac{1}{2}^m$ central star	19530n2909
6853	M27	◎	!! vB,vL,1E; Diam 8' x 5'; Mag 8; $13\frac{1}{2}^m$ central star; "Dumb-bell Nebula" (*)	19574n2235
----	I.4954	☐	F neby Diam 1' with 12[m] B-star	20028n2906
6885	20[8]	⦂⦂	B,L,pRi; Diam 20'; about 35 stars mags 6...11; incl 20 Vulp; Class E	20099n2620
6940	8[7]	⦂⦂	B,vL,vRi; cC; Mag 8, Diam 20' About 100 stars mags 9..... Class E (*)	20325n2808

DEEP-SKY VIEWS IN VULPECULA. Top: The "Dumb-bell Nebula" M27, a bright planetary nebula. Below: Star cluster NGC 6940. Lowell Observatory 13-inch telescope photographs.

DESCRIPTIVE NOTES

M27   (NGC 6853) Position 19574n2235, about 25' S from 14 Vulpeculae, or 3.3° almost due north from Gamma Sagittae, the brightest star in the little arrow of Sagittae. M27 is generally considered the most conspicuous planetary nebula in the entire sky for small instruments; with a total magnitude of about 8 it may be easily found in large binoculars. M27 was discovered by Messier in July 1764: "Seen well in 3½-foot telescope.... Appears oval and contains no star. Recorded on chart of comet of 1779... Diam 4'. " William Herschel thought it possibly "a double stratum of stars of a very great extent, one end of which is turned toward us", while John Herschel remarked on the "faint luminosity which fills the lateral concavities of the body and converts them into protuberances so as to render the general outline of the whole nebula a regular ellipse.." The popular name, "Dumb-bell Nebula", is derived from the appearance of "two hazy masses in contact" as T.W. Webb phrased it, with a narrower zone between; in a modern 6-inch telescope with low or medium powers it appears oval

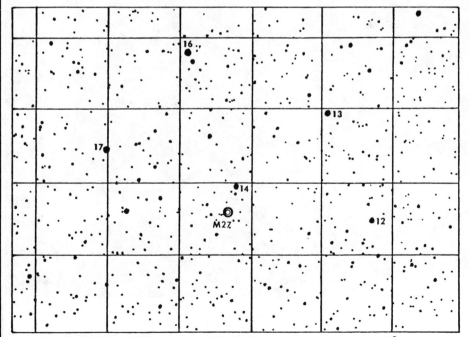

IDENTIFICATION CHART for M27. Grid squares are 1° on a side with north at the top; limiting magnitude about 9.

PLANETARY NEBULA M27 in VULPECULA. The "Dumb-bell Nebula" as it appears on a photograph made with the 61-inch astrometric reflector at the U.S. Naval Observatory.

or hazily rectangular with smoothly rounded corners. The
apparent dimensions are about 8' x 5', one of the largest
of the planetaries, and the first object of the type to be
discovered by Messier. It is, however, not one of the ring-
shaped or "annular" nebulae; the first one known was M57 in
Lyra, discovered by Darquier some 15 years later.

D'Arrest spoke of M27 as producing a "beautiful
appearance on magnification 95X.... very large and shining;
two objects blending into one another... In the more luminous S.p. portion is a brighter region placed eccentrically;
the N.f. part is of fairly even brightness..." Lord Rosse
thought to find some symptoms of resolution in the glowing
nebulosity, and J.E.Gore believed that some hint of a faint
spiral structure was present. Admiral Smyth called it a
"magnificent and singular object.... situated in a crowded
vicinity where field after field is very rich..... truly
one of those splendid enigmas which, according to Ricciolus
are proposed by God but never to be subject to human solution..." In Vol. XIII of the Lick *Publications*, M27 is
referred to as "One of the giants of the planetary class
and of great importance in theories of planetary [nebula]
structure because of the easy visibility of its intricate
details".

The large apparent size of this nebula classes it as
probably one of the nearest of the planetaries, though the
exact figure is not certain. C.R.O'Dell (1963) derived a
probable distance of about 850 light years, L.Kohoutek has
650, I.S.Shklovsky obtained about 490, A.Becvar in the 1960
edition of the *Atlas Coeli Catalogue* reported about 980,
while K.G.Jones gives the figure as 975 light years. If the
true distance is close to 900 light years, the actual size
is a little under 2½ light years, among the largest diameters calculated for any nebula of the type. The central
star, of magnitude 13½, has about ½ the solar luminosity,
if the adopted distance is reasonably correct.

As in all the planetary nebulae, the faint central
star is an abnormally hot bluish dwarf or sub-dwarf with a
spectrum resembling that of an ancient nova and a calculated surface temperature of about 85,000°K, one of the hottest stars known. Such a star appears to form the nucleus
of every planetary nebula which is near enough to be studied in detail. It is agreed that the planetaries have been

PROMINENT PLANETARY NEBULAE

NGC	OTH	CON	SUMMARY DESCRIPTION	RA & DEC
40	58[4]	CEPH	Mag 10½; Diam 60" X 40"; 11½ mag central star, spect O	00102n7215
246	25[5]	CETI	Mag 8½; Diam 4' X 3½'; 12^m 07 central star. Photo pg 651	00446s1209
650	M76	PERS	"Cork Nebula" Mag 11; Diam 140" X 70"; 16½^m central star. Photo pg 1436	01388n5119
1501	53[4]	CAML	Mag 12; Diam 55" x 48"; 13½ mag central star. Photo pg 336	04026n6047
1535	26[4]	ERID	Mag 9; Diam 20" X 17"; 11½^m central star in bluish disc	04121s1252
2392	45[4]	GEMI	Mag 8; Diam 40" with 10^m 08e central star; "Eskimo Nebula". Photo pg 941	07262n2101
3132		VELA	Mag 8; Diam 84" x 52"; 10^m central star. Photo pg 1175	10049s4011
3242	27[4]	HYDA	Mag 9; Diam 40"; 11^m central star in eye-shaped ring. Photo pg 1028	10224s1823
3587	M97	UMAJ	"Owl Nebula"; Mag 11; Diam 150"; 14^m central star. Photo pg 1999	11120n5518
6210	Σ5	HERC	Small bright disc Mag 9½; Diam 20" X 16"; 12½^m 06 central star	16425n2353
6543	37[4]	DRAC	Greenish disc Mag 8½; Diam 22" X 16"; 10^m 0-type central star; Photo pg 871	17588n6638
6572	Σ6	OPHI	Very bright bluish disc; Mag 9; Diam 15" X 12"; 12^m 0-type central star	18097n0650
6720	M57	LYRA	"Ring Nebula"; annular; Mag 9; Diam 80" X 60"; 15^m central star; Famous planetary; Photo pg 1165	18517n3258
6826	73[4]	CYGN	"Blinking Planetary"; Mag 9; Diam 25"; 11^m 0-type central star; Refer to note on page 1178	19434n5024
6853	M27	VULP	"Dumb-bell Nebula"; Mag 8; Diam 8' X 5'; 13½^m central star; fine large planetary; Photo pg 2118	19574n2235
7009	1[4]	AQAR	"Saturn Nebula"; Mag 8; Diam 25"; 12^m central star; Photo pg 191	21014s1134
7293		AQAR	Large faint annulus Diam 12'; "Helical Nebula" Photo pg 195	22270s2106
7662	18[4]	ANDR	Mag 8½; Diam 30"; Photo pg 158	23234n4212

formed by ejection of material from the hot central star, but the exact details of the process are uncertain. Current evidence suggests that the planetary nebula phenomenon represents that last activity of a dying star which is sinking toward the white dwarf state. The question of a possible relationship of these stars with the novae remains speculative.

The expansion of the Dumb-bell Nebula has been measured at about 17 miles per second through spectroscopic analysis, and an actual increase in diameter of about 1.0" per century has been detected by astrometric photograph measurements. If the expansion rate has remained fairly constant, an approximate age of about 48,000 years is found for this nebula, about 2½ times the "average age" of about 20,000 years deduced for typical bright planetaries. K.L. Cudworth of the Yerkes Observatory has determined that the central star of the Dumb-bell is very probably a physical double; the companion is a yellowish 17th magnitude star at 6.5" in PA 214°, and the components appear to form a true common motion pair with a projected separation of about 1800 AU. At the adopted distance, the derived absolute magnitudes of the two stars are about +6.2 and +9.7. The primary, which shows a virtually continuous spectrum, is the illuminating star of the nebula; its strong ultraviolet radiation excites the glow of the highly rarified gases, producing that eerie pale bluish fluorescence now attributed to the "forbidden radiation" of doubly ionized oxygen. The observer who spends a few moments in quiet contemplation of this nebula will be made aware of direct contact with cosmic things; even the radiation reaching us from the celestial depths is of a type unknown on Earth....

> " What we have learnt
> Is like a handful of earth;
> What we have yet to learn
> Is like the whole world..."

THE POETESS AVVAIYAR
(Ist CENT B.C.)

THE BRIGHTEST STARS

	NAME	BAYER DESIGNATION	MAG	SPECT	PAGE
1	SIRIUS	α Canis Majoris	-1.42	A1 V	387
2	CANOPUS	α Carinae	-0.72	F0 Ib	465
3	ALPHA CENTAURI	α Centauri	-0.27	G2 V	549
4	ARCTURUS	α Bootis	-0.06	K2 III	302
5	VEGA	α Lyrae	0.04	A0 V	1137
6	CAPELLA	α Aurigae	0.06	G8 III	261
7	RIGEL	β Orionis	0.14	B8 Ia	1299
8	PROCYON	α Canis Minoris	0.35	F5 IV	448
9	ACHERNAR	α Eridani	0.53	B5 IV:	888
10	HADAR	β Centauri	0.66	B1 II	553
11	BETELGEUSE	α Orionis	v 0.70	M2 Ia	1281
12	ALTAIR	α Aquilae	0.77	A7 V	205
13	ALDEBARAN	α Tauri	0.86	K5 III	1807
14	ACRUX	α Crucis	0.87	B1 IV	728
15	ANTARES	α Scorpii	v 0.92	M1 Ib	1655
16	SPICA	α Virginis	v 1.00	B1 V	2065
17	POLLUX	β Geminorum	1.16	K0 III	921
18	FOMALHAUT	α Pisces Austrini	1.17	A3 V	1485
19	DENEB	α Cygni	1.26	A2 Ia	755
20	BETA CRUCIS	β Crucis	1.28	B0 IV	729
21	REGULUS	α Leonis	1.36	B7 V	1057
22	ADHARA	ε Canis Majoris	1.49	B2 II	437
23	CASTOR	α Geminorum	1.59	A1 V	912
24	SHAULA	λ Scorpii	1.62	B1 V	1678
25	BELLATRIX	γ Orionis	1.64	B2 III	1301
26	EL NATH	β Tauri	1.65	B7 III	1827
27	MIAPLACIDUS	β Carinae	1.67	A1 IV	466
28	GAMMA CRUCIS	γ Crucis	1.67	M3 II	729
29	ALNILAM	ε Orionis	1.70	B0 Ia	1302
30	AL NA'IR	α Gruis	1.76	B5 V	949
31	ALNITAK	ζ Orionis	1.79	09 Ib	1305
32	ALIOTH	ε Ursae Majoris	1.79	A0p	1952
33	MIRFAK	α Persei	1.79	F5 IB	1404
34	DUBHE	α Ursae Majoris	1.81	K0 II	1949
35	KAUS AUSTRALIS	ε Sagittarii	1.81	B9 IV	1562

INDEX TO THE 25 NEAREST STARS

STAR	CONS.	DIST.	A Mag & Sp	B Mag & Sp	Mu"	PA °	RA & DEC	PAGE
PROXIMA CENTAURI	CENT	4.3	10.7 dM5e		3.85	282°	14263s6228	550
ALPHA CENTAURI	CENT	4.3	0.0 G2 V	1.17 dK1	3.68	281	14362s6038	549
BARNARD'S STAR	OPHI	6.0	9.5 dM5		10.29	356	17554n0424	1251
WOLF 359	LEO	7.7	13.6 dM6e		4.71	235	10541n0719	1071
LALANDE 21185	UMAJ	8.3	7.6 dM2		4.78	187	11007n3618	1980
SIRIUS (α C.MAJ)	CMAJ	8.7	-1.4 A1 V	8.65 DA5:	1.32	204	06430s1639	387
L726-8 (UV CETI SYSTEM)	CETI	9.0	12.4 dM6e	12.9 dM6e	3.35	80	01364s1813	641
ROSS 154	SGTR	9.5	10.6 dM5e		0.74	103	18467s2353	
ROSS 248	ANDR	10.3	12.2 dM6e		1.62	177	23394n4355	
EPSILON ERIDANI	ERID	10.8	3.7 K2 V		0.97	271	03306s0938	889
L789-6	AQAR	10.8	12.2 dM7e		3.25	46	22357s1536	
ROSS 128	VIRG	10.9	11.1 dM5		1.35	152	11452n0106	
61 CYGNI	CYGN	11.1	5.3 K5 V	5.9 K7 V	5.22	52	21044n3828	768
PROCYON (α C.MIN)	CMIN	11.3	0.3 F5 IV	11.0? DA?	1.25	214	07367n0521	448
EPSILON INDI	INDI	11.4	4.7 K5 V		4.69	123	21596s5700	1038
Σ2398	DRAC	11.4	8.9 dM4	9.7 dM5	2.28	324	18425s5930	869
GROOMBRIDGE 34	ANDR	11.7	8.1 dM2	10.9 dM4e	2.89	82	00155n4344	127
TAU CETI	CETI	11.8	3.5 G8 V		1.92	297	01417s1612	639
LACAILLE 9352	PSCA	11.9	7.4 dM2e		6.90	79	23026s3609	1487
BD+5°1668	CMIN	12.3	9.8 dM5		3.76	171	07247n0529	
CORDOBA 29191	MICR	12.7	6.7 dM1		3.46	251	21143s3904	
KAPTEYN'S STAR	PICT	12.7	8.8 dM0		8.70	131	05097s4500	1462
KRUEGER 60	CEPH	13.1	9.8 dM3e	11.4 dM4e	0.86	246	22262n5727	598
ROSS 614	MONO	13.1	11.1 dM6e		1.00	131	06269s0246	
BD-12°4523 (WOLF 1061)	OPHI	13.4	10.1 dM5e		1.18	123	16275s1232	1189

BIBLIOGRAPHY— A PARTIAL LIST OF WORKS CONSULTED

GENERAL STELLAR DATA

D.Hoffleit, CATALOGUE OF BRIGHT STARS, Third Revised
Edition; Yale University Observatory, New Haven, 1964.

A.Becvar, ATLAS COELI II KATALOG 1950.0; Ceskoslovenske
Academie, Prague, 1960.

W.Gleise, CATALOGUE OF NEARBY STARS; *Veroffentlichungen
des Astronomischen Rechen-Instituts,* Heidelberg, 1969.

R.Woolley, E.A.Epps, M.J.Penston, S.B.Pocock, CATALOGUE OF
STARS WITHIN TWENTY-FIVE PARSECS OF THE SUN; Royal
Observatory *Annals,* No.5, Greenwich, 1970.

C.W.Allen, ASTROPHYSICAL QUANTITIES, 3rd Edition; Athlone
Press, University of London, 1973.

R.E.Wilson, GENERAL CATALOGUE OF STELLAR RADIAL VELOCITIES;
Carnegie Institution of Washington, 1953.

Smithsonian Astrophysical Observatory STAR CATALOG;
Smithsonian Institution, Washington, 1966.

V.M.Blanco, S.Demers, G.G.Douglass, M.P.Fitzgerald, PHOTO-
ELECTRIC CATALOGUE: *Publications* of the U.S. Naval
Observatory, Second Series, Vol.XXI, Washington, 1968.

F.Schlesinger, L.F.Jenkins, GENERAL CATALOGUE OF STELLAR
PARALLAXES, Yale University Observatory, 1935.

Landolt-Bornstein; ASTRONOMY AND ASTROPHYSICS, Springer-
Verlag, 1965.

W.J.Luyten, A CATALOGUE OF 1849 STARS WITH PROPER MOTIONS
EXCEEDING 0."5 ANNUALLY, Lund Press, Minneapolis, 1955.

W.J.Luyten, AN ATLAS OF IDENTIFICATION CHARTS OF WHITE
DWARFS, *Astrophysical Journal,* Vol.109, No.3; 1949.

DOUBLE STARS

E.Crossley, J.Gledhill, J.M.Wilson, A HANDBOOK OF DOUBLE
STARS, Macmillan & Co., London, 1879.

S.W.Burnham, A GENERAL CATALOGUE OF DOUBLE STARS, Carnegie
Institution, 1906.

R.G.Aitken, NEW GENERAL CATALOGUE OF DOUBLE STARS, The
Carnegie Institution of Washington, 1932.

R.A.Rossiter, CATALOGUE OF SOUTHERN DOUBLE STARS, *Publications of the Observatory of the University of Michigan*
Vol. XI, 1955.

H.M.Jeffers, W.H.van den Bos, F.M.Greeby, INDEX CATALOGUE
OF VISUAL DOUBLE STARS: *Publications of the Lick Observatory,* Vol. XXI, 1961.

O.J.Eggen, THE NEAREST VISUAL BINARIES; *Astronomical
Journal,* Vol. 61, 1956.

R.G.Aitken, THE BINARY STARS; McGraw-Hill, N.Y. 1918.

BIBLIOGRAPHY   (Cont'd)

F.Holden, MEASURES OF DOUBLE STARS; *Journal des Observateurs*, Vol. 46, 1963.

W.H.van den Bos, MICROMETER MEASURES OF DOUBLE STARS; *Astronomical Journal*, Vols. 63, 64, 68; 1958- 1963.

C.E.Worley, A CATALOG OF VISUAL BINARY ORBITS; *Publications of the U.S. Naval Observatory*, Second Series, Vol. XVIII Part III; Washington, 1963.

C.E.Worley, MICROMETER MEASURES OF 1,056 DOUBLE STARS; *Publications of the U.S. Naval Observatory*, Second Series, Vol.XXII Part IV, 1972.

R.T.A.Innes, A SOUTHERN DOUBLE STAR CATALOGUE, Union Observatory, Johannesburg, 1927.

G.van Biesbroeck, MEASUREMENTS OF DOUBLE STARS; *Publications of the Yerkes Observatory*, Vol.VIII,Part VI; Vol.IX, Part II.

P.Couteau, P.J.Morel, ATLAS OF THE APPARENT ORBITS OF VISUAL DOUBLE STARS; Observatoire de Nice, 1973.

V.V.Kallarakal, I.W.Lindenblad, F.J.Josties, R.K.Riddle, M.Miranian, B.F.Mintz, A.P.Klugh, PHOTOGRAPHIC MEASURES OF DOUBLE STARS; *Publications of the U.S. Naval Observatory*, Second Series, Vol. XVIII, Part VII, 1969.

G.M.Popovic, FIRST GENERAL CATALOGUE OF DOUBLE STAR OBSERVATIONS MADE IN BELGRADE 1951-1971; *Publications de l' Obsvt. Astron. de Beograd*, 1974.

VARIABLE STARS

A.J.Cannon, SECOND CATALOGUE OF VARIABLE STARS; *Annals* of the Astronomical Observatory of Harvard College, Vol. LV, Cambridge, 1907.

L.Campbell, STUDIES OF LONG PERIOD VARIABLES; A.A.V.S.O., 1955.

J.S.Glasby, VARIABLE STARS; Harvard University Press, 1969.

J.S.Glasby, THE DWARF NOVAE; American Elsevier Publ. Co., N.Y., 1970.

F.B.Wood, EMPIRICAL DATA ON ECLIPSING BINARIES; Ch.19 in K.A.Strand "BASIC ASTRONOMICAL DATA", University of Chicago Press, 1963.

B.V.Kukarkin & others, GENERAL CATALOGUE OF VARIABLE STARS; Sternberg State Astronomical Institute, Moscow, 1969

C.Payne-Gaposchkin & S.Gaposchkin, VARIABLE STARS, Harvard Observatory, 1938.

L.Campbell, L.Jacchia, THE STORY OF VARIABLE STARS; Harvard Books on Astronomy, Blakiston Co., Philadelphia, 1941.

B.V.Kukarkin,(ed) PULSATING STARS; John Wiley & Sons, 1975.

BIBLIOGRAPHY   (Cont'd)

STAR CLUSTERS, NEBULAE, AND GALAXIES

J.L.E.Dreyer, A NEW GENERAL CATALOGUE OF NEBULAE AND CLUS-
TERS OF STARS; *Memoirs* of the Royal Astronomical
Society, Vol. XLIX, London, 1888.

J.W.Sultenic, W.G.Tifft, THE REVISED NEW GENERAL CATALOGUE
OF NON-STELLAR ASTRONOMICAL OBJECTS; University of
Arizona Press, Tucson, 1973.

K.G.Jones, MESSIER'S NEBULAE AND STAR CLUSTERS; American
Elsevier Publ. Co., N.Y., 1968.

H.Shapley, A.Ames, A SURVEY OF THE EXTERNAL GALAXIES
BRIGHTER THAN THE THIRTEENTH MAGNITUDE; *Annals* of the
Astronomical Observatory of Harvard College, Vol. 88,
No.2, Cambridge, 1932.

G.& A.de Vaucouleurs, REFERENCE CATALOGUE OF BRIGHT GALAX-
IES; University of Texas Press, Austin, 1964.
SECOND REFERENCE CATALOGUE OF BRIGHT GALAXIES; 1976.

G.de Vaucouleurs, SURVEY OF BRIGHT GALAXIES SOUTH OF -35°
DECLINATION; *Memoirs* of the Commonwealth Observatory,
Mt.Stromlo, Canberra, Australia, 1956.

S.van den Bergh, A RECLASSIFICATION OF THE NORTHERN
SHAPLEY-AMES GALAXIES; *Publications of the David Dunlap Observatory*, University of Toronto, 1960.

H.B.S.Hogg, A BIBLIOGRAPHY OF INDIVIDUAL GLOBULAR CLUSTERS;
Publications of the David Dunlap Observatory, Vol. I,
No. 20, 1947;   First Supplement, 1963.

H.Arp, ATLAS OF PECULIAR GALAXIES; *Astrophysical Journal
Supplement Series*, Vol. XIV, No. 123, 1966.

A.Sandage, THE HUBBLE ATLAS OF GALAXIES; Carnegie Institution of Washington, 1961.

H.D.Curtis, STUDIES OF THE NEBULAE; *Publications of the
Lick Observatory*, Vol. XIII, 1918.

J.L.Sersic, ATLAS GALAXIAS AUSTRALES; Observatorio Astron-
omico de Cordoba, Argentina, 1968.

L.Perek, L.Kohoutek, CATALOGUE OF GALACTIC PLANETARY NEBU-
LAE; Academia Publ. House of the Czechoslovak Academy
of Sciences, Prague, 1967.

E.E.Barnard, ATLAS OF SELECTED REGIONS OF THE MILKY WAY;
Carnegie Institution of Washington, 1927.

A.A.Hoag, H.L.Johnson, B.Iriarte, R.I.Mitchell, K.L.Hallam,
S.Sharpless, PHOTOMETRY OF STARS IN GALACTIC CLUSTER
FIELDS; *Publications of the U.S. Naval Observatory*,
Second Series, Vol. XVII, Part VII, Washington, 1961.

H.Shapley, GALAXIES; Harvard Books on Astronomy, Harvard
University Press, revised edition, 1961.

BIBLIOGRAPHY (Cont'd)

H.Shapley, STAR CLUSTERS; Harvard Observatory Monograph No
    2, McGraw-Hill, N.Y. 1930.
G.Alcaino, ATLAS OF GALACTIC GLOBULAR CLUSTERS WITH COLOR-
    MAGNITUDE DIAGRAMS; Universidad Católica de Chile,1973
F.E.Ross, M.R.Calvert, ATLAS OF THE MILKY WAY, University
    of Chicago Press, 1934.
P.Brosche, J.Einasto, U.Rummel, BIBLIOGRAPHY ON THE STRUC-
    TURE OF GALAXIES; *Publ. Astronomischen Rechen-Instit-*
    uts, Heidelberg, 1974.

OBSERVERS' GUIDEBOOKS

T.W.Webb, CELESTIAL OBJECTS FOR COMMON TELESCOPES, Vol.2;
    Dover Publications, N.Y. 1962
E.J.Hartung, ASTRONOMICAL OBJECTS FOR SOUTHERN TELESCOPES;
    Cambridge University Press, London, 1968.
D.H.Menzel, A FIELD GUIDE TO THE STARS AND PLANETS;
    Houghton-Mifflin Co., Boston, 1964.
J.Mullaney, W.McCall, THE FINEST DEEP-SKY OBJECTS; Sky
    Publishing Corporation, Cambridge, Mass. 1966
H.J.Bernhard, D.A.Bennett, H.S.Rice, NEW HANDBOOK OF THE
    HEAVENS; McGraw-Hill, 1941.
C.E.Barns, 1001 CELESTIAL WONDERS; Pacific Science Press,
    Calif., 1929.
W.T.Olcott, E.W.Putnam, FIELD BOOK OF THE SKIES; G.P.
    Putnam's Sons, N.Y., 1936.
A.P.Norton, STAR ATLAS & REFERENCE HANDBOOK; 16th Edition,
    Sky Publishing Corporation, Cambridge, Mass., 1973
A.Becvar, SKALNATE PLESO ATLAS OF THE HEAVENS; Sky Pub-
    lishing Corporation, Cambridge, Mass., 1958.

GENERAL HISTORY AND ASTRONOMICAL LORE

R.H.Allen, STAR NAMES AND THEIR MEANINGS: Stechert, N.Y.,
    1899, Reprint by Dover Publications, N.Y.
H.Shapley, A SOURCE BOOK IN ASTRONOMY; Harvard University
    Press, Cambridge, 1960.
M.E.Martin, THE FRIENDLY STARS; Harper & Brothers, 1907.
    Reprint by Dover Publications, N.Y.
A.M.Clerke, HISTORY OF ASTRONOMY DURING THE 19th CENTURY;
    Adam & Charles Black, London, 1893.
A.M.Clerke, THE SYSTEM OF THE STARS; Adam & Charles Black,
    London, 1905.
M.Proctor, EVENINGS WITH THE STARS; Cassell & Co., London
    1924.
L.C.Peltier, STARLIGHT NIGHTS; Harper & Row, N.Y., 1965.

THE CELESTIAL HANDBOOK-CONSTELLATION INDEX			
CONSTELLATION	PAGE	NORTON'S ATLAS CHART	SKALNATE-PLESO ATLAS CHART
ANDROMEDA	103	3	II, V
ANTLIA	160	8, 10	XIII VIII
APUS	163	16	XVI XIV
AQUARIUS	165	4, 14	XI XV VI
AQUILA	197	13, 14	X XI V
ARA	235	12, 16	XIV XV XVI
ARIES	245	5	VI II
AURIGA	253	5, 7	II III
BOOTES	297	11	IV IX I
CAELUM	313	6	XII
CAMELPARDALIS	314	1, 2	I II III
CANCER	337	7	VIII III VII
CANES VENATICI	353	9	IV III I
CANIS MAJOR	381	8	VII VIII
CANIS MINOR	447	7	VII VIII
CAPRICORNUS	452	14	XI XV X
CARINA	458	8, 16	XIII XVI XII
CASSIOPEIA	476	2, 3	I II V
CENTAURUS	535	10, 16	XIV, XIII, XVI
CEPHEUS	573	2	I V II
CETUS	621	4, 5	VI XII XI XV
CHAMAELEON	655	16	XVI
CIRCINUS	656	16	XVI XIV
COLUMBA	658	6	XII XIII
COMA BERENICES	661	9	IV IX VIII III
CORONA AUSTRALIS	693	14	XV XIV
CORONA BOREALIS	697	11	IV
CORVUS	716	10	IX XIV VIII

CONSTELLATION	PAGE	NORTON'S ATLAS	SKALNATE-PLESO ATLAS CHART
CRATER	723	10	VIII XIII IX
CRUX	726	16	XIV XVI XIII
CYGNUS	735	13	V I IV
DELPHINUS	817	13	XI X V
DORADO	833	15, 16	XVI XII XIII
DRACO	853	1, 2	I IV V III
EQUULEUS	875	13	XI
ERIDANUS	878	6	VI VII XII XVI
FORNAX	895	6	XII VI VII
GEMINI	905	7	III VII VIII
GRUS	945	4	XV XVI XII
HERCULES	950	11	IV X V IX
HOROLOGIUM	997	6, 15	XII XVI
HYDRA	999	8, 10	VIII XIII XIV IX
HYDRUS	1033	15	XVI XII XV
INDUS	1035	14, 15	XV XVI
LACERTA	1039	3	V I II
LEO	1047	7, 9	VIII III IX IV
LEO MINOR	1083	9	III VIII
LEPUS	1087	6	VII XII XIII
LIBRA	1101	12	IX XIV X
LUPUS	1111	12	XIV XVI
LYNX	1123	1, 7	III II I
LYRA	1131	13	V IV
MENSA	1179	15, 16	XVI
MICROSCOPIUM	1180	14	XV
MONOCEROS	1182	7, 8	VII VIII
MUSCA	1215	16	XVI XIV XIII
NORMA	1218	12	XIV XV XVI
OCTANS	1221	15, 16	XVI

CONSTELLATION	PAGE	NORTON'S ATLAS	SKALNATE-PLESO ATLAS CHART
OPHIUCHUS	1223	11, 12	X XIV IX
ORION	1271	5, 6	VII III II
PAVO	1351	15	XVI XV XIV
PEGASUS	1358	3	XI V VI II
PERSEUS	1394	5	II I III
PHOENIX	1453	4	XII XVI XV
PICTOR	1458	6, 16	XII XIII
PISCES	1464	3	VI XI II
PISCES AUSTRINUS	1483	4	XV
PUPPIS	1489	8	XIII VII XII VIII
PYXIS	1518	8	XIII VIII
RETICULUM	1525	15	XII XVI XIII
SAGITTA	1526	13	X XI V
SAGITTARIUS	1547	14	XV X XIV XI
SCORPIUS	1645	12	XIV X IX XV
SCULPTOR	1730	4	XII XV
SCUTUM	1741	14	X XV
SERPENS	1759	11	IX X XIV IV XV
SEXTANS	1794	9, 10	VIII
TAURUS	1798	5	VII II VI III
TELESCOPIUM	1889	14	XV XIV
TRIANGULUM	1894	3	II
TRIANGULUM AUSTRALE	1904	16	XVI XIV XV
TUCANA	1907	15	XVI XV XII
URSA MAJOR	1922	1, 9	III IV I
URSA MINOR	2006	1	I IV
VELA	2027	8, 10	XIII XVI XIV
VIRGO	2042	9, 10	IX VIII XIV
VOLANS	2110	16	XVI XIII
VULPECULA	2111	13	V X XI

Since the lists of objects in the *Celestial Handbook* are all arranged in a definite logical order, the author feels that the preparation of a *complete* index to this book would serve no real purpose. There are, however, many celestial objects which are often referred to by a common or popular name (The Whirlpool Galaxy, The Swan Nebula) as well as objects which are usually designated by the discoverer's name or number (Groombridge 1830, Ross 614, Kapteyn's Star). In such cases the name alone does not identify the constellation in which the object is located. The following table has been prepared as a quick page reference to these objects.

Carl Sandburg has told us that Abraham Lincoln, in his law office in Springfield, kept an envelope marked *"If you can't find it anywhere else, look in here."* The same policy should guide the user of this Index.

INDEX TO GENERAL TOPICS (Cont'd)

INDEX TO TABLES OF DATA IN THIS HANDBOOK